Reports of the United States Commissioners to the Paris Universal Exposition, 1867
by United States. Commission to the Paris Exposition

REPORTS

OF THE

UNITED STATES COMMISSIONERS

TO THE

PARIS UNIVERSAL EXPOSITION, 1867.

PUBLISHED

UNDER DIRECTION OF THE SECRETARY OF STATE BY AUTHORITY
OF THE SENATE OF THE UNITED STATES.

EDITED BY

WILLIAM P. BLAKE,

COMMISSIONER OF THE STATE OF CALIFORNIA.

VOLUME I.

WASHINGTON:
GOVERNMENT PRINTING OFFICE.
1870.

CONTENTS.

PARIS UNIVERSAL EXPOSITION, 1867.
REPORTS OF THE UNITED STATES COMMISSIONERS.

INTRODUCTION,

WITH

SELECTIONS FROM THE CORRESPONDENCE

OF

COMMISSIONER GENERAL BECKWITH AND OTHERS,

SHOWING THE

ORGANIZATION AND ADMINISTRATION

OF THE

UNITED STATES SECTION.

WASHINGTON:
GOVERNMENT PRINTING OFFICE.
1870.

INTRODUCTION TO THE REPORTS OF UNITED STATES COMMISSIONERS.

Among the most instructive developments of modern civilization are those international exhibitions which, commencing in London in 1851, under the inspiration and auspices of the late sagacious and public-spirited Prince Albert, have been succeeded by more and more extended and comprehensive ones, closing in the Universal Exposition held at Paris during the summer of 1867. The projectors of this great international reunion, after mature study of preceding exhibitions, evolved a programme which embraced in its scope the productions and results of every industry, art, and science, as well as their processes and methods of operation. This was done not merely for purposes of competition and the distribution of prizes, but also, and more especially, with the object of passing in review, under the scrutiny of the most accomplished experts and men of science, all of the fruits of the skill, industry, and inventive and artistic genius of every nation, in such a manner that the exact condition and the comparative merits or defects of the industrial development of each nation and of each description of article or process could be set forth; the progress which such examination indicated, measured, and explained; and the highest standards of excellence be placed within the reach of all by means of carefully prepared reports.

From the commencement of the industrial epoch which dates from the London Exhibition of 1851 the profound significance and value of such exhibitions have been realized by the people and governments of the civilized nations. Their beneficent influences are many and wide-spread; they advance human knowledge in all directions. Through the universal language of the products of labor the artisans of all countries hold communication; ancient prejudices are broken down; nations are fraternized; generous rivalries in the peaceful fields of industry are excited; the tendencies to war are lessened; and a better understanding between labor and capital is fostered. It is gratifying to note that these great exhibitions are planned and executed in the interests of the mass of the people. In this last instance those industries, products, and organizations designed to promote the material and moral well-being of the people were made prominent, and the underlying animating spirit and impulse of the whole plan were for the advancement, prosperity, and happiness of the people of all nations. One of their most salutary results is the promotion of an appreciation of the true dignity of labor, and its paramount claims to consideration as the basis of national wealth and power.

Such exhibitions have become national necessities and duties, and as such it may be expected that they will be repeated again and again hereafter.

The programme laid down by the French Imperial Commission under the presidency of Prince Napoleon, charged with the preparation and management of the Exposition, made it an absolute condition for the admission of exhibitors from any country that the government of such country should first accept the invitation and assume the responsibility of forming the exhibition of its section; and in that event, suggested the appointment of some competent person to be intrusted with the general supervision of the business on its behalf, and to communicate with the Imperial Commission.

On the 27th of March, 1865, the government of France, through M. de Geofroy, their chargé d'affaires residing at this capital, invited this government to participate, upon the terms above indicated, in the Exposition. The considerations which have been hereinbefore mentioned, and their special application to this country at a period when it was concluding its repression of a formidable rebellion, made it peculiarly desirable that the United States should not hold aloof from such an assemblage.

Mr. Bigelow, then minister at Paris, was accordingly, on the 5th of April, 1865, instructed to inform the minister for foreign affairs of France that President Lincoln regarded the proposed Exposition with great favor, as well because of the beneficent influence it might be expected to exert upon the prosperity of the nations, as of its tendency to preserve peace and mutual friendship among them; that what the executive government could do, by way of concurrence in the movement, would very cheerfully be done, but that that was as far as the President was able to proceed without special legislative authority, for which application would be made to Congress when it should next meet. Mr. Bigelow was at the same time requested to act temporarily as a special agent for this government in the premises.

Mr. Bigelow, by a letter of the 17th of April, 1865, recommended the appointment, as Commissioner General of the United States, of N. M. Beckwith, esq., a citizen of the United States, then residing in Paris, of whose eminent qualifications for the post the Department had most satisfactory proofs.

Accepting the onerous duties of that office, without compensation, Mr. Beckwith entered upon them with an activity, zeal, intelligence, and executive ability to which, with the assistance of other commissioners, is mainly due the measure of success that, notwithstanding unlooked-for and frequent impediments, was attained by the United States Section in the competition for awards and in the instruction and general benefits derived by the nation from the Exposition.

Under these circumstances I perform a pleasing duty in placing on record the grateful acknowledgments of this Department, and I venture to express a hope that Congress will signify in some public manner its sense of services of a most responsible and arduous character, rendered not only without compensation, but involving many expenses incidental to the position which would not otherwise have been imposed upon Mr. Beckwith.

For an exposition of the nature of these duties and the manner in which they were discharged, and for many terse philosophical commentaries incident to them, I refer with pleasure to the appended extracts from the official correspondence of the Commissioner General with the Department.

On the 9th of October, 1865, a general agency was organized at New York under the direction of J. C. Derby, esq., who, taking counsel of competent committees specially qualified to advise him in the selection of products belonging to each group and class, adopted prompt measures to make known to producers the inducements which existed for taking part in the Exposition. Circulars and pamphlets giving full details of the plan and organization were prepared by the Department and circulated through every available channel in every State and Territory.

Special acknowledgments are due to Mr. Derby and to Mr. William C. Gunnell, chief civil engineer and architect, Mr. A. P. Mulat, engineer and architect, and to the other gentlemen connected, from time to time, with the New York agency, for their co-operation with Mr. Derby in his efforts to bring out a representative exhibition in the United States Section which would fitly indicate the condition and resources of the country. These efforts, considered with reference to the spirit of the debates in Congress, and to the delay caused by hesitation to make the necessary appropriations, may be regarded as remarkably successful; and although the United States Section did not contain such a collection of products as would constitute anything like a proper or just basis for estimating the industrial or natural resources of the United States, the summary of prizes cannot but be regarded as highly gratifying to the country.

To the advisory committee, and to the Chamber of Commerce of New York, for the effective measures adopted by that influential and public-spirited organization to promote the success of the movement in the United States, the department and the country are much indebted. Messrs. Samuel B. Ruggles, Elliot C. Cowdin, and Professor Charles A. Joy presented the subject for consideration and labored with commendable and efficient zeal in awakening a proper appreciation of it in the public mind throughout the country.

The light thrown upon the subject by the very able speech of the Hon. N. P. Banks in the House of Representatives did much to promote the enthusiasm of the people and a greater appreciation of the importance of the Exposition.

The general charge of the preliminary correspondence with the New York agency, and with the co-operative committees, was early confided to Mr. Henry D. J. Pratt, to whose desk in the diplomatic branch of the

Department the subject pertains. The efficient manner in which this duty has been performed invites special acknowledgment.

Commissioners Bowen and Reynolds, of Illinois, and State Commissioner Gotthiel, of Louisiana, made special efforts and obtained valuable contributions from their respective sections of the country.

In California, and upon the Pacific coast, through the exertions of the citizens of San Francisco, the Chamber of Commerce, the State Commissioner, I. N. Hoag, secretary of the State Agricultural Society, and a committee of the citizens of Nevada, a very satisfactory representation of the products of that portion of the country was added to the Exposition, and contributed largely to the value of our national representation.

In order to present a comprehensive and connected view of the progress of the executive administration of the Exposition intrusted by the Department to the Commissioner General, as well as to show the difficulties and the nature and details of the labor required for the proper conduct of a participation of the country in such great international displays, I present extended selections from the official correspondence of the Commissioner General and others, which, while giving a historical epitome of the relation sustained by the United States to the whole Exposition, will serve as a general introduction to the valuable series of special reports by the United States Commissioners and scientific experts. These reports constitute a valuable portion of the fruits of the participation in the Exposition by the United States, and present to the people of this country much useful and instructive information concerning the practical arts, and constitute a novel and profitable class of public documents, the tendency of which will be to expand and improve manufactures and arts, and increase the application of scientific principles and discoveries, which, so far as they cheapen the transformation of raw materials to articles for the use of man, or improve their quality, increase the wealth of the nation and lighten the burdens of taxation.

The editorial care and direction of the publication of these reports have been intrusted to Professor William P. Blake, of California, who attended the Exposition as Commissioner from that State, and was one of the scientific experts selected by the Commission. I feel very sure that Congress, and the general public, will sustain me in the opinion that this important responsibility has been discreetly and faithfully discharged.

WILLIAM H. SEWARD.

MARCH 3, 1869.

CONTENTS.

I.

ORGANIZATION AND FORMATION OF THE EXPOSITION.

Preliminary correspondence—The invitation to the United States—Suggestions for the organization—Appointment of an agent in New York—Importance of the Exposition to the United States—Notices to persons intending to exhibit—Selection of products and the allotment of space—Letters from the Commissioner General to the agent in New York—Grouping and classification adopted by the Imperial Commission—Transmission of plans of the United States Section—The advisory committee in New York—Resolutions adopted by the Chamber of Commerce of New York—Efforts to obtain extension of time—Transfer to Paris of the labor of apportionment of space—Motive power—Proposed exhibition of costumes and of aboriginal races—Exhibition of heavy cannon and munitions of war—Society of International Travel—Completion and opening of the Exposition—The opening on the first of April—Condition of the United States Section at the opening.—pp. 9–75.

II.

THE PROGRESS AND CLOSE OF THE EXPOSITION.

Scientific Commission; the importance of obtaining the assistance of professional and scientific persons to study and report upon the Exposition—Reports upon the progress of letters and science in France—The organization and duties of a scientific commission—Commission upon weights, measures, and coins—International Exhibition of measures, weights, and coins—Preparation of the catalogue of the United States Section and publication of statistics—Field trials of agricultural machines at Billancourt—International jury and its organization—New order of awards—Apportionment of jurors to the United States—Work of class juries—The distribution of prizes—Honorary distinctions—Exhibition of medals and diplomas—Prizes for reaping and mowing machines—Condition of the industrial arts indicated by the awards—Commission of the United States—Regulations issued by the Secretary of State—Meetings and proceedings of the Commission—Close of the Exposition and delivery of products—Cereals collected by exchange—Minerals donated to various institutions and letters received in reply.—pp. 77–139.

III.

THE ACTION OF CONGRESS—ESTIMATES, APPROPRIATIONS, AND EXPENDITURES.

Joint resolutions passed by Congress—Estimates by the Commissioner General of the cost of the Exposition—Estimates in detail for transportation, unpacking, installation, guarding, linguists, foundations, and fixtures for machinery, decorations, cases, storage, legal expenses, &c.—Estimate of expenses of Scientific Commission—Discussion of the amendment proposing to strike out the provisions for the payment of a part of the appropriations in coin—Report of the advisory committee upon the necessity of further appropriations—Expenditures, report from the Commissioner General—Report from the agent in New York.—pp. 141–158.

IV.

THE PUBLICATION OF THE REPORTS,

Statement of the authority under which the reports have been printed—Publication in a separate form and reasons therefor—Grouping of the reports in volumes—List of the reports by title, arranged according to subjects—Alphabetical list of the authors of reports.—pp. 159–163.

V.

CLASSIFICATION OF THE OBJECTS EXHIBITED AND GENERAL INDEX.

The classification of objects adopted by the Imperial Commission—Its comprehensive and exact character—Its value as an index to the Exposition and to human industry in general—Enumeration of objects in each group and class and references to the reports.—pp. 165–181.

LIST OF UNITED STATES COMMISSIONERS.

ORGANIZATION AND ADMINISTRATION OF THE UNITED STATES SECTION.

I.

ORGANIZATION AND FORMATION OF THE EXPOSITION.

PRELIMINARY CORRESPONDENCE—THE INVITATION TO THE UNITED STATES—SUGGESTIONS FOR THE ORGANIZATION—APPOINTMENT OF AN AGENT IN NEW YORK—IMPORTANCE OF THE EXPOSITION TO THE UNITED STATES—NOTICES TO PERSONS INTENDING TO EXHIBIT—SELECTION OF PRODUCTS AND THE ALLOTMENT OF SPACE—LETTERS FROM THE COMMISSIONER GENERAL TO THE AGENT IN NEW YORK—GROUPING AND CLASSIFICATION ADOPTED BY THE IMPERIAL COMMISSION—TRANSMISSION OF PLANS OF THE UNITED STATES SECTION—THE ADVISORY COMMITTEE IN NEW YORK—RESOLUTIONS ADOPTED BY THE CHAMBER OF COMMERCE OF NEW YORK—EFFORTS TO OBTAIN EXTENSION OF TIME—TRANSFER TO PARIS OF THE LABOR OF APPORTIONMENT OF SPACE—MOTIVE POWER—PROPOSED EXHIBITION OF COSTUMES AND OF ABORIGINAL RACES—EXHIBITION OF HEAVY CANNON AND MUNITIONS OF WAR—SOCIETY OF INTERNATIONAL TRAVEL—COMPLETION AND OPENING OF THE EXPOSITION—THE OPENING ON THE FIRST OF APRIL—CONDITION OF THE UNITED STATES SECTION AT THE OPENING.

In order to give a history of the participation of the United States in the Exposition at Paris in 1867 selections are presented from the official correspondence of the Commissioner General and others, particularly such dispatches and inclosures as show the organization and progress of the exhibition made by this country. Some of the earlier correspondence was in part transmitted to Congress by the President, December 11, 1867, and was published in a series of pamphlets for general distribution.[1] A portion of this earlier correspondence is here reproduced

---

[1] The early publications setting forth the progress made from time to time in preparing the Exposition appeared at intervals until the time of opening in 1867. These publications, in the order of their issue, were entitled as follows:

1. Message from the President of the United States, December 11, 1865, transmitting a report from the Secretary of State concerning the Universal Exposition to be held at Paris in the year 1867. 8vo, pp. 58.

2. Supplemental circular relative to the Paris Universal Exposition of 1867: Proceedings of the Chamber of Commerce of the State of New York. Washington, Government Printing Office, 1866. 8vo, pp. 14.

3. Speech of Hon. N. P. Banks, of Massachusetts, upon the representation of the United States at the Exposition of the world's industry, Paris, 1867. Washington, D. C., Mansfield & Martin, publishers, 1866. 8vo, pp. 24.

4. Second supplemental pamphlet, Paris Universal Exposition of 1867: Details of organization. Washington, Government Printing Office, 1866. 8vo, pp. 64.

5. Third supplemental circular respecting the Paris Exposition of 1867: Importance

in connection with the later letters, in order to give a connected view of the organization, progress, and general administration of the United States Section of the Exposition.

In a message from the President transmitting to the Senate and House of Representatives a report from the Secretary of State concerning the Universal Exposition to be held at Paris in the year 1867, the subject was commended to the early and favorable consideration of Congress. The letter of the Secretary of State, dated December 11, 1865, was as follows:

"The Secretary of State has the honor to submit a copy of correspondence between the Department of State and the minister of France upon the subject of an invitation extended by the government of France to that of the United States, to take part in a proposed Universal Exposition to be held at Paris in the year 1867; also a copy of correspondence between the department and the minister of the United States at Paris, and other papers, explaining the nature and magnitude of the Exposition, the general utility of such exhibitions, and the measures which it has been found expedient to adopt, subject to the approval of Congress, in order to secure for the United States the advantages of participation by their citizens in the Exposition.

"It being necessary that the Imperial Commission at Paris should, to enable them to carry out their programme of arrangements so far as it relates to the United States, be notified, without delay, of the decision of this government, it becomes important for Congress, at the earliest practicable moment, to adopt such proceedings as in their judgment may be best calculated to meet the requirements of the occasion.

"Special attention is invited to the copy of a letter of the 16th ultimo from N. M. Beckwith, esquire, the Provisional Commissioner General of the United States at Paris, which is appended to one of the same date from Mr. Bigelow, and which clearly explains the importance of prompt action.

"From the correspondence it will appear that the selection of the officers hereinafter named, subject to the approval of Congress, was an indispensable preliminary for any participation by the United States in the Exposition, namely: John Bigelow, esquire, (the minister of the United States at Paris,) special agent of the United States for the Exposition, (without extra compensation for that service;) N. M. Beckwith, esquire, Commissioner General of the United States, (without compensation;) Monsieur J. F. Loubat, Honorary Commissioner of the United

---

of prompt action, &c.   J. C. Derby, general agent for the United States.   Washington, Government Printing Office, 1866.   8vo, pp. 71.

6. Senate Ex. Doc. No. 5, 39th Congress, 2d session: Message of the President of the United States, communicating, in compliance with a resolution of the Senate of the 19th December, 1866, information in respect to the progress made in collecting the products, and the weights, measures, and coins of the United States, for exhibition at the Universal Exposition at Paris in April next.   8vo, pp. 52.

States, (without compensation;) J. C. Derby, esquire, general agent in the United States, resident at New York.

"It will also appear that such appropriation for the payment of necessary expenses as may be made will be a judicious outlay, from which large returns may be confidently anticipated in effects upon the national revenues and resources, by tending to expand the demand for our productions, by attracting for the development of our latent wealth reenforcements of labor and capital, and in the collection and diffusion of useful knowledge, of the improved applications of science to agriculture, manufactures, and art, through the results of the reports of the general scientific committee. The moral influence, moreover, of a just and liberal illustration of the vitality and progress of this nation, at such an international gathering, so soon after a great civil war, ought not to be overlooked in the consideration of this subject."

PRELIMINARY CORRESPONDENCE.

THE INVITATION TO THE UNITED STATES.

The following is a translation of the letter addressed to Hon. William H. Seward, Secretary of State, by L. de Geofroy, minister of France to the United States, and dated at the legation of France to the United States, March 27, 1865:

"By two decrees, dated June 22 and the 1st of last month, the Emperor has ordered that a Universal Exposition of the productions of agriculture, manufacture, and the fine arts should be opened at Paris May 1, 1867. Another decree, also issued February 1 of this year, and published in the Moniteur the 21st of the same month, has placed this grand international solemnity under the direction and supervision of a commission, the presidency of which has been confided to his Serene Highness Prince Napoleon.

"Such a selection bears too high testimony to the importance which the Emperor attaches to the success of this Universal Exposition to leave any need to dwell upon it. As to the commission, it is composed of several of his Majesty's ministers, of high functionaries of state, as well as of the most competent of notable individuals.

"The government of his Majesty charges me to give notice, officially, of these aforesaid decrees to the cabinet of Washington, to invite its valuable concurrence, and to designate an authority with which the Imperial Commission could have a direct understanding.

"It would also be of advantage, to avoid all loss of time, that the government of the United States should make choice at Paris of an agent who would be specially delegated to be near his Serene Highness the Prince Napoleon.

"This mode of procedure is the most suitable channel, and the speediest, to convey to the knowledge of the Imperial Commission the wishes of the exhibitors from abroad.

" The government of his Majesty would attach a high value to being informed as early as possible of the result of the steps I am charged to take, which have an exceptional character of urgency.

" The objects sent to the Exposition will be received, in effect, in a palace constructed for the occasion of this solemnity, and the size of which should meet the actual need of the exhibitors of all nations. But that the general arrangements and plans which shall be adopted may be in relation with the claims for space which will be preferred, it will be necessary that the Imperial Commission should know, with the least delay, what states will take part in the Exposition, and how much space each would desire to obtain.

" In ending the letter he has written to me on the subject, the minister for foreign affairs adds that he is gratified to hope that the government of the United States will show a disposition to facilitate, so far as it is concerned, the success of the work confided to the Imperial Commission. It is too enlightened not to appreciate the advantages of these solemnities, at which nations contract new ties, collect useful and mutual lessons, and thus assure the development of their prosperity."

A copy of this note was transmitted, April 5, 1865, by Mr. Seward to John Bigelow, esq., minister of the United States to France, with the following letter:

" I give you, for your information, a copy of a note which I have recently received from M. de Geofroy, chargé d'affaires of the Emperor, concerning a projected Universal Exposition of productions of agriculture, manufactures, and the fine arts, to be opened at Paris on the first day of May, 1867, under the direction and supervision of a commission in which his Serene Highness the Prince Napoleon will preside.

" You will inform M. Drouyn de Lhuys that the President of the United States regards the project thus described with great favor, as well because of the beneficent influence it may be expected to exert upon the prosperity of the nations as of its tendency to preserve peace and mutual friendship among them.

" The Prince Napoleon is most favorably known on this side of the Atlantic, and his connection with the Exposition will increase its proper prestige in the eyes of the government and people of the United States.

" What the executive government can do by way of concurrence in the noble purpose of his Majesty will, therefore, be very cheerfully done. The design and arrangements will be promptly promulgated. For the present you will confer with M. Drouyn de Lhuys, as a special agent of this government, and will bring yourself into near relations with the Prince.

" This is as far, however, as the President is able to proceed without special legislative authority. Application for that authority will be made to Congress when it shall have convened. In the mean time this department will receive and give due attention to any suggestions which the government of France may desire to offer, with a view to a complete success of the contemplated Exposition."

Mr. Bigelow also addressed Mr. Seward upon the subject, as shown by the following extract from a communication dated at the legation of the United States, Paris, April 12, 1865 :

" I presume you have already received official notice of the Universal Exposition which it is proposed to hold in Paris in the summer of 1867, coupled with a request that the ingenuity and enterprise of our people should be represented in it. That you may lack none of the elements in my possession which are necessary to determine the true policy of the United States in reference to this Exposition, I will state what has occurred at this legation in connection with it.

" On the 18th of last month I received a note from Prince Napoleon, president of the Imperial Commission, inviting me to confer with M. Le Play, Commissioner General of the Exposition, in reference to a proper representation of the United States on the occasion, to which his Imperial Highness professed to attach much importance. Early in the following week M. Le Play called upon me at the legation, and since then I have had a second interview with him at his office. He seemed anxious to know, in the first place, if my government would feel an interest in having the ingenuity and skill of the country represented at the Exposition. I ventured to express to him my decided conviction that it would ; that in 1867 we all hoped and believed grim-visaged War would have smoothed his wrinkled front in the United States, and the arts of peace would have resumed their accustomed supremacy, in which case an opportunity of seeing, at a glance, what progress the whole world had made in the arts of civilization during the preceding five or ten years, and also of showing to the world what we ourselves had accomplished, would unquestionably be highly prized by my countrymen.

" M. Le Play seemed highly gratified by this assurance. He said the Prince president had been very much astonished by the marvels of ingenuity and skill which he had observed in the United States, and was anxious to have them more known and appreciated in France.

" M. Le Play, with the utmost delicacy, suggested that it would be desirable that our government should place the direction of its representation at the Exposition in the hands, and, as far as possible, under the absolute control, of some person worthy of the trust, through whom the exhibitors, or their agents, and the central commission, might communicate as occasion required. He spoke of this arrangement as likely to obviate some of the inconveniences which the commission experienced at the exposition of 1855. On that occasion nearly every State had its separate commissioner, subordinated to no central authority. Infinite confusion, and a great deal of dissatisfaction on both sides, were the inevitable consequences. M. Le Play, who was also commissioner general of the exposition of 1855, seemed to think it highly desirable that some trusty and competent person be invested with exclusive authority to communicate officially with the central commission, and to require the several State commissioners or agents to communicate through him as the proper agent or representative of the whole nation, just as on all political matters they would communicate through its diplomatic agent. I told M. Le Play that I concurred entirely with him in this suggestion, and should not fail to recommend it to my government, though, as an appropriation for money would be necessary to give such a commissioner his proper efficiency, the suggestion had come too late, I feared, for as early action as would be desirable. Congress having adjourned, no money could be appropriated by the government for this purpose before next winter, and it was, therefore, impossible for me to say in what way my government might find it convenient to manifest its interest in the objects of the Exposition before that time. M. Le Play seemed to regret the delay, which he feared might prejudice the interests of our representation in this wise: It is proposed to appropriate the Champs de Mars to the Exposition. A vast building is to be constructed in the center of this beautiful space, which embraces about one hundred and fifty acres; and around the edifice, at a proper distance, groups of houses, or small villages, will be constructed and furnished to represent the domestic habits and characteristics of different nations. This will probably be the greatest novelty of the Exposition, if successfully executed, and nothing will be neglected by the Prince president, who has his heart very much in it, to make it a success. The plans for the structures necessary to the development of this feature ought to be matured without delay, and for that purpose there is immediate need of a commissioner to advise with in regard to the United States. I suggested that perhaps the President might take it upon himself to name a commissioner now, and define his duties, leaving it to Congress, when it meets, to fix his compensation, if he is to be paid, and, in any case, to supply him with the funds required in the proper execution of his duties.

"He seemed to think that the sooner such a person should present himself here the better, and at the same time gave me to understand that an office would be provided for him in the Palais de l'Industrie, beside his own, and all the architects and personnel of the commission would be at his disposal.

"M. Le Play further informed me that it is the present intention of the Imperial Commission to assign about six times the space to exhibitors from the United States which was assigned to them in 1855. This is to be independent of the space occupied by the outside structures, which will doubtless be in proportion.

"When this subject began to occupy my attention, I consulted Mr. N. M. Beckwith, a very intelligent American gentleman, at present residing in Paris, who had been one of the commissioners at the New York Exposition of 1853, and who was also more or less in the councils of the American exhibitors at the Exposition of 1855. His experience and good judgment led me to attach great value to his opinion in regard to the proper mode of turning the Exposition of 1867 to the best account, and I requested him to give me his views in writing. He has been good enough to do so, and I have taken the liberty of annexing them to this dispatch.

"So far as I have any well-defined opinions upon the subject, they lead me to approve of the suggestions of Mr. Beckwith. I think, however, the success of the whole thing depends mainly upon having a competent central commissioner. He should be a man of high character; reasonably familiar with the great sources of our national wealth; accustomed to organize and employ the labor and talents of others; thoroughly acquainted with the French people and their peculiar modes of organizing their industry; and, above all, he should be conversant with their language, without which all other accomplishments would be nearly valueless."

SUGGESTIONS FOR THE ORGANIZATION.

The following is an extract from the letter of Mr. Beckwith to Mr. Bigelow referred to above. It is dated at Paris, April 3, 1865:

"In continuation of our conversation about the International Exposition, permit me to add a few words.

"The value of French exports last year was five hundred and eighty-one million dollars, and shows an increase of fifty-one per cent. in four years.

"This growth of the external commerce is but the index of the greater growth of internal commerce, resulting from the increased productiveness imparted to labor, skill, and capital; and the increased productiveness is traceable in details directly to the application of the sciences to the industrial arts.

"If it be true that civilization was led in most countries for a long period by a few men of genius skilled in political science and literature, it is not less true that the men of physical science have at length come to their aid.

"The geologists, naturalists, chemists, mineralogists, inventors, and engineers are now directing the labor of the world with a success never before attained.

"As the intellectual domination of the material world increases, the hardships and barrenness of toil diminish and its products multiply; and while political science emancipates the enslaved races, physical science enslaves the elements and forces of nature and emancipates mankind.

"In this great movement the largest benefits will fall, with the largest markets in the world, to those who make the best provision for the development and diffusion of the practical sciences as applied to industry.

"No nation produces within itself all these in perfection, nor keeps up with the daily progress in them; but those are most advanced in the race who adopt the best methods of collecting and disseminating the progressive knowledge resulting from the studies and labors of all.

"Among the methods for this purpose, international assemblies and exhibitions are increasing in numbers, in frequency, and in importance.

"A knowledge of many of the useful and successful combinations of science and industrial art cannot be conveyed in words; they must be studied in models and specimens, which display at once the combinations and effects, the modes and results.

"These being the products of many localities and many countries, bringing them together facilitates their study, and affords, at the same time, the opportunity of careful and accurate comparisons, without which no study is complete.

"The utility which experience ascribes to this method is indicated in France by a comparison of the provisions made for the exhibition of 1854 with those making for 1867.

"The first was entered upon timidly, the government relying chiefly on private capital and enterprise, on which the labor and risk were thrown.   The latter has been taken up boldly as a business of state, and projected on a larger scale, contemplating an expenditure of twenty millions of francs, of which twelve millions are to be supplied from the public funds, leaving eight millions as the probable contribution of visitors.

" The United States have never participated in these assemblies to the extent naturally suggested by their interests, intelligence, and enterprise, nor derived from them the benefits they might have done.   I attribute this to the want of a suitable organization of the movement, to the want of timely information on the subject, and provision for the transportation, placement, and proper exposition of objects, and to the absence of the necessary co-operation of the government in aid of the exhibition.

"First. The first step toward a proper organization is indicated by the regulations of the Imperial Commission, which require the governments intending to co-operate to appoint a commissioner, duly accredited to the Imperial Commission, which commissioner will have charge of the business belonging to the country whose government appoints him.   It is necessary for the commissioner to be in constant communication with the Imperial Commission, to enable him to lay before the exhibitors early information of the plans and designs as they are developed during the whole progress of the formation of the Exposition.

"Second. The commissioner will require an agency in New York, to centralize the movement in the United States, to communicate with exhibitors and impart to them the requisite information in detail, and to facilitate in general the movement.

"The commissioner will also require (at a later period) the assistance of a committee, composed, first, of the professional and scientific persons whom the government should appoint to study and aid in preparing a suitable report of the exhibition, to be subsequently published; second, of the agents appointed by different States, or associations, and such other persons as the commissioner may find necessary to aid in the general work.

"*Remark.*—The agent in New York, and the professional men the government may appoint, should be paid; all others should serve without pay. The agent should select his own local committees or assistants, and so distribute them throughout the States as to render the movement active and efficient.

"This organization, completed in smaller details, is the simplest and the least that will answer the purpose, and I feel no hesitation in expressing the conviction that nothing will be done on a scale worthy of the country, and with the completeness requisite for public benefit, if the government does not take the initiative in the manner and to the extent here indicated.

"It is obviously necessary that the organization should conform to the plan of the Imperial Commission; and it is equally obvious that in a movement of this kind, where there is no authority, and no corresponding responsibility, (which can only emanate from the government,) there is not likely to be the order, co-operation, and unity requisite for efficient management and useful results.

"If the government decides to inaugurate the business in this way, the monetary provision required from Congress will, doubtless, be readily made. The country which taxes itself and appropriates more public money to education than all other countries will readily aid its men of the industrial sciences and arts to be present with the evidences of their skill in an assembly of nations where all contribute for the improvement of all, and from which none can retire without benefit.

"The diffusion of knowledge is in proportion to the numbers brought in simultaneous contact with its sources and with each other; and the more numerous the objects assembled, the more numerous the exhibitors and visitors brought together, the better will be the results."

On the 2d day of August, 1865, Mr. Bigelow again addressed the department upon the organization of the Exposition. The dispatch is given entire, but the inclosures are omitted:

"At a recent interview with M. Le Play, the Commissioner General of the Universal Exposition of 1867, he informed me that the Imperial Commissioners had finally fixed upon the Champs de Mars for the site of the Exposition, and had proposed to reserve for the United States 3,346 square metres of space within the edifice, with the privilege, if we required it, of some 1,600 metres lying adjacent and not yet appropriated. The map which accompanies this dispatch, and marked inclosure No. 1, will show the manner in which this space is distributed, and the proportion which the aggregate bears to the allotments made to the other powers.

"M. Le Play wished to know what assurance I could give that we would occupy so much space. I replied that, unfortunately, this subject was not brought to the attention of my government until after the adjournment of Congress, which does not meet again until December next; that the amount of space we should require would depend very much on the liberality of its appropriations, the executive government having no

2 P E

funds or credits available for such a purpose. I also read to him from your dispatch, in which I was designated as 'special agent,' the expressions of the interest which our government took in the Exposition; directed his attention to the important changes in our domestic affairs since that dispatch was written, all calculated to favor our participation in the Exposition; and I concluded by expressing my personal conviction that the United States would make good use of all the space that had been allotted to it, and that no effort would be wanting, on my part, to secure such a representation as would be creditable to my country.

"Further than this I told him I could not go; for though I believed that any recommendation which the President might make upon this subject to Congress would receive its approval, I could give him no stronger assurance of it than my personal conviction. I urged the Commissioner General, at the same time, to let me have the detailed plans of the Imperial Commissioners at as early a moment as possible, to submit to my government, that no time should be lost, on the one hand, in preparing a programme for the action of Congress, and, on the other, in taking steps to ascertain the disposition and requirements of exhibitors.

"About two weeks after this interview I received from M. Le Play two communications. Of the first, inclosure No. 2 is a copy, and inclosure No. 3 is a translation; and of the second, inclosure No. 4 is a duplicate, and inclosure No. 5 is a translation. By inclosures Nos. 2 and 3 it will be observed that the Imperial Commission has felt constrained, in consequence of my inability to give the Commissioner General more definite assurances, to reduce our allowance of space room from 3,346 to 2,788 square metres.

"I have as yet made no reply to this communication, for I have none to make. Though the commission has left us about nine times the space that we occupied in 1855, still I reget the reduction, so firmly persuaded am I, should the opportunity be fairly presented to our people, that the proportions which this Exposition is destined to take in the eyes of the world within the next twelve months will render it much more difficult to limit our contributions to the larger space than to fill it creditably.

"Inclosures Nos. 4 and 5 embrace the general regulations and the system of classification adopted by the Commission. For the translation of the classification I am indebted to Mr. Beckwith, who has consented to act in the capacity of a special commissioner, under a power derived through me, as the special agent of the United States. In a note which accompanied this translation, Mr. Beckwith says: 'If the government would publish the classification in the newspapers, they would thus probably reach every individual in the United States interested in the subject. The classifications, like a carefully written chapter of contents, comprise more information as to the scope, limits, character, and objects of the Exposition, than could be given in any other form in an equal space. They suggest, of themselves, much of the information most useful and most desired by the public at this stage of the enterprise, which renders it important that they should be published and distributed without delay.'

"I concur entirely in this recommendation, for the reasons to which I shall refer more at length presently. If *our* people are to participate in this Exposition, no time should be lost in supplying them with the means of knowing how they may do so to the best advantage, and for that purpose they must study the regulations and systems of classification patiently and thoroughly. They may do that profitably, whether they finally exhibit or not, for they will there find probably the most complete classification of the products of human industry and art anywhere to be found in print.

"There are some features of the regulations to which it is proper that I should invite your attention at once. I may have occasion to trouble you about some of the others at a later day.

"The Exposition is to open on the 1st of April, 1867, and to close on the 31st of October of the same year. The foreign commissioners are to be notified of the *space allotted* to their respective nationalities before the 15th of August instant, after which, I am given to understand that it will be impossible to make any material changes in that regard. All applications for admission, with a description of the articles to be exhibited, must be presented before the 31st of October, 1865, prior to which time also a plan or chart of the uses to which the space will be put by each nationality respectively must be made by the foreign commissioners, on a scale of $0^{m}.002$ per metre, and sent to the Imperial Commissioners.

"Detailed plan of articles, and their distribution in the space assigned them, must be furnished on the same scale by the foreign commissioners, as well as materials for the official catalogue, before the 31st of January, 1866.

"It thus appears that within the next six months, and before any action is likely to be taken by Congress, the Imperial Commission must know not only precisely what articles will be offered for exhibition, but they must have an accurate plan of their distribution. How far these regulations may be relaxed, and the time extended, will depend upon circumstances; but, from the nature of the case, it is impossible that they should be relaxed so as materially to relieve American exhibitors, for the reason that the plan of the Exposition requires a peculiar disposition of the articles, from which any serious departure is impracticable. This plan is explained in a communication from Mr. Beckwith, of which inclosure No. 6 is a copy, and to all of which I invite your attention.

"It may, therefore, be assumed that to wait for the action of Congress before organizing the American department of the Exposition of 1867 is equivalent to an abandonment of all profitable participation in it. All the plans must be laid, and the chief expenses incurred, if not made, before Congress can be heard from.

"Should our country people, however, attach to the privilege of sharing in the Exposition anything like the value which is attached to it by the people of Europe, it ought not to be difficult to find capitalists willing to anticipate the action of Congress by requisite advances of means whenever the government shall submit to them a plan or line of policy which it is prepared cordially to recommend to Congress and the public.

"I trust that in the documents which I have already transmitted, with those which accompany this communication, the government will find all the information it will require to fix, without delay, upon the policy it ought to pursue.

"Before closing this communication, there are one or two other features of the regulations to which it is my duty to invite your attention.

" By article 5 it is provided that all communication between foreign exhibitors and the Imperial Commission shall take place through the commissioners of the respective countries, and in no case will they hold direct communication with the exhibitors. For this purpose foreign commissioners, if there are many, are invited by article 6 to appoint a delegate, as soon as possible, to represent them near the Imperial Commission.

"These provisions are designed to meet the inconveniences which have heretofore resulted from a multiplicity of commissioners, who were often exhibitors, and to concentrate the practical cares of managing the Exposition in the hands of persons specially selected for the duty, and who, by a careful study of its plan and familiarity with every stage of its growth, are best qualified to promote its success. These regulations also tend greatly to simplify the organization through which our government will have to operate. With an appropriation sufficient to pay such portion of the expenses of transportation as it may conclude to assume, and other allied expenses, (I would recommend that it assume the charge of all articles at tide-water in the United States until they are returned, those sold during the trip to pay their own charges,) and with two commissioners, one to reside in Paris and the other in New York, properly qualified for their duties, the official or governmental organization would be, for the present, and for the next eighteen months at least, complete. This subject is more fully developed by Mr. Beckwith in inclosure No. 6, to which, for the present, I content myself with inviting your attention, as presenting what seems to me the simplest, the most economical, the most harmonious plan of operation that I can imagine, and one open to fewest objections, and most certain to work successfully. I think it would be wise to take measures to avoid, as far as possible, any representation by States at this Exposition, for the Imperial Commission never know what relative value to attach to such commissioners, and the result of such a representation here would be, as it has always been before, that the whole national character of our part of the Exposition would be sacrificed to the interests of a few sharp-witted speculators who might chance to know best how to turn the inevitable confusion and disorder that would result to their own account.

"When the Exposition is ready to open, it will be proper for the United States to be represented by a very different and more numerous body of men, who, by their knowledge and accomplishments, are qualified to describe in popular language the novelties with which the Exposition may abound. It is from the labors of such men as these that the country ought to derive its chief advantages from such an Exposition, but such men are not apt to be qualified nor to have the leisure or taste for any of the labor which precedes the opening or which follows the closing of the Exposition.

"In France it is provided that the Imperial Commission shall organize in each department what it terms departmental committees, whose duties, among others, it will be to create a commission of savans, agriculturists, manufacturers, master-workmen, and other specialists, who should make a special study of the Exposition, and prepare and publish a report on the various applications which may be made in their department of the information they may gather. To meet at least a portion of the expense of this work, private subscriptions are authorized to be opened in the several departments.

" Something similar should be done by our people and government; and in the selection of candidates for such work, no pains should be spared to select the most capable from among the class of men who have enough of our own skill and resources to determine what is new and worthy of transplantation to the United States. This work will be done for the nations of Europe by their ablest men, for thus only are the important lessons of the Exposition to be perpetuated and diffused. I hope we shall not disregard their example. In making choice of men for this labor our academies of art and design, our agricultural societies, our mechanics' institutes, and other literary and scientific societies, might possibly be consulted to advantage.

" With no other apology for these somewhat perfunctory suggestions than my desire that our country may not only appear to advantage at the Exposition of 1867, but that its artists and artisans may profit by the unexampled opportunity for instruction which it will present, I remain, sir, with great respect," &c.

APPOINTMENT OF AN AGENT IN NEW YORK.

Mr. J. C. Derby, the United States dispatch agent, New York, having consented to act as the agent for the Exposition in the United States, he was instructed by the department as follows, under date of October 9, 1865:

" SIR : Having been informed of your willingness to act as the agent in the United States for the Paris Exposition for 1867, I inclose for your guidance and information a copy of a pamphlet prepared and published by this department, and which contains the dispatches of Mr. Bigelow relative to the conditions upon which citizens of the United States can participate in the Exposition. The limited period allowed for applications to be filed was, on the 2d of September, pointed out to Mr. Bigelow, and he was requested to inform the Imperial Commission that an extension of the time would be gratifying to this government; and on the 21st of that month his attention was again called to the importance of such an extension of time as would enable all of our citizens, who are so disposed, to unite in the Exposition so far as the space assigned will permit.

"Your attention is particularly invited to the suggestions made by Mr. Beckwith, in his letter of the 30th of July, printed on page 26 of the pamphlet, and to Mr. Bigelow's remarks on page 7 of the same.

"Two thousand copies of the pamphlet have been distributed, a number having been sent to each of the governors of States and Territories, and a number having been sent to various other quarters where they would be likely to reach parties interested. Seventy-five copies, which remain on hand, will be forwarded to your address without delay, for such disposition as you may think proper. Whenever the result of the application for extension is known here, you will be informed of it."

Mr. Bigelow, at Paris, was informed of this appointment by a letter of the same date, from the department, (dispatch No. 284,) and of which the following is an extract:

"SIR : With reference to the correspondence which has taken place upon the subject of the French Universal Exposition for 1867, I have to inform you that J. C. Derby, esq., the dispatch agent of the United States at New York, has been selected, and has consented to act as the agent for the Exposition in this country. I will thank you to request Mr. Beckwith to enter into correspondence with him as to the steps which it may be advisable for him to take in that capacity.

"With regard to the extension which you have been requested to ask for of the time for filing applications of our citizens to become exhibitors, I would suggest that, if it should be found that the Imperial Commission is unable formally to accede to the proposed change, you will request Mr. Beckwith, when he prepares the general plan of organization of our branch of the Exposition, required according to the programme on the 31st of the present month, to make such allowance as his judgment may dictate for additional machinery and articles for which it may be expected subsequent applications will be made."

The following is Mr. Bigelow's reply, (October 27,) together with a communication from Mr. Beckwith, under date of October 26, 1865 :

"SIR : I have the honor to acknowledge the receipt of your dispatch No. 284, with an inclosure, by which I am advised of the appointment of J. C. Derby, esq., of New York, as agent for the French Universal Exposition of 1867, to reside in the United States.

"I also have the honor to inclose a copy of a letter this day received from Mr. Beckwith, Commissioner of the Exposition for the United

States residing at Paris, from the tenor of which it would appear desirable that Americans wishing to exhibit should be notified as soon as possible to send in their applications with specifications to Mr. Derby, instead of sending them to Mr. Beckwith. The reasons for giving this direction to the applications are sufficiently disclosed in Mr. Beckwith's note. I would suggest, also, that exhibitors be notified at the same time to make their applications as soon as possible, that the New York commissioner may have time enough to make his selections, allotments of space, drawings, &c., and transmit them to the Commissioner at Paris before the 31st of January.

"It may be also desirable that the public be prepared in some way, either in this notice or otherwise, to expect that it will be the endeavor of the commissioners to secure as complete a representation of the art and industry of the United States as possible, and for that purpose it will be necessary for them to make selections of representative articles in every class or group, rather than accept many specimens in the same class, whatever may be their merit. As the space will be limited, it is as well that this guiding principle of having a complete Exposition, if we are to have any, should be known early, both to aid in bringing about such a desirable result, and to prevent needless disappointment.

"It is to be presumed that the Army and Navy Departments have some novelties appropriate for this Exposition; if so, it is needless for me to say that anything coming from those quarters would be likely to command special attention."

Mr. Beckwith to Mr. Bigelow.

"PARIS, *October 26, 1865.*

"DEAR SIR: In conformity with the instructions of the Secretary of State which you communicated to me, I have to-day placed myself in correspondence with J. C. Derby, esq., agent, New York. I have prepared for him—

"1st. A general letter placing before him the present state of that part of the business of the Exposition of 1867 which he will have first to take up.

"2d. The loss of time consequent upon the necessity of waiting for the action of Congress renders it necessary to transfer to New York the work of dividing the ground among exhibitors, (as suggested in my letter to you of the 30th July, published,) where preparation can be made pending the needful legislation, to complete the work of distribution in a brief space of time afterward.

"I have, therefore, transmitted to Mr. Derby eighteen letters, comprising all the applications for space in the exhibition which I have received to this date. I have desired him to place the letters on record as a part of the applications to be considered in making the distribution of ground, and I have in conformity advised the writers that they will receive from Mr. Derby, in due time, definitive advices of the result of their applications.

"I shall now prepare as early as possible the plans and drawings by which Mr. Derby will be governed in making the allotments, and shall point out to him the manner and extent to which he can alter these plans to suit circumstances without departing from the general order to which all conform. These documents will be accompanied by explanatons and information which will, I hope, render the work easy.

"I would now suggest the expediency of a notice, authorized by the government, requesting all who wish to exhibit, and have not made applications, to send in their applications to Mr. Derby, with a limit of time in the notice beyond which no applications can be received.

"The work will be so far advanced by this method, I trust, that by the time the needful legislation is finished the allotments can at once be made, and the plans, catalogues, and reports sent forward, so as to be returned to the Imperial Commission within the extended time they will be able to allow us.

"I beg to call your particular attention to the importance of the allotments of ground; this, in reality, is the formation in embryo of the Exposition.

"The selections of products will be limited in quantity to the area they are to occupy, but in variety and character they should comprise a full and fair representation of American products, industry, arts, and science.

"To make these selections and the allotments of space for them is the work which now devolves on Mr. Derby, and for the selections it is not probable that any one man could be as competent as several, each chosen for his knowledge in different departments.

"When the applications are all in, and the work prepared, the selections and apportionments, which must proceed together, will occupy but little time.

"The attention of the government, I trust, will be given to this, and suitable persons invited to assist Mr. Derby for a brief period in this important part of the work."

In regard to the extension of time requested of the Imperial Commission, Mr. Beckwith wrote to Mr. Bigelow November 16, 1865:

"DEAR SIR: The observations relating to the action of Congress in regard to providing transportation for the Exposition, contained in the article annexed to the circular of the Department of State of the 18th November, leave the impression that there is no occasion for the immediate decision of Congress on that subject, and as no other subject is named requiring early attention, the inference naturally suggests itself that there is none.

"I cannot doubt, however, that your dispatches and my letters have presented the real situation, which requires an early decision, and that this will appear in the communications of government to Congress.

"The application for time (which was granted) related only to the report due on the 31st October. That report was preliminary, and admitted of subsequent modifications, and delay in regard to it was not of great moment, but the important report called for on the 31st January next is final in regard to that part of the work. It includes the allotment of ground and formation of the Exposition, (in embryo,) leaving but the subsequent labor of bringing it to maturity. This report cannot be made until after the action of Congress.

"All that has been done is provisional and contingent on the future decision of the government; but to make the report in question we must abandon contingencies, and enter upon positive engagements with the Imperial Commission and with exhibitors. The early decision of Congress is therefore indispensable to avoid further delay and another appeal for more time.

"There can be no doubt of the readiness of the Imperial Commission, and of the Emperor, to grant all the delay possible, without interrupting seriously the general progress of the work; but how far a delay of the important report alluded to would embarrass the general movement I am unable to judge.

"All that the Imperial Commission has said on the subject is, that the work is well advanced; that we are the only nation now in arrear, and they hope, and appear to expect, we will soon be able to make up lost time. I am the more anxious to have the present state of the business clearly understood, because, after the action of Congress, we shall need all the delay we can obtain.

"There is a good deal of work to be done in New York, which has been presented in ample detail to Mr. Derby, but the work cannot be done till after the decision of Congress, and if forced to be done hastily, cannot be well done."

IMPORTANCE OF THE EXPOSITION TO THE UNITED STATES.

Mr. Beckwith to Mr. Bigelow.

"PARIS, *November* 23, 1865.

"DEAR SIR: In proposing the Exposition of 1867, the French government represented its chief object to be a collection of the useful products of all countries for the purposes of comparison and the study of the methods and processes connected with the production and fabrication of the objects collected, and that this end would be attained in proportion to the variety and universality of the collection. National exhibitions thrown together by the spontaneous action of producers never have the character of universality desired. Producers who are most active, or who act most in the spirit of sparing no expense in advertisements to increase sales and profits, come forward, while many whose products are equally desirable, and perhaps more instructive, have no occasion or no disposition to make use of the method, and they

do not appear.   Such collections are defective, and to that extent
failures.

"The course adopted by the French government on this occasion
differs from that of preceding attempts, and is expected to have better
results.   Invitations to co-operate are limited to governments, and the
respective governments are solicited to undertake the work for their
respective countries, giving to their exhibitions the arrangement provided in the general programme, which will bring them all in harmony
with each other.   Governments thus co-operating, it may be usually expected, will adopt each for itself the local measures necessary to prevent a
partial exhibition and to secure a collection more universal and fairly
representative of the country in every department of national and industrial products.   In this connection you will appreciate the importance
which attaches to the distribution of the ground to exhibitors, because
that comprises the formation of the Exposition and determines its character.

"I consider it superfluous to develop and discuss the direct advantages
of international exhibitions in general, or of this one in particular, to
the United States.   They present themselves to intelligent minds, and,
fortunately, we have no others to present them to.

"Those who are familiar with the industrial products of England
(and who are not ?) are aware that their prominent qualities are strength,
solidity, and utility ; that those of France have always been remarkable for beauty and taste.   They cannot have failed to observe, also,
since the epoch of international exhibitions, the rapid improvement of
English products in graceful forms, beautiful combinations of colors,
finer designs, and superior taste, while those of France rise equally in
the important elements of strength, durability, and fitness.   Similar
observations apply in an eminent degree to Belgium, which learns and
combines from both ; and the same may be said in some degree of other
surrounding nations.   Nor is this surprising.   Inventions, combinations,
discoveries, improved methods and processes, spring to light simultaneously in many fertile minds, and in many localities of all countries,
but the knowledge is slow in spreading itself into general use.   Its
diffusion is quickened by international gatherings and exhibitions.
But on this occasion there are indirect considerations which invite us
with unusual urgency to co-operation.

"No one is more sensible than yourself of the deficiency of exact information in Europe in regard to America previous to the rebellion, in
a political, literary, and moral sense, in a physical, geographical, statistical, financial, industrial, scientific, and productive sense, and in
every sense.   It was obvious at every step, everywhere, and among all
classes, and it suggested an incredible indifference, unaccountable to
those not acquainted with the causes of such deficiency.   The events of
the last four years have made the United States more known than all
the events of their previous history.   Their magnitude, their resources,

and their strength are now acknowledged. The strong impression produced is pleasing or unpleasing, according to the sympathies or aversions of classes and interests, but none deny the presence of a great power, and its advent is acceptable and hopeful to the masses of the numerous peoples.

" Emigration of the productive and industrial classes from Europe to America is an acknowledged source of prosperity, and has long received the encouragement of the government.

" An exhibition of the products of America in the center of Europe, well selected, and complete enough to be national, showing the mineral and agricultural resources, the state of manufactures, the varieties and quantity of machinery, and the condition of the industrial arts in general, would, in my judgment, produce an impression of surprise analogous to that produced by the disclosures of the war. The strongest impression would naturally fall on the mind of the most intelligent portion of the productive classes, who are most appreciative in this sense, and have the best means of being informed. This is the class of skilled labor and of practical knowledge, whose emigration is highly desirable, but who are slowest to risk the change. They would see and judge for themselves of materials and resources and products; of the existing conditions and opportunities open to them to better their condition in life.

" Financial organizations under the patronage of the French government (a plan of which I have sent Mr. Derby) are now forming to aid the class of operatives in question to assemble from all parts of Europe to be present at the Exposition and to remain and study it. The concourse will be large, and they are the practical students of exhibitions.

" We can participate in the benefits resulting from this, and I do not think it chimerical to suggest that an American exhibition, well selected and really national, viewed merely in its economical aspect, is desirable, and would return to the treasury, by increased immigration and augmented revenues, more than its cost, however liberal the provision of Congress.

" The United States are the only nation of importance which has yet to express itself definitively on the subject, and a lively interest attends the action of Congress, not only on account of its bearing on the Exposition, but as an expression of its appreciation of the object and enlightened spirit of the undertaking."

NOTICES TO PERSONS INTENDING TO EXHIBIT.

Soon after the organization of the agency in New York, Mr. Derby issued a revised and enlarged edition of the official pamphlet, giving information to the public and directing attention to the importance of the proposed Exposition. He also issued a circular letter to the manufacturers, mechanics, inventors, producers, engineers, architects, artists, and scientific and educational organizations of the United States, of which the following is a copy, and sent one hundred copies to the governor of each State and Territory:

"NEW YORK, *November* 23, 1865.

"The undersigned, having been appointed by the Secretary of State to the above-named agency, and being desirous of the co-operation of his countrymen in his efforts to make as complete, interesting, and creditable as possible the representation of this country at the great Exposition, adopts this method of conveying to them information and suggestions upon the subject.

"In compliance with a request made through our minister at Paris, the time for filing applications from the United States has been so far extended that all which reach the undersigned before the 1st of January next will be in season. When examined and considered, the decisions will be duly made known.

"Parties wishing to exhibit are requested to apply immediately to the undersigned for correct forms of application and instructions, inclosing postage stamps for reply.

"Articles accepted should be delivered at New York prior to January 31, 1867.

"Accepted articles will be shipped from New York to Paris and returned at government expense, provided the necessary action of Congress obtains.

"To prevent unnecessary trouble, it should be understood that it is a primary object to make the representation of the United States as complete as possible in all the classes and groups enumerated in the programme published in the official córrespondence, and that it will therefore be necessary to select representative articles in every class or group rather than accept an excess of any one class.

"In order to secure the universality of character above indicated, it is suggested that in each city or neighborhood those classes of manufacturers, artisans, and others who produce articles for very general use or consumption, should, without any delay, agree among themselves as to the specimens for which space should be applied for.

"Every effort should be made to bring forward new and useful mechanical inventions, combinations, and fabrics, and pains should be taken to have all articles neatly and thoroughly finished and prepared for exhibition.

"As the decisions, report, and plan of arrangements from the undersigned must reach Paris prior to the 31st of January next, it is very desirable that all applications should be sent in as much earlier than the 1st of that month as may be practicable."

SELECTION OF PRODUCTS AND ALLOTMENT OF SPACE.

THE COMMISSIONER GENERAL TO THE AGENT IN NEW YORK.

"PARIS, *October* 26, 1865.

" In conformity with the directions of the Secretary of State, I have the pleasure to address you on the subject of the Universal Exposition of 1867.

" The information which has doubtless been sent you by the Department of State, and the publications from the same source, will have placed the business before you in its present state. This relieves me from the necessity of further preface, and I take up the subject at the point where the publications leave it.

" The delay which has unavoidably occurred in organizing and initiating the work has rendered impossible a strict compliance with that part of the Imperial Regulations (article 7) which calls for a report and specifications on the 31st of October, 1865.

"An application now pending has consequently been made for an extension of time, and there is reason to expect that it will be granted. But as nothing definite can be done by us until authorized by Congress, and as the action of Congress is still distant and the period of its decision uncertain, it becomes the more necessary to be prepared to act rapidly when the time arrives. In the ordinary course it would have been practicable to receive in Paris all the applications of exhibitors and to make the allotments of ground here, but this would require six weeks or two months more of time, and, in view of the time already lost and still to be lost, it is necessary to transfer this part of the work to your side. The work, therefore, which will first come before you will be as follows:

"First. A public notice for all applications (not previously sent in) to be sent in to you within a limited period.

"Second. When this time has elapsed, and the applications (including those made here, which I shall send you) are all in, the selections will have to be made, which will form the exhibition, and the total quantity of products accepted will be limited to the total area provided for them.

"Third. The apportionment of ground to each exhibitor, to which his name and number will be attached, to his locality in the place designated.

"Fourth. The classification of products and placement in conformity with the general plan of the Imperial Commission, and drawings, to a scale of $0^m.002$, which will exhibit the distribution made, together with the plan and arrangement of tables, cases, and fixtures of all kinds for the exhibition of the products.

"Fifth. Plan of the fixtures required, and statement of the force needed for machines in action.

"Sixth. Catalogues of names of exhibitors, with their numbers, and catalogues of the objects to be exhibited, as described in article 7 of the Imperial Regulations.

"I am preparing, and shall transmit to you as early as possible, the drawings and detailed plans which you will need in the distribution and apportionment of ground; and I shall indicate the extent to which the Imperial Commission will allow the remodeling of these plans to suit your particular requirements, without breaking up the general plan conformed to by all; and I will add such other information and explanations as will, I hope, make this part of the work easy.

"The most important part of the labor will be the selection of products, selection being necessarily limited to the area for exhibition, and the governing idea being a fair and, as far as possible, a full representation of national products.

"For this labor not any one man probably would be as competent as several, each of whom would be better informed in some particular department.

"I am not informed as to whether the government will provide for this, or whether it will be left to you, nor is it of much moment, provided the right thing be done. But you will observe the importance of the selection, *which, in fact, is the real formation of the Exposition,* and its completeness and value will depend on the knowledge and judgment displayed in this department.

"The preceding will serve to inform you of the work you have to prepare for."   *   *   *   *   *   *

"No. 4.]   "PARIS, *November* 1, 1865.

"DEAR SIR: The leading object of the French government in undertaking the Exposition of 1867 is indicated in the method adopted by the Imperial Commission for the purpose of forming the Exposition. The principal motive of producers in exhibiting may be to advertise the qualities and value of their products, thus augmenting sales and profits. But these considerations are only collateral and secondary with the government. The primary object is an opportunity for the comparison of products and the study of processes by which the knowledge that multiplies products, improves their qualities, and diminishes their cost, is diffused.

"For this purpose, it is obvious that the Exposition should be 'universal;' that is, it should comprise specimens of the useful products of the universe. To give to the Exposition, as far as possible, the character of universality, the method of forming it, suggested by experience, and adopted by the Imperial Commission more fully than in any preceding exhibition, is the following: All useful products are first divided into groups, and the groups divided into classes. The ground on which the products are to be exhibited is then divided into compartments corresponding to the groups and classes, and these compartments are, in due course, to be filled with their appropriate objects. By this method of proceeding, the Exposition will of necessity have the character of universality intended.

"An examination of the grouping and classification which have been published will show that, however diversified and different the products of different countries and climates, they will all find a place in the different classes, while no country of any extent, probably, will be found destitute of products suited to each class.

"In dividing the ground, the importance of some products as compared with those of the same country is not overlooked. The more important should have a corresponding representation, which, in general, implies a larger space.

"This is provided for as follows: The divisions suited to the products of France and adopted by the Imperial Commission are represented as a model. But discretion is reserved to the commission of each country to remodel this plan and adapt it to their own wants, which is only limited by the skeleton or autonomy of the general plan, which requires all groups and classes to be preserved, and precludes any from being entirely obliterated.

"The property of this provision may be explained in this manner: All countries, for example, produce clothing; but the makers of clothes in our country might not feel much interest in exhibiting their work in another country, with a view to markets, where differences of climate, of race, and of habits are against them.

"There is, however, no product of labor more important, none in which human skill has been more universally, nor to which science and art have been more elaborately, applied in the conversion of raw material, in the adaptation of garments to climates, to particular uses, and to the various conditions of life, and for the comparisons necessary to an appreciation of the best qualities of each; collections of native costumes or clothing from all countries are equally desirable and valuable.

"The method thus carried out will obviously produce the conditions desired—facilities of comparisons and the studies of processes relating to products of greater importance, and to those of less importance to the products of one locality, as compared with those of another in the same country, and to the products of all countries compared with each other.

"The Exposition will at the same time be, to a large extent, an advertisement of products for the direct interest of producers.

"My chief purpose in this brief explanation of method and object is to call your attention more pointedly to one of the topics in my letter of the 26th of October,[1] viz: The allotment of ground to exhibitors.

"The allotment of ground is the formation of your exhibition; when this is complete your exhibition (in embryo) will be completed.

"The success of its representative character, in a national sense,

---

[1] The letter here referred to is published in the official pamphlet, second edition, page 37, and with these papers, pages 23–24.

depends, therefore, in the knowledge and judgment displayed in the allotments, because that determines at once the variety of products to be displayed, and the quality and importance ascribed to those selected for exhibition in each department. A right understanding of the views of the French government in regard to the Exposition, in which the United States are invited to co-operate, and the importance which attaches to the allotments, will, I hope, excuse my having returned to this subject and dwelt so long upon it."

"No. 9.]   "PARIS, *November* 8, 1865.

" DEAR SIR: I have alluded in previous letters to the great import-ance attached by the Imperial Commission, not only to the exhibition of useful products, but to the exhibition of the methods and processes by which these objects are produced.

"Extensive preparations will be made in the Palace and in the Park to exhibit machinery in action, accompanied by the persons usually em-ployed with it, displaying at once its method of action and its products.

"Great efforts are also making to bring together and exhibit groups of families of persons of all nations usually employed in the industrial arts, whether carried on by mechanical means or by the use of a few tools and implements combined with manual labor and skill, dressed in their native working costumes, installed in their usual habitations, or those resembling them, and fabricating the objects they exhibit.

"The interest and importance which the Imperial Commission ascribes to the exhibition of methods and processes, the scope intended to be given to this department, the police, sanitary, and other peculiar pro-visions requisite, and the general co-operation which is invited, are set forth in the document hereto annexed. It comprises thirty-two pages, chiefly in lithograph and partly in manuscript. It has not yet been pub-lished, and is incomplete. The plan is developed day by day, under the study of the Imperial Commission, aided by the suggestions of others, which are invited and frequently adopted.

" I send it in the imperfect form, because I think it sufficiently devel-oped for your purposes, and no more time should be lost in presenting it for your consideration and that of the persons with whom you will doubt-less advise in forming the exhibition.

"The programme, you will observe, includes all nations and nation-alities, civilized and uncivilized, among whom industrial arts exist; and there are few people without them.

"Doubtless the greatest variety and number of these industrial groups will come from Oriental nations, which are little advanced in the science of mechanics, and destitute of the great combinations of capital and skill embraced in large manufactories. Industrial art among them is still confined to the family circle; but their products are abundant in variety and quantity, frequently excellent in quality, often of great beauty, and in the important elements of utility and cost they still hold in check and nearly control the great markets of the East, exposed to the competition of the best fabrics of Europe and America.

"But the Imperial Commission does not limit its Exposition to the East; it hopes for similar exhibitions from North America and from South America; and I am desired to bring the subject to your particular attention.

"The programme is comprehensive in the scope of industries it proposes to exhibit—workers in metals, in glass, in chemicals, in wood, in leather, in all materials; hand-spinning, weaving, and embroidery, machine sewing, machine shoemaking, knotting of fish-nets, twisting of fish-lines. No industry will be out of place, even to a group of red Indians making pipes, bows, wampum, feathers, or baskets. These last, indeed, would be among the most unique and interesting objects you could send. They would add a valuable feature to the ethnological elements which the many nationalities assembled, with their peculiar habits, manners, industries, and character, are expected to display, and which subject the French Scientific Commission has been particularly directed to study.

" However uninteresting a group of red men may be in America, few objects would be thought more interesting in Europe; while similar groups brought from the East may afford subjects equally curious and instructive to Americans."

" No. 10.] "PARIS, *November* 8, 1865.

"DEAR SIR: The special committee (French) on admissions, Class No. 93, on habitations combining cheapness, health, and comfort, have published the document annexed.

"Ground in the Park is appropriated for this purpose, and great importance is attached by the Imperial Commission to the exhibition of rural habitations from all countries. It is suggested, also, that the furniture adapted to them, being on exhibition, may be placed in them, and that they may be inhabited by the families or groups of persons alluded to in my letter No. 9, and the documents attached to it.

"The impression prevails that we produce in America model houses of iron, combining many useful qualities and adapted to many localities; also model houses of wood, comprising similar qualities in a higher degree—such houses as are shipped to California, &c. But great interest attaches to the exhibition of rural habitations, of whatever material, adapted to all classes of laborers and every grade of fortune, including the log-houses of remote settlers and those of the transitional condition, from a humbler to a higher state of prosperity and comfort corresponding with the use and development of condition and wealth in settlements of rapid growth, in which no country can compare with America. A row or group of this kind would speak strongly to the eye and the mind. It would contrast strongly with corresponding groups from different parts of Europe and the East, where characteristics are immobility and poverty—no growth, no change. Habitations of this description are typical

3 P E

of the moral and physical condition of the great bulk of the population of all countries; they indicate the degrees of intelligence, thrift, and prosperity among them, and would be objects of interest and instruction to the great emigratory classes, as well as to the philosopher and economist."

"No. 17.]   "PARIS, *November* 10, 1865.

"I inclose an application, in behalf of the State of Michigan, for twenty-one thousand feet of blank space in the Exposition. I also inclose my reply to M. d'Aligny, which please read and forward to him. I suggest the expediency of your publishing in the newspapers as advertisements, or otherwise, a notice to applicants, comprising the observations I have made to M. d'Aligny on the necessity of an exact description of each article to be exhibited, without which you never can complete catalogues for the report on which the Imperial Commission compiles its publications. The description is indispensable, and in the outset I have assigned some explanatory reasons for this method of proceeding, which you will probably agree with me are requisite to satisfy applicants and induce them to comply with the requisitions.

"You are doubtless conscious how strongly we Americans are disposed to revolt at everything chalked out for us, and how inclined to think we could make it better, (and perhaps we could,) and therefore take our own way about it.

"But the question of better or worse does not arise in this case; it is merely a question of method or no method. Any method that all follow is better than none; and as it belongs to the Imperial Commission to lay down the method, it belongs equally to us to follow it, and we cannot get on without. You had better be firm on this at the outset; you will have to come to it, and it will save time."

"No. 21.]   "PARIS, *November* 27, 1865.

"DEAR SIR: I beg to hand you with this a number of drawings, six in all, numbered 1 to 6; they develop plans of that section of the Exposition Palace appropriated to the United States, and are accompanied by detailed explanations of each drawing, which document is numbered 77. Explanations of this kind seldom appear as clear to the reader as to the writer; many details which are present in his mind, and fill up the outlines, are omitted in the description from a feeling that they will suggest themselves, and that a record of them is superflous, and would only make the description tedious and obscure, rather than clear.

"But the plans and explanations will, I hope, be found sufficient to enable you to make the distribution of groups and classes, and the allotments of place to exhibitors with facility, and free from error.

"At all events, if you find my details defective, I must refer you to the French plan, No. 1, which I send you; it is all I have had to work from, and I hope you may find the study of it more interesting than I do.

"The plans herewith relate only to the Palace; nothing is said of the Park, nor of the three groups (VIII, IX, X) and twenty-two classes which belong to the Park. I shall return to this subject as soon as the Imperial Commission makes up its mind on it, and decides on the distribution and manner of occupying it.

"No definite apportionments of ground in the Park to nationalities have yet been made. All are told they can have what they want, but I imagine there is some difference of opinion as to the manner of occupying the grounds. The Imperial Commission is, therefore, inviting from the foreign commissions suggestions as to how much ground they want, and how they wish to employ it. Doubtless in a few days the plan will be settled, and the appropriations made, to be occupied in conformity with the ground-plan, which will be promulgated.

"I think you will find we have less room in Group VI in the Palace than we require. My impression is, we ought to occupy twice as much room as we have in that department. The United States are not so strong in products of the other groups as in those of the sixth, and they are of a kind that require room. But the plan of the building does not admit of giving us a larger portion of room in that group; it is the same as falls to other nations; but the products of other nations do not demand so much room in that department.

"I have, therefore, proposed, in writing to Washington, if it should be the opinion of your committee also, to supplement the ground of the Group VI in the Park, provided the Imperial Commission will consent to the requisite modification; and my present impression is, they will do so, though they have not yet given me a definite answer. You had better, therefore, as soon as you are ready to do so, express your opinion to the government on this subject, and inform me also of your views.

"I shall not wait, however, for the advice, but secure the ground conditionally if I can, but I wish to hear from you as soon as possible in regard to it.

"If we occupy a space in the Park with objects of Group VI, it will necessitate the construction of a building suitable to the purpose at our expense; but I think we shall not hesitate about that if we want it, nor do I imagine Congress will hesitate.

"The government of the United States will not be satisfied to undertake an exhibition of the produce of the country on a diminutive scale, nor permit it to fall short and be deficient for want of room, nor on account of the additional expense this may involve. Belgium is in a similar situation, and has resolved to supply the room she needs by building on the Park, which, I have no doubt, will be permitted; but if we find we can do without it I shall be glad of it.

"This proposal does not affect the arrangements to be made for Groups VIII, IX, X, which belong to the Park, the provisions and allotments for which are now delayed by the Imperial Commission."

"No. 22.]   "PARIS, *November* 29, 1865.

"DEAR SIR : I am favored with yours of the 13th instant, which reached me last evening, and I take due note of your observation.

"You will by this time have acquired a good idea of the work to be done, and papers I send you by this mail will complete your impression of the best way of doing it. It is necessary to appreciate the difference between an irregular and defective exhibition, which characterizes itself by spontaneous movements without concert of producers, and an exhibition formed by the state, which should be well selected, classified, and complete in all its parts.

"You desire to know how long it will be safe to continue to receive applications, and the date of the latest mail which will reach here in time.

"It would be easy to reply to those inquiries if we could be governed solely by the demands of the imperial programme for January ; but this is impossible. We must be governed by the requirements of the programme, taking the risk of failure ; there is no other way.

"The first thing you have now to do is to sketch your plans of the ground for Groups II and V, (see my explanation of plan 3.) The second is to decide the space you will give (or thereabouts) to each class of objects in the respective group, and mark out the space in conformity ; and the third is to select from your applications the most representative and suitable products, and form and file the groups and classes laid out in your plan.

"When this is completed your exhibition will be formed. This work requires knowledge of products, judgment, and care ; it cannot be hastily done and well done, but it can doubtless be accomplished, and your plans drawn and catalogues made while Congress is deliberating.

"There will then remain but little to do after the decision of Congress but to announce to applicants the result of their applications. This announcement will constitute the definitive allotments of ground to exhibitors. It forms the contract between the exhibitors and the government, and between the government and the Imperial Commission, and cannot be made, of course, till authorized by the government, neither on your side with the applicants, nor on my side with the Imperial Commission. You will doubtless have the work so far advanced, in the form indicated, by the time you receive the orders of the government, that you can close up the part necessary to the report of January 31 in very brief time.

"This is all you can do in advance, and you must be governed by the movements of Congress up to that period ; you cannot be governed by the requirements of the imperial programme. If we keep up with the action of Congress, (which we must do,) and still the business from the delay of Congress falls behind and finally fails, we shall have done all we could do. As you will see the movements of Congress you can shape your own by them ; but should Congress decide sooner than I anticipate, you must still take time to do the work in a proper manner. We were not authorized to begin sooner, and it would be a mistake to close the work prematurely, half done or badly done.

" It is easier to find a reasonable and acceptable excuse for taking the time absolutely necessary than to apologize for imperfect and bad work when it appears.

"As soon as it is decided that Congress will pay the expenses you will have applications enough, which will enable you to fill up the groups and classes, and form the exhibition in a more complete manner than you could otherwise obtain. But the work of filling up the groups and classes will doubtless involve some negotiations with exhibitors, and will inevitably require time, and the work should not be slighted; dispatch will depend on the skill and competency of your assistants. But I have no doubt of your being able to keep up with Congress, and you can judge better than I can when and what notice, or if any notice, to close the acceptance of applications, is necessary.

" If anything occurs to make it necessary to be more positive in regard to dates, I will of course advise you. All I can now say is, you have the programme and knowledge of the situation, and have only to use the greatest dispatch compatible with the circumstances and with the work which must be done.

" I have no doubt from what I have since heard of Mr. Evans,[1] he is qualified for the work I suggested, and will be very useful to you. You cannot get on without a competent man in that department, and if he is competent I may want him here. If Congress does not refuse to undertake the work, it will not refuse to pay the cost of it; if it does refuse, there is an end of it. No provision is made for the reception of the produce of any foreign country not presented by the government of the country, nor is it likely, under the circumstances, that producers would be willing to appear in any other way even if it were practicable, which it is not."

ADVISORY COMMITTEE.

Some of the citizens of New York who felt a deep interest in the proper representation of our country in the Universal Exposition, at the request of Mr. Derby, and with the approbation of the State Department at Washington, organized an "advisory committee" to assist Mr. Derby, particularly in making selections from the numerous applications for admission of products. The services of this committee were gratuitous. It consisted of ten members, one for each of the ten groups as set forth in the programme of the Imperial Commission. The aid of experts in each group or class was obtained, and great efforts were made, not only in the city of New York, but elsewhere, by journeys through portions of

---

[1] Mr. F. W. Evans, of Boston, Massachusetts, chairman of the advisory committee upon Group VI, who died while engaged upon the work of organization. Mr. Evans was educated as an engineer at the *École Centrale*, Paris, and his untimely death caused a great loss to the work.

the United States, to secure a full representation of the multiform and various products of the country. They had the benefit of the active co-operation of several State Commissioners, of many societies, and of private individuals, and occasionally had the opportunity of consulting with some of the Government Commissioners.

At a meeting of this advisory committee held in the city of New York, on the fourth day of December, 1866, the secretary, Professor Charles A. Joy, was directed to prepare an abstract of the minutes of previous meetings, and to state what further measures would, in the opinion of the committee, be required in order to carry on the work to a successful completion. From the report presented in conformity with these instructions, it appears that early in January of 1866 the following communication was addressed to Mr. Derby by the committee:

"Your communication of the 19th instant, informing us that 'upon consultation with prominent citizens interested in the growth and development of the resources of our country,' we had been designated as a committee to aid you in the selection of proper articles for exhibition in Paris in 1867, has been duly received; and after a brief consideration of the subject, and in compliance with your request, we beg leave to submit the following suggestions:　　　　　　　　　　　　　　　　＼

"This is the first time that the government has proposed to take part in a foreign exhibition. Hitherto the representation has been by individual effort and without system, and has been in no sense national.

" It is now incumbent upon those having the matter in charge to take prompt, efficient, and comprehensive action, to insure a creditable display of the products and productive capacity of the United States; and if, in consequence of the shortness of time and of inadequate appropriations, it is found that the work cannot be properly done, it would be better for the nation to be excluded from the Exposition than for us to send forward a defective and partial exhibition, which will be neither useful nor respectable, nor in any way representative of the products of the country.

"The representation of the United States at the Exposition of 1867, that would be satisfactory to its government and its people, and worthy of effort and expenditures, would be one that furnishes its representative products in each of the several classes as set forth by the Imperial Commissioners, so far as they are known to exist in this country.

"In our judgment, even if the time were not short, there would be great difficulty in undertaking to obtain these products by an appeal for voluntary offers; but under present circumstances, and expressly in reference to that of time, it is not to be expected that such measures will effect the desired representation, and that therefore recourse must be had to very different means in order to insure the end in view.

"First. As it appears to us, it is necessary that it be made known to the people of the United States that it is the intention of the government, in view of great and important national considerations, to take the necessary measures, with the co-operation of its citizens, to have the products and productive capacity of the country fairly represented at the Exposition of 1867.

"Second. That the government will furnish all the transportation necessary from the seaports of the United States to Paris and back; that it will provide agents to receive, take care of, and return the products furnished; and that it will empower a suitable commission to apply for and receive applications in such detail as may be necessary for selection, and finally to determine what articles are to be asked for, obtained, and forwarded, and that, in defining the duties of such commission, it shall be specially provided that the best products of the several kinds shall be selected, and where there are numerous producers of the same class of products of the same degree of excellence, care shall be taken to apportion the articles among as large a number of producers as possible.

"A publication of this intention of the government, accompanied by an appeal in the proper spirit and language, and setting forth clearly what is asked for of the producers, and, impressively, the principle of fairness and impartiality that will be required of the commission, would, it appears to us, meet with a response which would enable the commission to perform its part.

"To some extent the commission might find it necessary to make special application to obtain creditable products. It would be of great service to such commission to have copies of the catalogues of the expositions of 1851, 1855, and 1862, in Europe, and of 1853 in the United States.

"The government of the United States ought to be a contributor, as is the case with foreign governments. It could order the whole of the larger parts of an engine for a war steamer to be set up in Paris, as a fair indication of our capacity in that class of production.

"Should the action of the government and of the producers of the United States be of the character briefly set forth, it is evident that no small space at the Exposition will be required; and we deem it necessary to remark that, in view of such action, the spaces occupied in the expositions of 1851, 1855, and 1862, under entirely different influences, afford no proper basis of conclusion as to the allotment required in 1867.

"Not having before us any estimate of the expenditures required for the participation of the United States in the manner proposed, we can hardly with propriety name any sum; but in our view of the urgent need of very prompt action, we deem it proper to say that it appears to us that not less than $300,000 should be placed at the command of the appropriate department, from which the commission would receive its powers and instructions, and to which it would make application for such funds as may be necessary to perform the work intrusted to them.

"In this communication we have aimed to present, in a summary manner, the views which we have formed. Of course, very much remains to be considered and decided.

" If in the future proceedings it is thought that we can be of service, we shall be happy to meet you and to render such aid as may be in our power.

" For the advisory committee: [Signed by] Horatio Allen, Samuel B. Ruggles, Frederick Law Olmsted, Charles A. Joy, sub-committee."

An estimate of expenses was prepared in conformity with the above letter.

" Mr. Ruggles and Mr. McElrath, of our committee, repeatedly visited Washington to urge upon Congress the necessity for immediate action.

" They, with others, addressed public meetings and published articles in the papers of the day.

" It was not until the 5th of July last, more than a year after the attention of the government had been first called to the subject, that any appropriations were made, and those then made were quite inadequate in amount.

" The United States agency has therefore labored under disadvantages not experienced in other countries.

" The uncertainty which prevailed to some extent in Congress, in the peculiar condition, at a certain period, of our public relations with France, whether the United States would participate at all in the Exposition, and the consequent delay in the passage of the appropriation, rendered it impossible, at an early day, to arouse the national spirit to the extent that a different state of facts would undoubtedly have witnessed.

" Notwithstanding these untoward circumstances, a very considerable work has been done, and much more can be accomplished if immediate additional aid be rendered by Congress.

" The highest interests of the nation evidently demanded the utmost efforts of your committee to stimulate the country without delay to a full exhibition of its products, notwithstanding any temporary inadequacy of the appropriations.

" They have proceeded under the conviction that Congress, when fully acquainted with the magnitude of the subject, and its consequent necessities, would make any necessary increase in the appropriations.

" At the meeting of the Advisory Committee, December 4, 1866, the respective chairmen of the ten groups submitted full reports of what they had been able to accomplish up to that date."

Mr. William J. Hoppin, chairman of the committee for Group I, embracing works of art, &c., reported that a general invitation to participate in the Exposition had been addressed to artists and others, and extensively circulated in the newspapers. A committee charged with the duty of selecting was organized from among the owners of private galleries and familiar with the condition of art in this country.

" They adopted the rule to accept, if possible, only the best things we have done since 1855, and this rule necessarily excluded some interesting and creditable works, which, if the competition were among ourselves, and not between the United States and foreign nations, would probably have been admitted. Want of space also compelled the exclusion of some valuable productions.

"It was determined to give great predominance to landscapes in our selection, because this was the department in which the American school of art has gained most distinction.

"In obedience to these rules, the art committee endeavored to decide which were the best pictures that had been painted by the leading men within the last ten years, and then to obtain these works by direct applications to their owners.

"In sculpture the same general rules of selection prevailed, and some of our best productions will be sent to Paris.

"The owners of these works of art expect no private advantages from this enterprise, and are willing, for no other motive than to increase the fame of the artists and the credit of the nation, to submit to the absence of their treasures for nearly a year, and to the risk of their possible loss. It therefore seems no more than reasonable that Congress should make an additional appropriation for return freight, premiums of insurance, and the necessary expense of an agent or custodian.

"The value of the works of art thus contributed and loaned by these individuals for the public benefit is at least $150,000, and it would be exceedingly unjust and ungenerous if, in addition to the sacrifices made by them, they should be called upon to pay the charges indicated above."

Professor Charles A. Joy, chairman of Group II, embracing books, proofs and apparatus of photography, musical instruments, medical and surgical instruments, mathematical and philosophical instruments, &c., reported:

That there were 147 applications for space; of these 50 were withdrawn voluntarily or rejected, leaving 97 producers to whom space was assigned.

Mr. Samuel B. Ruggles, chairman of the committee upon Group V, and commissioner, reported:

"The chairman of this group, soon after his appointment by the government in July as one of the ten professional commissioners, for the purpose of securing adequate action by the country personally visited all the States from New York westward to Minnesota and Iowa inclusive, explaining the importance of the Exposition to the interests of the various portions of the United States.

"In these efforts, and especially in the northwestern States, he was actively and efficiently aided by two of his associates in the commission, Mr. James H. Bowen, of Chicago, and Mr. Henry F. Q. d'Aligny, of the Upper Peninsula of Michigan, and also by the zealous co-operation of several of the commissioners appointed by the several States, including Mr. J. L. Butler, of Missouri, Mr. J. P. Reynolds, of Illinois, and Mr. J. A. Wilstach, of Indiana."

* * * * * * *

The magnitude in number and in bulk of the contributions in Class 40 of this group, the products of mines and metallurgy, rendered it necessary to select only the most important and characteristic portions. It was therefore necessary to call in the aid of experts, not only to make the necessary selections, but to classify, label, and properly pack in boxes the specimens to be sent, and for that purpose to procure suitable rooms and several skilled assistants.

This labor for a portion of the collections sent to Paris from New York was performed chiefly by and under the direction of Professor Thomas Egleston of the School of Mines of Columbia College, in the city of New York.

Mr. F. W. Evans, chairman of Group VI, embracing machinery, &c., reported:

"The committee on Group VI was organized in July last, as soon as the action of Congress rendered it certain that the articles accepted could be sent. They had to select from about five hundred applications, and their aim has been to fill up the space allotted to them with representative articles for each class, paying no regard to priority of application, and taking care that every branch of manufacture and of industry comprised in this group should be represented.

"In order to do this, the space being limited, the committee had first to decide on the relative amount of ground to be allotted to each class, and then to fill up such space with the representative articles corresponding. This part of the work required careful study, much correspondence, and some travel, in order to see and understand, so as to decide knowingly on the merits of the articles for which space was demanded.

"Some of the best articles not being forthcoming, the committee deemed it advisable to solicit their representation, especially when such exhibition would necessarily entail great expense upon the owners. And it is to be regretted that it was not in the power of the agency to furnish material aid for some of the manufacturers of expensive and complicated machinery, whose exhibition would confer lasting honor upon the mechanical skill of the country without any immediate pecuniary benefit to the owners."

Mr. W. S. Carpenter, chairman of Group VIII, embracing animals and specimens of agricultural establishments, reported:

"That, under the prohibition by the minister of the interior in France, in view of the danger from the prevalent cattle plague, it was found impracticable to send live animals to the Exposition. The few articles applied for were transferred to Group VI."

In conclusion, the committee reported as follows:

"From the preceding abstracts of the reports of the chairmen of the ten groups some idea may be formed of the amount of work that has been accomplished by your committees in the limited time at their disposal. There have been about twelve hundred applications for permission to exhibit products. Some of them were made in the name of States, and cover a large number of individuals. The number of persons directly interested in the Exposition amounts to several thousands.

"The money value of the articles to be exhibited cannot be stated with accuracy. It would be difficult to form a just estimate; but as only choice articles have been accepted, it can safely be put down at many hundred thousand dollars.

"Many products, the exhibition of which would have proved highly advantageous to the country, were practically excluded for the reason that there was no provision for return freight. To send them to Paris was, in some instances, equivalent to giving them away.

"There has been much enlightened patriotism displayed on the part of exhibitors. Many of them have expended large sums of money for the purpose of showing to the world what we can produce, and western railroad companies have liberally offered to carry freight for the Exposition free of charge.

"Your committee having been familiar with all the details of the work from the beginning, knowing what has been accomplished and how much may yet be done, are in the condition to state what further sums are required to maintain the credit of our country in participating in this world-wide enterprise.

"To sum up these necessities, there is urgent need of an immediate additional appropriation of one hundred thousand dollars to save the property of exhibitors and to complete the work begun.

"The enlightened citizens who have loaned their valuable works of art must be secured from pecuniary loss on freight and insurance; the expense of collecting, assorting, selecting, and labeling ores and minerals, and of publishing concise statistical statements of the extent and value of our mineral lands, ought to be defrayed by the government, with additional appropriation for return freight of suites of specimens which institutions and individuals are willing to loan for the Exposition.

"In the department of machinery the sum at the disposal of the agency is altogether inadequate; there is an absolute necessity for motive power in the supplementary building in the Park, or a very large class of exhibitors will be deprived of the opportunity of showing their machines in motion, and a considerable addition to the transportation fund is required to enable the agent to forward some of the most important machines yet offered.

"The fund is also inadequate for inland transportation in France and return of the packages to the seaport; also for the care of them in Paris and the necessary services of agents and interpreters.

"There is not sufficient money to defray the necessary expenses of the agent in New York, and it is safe to say that, but for the gratuitous aid received from persons not officially connected with the Exposition, and the meager salaries accepted by yourself and others, the work would have been seriously interrupted.

"In the original plan of organization, prepared by the secretary of this committee, provision was made for the appointment of ten commissioners to report the scientific results of the Exposition, and it was proposed to give each commissioner authority to employ the necessary assistants. They should also be authorized, as a body, to appoint a secretary to keep and preserve proper records of their proceedings and their correspondence, and to provide rooms at Paris for meetings and business, with the necessary incidental expenses.

"Adequate provision should also be made for the expense of collecting and exhibiting the weights and measures, and especially the coins of the United States, reaching back to our colonial era, to properly prepare for the international discussion invited by the French commission of the very important question of a common unit of money for the use of the civilized world. The successful establishment of a coinage of uniform weight and fineness, and common to all the nations of the world, would annually save hundreds of thousands of dollars to the citizens of the United States.

"For the necessary objects above specified, your advisory committee are of opinion that an expenditure of fifty thousand dollars by the professional commissioners will be necessary, and should be appropriated by Congress.

"It should be considered, moreover, that the task which has been assigned to these ten commissioners, of preparing a report or series of reports upon the Exposition, and upon the several departments of industry which will be represented in it, is one which, for its proper execution, will require a species of assistance for which no provision has been made in the resolutions under which they have been appointed. In order that such reports may subserve the purpose intended of promoting the advancement of the arts of industry in the country, and thus contributing to the national wealth, they should exhibit not only the present condition of each department, but also some sketch of its history, and some account of the progressive steps by which it has reached its present state of perfection. They will consequently require a large amount of special study and of correspondence or personal communication with the scientific and practical men of other countries.

" For the intelligible presentation of the results they will require to be illustrated by numerous drawings and diagrams, exhibiting the constructions, apparatus, and machinery employed in the various processes which they describe. The purely mechanical labor of digesting the literary material thus collected, and of preparing the illustrations necessary, would be more than sufficient to occupy all the time of the commissioners, were not their proper task a higher one than that of mere historians. If their labors are to be practically useful, they must be free to study, discuss, and criticise the objects and processes upon which they report, to bring into clear relief whatever is most meritorious in each, and to point out the particulars in which improvement is still to be desired, and the directions in which it may be sought. They should, therefore, be authorized and enabled to employ such artistic and professional assistance as may relieve them of that portion of their work which they could only perform in person, to the great prejudice of the final value of their reports.

"The necessity of providing the commissioners with such assistance was early perceived and pointed out by Professor Joy in a letter to yourself published by Congress and by the Commissioner General of the United States in Paris.

"In a communication addressed to the Secretary of State under date of 31st January, 1866, Mr. Beckwith, with the intelligent forecast characterizing all his official communications, remarks: 'The resolutions presented to Congress on the 21st of December proposed appropriations for a scientific commission of ten members, corresponding to the ten groups of products. *But this number, unassisted, will not be sufficient.* It will devolve upon them not only to make the requisite studies and reports, but also to serve on international juries. The latter service, though requiring much time, will afford the best opportunities for information resulting from the investigations, experiments, and discussions of the juries. But they will not be equal *to the work without assistants*, and they can be obtained at a moderate cost. The services of scientific and professional assistants can be engaged, whose special studies, colloquial knowledge of continental languages, familiarity with the continental nomenclature of the sciences and industrial arts, together with their personal acquaintances, access to sources of information and works of authority and local knowledge in general, will render their services as assistants highly efficient. The scientific commission thus supplemented will be equal to the work required of it, and more useful labor can be accomplished in this way at less cost than in any other way.'

"The Advisory Committee have reason to believe that the several governments of Europe which have resolved to participate in the Exposition have not been in any case unmindful of this important provision. Our professional and scientific commissioners cannot but deeply feel the disadvantage under which they must necessarily labor, unless Congress shall see fit to concede to them the same aid in the execution of their task as will be enjoyed by their fellow-commissioners from other lands.

"By reference to the early correspondence between the Commissioner General and the minister of the United States in Paris, it will be seen that the appropriations already made by Congress fall short by more than forty thousand dollars of the sum estimated by the Commissioner General as the very minimum necessary to secure for our country a creditable representation at the Exposition, and very much further below what he thought desirable. These estimates were made with a perfect knowledge of what other governments were doing, and could have been dictated solely by a patriotic desire, not only to secure to our country all the important advantages which may be made to flow from this great international comparison of industries, but also to see her honorably sustaining her part in this most generous of rivalries.

"His estimates will be found in a published correspondence, in a letter addressed to Mr. Bigelow under date of November 22, 1865, and it will be seen that all the additional appropriations asked for by the undersigned might be made without transcending the limits assigned by him, and which the necessities of the case, as they have developed themselves, have shown to be too low.

"The Advisory Committee beg leave further to submit that the provision of the joint resolution of Congress making an appropriation of a certain definite sum for the purpose of defraying the personal expenses of the ten commissioners while engaged in the discharge of their duties, might with propriety be modified.   While these professional men may desire to derive no pecuniary advantage from their connection with the commission, it cannot be proper or just that they should suffer positive pecuniary loss.   Their services, if properly performed, cannot fail to be of material benefit to the country.   If worth having, they are worth paying for.   Their terms of service, including the time occupied in going and returning, extend over a period of eight months.   A moment's consideration is enough to show that the cost of a voyage to France, out and back, and the necessary expense of living for such a length of time in a foreign capital crowded with visitors, and at prices greatly enhanced, are most inadequately met by the appropriation in the joint resolution. It would surely be more just, and far more consistent with the dignity of the nation, that provision should be made for the payment of the actually necessary expenses of the ten commissioners, to be duly audited on proper vouchers by any appropriate officer of the government.

" In conclusion, and in view of the preceding facts and considerations, we, the undersigned, are of opinion that the pecuniary means now at your disposal are quite inadequate to the requirements of the various industrial and public interests of the country.

"We therefore respectfully recommend that you make immediate application to Congress for an additional appropriation, amounting in the aggregate to one hundred and fifty thousand dollars.

"The total expenditure would even then fall considerably short of the proportionate expenditure by most of the countries represented in the Exposition, but it would enable the United States to maintain to a fair extent its just rank in this great concourse of nations."

This report was addressed to Mr. Derby, and was signed by the chairmen of the admission committees of the ten groups.

RESOLUTIONS ADOPTED BY THE CHAMBER OF COMMERCE.

A special meeting of the Chamber of Commerce of New York was held on Friday, January 12, 1866, to hear the report of the committee, consisting of Mr. Samuel B. Ruggles, Mr. Denning Duer, Mr. George Opdyke, Mr. J. S. T. Stranahan, and Mr. Elliot C. Cowdin, in relation to the Universal Exposition of Industry to be held in Paris in 1867; President A. A. Low in the chair. The Hon. Samuel B. Ruggles, in behalf of the committee, reported the following resolutions for adoption:

"*Resolved*, That the Chamber of Commerce of New York have learned, with profound satisfaction, that the government of the United States has accepted the invitation of the government of France, to unite with the other governments of the world in the Universal Exposition at Paris, in April, 1867, of the products of each; and will confidently rely on the intelligence and liberality of Congress to make timely and adequate appropriations for exhibiting the products of the American Union on the proposed occasion, in such a manner and on such a scale as shall maintain its just rank among the civilized nations of the earth.

"*Resolved*, That in view of the well-considered action of the French government calling upon all its departmental authorities, including the Chamber of Commerce, boards of trade, and academies of art, to co-operate, within the proper limits of their authority, in the enlightened design of fully displaying the products of France, the Chamber of Commerce of this the principal national city of the United States feel called upon to exert whatever influence they may possess with their fellow-citizens throughout the Union to induce them promptly to furnish to the proposed Exposition, in the most liberal manner, such specimens of their products of industry or art as may elevate our national character; and to secure more effectually this object, they do now invite appropriate action on the part of the other chambers of commerce and boards of trade of our country.

"*Resolved*, That it be referred to a committee of five members of this chamber, to invite the attention of the chambers of commerce and boards of trade in the different cities of the United States to the peculiar national importance, both political and financial, of the proposed Exposition, in exhibiting to the governments and the peoples of Europe the natural and industrial resources of the American Union, now happily restored in its full constitutional authority."

Mr. Ruggles supported the resolutions with eloquent and appropriate remarks, after which Mr. Cowdin addressed the chamber on the subject.

The resolutions were unanimously adopted, and the committee authorized to forward them to Congress, and also to the various chambers of commerce and boards of trade throughout the country.

EFFORTS TO OBTAIN EXTENSION OF TIME.

Mr. Beckwith to Mr. Seward.

"PARIS, *January* 25, 1866.

" I beg to acknowledge receipt of your letter of the 30th December, advising reception of my communication of the 15th, and to thank you for the attention it had received.

"I think you will desire to be informed exactly of our relations to the Imperial Commission in regard to the extension of time, and I take leave to annex hereto copy of my letter to Mr. Derby of the 24th instant, which contains at once a statement of the situation and my advices to him in conformity therewith, for his guidance."

Mr. Beckwith to Mr. Derby.

"PARIS, *January* 24, 1866.

"DEAR SIR : I am favored with your letter of the 5th instant, No. 15, and am gratified and encouraged by the good spirits in which you write.

"The action of the Chamber of Commerce will undoubtedly receive the favorable consideration of Congress, and if followed immediately by the concurrent action of other chambers, the movement will have still more weight.

"The decision of Congress is vital, and if the Exposition is worthy of their attention, it deserves immediate attention, before it is too late.

"My letter of the 23d of December, No. 31, will have relieved you, I trust, from immediate anxiety in regard to time for filling up classes, and enabled you to go on with the work without interruption.

"I regret that I cannot make the extension of time definite and name the utmost limit that can be obtained.

"But I feel no hesitation in suggesting that it will not exceed three months from the 31st instant, and this is more than I think the Imperial Commission would now consent to.

"You should not, therefore, undertake anything on a scale which can-not be brought to maturity so as to enable you to complete the list and catalogue within this time.

"We must bear in mind that near ten months have elapsed since the proposal of the French government was communicated to the cabinet at Washington, that Congress assembled early in December, that the French government has not yet been informed that the United States will take part in the Exposition, and that we are not yet authorized to make defin-itive engagements with the Imperial Commission.

"The embarrassments resulting from this delay are not mitigated by their being unavoidable. We are, therefore, not in a favorable situation for asking the Imperial Commission to put themselves to further incon-venience. We should be able, first, to report to them the favorable action of Congress, which would carry the assurance that the changes we ask them to make would not be made in vain.

"Neither is it expedient for us to move in this prematurely, and pre-sent to the Imperial Commission occasion to act under circumstances so unfavorable to us as to be likely to result in giving us less time than we may expect at a later period.

"The date when the second report will be due (31st instant) has not yet arrived. Previous to that the Imperial Commission cannot act on its own impulse. It is therefore safe to wait and leave the initiative to them, and, as they know our situation and are disposed to favor us, they will not move in the matter till they are obliged to. When they call on us to report I will respond and make the best terms I can for time; but before this event occurs I fully expect the action of Congress will change the situation and make it more favorable.

"The Imperial Commission is well disposed to aid us in regard to time ' as much as it can,' and I observe that ' it can ' a little more, if we help them to help us, by leaving to them for the present the difficulty of preventing us from helping ourselves.

"I have fully explained the situation to Mr. Bigelow, and we are of one opinion on the subject.

"I have also discussed it with M. Le Play, and have informed him that I shall at present leave the initiative to him, but that we cannot afford to cut the work short at this stage and spoil it, and must assume that as much time as possible will ultimately be granted.

"His replies are made under the reserves which comport with his relations to the Imperial Commission, but he appreciates the situation, and is satisfied with the course which, for the present, I propose, and this leaves me no uneasiness on the subject.

"Be assured that I shall obtain as much time as can be obtained, which, I think, will in no case exceed three months, and shall, in so doing, preserve a good understanding with the Imperial Commission, which is all that the situation requires—at all events, all that it admits of.

"The time we have lost cannot be recovered nor wholly made up from the future, and we shall suffer some inconvenience from it.

"The gentlemen you name as having come to your aid will be of great service to you, and I am happy to learn that they are willing to lend their influence and co-operation.

"The estimates of cost of a suitable building in the Park, returned to me, were so unsatisfactory that I abandoned the idea of producing them as a basis for appropriations.

"I subsequently obtained estimates from the architects of the Palace, who compute the cost of what we shall require at about five dollars per square yard of ground covered by a building, and I have reported this estimate to the Department of State."

Mr. Beckwith to Mr. Seward.

"PARIS, *April* 29, 1866.

"SIR: The Imperial Commission has thus far assented to the delays I have been obliged to ask for, which it was possible to grant, without arresting the progress of some portion of the works in the Champ de Mars.

*　　　*　　　*　　　*　　　*　　　*

·   "It is necessary to the progress of the works in the Champ de Mars that we should now decide and agree definitively to occupy the ground reserved for us or relinquish it, that it may be prepared for other uses.

4 P E

" The annexed letter on the subject, addressed to me by the Imperial Commission, sets forth fully the situation of the Park business, and indicates the necessity of an early decision.   The letter is accompanied by a plan showing the ground reserved.

*   *   *   *   *   *

" I have, therefore, asked for the further delay requisite to make this communication, and that I may be exactly informed on the subject before surrendering the ground, if it be not possible to retain it."

*   *   *   *   *   *

" PARIS, *June* 1, 1866.

" I annex hereto a letter from the Imperial Commission.   *   *   *   It sets forth very clearly the progress and present situation of the preliminary work for the Exposition, and the necessity for proceeding on our part, and fixes the 30th of June for the delayed reports due from us.

" I am not without hope that the action of Congress will have enabled Mr. Derby to proceed, and that he will be able to close up and send in his report by the time named."

M. Le Play to Mr. Beckwith.[1]

" PARIS, *May* 26, 1866.

"I have the honor to remind you that, in accordance with article seven of the general regulations, the foreign committees are requested to furnish a plan of places to the scale of 0.002 of a metre, indicating by group and class the space assigned to each exhibitor, with the exhibitor's name, previous to the 31st of January, 1866.

" It is now four months since the expiration of the time, and the Imperial Commission, not having received that document, needs the information, in order to complete the construction of the general plan.

" The Imperial Commission leaves the foreign commissions free to arrange their articles as they please, within the space allotted to them, provided the principles of general classification are regarded; but certain measures must be considered together, by comparing the plans of the different states, and settled jointly, for the following reasons:

" Each nation, as you know, is separated from its neighbors, on one side by a partition put up by joint expense, and on the other by a passage-way.   In regard to this partition, the architects wish to make a certain number of doors in it, to afford a free circulation.   The Imperial Commission offers its kind mediation to arrange this communication between neighboring states; but this cannot be effected without an exact knowledge of the mode of location adopted by each party, and this information is indispensable, in order to agree upon the height of the partition, the position and dimensions of the screens, &c.

" Along the passage-ways separating two joining nations the different commissions intend to erect ornamental fronts of a national style of architecture.   Two opposite fronts, though differing essentially in their

---

[1] Translation.

general appearance, must have a similarity of construction, which cannot be determined upon without seeing the plans.

"I have the honor to inform you that the Imperial Commission has appropriated the seventh gallery for particular public purposes, as post and telegraph offices, police station, fire-engine, talking-room, water-closet, dressing-room, &c. The commission is now ready to begin this work, but must first know the plans of exhibitors near the locality.

"The eating-rooms and such places in the foreign department are to be constructed by national workmen appointed by the different commissions; but, in case it is neglected, the Imperial Commission will have the work done by Frenchmen; it is, therefore, absolutely necessary to have the plans and specifications to complete the work of this branch.

"Section seventh of the gallery, lighted at night and open to the public, must be separated from the Palace, which will be closed at sunset. The Imperial Commission is now arranging this department, and, as it wishes to pay due deference to the plans of the foreign commissions, the construction of that portion of the edifice will be put off till the 30th of June, hoping to get the necessary information by that time.

"The buildings in the Champ de Mars are advancing, and in two months a large part of the Palace will be done. Exhibitors should take possession as soon as possible.

"I have already mentioned that it is desirable that foreigners should have their show-cases made at home, so that everything may accord in nationality; yet everything of the kind can be made here, by Frenchmen, if exhibitors prefer it. As the opening of the Exposition approaches, workmen will become more scarce, and they will raise their prices for labor; so it is better to have everything done at once. A strike among the workmen might, moreover, cause some delay toward the last.

"For these many reasons, I beg you to forward to me the plans for the United States by the 30th of June. Of course a modification of the plans can be subsequently made. Send me also a plan of the houses to be erected in the Park, and the trees to be planted by the United States Commission in the allotted space, as announced on the 25th of April last."

RE-TRANSFER TO PARIS OF THE LABOR OF APPORTIONMENT OF SPACE.

Mr. Beckwith to Mr. Derby.

"PARIS, *October* 11, 1866.

"I beg to acknowledge receipt of your favors of the 22d, 24th, and 25th; Nos. 45, 46, 47, 48, and 50. No. 49 has not been received.

"I have also received the lists of applicants in the different classes of Groups II, III, IV, V, VI, VII, VIII, IX, and X, and the supplementary lists of applicants and withdrawals, named in the above correspondence.

"I will reply more fully to your letters herein acknowledged in a short time.   I cannot express my surprise and my embarrassment arising from the incompleteness of these reports.

"With the exception of the Palace portion of Group VI, which is well formed, nothing definitive has been done.

"In my letter, No. 114, I referred back to your committee the work of completing the formation of Group VI, in the annex.

"I beg now to cancel that reference, as it is impossible to wait for the work to be done on your side.

"All the other Palace groups are equally unformed, and there is now no possibility of avoiding a complete failure of our exhibition but for me to undertake the formation of the groups myself.

"I shall have to estimate the space for each product, place it, allot the space to applicants, make the plans for structures, in that conformity, from the catalogues which must be an exact index of this, and report them to the Imperial Commission for the great catalogue which is now printing and will be a finality, and then proceed to construct the installation to correspond with these arrangements.

"To do this, I have procured the best aid I can obtain, and we are engaged upon the work day and night.

"The applications will not fill some of the groups; in others they will be greatly in excess of the space.

"Those for Group VI far exceed the ground and the money, both on your side and on this side, and must be greatly reduced.

"As soon as it is possible I will send you a list of the products to be received, and a separate list of those which cannot be received, and request you to advise both parties of applicants in that conformity.

"This will be definitive, as the catalogues will be printed, and the works constructed, to correspond with this distribution.

"I make these observations with the utmost reluctance.   I am convinced of your attention, and zeal, and earnestness, and I know you have had difficulties.

"But the work thus thrown upon me forces me to undertake it myself, and accomplish it as I best may, which requires an explanation of what I am doing; or to abandon the Exposition, which would be a dereliction of duty that is impossible."

Mr. Beckwith to Mr. Seward.

"PARIS, *October* 11, 1866.

"SIR: I am under the necessity of reporting to the department the present state of the Exposition.

"The work of receiving applications, allotting space to applicants, and making plans and catalogues in conformity, upon which the necessary structures to receive the products could be made in advance of their arrival, was committed to Mr. Derby about twelve months since, and he was recommended to form a suitable committee to advise and assist him.

"Mr. Derby reported in due course that he had formed a board of able assistants, and would proceed with the work as rapidly as possible.

"The inaction of Congress caused delays, and I obtained corresponding extensions of time, which were protracted to the last moment compatible with the possible execution of the preliminary work on this side.

"I have now received from Mr. Derby the reports of what has been done; but with the exception of the formation of a part of Group VI, nothing definitive has been done.

"There have been no allotments of space to exhibitors in any of the other groups; the products have not been placed in them, the space they will occupy has not been ascertained, consequently there are no plans of the structures required, nor any catalogues, forming the index to this work, to be reported to the Imperial Commission.

"A portion of Group VI, in the Palace, has been formed, and it is well done. But the other seven or eight groups are unattempted; the ground is vacant, and presents only imaginary sketches of *pro forma* plans, similar to those which were sent from this as models nearly a year since.

"In place of all this work I have received nothing but lists of applicants, and of their products, copied from their applications, and arranged in classes.

"But the space these products will occupy is unknown; the space required by the applicants is not named; and with the exception of Group VI, the applications themselves have not been sent—nothing but the brief lists of names and products, as above stated.

"I have neither allotments, plans, nor catalogues, nor the elements of which to make them in a proper manner.

"The Imperial Commission is now printing the great catalogue, and pressing for mine, which has been promised, but I have none to report, and the structures must soon be begun or they cannot be made.

"There remains but one possible way of avoiding a complete failure of our exhibition.

"I must undertake myself to estimate the space each product will occupy, with the allotments of ground to applicants, form the plans of structures to correspond, compose the catalogues in this conformity, and report them to the Imperial Commission for publication, and proceed to make the necessary structures, as I best may, on the slender information above described.

"I have not any doubt that it is my duty in this emergency to adopt this course, for no other but failure is possible; and having solicited the aid of the most capable persons within reach, we are now engaged upon it day and night, and shall be able to report it in a few days to Mr. Derby for his guidance in advising applicants of the result of their applications, and in collecting and forwarding the products.

"Some of the groups will not be quite filled, but in others the applications are greatly in excess of the space or of the provisions of Congress, for the expenses, and large numbers will be excluded.

"This I am most desirous of avoiding, because it will give disappointment and dissatisfaction to applicants, but it is not possible to avoid it.

"It is not in my power to make this brief and accurate statement of the situation without appearing to reflect on the work of Mr. Derby.

"But that is not my desire; on the contrary, I am convinced of his attention, his zeal, and his earnestness, and that his failure in completing the work placed in his hands is owing to his inability to obtain the requisite assistance, from some cause which he can probably explain, but which is unknown to me."

NECESSITY OF EARLY INSTALLATION OF MACHINES.

Mr. Beckwith to Mr. Derby.

"PARIS, *November* 4, 1866.

"I beg now to recall your attention particularly to the documents accompanying my letter of the 24th of June, published by you, page 45, [Third supplemental circular,] article 18, as follows:

"'Between the 1st and 14th of April each class jury of Groups II, III, IV, V, VI, and X will examine the products and class the exhibitors deserving prizes, without distinction of nationalities.'

"This important work will be completed within the first fourteen days after the opening of the Exposition, and the reports thus made will form the basis on which the awards will be made.

"The time allowed appears, at first sight, short, but there will be sixty-eight separate juries, which is one jury on each class in these groups, and they will work separately and simultaneously.

"The labor being thus divided, the time will be ample.

"My object at present is to remind you that we have designated between sixty and seventy machines to be installed and put in motion in Group VI; and if this labor be not completed, and the machines in full and perfect action at the opening on the 1st of April, they will lose their chance of favorable reports from the juries, and consequently of the awards which their qualities, displayed in action, might command.

"Machinists will appreciate the labor which is requisite to place and adjust in good working order so many machines, and that this cannot be done but by the concurrence of many persons within the time that remains for it.

"I have already stated to you in previous letters the defects in the information required for foundations which should be laid before the frosts set in, and have only to repeat my hopes that the necessary information will arrive in time.

"I wish now to repeat also, and it should be made known to the owners of each of the machines, that no preparation can be made in advance for the transmission of steam by separate steam pipes, nor of force from the main shafts to the respective machines.

· "These transmissions and the structures they may require will be at the expense of the owners of the machines respectively.

"The machines to be operated should, therefore, be sent forward as early as possible, and the machinist who is to set up and work the machines, or each machine, should come with it, prepared to complete the work at once, and to defray the expenses each of the machines may require.

"If the owners of the machines do not respond with alacrity to this request there will be lamentable defects in this department at the opening, and it is the department in which our strength lies—where we shall be successful, if anywhere.

"If any of the parties whose machines have been designated for action are not prepared to do the needful in good time, I beg to be notified of this at once that other machines may be substituted, if possible.

"I will thank you to communicate the substance of this letter to each of the parties interested as early as possible.

"I have already been notified informally that a portion of Group VI will be delivered to me in a few days, on which I can commence work, and I expect shortly the delivery in form. "

MOTIVE POWER.

Mr. Beckwith to Mr. Derby.

"No. 14.]   "PARIS, *November* 8, 1865.

"DEAR SIR: Class No. 52, in Group VI, comprises machines and apparatuses suited to the uses of the Exposition.

"The plan of the special committee to which the most of this work is assigned is to supply motive power to the Exposition, as far as practicable, by using the machines exhibited.

"The arrangements for steam power are as follows :

"The machines and apparatuses to be moved by steam power belong to Classes 47 to 66, Group VI, and will occupy the great gallery (hall) forming the outer circle but one of the Palace.

"The furnaces and generators will be placed in the Park, outside the walls of the Palace, in a circular line, parallel with the wall, but at equal distances from each other, to correspond with the different localities within the Palace requiring steam.

"This service will be divided into fourteen sections, organized and worked separately.

"The force will be transmitted to shafts in gallery No. 6 ; the shafts will extend in polygonal lines, yielding to the curve of the gallery, and transmitting the force to various machines to be moved.

"It is proposed by the commission to supply requisite motive power by letting the work in sections to contractors *a forfait*, (by the job)

"The annexed document in lithograph presents the conditions and bases on which the commission invites the offers of contractors, and they engage to give a preference to the contractors belonging to the nationality to which the contract may apply.

"It may be doubtful if any of our good engineers happen to be familiar enough with the elements of such a contract, such as the cost of material, fuel, labor, living, &c., in Paris, to enable them to make safe estimates and offers; and equally doubtful whether their present employment is not more remunerative than any they would be likely to obtain here, in competition with lower wages, permanent residence, and better knowledge of the situation. But there may be those who may be able to see their interest in it, and, in conformity with the inventors of the plan, and the wishes of the committee, I submit the matter to your consideration."

"No. 39.]   "PARIS, *January* 16, 1866.

"I had the pleasure to address you this morning, and have received this evening your favors of the 22d of December, No. 10, and of the 23d December, Nos. 11 and 12.

"No. 10 relates to the efficient measures you propose for disseminating the information therein alluded to, and refers to the difficulty of engineers in offering to supply motive power for machinery in the absence of specific information regarding the price of labor, fuel, board, and other elements of cost.

"I had foreseen this difficulty, but not the means of obviating it.

"I have sent you all the documents and all the information on this subject provided by the engineering department. They consider it in the province of contractors themselves to make the investigations on which their offers must be based. It is an object with the department, in adopting the contract method, to divest itself of the labor and responsibility of the estimates and of the fluctuations of market prices which fall to the side of the undertaker.

"1 will make further inquiries in other quarters, being desirous of having the motive force supplied by our own engineers, but I have not much expectation of being successful in the inquiries because the subject requires the investigation of a practiced engineer, whose researches can be relied upon as the basis of contract.

"I have no authority to employ an engineer for this purpose. Indeed, the first step of a contractor should be to make or provide the means of such investigation for himself, as that is a part of the labor and expense intended to be thrown on him and is implied in his contract.

"The general disposition of the apparatus for the motive force you will find, I trust, sufficiently indicated in the cahiers I sent you, and as the American section will be operated by itself you will have in your own hands the elements for computing the aggregate force required, the velocities, &c., for it is upon the elements to be supplied by you that the Imperial Commission itself would have to make those estimates.

*   *   *   *   *   *   *   *

"No. 41.]   PARIS, *January* 22, 1866.

" Referring to my letter of the 16th instant, No. 39, I have not been able to obtain the information requisite as to the cost of materials, &c., on which a contract could be safely made for the supply of motive power in Group No. VI. But I have made an (verbal) understanding with the *chef de service* in the engineering department, by which he agrees to pay an American contractor the average price paid to French contractors for similar work.

" This is the only basis for a contract which I can give you, and, from the nature of the case, I imagine that this method will be followed by other nations who may wish to have their own engineers employed, but who will have the same difficulties in obtaining local information as to cost of elements.

"If, therefore, you can arrange with a respectable and responsible party, in whom you have confidence, who wishes to exhibit his machinery, and is desirous of working it for the supply of motive power on the terms above named, please do so.

"The arrangement on your part will be provisional, and you will transfer the contractor to the Imperial Commission to complete his contract. He will be their employé, and under their orders, and will receive his pay from them, but you can assure him the contract upon the basis above named.

" The nature of the service to be performed, the apparatus to be supplied, the structures to be made at his expense, the hours of work, the prolongation or abridgment of time, and all the general conditions and regulations applicable to the contract, and binding upon both parties, are set forth in the document accompanying my letter No. 14, of November 8th, p. 55, with all which conditions the contractor should first make himself acquainted.

" You will be able also to inform him pretty nearly as to the amount of motive power you will require. This is of moment because the outlay and preparatory expenses of the contractor will be as much nearly for the supply of a small force as for a larger one, while the pay will be in proportion to force. If, for example, you want thirty horse-power, and the price is $100 per horse for the season, (which perhaps is not a bad guess as to probable offers,) the contract money would amount to $3,000, and for sixty horse it would be $6,000, while no such increase of cost in fixtures or structures would occur. It is also for the contractor to consider that he must arrive in advance, complete his contract, and see that he has his apparatus in order for work in time; the days get short and weather bad, and work expensive late in the season. I should think October would be as late as it would do to arrive here and commence the placing of apparatus.

" I have only to add to these observations that the Imperial Commission is now engaged in making contracts, and is desirous of being informed, as early as convenient, whether or not you will provide a contractor, and I have informed the commission that, I think, within a fortnight after you receive this letter you will be able to satisfy yourself on the subject, and will advise us in conformity."

Mr. Beckwith to Mr. Seward.

"PARIS, *May* 6, 1866.

"SIR : I had the honor to address you on the 29th April, transmitting a letter from the Imperial Commission on the subject of the ground which we propose to occupy in the Park.

"I now transmit another letter from the Imperial Commission on the subject of motive force, dated the 3d instant, and received this morning.

"I beg to state briefly that the method adopted for supplying force for machinery is by separate contracts for each national section.

"Each nation may employ its own engines and engineers, and, for the force thus furnished, six hundred francs per horse-power will be paid by the Imperial Commission, or the nations may decline furnishing the force they require, and leave it to the Imperial Commission.

"An excellent opportunity is thus presented without expense to the exhibitor to display the qualities and results of his engine-boilers and apparatus.

"I transmitted to Mr. Derby early in November the general plan and conditions, (which have been printed and published in the United States,) and desired him to advise me in due time whether or not he would furnish a contractor for the motive force, and if not, to inform me of the amount of force he would require, that I might request the Imperial Commission to supply it.

"On the 8th April, at the request of the Imperial Commission, I applied to Mr. Derby again, informing him of the necessity of immediate decision.

"But owing, I doubt not, to the delays in Congress, Mr. Derby has not been able to arrive at any decision, and I am without information on the subject.

"The Imperial Commission now calls on me (in the annexed letter) to enter into a contract with them to furnish the motive force which we may require, or to decline it definitely, and, in so doing, inform them what amount of force we will need, that they may contract for it, and proceed to construct the necessary works. They remark, also, that if I cannot comply with either of these demands, the works in general must not the less go on, and they cannot be responsible after the present notice for the inconveniences which may result to us from further delay in this department.

"I have concluded not to reply to this letter until the last moment which M. Le Play will concede to me, and if advices do not arrive to relieve me from the embarrassment, I must then surrender the privilege of our exhibitors to furnish their own motive force, and request the Imperial Commission to supply it.

"This is the only course that appears open to me, but it is not likely to result very satisfactorily. I must assume the amount of force we shall need. If I fix it too high, and the Imperial Commission make the contract in conformity, and commence the construction of furnace, chimney, steam-pipes, &c., for a larger force than we shall need, they will have to compromise subsequently with the contractor, or pay him for wasted force, and in either case they will suffer some loss which they will probably ask me to pay. If, on the other hand, I fix the amount too low, we shall be without the requisite force.

"I feel bound to acknowledge in this connection the continued disposition of the Imperial Commission to yield all the delay that is possible. But we are now on the fourth month of delay, at our own special request, and I am aware that the works on the Champ de Mars have reached a stage which requires the question of force to be settled.

"It is also evident that similar questions will continue to arise in pretty rapid succession which will not admit of further delay."

M. Le Play to Mr. Beckwith.[1]

"PARIS, *May* 3, 1866.

"MONSIEUR LE COMMISSAIRE : The Imperial Commission has recently settled the details of the organization of the mechanical service; they have approved the contracts made with the furnishers of force, and the general dispositions for the transmission of the force.

"It is, therefore, indispensable that, without loss of time, the committee of the United States of America proceed to a similar work, which the information contained in this letter will enable you promptly to complete.

"You have already learned, from reading the third instruction, (of which I send herewith another copy,) that the general transmission is made by two [parallel] shafts, distant from each other $4^m.71$, elevated $4^m.36$ above the ground, and communicating movement to each other. The shafts are $0^m.29$ in diameter, forming polygons of which the sides are $13^m.8$ in mean length, producing an angle between them of about $5°$. The revolutions for the French section will be one hundred per minute, but the American section having no connection of movement with neighboring sections, you can choose yourself, according to your wants, the velocity which seems to you most advantageous.

"I pray you only to recollect, in determining the velocity, the fact that the general arrangement will not admit of *poulies* (wheels on the shafts) of more than $1^m.00$ in diameter.

"This general transmission thus suspended is very expensive, costing not less than six hundred and fifty francs the running metre. It is, therefore, of great importance to reduce the length of the shafts as much as possible. In the French section the movement is supplied to about

---

[1] Translation.

one-third the length of the Gallery VI. It is confined to certain localities, leaving others without motive force; and finally, in regard to
certain localities which require but feeble force, we have provided it, not
by transmission direct from the main shafts, but by one of the three
following methods :

"1. By special motor.

"2. By a small secondary shaft in rear.

"3. By a shaft under ground.

"I hope the Commission of the United States will adopt the same
principle to regulate the installation of their machines.

"Not having yet received definitive advices of the arrangements they
intend to adopt, and being unable to wait for full advices before ordering
the supports and shafts of which the execution requires a great deal of
time, I think it necessary to fix upon a plan of placing them analogous
to that adopted in the French Section.

"The plan hitherto annexed indicates the position of the shafts (on
this hypothesis) in your section.

"The transmissions will occupy a *travée* of 14^m.00, and will have thus a
double length of shaft, say 28^m.00. It would seem that this should be
sufficient for your wants; if not, or if you wish to substitute the *travée*
indicated by another, which you find more convenient for your installations, or, finally, if you think you will not have need of this length, I
pray you to inform me immediately, in order that I may consider it while
there is yet time.

"If any apparatus which ought to move be placed outside of this
travée, it will receive its force from one of the three methods above named,
which you can choose and apply in each particular case.

"A platform, supported independently of the transmissions, 4^m.00 in
breadth and 5^m.15 in height, will extend continuously (except, perhaps,
across the great entries) the whole length of Gallery VI, (*des arts usuels.*)

"This will serve as a promenade for visitors, who will find in the
salons garages (enlarged spaces with seats) in the middle of each sector a place of rest, where they can sit and enjoy the spectacle of mechanical activity displayed at their feet. Certain exhibitors of objects of
great height, which occupy two stories, expect to derive great benefit
from this platform by carrying a passage from it to their second story.
I allude particularly to some exhibitors of agricultural machines, sugar
apparatus, light-houses, organs, &c. Similar arrangements might be
adopted in your section, which would render its appearance more impressive.

"It will be indispensable to regard the supports of the platform in
placing your apparatus.

"The general plan herewith indicates exactly the places of the supports.

"These arrangements being well defined, it remains to consider those
which belong to the furnishing of the motive force.

"All the contracts that have been made with French undertakers for our French Section have been made on the basis of six hundred francs per effective horse-power, measured on the shaft. This sum serves equally for base in our contracts with England and Belgium, and the same should be adopted by you, if, in conformity with my preceding communications, you have organized yourselves your mechanical service with contractors of your country. This sum includes also the furnishing and placing completely of the furnace, boiler, engine, transmission, construction of the building for the boilers and furnace, the chimney, the steam-pipes, and the passage in which the pipe is laid, the combustibles, and the persons required for the apparatus. It is also understood that all these materials remain the property of the contractor after the exhibition.

"I send you herewith a form of contract which indicates the principal conditions of these agreements made directly between the Imperial Commission and the foreign commissioners themselves, and not with those of their countrymen whom they choose for contractors.

"This *pro forma* contract presents some blanks which should be filled up, and of which the most important relates to the motive force, and consequently to the amount to be paid by the Imperial Commission for the force.

"My previous communications on the subject of the mechanical force necessary to your section having remained thus far without response, I cannot fill up the blanks, and I renew my entreaty to be informed the most promptly what is possible in this respect. This force once fixed as exactly as possible, will indicate the sum to be paid, by multiplying the number of horse-power by six hundred francs. But the sum thus calculated will be the maximum, and subject to proportional reductions, if by dynamometric observations the power actually furnished be less than the amount named as a basis of calculation.

"To aid you in completing the organization of your mechanical force, I hope to be able to place at your disposal some machines, portable or fixed, exhibited by French contractors, but on condition that you inform me as soon as possible what machines you may have need of.

"Finally, (and if you have any objections to make I shall be obliged if you will make them at latest before the 15th instant,) your section will comprise arrangements for the general transmission of force, to the extent of fourteen metres in length, corresponding to double that length of shaft.

"It is desirable, in conformity with the example of Belgium and England, that the United States should agree with the Imperial Commission to furnish the motive force necessary to their section, reserving to themselves to come to an understanding with their own national contractors afterward. Thus, as I have explained to you in my various communications relative to this object, the Imperial Commission thinks that all considerations unite in favor of making this method general; in this case there will be occasion for a contract analogous to the outline of agreement which I send you.

"If, on the contrary, the Commissioner of the United States does not think himself able to agree to this, it will be indispensable for him to advise me immediately, and to indicate the motive force that will be required to enable me to proceed in his place to prepare the necessary mechanical constructions in his section. The French contractors are about to commence the construction in the Champ de Mars of their buildings for furnaces and boilers, passages for steam pipes, &c. A longer delay in deciding for your section will tend to compromise the work that is requisite for it, and the Imperial Commission must decline the responsibility from this time for the consequences which further delay may entail.

"In the expectation of a prompt response to my communication, I pray you, Monsieur le Commissaire, to accept the assurance of my distinguished consideration."

Mr. Beckwith to Mr. Seward.

"PARIS, *November* 27, 1866.

"SIR: It is my desire and effort to occupy the attention of the Department as little as possible with details, but some of them should be brought to your notice in passing, that they may be understood.

"The regulations and formalities by which the Imperial Commission conduct their work are applicable to all nations alike, and we must conform to them, or we cannot proceed. The more we show a disposition to reconsider what has been done and go back to change it, or propose methods which we may think better, but which are not in accord with their methods, the more we come in conflict and embarrass the work.

"To avoid this result at this late date is of great importance, and in the endeavor to do this I have several times of late been obliged to place myself in apparent opposition to the proposals from New York, even when I should cordially agree with the object, if it were practicable in the way proposed. This pressure arises from particular interests, which might have been more fully accommodated at an earlier period if they had come forward, but which it is now more difficult to satisfy.

"The contract for motive force was kept open, at my request, until it became so embarrassing to the Imperial Commission that they notified me I must close it, or sign a contract which they sent me, agreeing to supply the force myself and commence at once the structures. Being unable to comply with this request or to present a contractor acceptable to the Imperial Commission, I abandoned the attempt, and called on them, on the 13th of July, to provide the requisite force, in conformity with the general regulations, of which I duly notified Mr. Derby and the department.

"Mr. Derby writes on this subject, on the 9th November, 'that there is much feeling, among those interested in machinery, about motive force in our section, and they think we ought to have had our own engine and engineer.'

" To this I replied as follows: ' That is precisely my feeling; I agree with them; and when that contract on fair terms was presented month after month, without takers, and I was persuading the Imperial Commission time after time to keep it open, and still nobody offered, I was disappointed. The result of this delay was that the works went on, and when I was called on finally to close up I was obliged to pay a considerable sum extra to get the power you required, because the preliminary work was too feeble in structure and had to be done over; and, as this was owing to our delay, I was compelled to yield or go without the force. I surrendered this business from necessity, with a feeling of disappointment and chagrin; and I might use a stronger expression, for I fully believed our people would take that contract freely, and relied on it, and suffered for my mistake. Therefore I have no more to say on that subject but this : feelings which are not strong enough to lead to action are of no value; if our machinists feel sufficient interest in it to buy out the contractor, they can do so, and if not, not.'

" Mr. Derby writes again, on the 13th instant, as follows: ' If you will propose to the French contractor for the motive power of the American Section that we will furnish our own power at our own expense, and at the same time allow him to draw his contract money from the Imperial Commission just as if he furnished it according to contract, the money will be supplied by parties here for furnishing this power, as it is considered of the greatest importance, not only by exhibitors but by leading men in this country, that this power should be furnished by an American contractor, and that an American engine and boiler should be used for that purpose. If the French contractor has already constructed buildings for boilers, &c., and put up the shaftings or supports for it, these can be used by the American contractor. If he has not, we will furnish them from this side; i. e., at our expense. As I have heretofore advised you, there is much feeling here upon this subject—which will not be diminished when the Exposition opens to the view of Americans in Paris—of American machinery propelled by a French engine and French engineer.'

" To this I have replied by this day's mail as follows: ' Referring to the remarks of your letter No. 78 relative to motive force, the subject will perhaps be made clearer by restating the conditions. It is incumbent on the Imperial Commission to furnish motive force, and they retain the entire control of the force. They proposed to accept a contractor for our section, presented by us, provided the contractor would accept of their terms, by which he would become responsible to them, receive his pay from them, and be entirely under their control. By that arrangement we would continue to look to the Imperial Commission for force, as if we had not presented the contractor; they would take the risk of the contract, and if the machine broke down or any other accident disabled it, the Imperial Commission would be bound to supply its place to us at their expense, they settling with the contractor. The same conditions exist, whether the contractor be presented by us or not. These are not our terms, but those of the Imperial Commission, and they are applicable to all foreign nations. We were unable to nominate a contractor in time, as you are aware, and the Impérial Commission made a contract with another contractor. We have never had any control of this contract, nor can the Imperial Commission recall it; it is the property of the holder. He may sell it if he can find a buyer, provided always that the other contracting party—the Imperial Commission—will accept the buyer in place of the seller. Therefore any party wishing to make this contract must buy out the holder and agree with the Imperial Commission to accept him in place of the seller, and enter into a new contract in that conformity. With this change we have nothing to do, except to oppose it or promote it, according to our interest, as far as our influence may go. Now, I shall be extremely glad to have an American contractor and engine in place of the one we have; it is what we ought to have, and I am ready to do all I can to effect this change, provided always that the new contract will be equal to our wants. But I cannot propose the canceling of the existing contract, which, if accepted, would leave us at this late date without a positive contract for force; nor would the Imperial Commission listen to such a proposal; neither can I become myself the contractor, which would, in effect, be my position by your proposal. The new contractor must come forward and negotiate for himself; he must agree with the holder on the terms of sale, and till this is done nothing can be done; he must then agree with the Imperial Commission to accept him as a substitute for the other, and enter into the obligations and responsibilities which they require of all contractors. I will help him in this as far as I can, provided always his offer is equal to our wants and compatible with the general interest of our exhibition, which it is incumbent on me to look after. I think it best, therefore, for me not to make the proposal you suggest, and would recommend your contractor not to begin with such a proposal, because it would come to nothing either with the holder or with the Imperial Commission. The holder is a machinist of reputation and wealth, who wishes to exhibit his machine, and cares very little for the pay. I do not think he would listen to a proposal to give up his contract and continue to draw his pay; I think he would refuse it; at the same time, if the case were properly stated, and he were asked to name his terms, he might name terms more moderate than the buyer is ready to offer. These are my impressions, but I cannot undertake this negotiation; it is the business of the new contractor, and I shall be glad if I can help him in it in the way I have suggested, and glad if it succeeds.'

"It will be readily seen that I cannot propose the canceling of the existing contract and substitute nothing in its place but a vague understanding that parties who are not yet named will come forward and make another contract. The Imperial Commission would not consent to this, and if they did it would only deprive our exhibition of the certainty it now has of sufficient force, and leave the common interest to the uncertainties of an incomplete engagement not reduced to the forms of business which secure fulfillment.

"It is for the interest of the exhibition to have the new contract perfected before the old one is relinquished, and it is incumbent on those who are directly interested, and desire to profit by the change, to come forward and complete it in advance.

"I think I should jeopardize the general interest of our exhibitors in consideration of the particular interests of contractors if I acted otherwise, and my object in this communication is to explain this situation.

"The pressure from particular interests at this stage naturally increases, and the numerous letters which I receive direct from parties themselves are now embarrassing.

"I shall endeavor to satisfy each as far as is compatible with the common interest of our exhibition, which should be kept uppermost; but I cannot deviate from that, unless in particular cases, which may be referred to you, you shall think me mistaken and direct me to act otherwise."

PROPOSED EXHIBITION OF COSTUMES AND OF ABORIGINAL RACES.

Mr. Beckwith to Mr. Derby.

"No. 12.]   "PARIS, *November* 8, 1865.

"DEAR SIR: The annexed publication is from the special committee on costumes, Class 92, and indicates the method adopted in France for perfecting that part of the Exposition.

"The peoples of Western Europe descend from successive invasions of numerous races which settled in various localities, holding comparatively small intercourse with each other previous to the epoch of railways, and preserving, consequently, great variety of dialects, habits, manners, and costumes.

"These characteristics are suggestive, not only of differences of origin, but of the influences which tend to preserve or create the differences in question, such as peculiarities of climate, soil, geographic configuration, occupation, &c., in localities but little removed from each other.

"The difference of origin and the better means of communication in America, the uniformity of institutions, the diffusion of a common literature, the superior intelligence, and the homogeneous character of the nation, tend alike to preclude the preservation or growth of similar local distinctions, while the brief history of the country, from its settlement, embraces too short a period of time for the modifications of character and development of local differences, which it is becoming the fashion to ascribe, with or without reason, to the powerful influence of the elements.

"I doubt if you will be able to make a collection of native costumes that will be very interesting or instructive, either in a historical or an ethnological sense."

5 P E

Mr. Beckwith to Mr. Seward.

"PARIS, *September* 19, 1866.

"The project of bringing together at the Exposition groups of aboriginal races from different quarters of the globe may appear at first adapted merely to gratify the curiosity of the multitude.

"But, however legitimate such a wish might be, the project includes a higher object.

"The interesting researches which relate to the natural history of man, it is well known, are now pursued with great zeal, and are pushed back to periods long anterior to the commencement of the historic period.

"The elements of these researches include careful studies of the physiology of races, of the habits and manners of existing races, of languages living and dead, and of fossil remains.

"The persons most occupied with these inquiries are seldom men of fortune, and rarely travelers, but they are usually men of small means, devoted to special pursuits, which they follow with untiring zeal, depending, to a great extent, for the material facts on which their generalizations are based, upon the hasty and often superficial observations of unscientific travelers and upon accidental discoveries.

"Bringing together specimens of races, as proposed, will present a rare opportunity for the linguists, the sinologues, the ethnologues, the physiologists, &c., to perfect and verify their theories—to correct them or to originate new ones—an opportunity which most of them have never enjoyed, nor could in any other way.

"The American Indians, as regards their physical qualities, their moral and intellectual qualities, their present condition, their obscure past and more obscure future, are unquestionably among the most interesting of the early races of man.

"Their gradual diminution is considered by some as the evidence and effect of that law which they contend governs the animal kingdom, in conformity with which the lower precedes the higher, and is in turn exterminated by it. From this it is argued by one party that civilization spreads only by extermination, while their opponents maintain that all races are capable of civilization and preservation, and that extermination results only from the ignorance and consequent enmity of races.

"But, whatever the causes of decay, the fact is obvious that the aboriginal inhabitants of America are diminishing, and it may be doubted whether it is in human power to preserve or even to prolong their existence.

"The journals from Washington just received contain the legislation of Congress, Document No. 157, relating to certain tribes of Indians.

"The pains taken to introduce among them the arts and habits of civilization is remarkable. Oxen, horses, plows, hoes, axes, log-chains, saw-mills, grindstones, spades, farming implements of all sorts, and domestic utensils, are not only provided for them, but white persons of both sexes are sent among them to teach them the uses of these things and the habits of a higher life.

" The consideration and care of the government and people of the United States for these ancient races are beneficent and even parental. But this fact is little known in the world, and we are frequently reproached with pursuing a cold and cruel policy toward the Indians.

·' A better understanding of this subject would relieve us from these reproaches and justify the policy of the government and nation, by showing that it is eminently humane and wise, and really up to the level of the highest civilization of the age.

" The history of this policy and its effects, carefully studied, would also throw great light on the ethnological question to which I have alluded, touching the destiny of races as affected by human laws and by laws which are higher than those of human origin.

" If I could succeed in adding a group of Indians to the assembly of races which it is hoped will be brought together at the Exposition, I think it might give rise to inquiries and researches which, in a scientific sense, would be interesting and useful, and in a political sense would tend to diffuse a knowledge of facts in every way creditable to the government and the country; and I am not without hope that you may think the subject of sufficient interest to bring it again to the attention of the Secretary of the Interior."

EXHIBITION OF HEAVY CANNON AND MUNITIONS OF WAR.

The two letters following, from Mr. Beckwith to Mr. Seward, explain the absence of an exhibition by the government of materials of war in the United States Section :

Mr. Beckwith to Mr. Seward.

"PARIS, *April* 19, 1866.

" SIR : The fabrication of heavy cannon and materials of war in general being, to a large extent, the work of government, the Imperial Commission omitted articles of this kind in forming their catalogues for the Exposition.

" But the nations most advanced in products of this description, England, Prussia, Belgium, &c., have expressed a desire to exhibit them, and the Imperial Commission has resolved to add them to the catalogue.

"The French government will, therefore, form for itself in the Park a separate exhibition, comprising all descriptions of materials of war, and other similar exhibitions will be formed by other governments or manufacturers, or by both.

"An exhibition of this kind by the United States, through the cooperation of the Navy and War Departments and manufacturers, might be made with great effect, and a place could be provided for it in the Park, under the same roof where I propose to supplement Group VI, alluded to in my previous letters.

" The additional expense this would involve would not be large on this side, and the cost of the proposed building could, I think, be kept within the sum I have named for that purpose, which Congress appears disposed to provide.

" A branch from the railway which encircles Paris will be laid to connect with the Park, which will facilitate the transport of heavy objects, and suitable machinery for handling and placing them will be provided.

" A collection of war materials would add great attractions to our exhibition, and undoubtedly be highly appreciated.

" I have requested Mr. Derby to apply to you for information, and I beg your favorable consideration of the subject."

Mr. Beckwith to Mr. Seward.

" PARIS, *May* 31, 1866.

" SIR: I beg to acknowledge receipt of your letter of the 11th instant, referring to mine of the 19th April, on the subject of an exhibition of materials of war.

" Your letter includes a copy of the observations of the Secretary of the Navy on the subject, in which he remarks that he is 'aware of no benefit that would accrue to our government or country from an exhibition of specimens of our ordnance in Paris,' from which I infer that I must have failed to present the subject in the light which I intended.

" It has been the occasional custom of the United States government, and it is the constant custom of European governments, to dispatch commissioners to different countries to study and report upon the progress and condition of the materials of war.

" These inquiries are attended with great expense, on account of the extended journey they require. The inquiries are in themselves difficult and the results imperfect, owing to the objections and obstacles often thrown in the way of them, and the reports are defective, which result from such hasty and imperfect studies without the means of comparison.

" The Imperial Commission omitted this subject in its original programme, but England, Prussia, and Belgium, countries among the most advanced in products of this kind, thought the occasion should not be neglected for bringing together collections of the most improved and advanced materials of war from all countries, which would present at once the best possible opportunity for the study and comparison of them without obstacles.

" At their suggestion the Imperial Commission reconsidered the subject, and resolved to provide for such an exhibition.

" The French government concurred in this view, and the result will be national exhibitions of the best war materials of the countries above named, in which each will exhibit not for its own especial benefit but for the mutual common benefit, which accords with the spirit and meaning of the entire Exposition of 1867.

"I feel that I should apologize for intruding the subject a second time on your attention, but I am not without hope that the Secretary of the Navy and the Secretary of War may be willing to reconsider the matter in the light now presented.

"If those departments could be induced to contribute to the Exposition, and send a competent officer to study and report upon it, (of whom there must be many who would accept the commission without expense,) they could not fail, I think, to obtain more complete and valuable information than they could get in any other way of the quality and condition of the materials of war of every kind in all countries where great attention and skill are applied to the production of them."

SOCIETY OF INTERNATIONAL TRAVEL.

The following letter from Commissioner Beckwith to Mr. Derby, dated Paris, November 8, 1865, explains the organization and objects of "The Imperial Society of International Travel:"

"Many persons engaged in agriculture, manufactures, and various industries will desire to visit the Exposition for the purpose of studying it in connection with their particular interests. It is likely also that many of those persons whose studies would produce practical and useful results may not be able to afford the whole expense which it involves. The annexed publication emanates from an association collateral to the Imperial Commission, founded on a capital of $100,000, for the purpose of aiding the class of persons in question to visit the Exposition by means of contracts in their favor at reduced prices, with railways, steam navigation companies, hotel-keepers, &c. The articles of association and method of proposed operation are described in the annexed pamphlet.

"I send it merely as a suggestion, which some ingenious and well-disposed person may embrace, to originate a similar organization if thought useful and requisite on our side."

The object of the society is:

1. To make arrangements with railway companies, steamship companies, and others, in regard to running trains and making trips at reduced rates, from the principal towns of France, Algiers, and from foreign countries, for the express purpose of transporting the working classes, farmers, and mechanics, to the Universal Exposition of 1867, at Paris.

2. To enable all these persons to reach, in a safe and easy manner, the great manufacturing and agricultural centers.

3. To furnish them with all kinds of information, through the agency of competent persons, attached to the special service of the administration.

4. To provide for them capable interpreters.

5. To direct them to vacant apartments, and, in certain cases, to supply board and lodging for travelers at Paris, or in other places.

The society will base all its operations upon a moderate tariff, within the reach of all.

It will make arrangements with railway companies, so that travelers of all classes coming by the ordinary trains can procure, at starting, a certificate allowing them full possession of all advantages offered by the society.

The directors of the society, according to the wish expressed in article five of the regulations of the Imperial Commission for the Universal Exposition of 1867, at Paris, will provide for the running of third-class trains, specially intended for farmers, overseers, workmen, and mechanics. They will place themselves, as soon as possible, in communication with prefects, sub-prefects, mayors, heads of institutions, presidents of chambers of commerce, corporations, &c., and with the ministers of foreign powers, for the purpose of soliciting their valuable assistance and advice in regard to the best method of making known the conditions of this way of traveling, and the manner of receiving the sum to be paid, by means of small weekly installments. For this purpose, the society will establish in each department an agency, having power to appoint sub-agents in all towns and villages, who will be provided with books containing small printed receipts to be given in exchange for each payment of fifty centimes or one franc.

Upon the first page of this book will be printed an extract from the regulations, as follows:

A.—These books are not transferable unless notice has been previously given to the agent of the administration.

B.—The sums collected in each department will be paid in, every week, to the receiver of finances, or to some person of equal responsibility.

C.—Each holder of a book, by giving notice ten days in advance to the departmental agent, will be reimbursed for all sums he may have expended, except the premium of two francs, payable by each book, and a reserve of three per cent. intended to cover the expenses of printing and of commission to the agents and sub-agents.

D.—Members of workmen's societies, or even of workshops, can, if they wish, form companies and make direct contracts with the society for their journey and sojourn in Paris.

The receipts will be distributed through all the towns and villages, and it will be easy at any time for any person wishing to visit Paris in 1867 to purchase one or more of these receipts, according to the expense of his ticket and of his sojourn in Paris, if he desires it.

This arrangement will give an opportunity to persons interesting themselves in social and universal progress of purchasing these receipts in any place, and of disposing of them where and when they wish.

At the railway terminus in Paris persons in the society's employ will be constantly stationed to furnish gratuitously any information desired by travelers of all classes. These persons will be provided every day with lists of apartments, unengaged chambers in hotels, furnished houses, and all other particulars.

In short, the persons in employ of the society should endeavor to be useful in every way to the stranger, and to make his sojourn in the capital as agreeable as possible.

The office of the society will be open day and night for the reception of travelers.

A hospital will be prepared, under the direction of a physician, with an apartment for ladies.

COMPLETION AND OPENING OF THE EXPOSITION.

Mr. Beckwith to Mr. Seward.

"PARIS, *January* 21, 1867.

"SIR: The dates fixed by the imperial regulations for placing the products which are to form the Exposition are as follows:

"The structures in the Palace and the Park to be completed by the 1st of December; the show-cases, tables, and fixtures of all kinds to be placed before the 15th January; the reception and unpacking of products to commence on the 15th January, and to terminate on the 10th March, after which no more will be received. The products to be arranged for exhibition between the 11th and 28th March; the 29th and 30th are allowed for cleaning and sweeping, and a general inspection on the 31st will take place preparatory to the opening on the 1st April.

"The latest notice on this subject which I have received from the Imperial Commission is dated the 12th instant, reminding me that the above regulations will be adhered to; that the Emperor will inspect the Exposition between the 28th and 31st March; and that the opening will take place on the 1st April, without fail.

"The dates for finishing the structures which we had to make, and for commencing the introduction of products, (15th instant,) being past, I now propose to report the situation of our work.

"PALACE.—I have completed the flooring of Groups II, III, IV, V, in the Palace, and laid out upon them the plans in conformity with which the installations (fixtures) are to be made and placed.

"In Group VI one part of the floor is being laid, and will soon be finished; and in the other part of the same group the foundations in masonry are in progress for machines, of which plans of foundations have been sent me, upon which I could construct in advance; but all the necessary plans have not yet reached me. The concrete in Groups I and VII, laid by the Imperial Commission, will be sufficient in those groups, and answer in place of wood floors.

"PARK.—The annex in the Park will be about three hundred feet in length, and nearly thirty-four feet in breadth. The frame of this building is erected, and the covering commenced; this, by contract, should have been completed on the 15th January, but the tempestuous weather which set in on the 2d January, and severity of the cold which still continues, have retarded this work; the material for the covering and the flooring is prepared and ready to be laid, and a very short period of milder weather will enable me to complete this building.

"With respect to the buildings to be erected in the Park—two houses, one school-house, and a bakery, to be sent from the United States—the information sent me is not such as to enable me to prepare the ground for them, and there is likely to be some delay in consequence after their arrival.

"The contracts for the installations (tables, show-cases, shelves, frames, partitions, and other fixtures) in Groups II, III, IV, V, and VI, require the completion and delivery of this work by the 31st instant; but I have been obliged to extend the time for a part of it to the 9th February.

"The preparation of the walls in Group I, for the reception of pictures, is nearly completed; and I rely upon being in a condition to commence the reception of products in the Palace from the 25th instant to the 30th instant, and to commence the unpacking and placing throughout the Palace and annex by the 10th February.

"Most of my contracts for the more expensive work have been made in Belgium, at lower prices than I could obtain in Paris, and where circumstances admit of more reliance on punctuality.

"I ought not to omit to state in this connection that the backward and still incomplete condition of the catalogues has compelled me to undertake and carry on the expensive part of the work in question under great disadvantages.

"Taking the preliminary catalogues and allotments which I transmitted to the department on the 24th October as a basis, I have been obliged to make the contracts for the construction of the fixtures in that conformity, as being likely to be pretty nearly what would prove to be in the end necessary.

"But as there have been many changes in those lists of products and allotments of space, and these changes are still going on, it is not unlikely that when the products and the fixtures come together they will not in all cases fit each other.

"I am liable to find a space for which I have prepared an expensive show-case occupied by a stove, or another space for which I have prepared a table, appropriated to products requiring a different method of installation for exhibition, &c.

"This contingency results inevitably from carrying on simultaneously two distinct works, one of which (the catalogues) should precede the other, by which method alone the fixtures could be made in advance to fit the products when they arrive.

"The incongruity between the products and the installations prepared for them, to whatever extent it may be found to exist, will cause further delays, probably considerable waste or expenditure of money that might have been avoided, and can hardly fail to render it impossible to place and expose the products in all cases in the way and manner desired by the exhibitor, and intended. Some changes and disappointments from this source may become unavoidable, and give rise to dissatisfaction and complaints from exhibitors thus disturbed, and who perceive no cause for it but what appears to them very bad management.

" But it is obvious that if the construction of the fixtures had been delayed for the completion of the catalogues, (not yet completed,) such delay would have been equivalent to an abandonment of the Exposition, and it will require unceasing efforts, as it is to bring the products and the fixtures together, however incongruous their condition, in time to prevent their exclusion from the Exposition.

*　　　*　　　*　　　*　　　*　　　*　　　*

" I have not yet been able to report any of the catalogues to the Imperial Commission. Their urgency increases daily and their hopes have been fed by the continued advices above quoted, each of which in succession seemed to indicate that but little remained to do, and that the final report might be fairly expected by the following mail.

" But the result is, I regret to say, that the Imperial Commission has at length become impatient. They have received my representations of late with apparently diminished confidence, and have now given me final notice that if my manuscript catalogues are not delivered to them by the 25th instant for publication, the Exposition will open on the 1st of April without them.

" I still hope to avoid this result; it would place our exhibitors at great disadvantage, and I look with increasing anxiety for the final reports from the agency at New York."

THE OPENING OF THE EXPOSITION.

Mr. Beckwith to Mr. Seward.

" PARIS, *April* 2, 1867.

" SIR : The Exposition was opened yesterday, the 1st of April, in conformity with the regulations published by the Imperial Commission.

" Work in the building and the Park was suspended for the occasion, the doors were opened to the public, the attendance was numerous, and the weather was brilliant.

" The diplomatic bodies and the other invited guests were assembled in the interior gallery appropriated to the fine arts. The national commissions were stationed, each in its own section, on the elevated platform which runs through the great gallery of machinery comprising the larger circuit of the building.

" His Majesty the Emperor and the Empress arrived at 2 o'clock p. m., accompanied by the chief officers of state, several ladies of the court, the Imperial Commissions, and a numerous suite of functionaries connected both with civil departments and with the Exposition.

" The imperial cortege on arrival ascended the great platform or promenade, and made the entire circuit of the building, the various national commissions being presented in succession by the minister of state, vice-president of the Exposition, to the Emperor and the Empress as the cortege arrived at the different sections.

" The national commissions then repaired to their respective sections in the gallery of fine arts and joined the invited guests. The imperial cortege descended from the platform and made the tour of the gallery of fine arts, their Majesties saluting the audience as they passed, receiving in return their cordial greeting.

" The imperial cortege then retired by the great door opposite the one by which they had entered, the Exposition was declared to be open, the barriers and guards removed, and the avenues left free to the circulation of the multitude."

CONDITION OF THE UNITED STATES SECTION AT THE OPENING.

Mr. Beckwith to Mr. Seward.

" PARIS, *April* 3, 1867.

" SIR: I beg to state briefly that at the opening of the Exposition the structures in our section were nearly completed and the placing of the products about half finished. Many of them are still on the road between this and Havre, which has been greatly clogged by accumulation beyond the means of rapid transport.

" Fully a month will be required to complete the work, and this observation is applicable to every national section of importance, including the French section.

" Very little machinery was ready in any section for movement, though a few machines in some sections were put in motion for effect.

" Three or four of our machines, under charge of Mr. Pickering, were belted and shafted ready for work, but the Imperial Commission were not ready to supply us with steam or water, and the machines did not run.

" Each nationality has been urgent in pushing forward its work for the opening, in which anxiety I participated, and increased the number of workmen, employing one set during the day and another for the night till five in the morning for a short period.

" The natural anxiety in my section was sufficient, and the movement was overdone by the severe pressure of the Imperial Commission. This caused an accumulation and a clog which retarded instead of hastening the work.

" The contracts for transport, cartage, carpenters' work, masonry, decoration, &c., all broke down, new contracts were made, wages were doubled, the men became masters, and with this accumulation of force and expense the work went slower every day.

" With the business in this train many of our exhibitors arrived, anxious to find their products and get them in place; but destitute of any knowledge of the situation, ignorant of the regulations, and to a great extent of the language, they have met with difficulties and delays they did not look for, and have shown some dissatisfaction.

" But time and patience will remedy this, reasonable grounds of complaint—if such exist—will be removed, and imaginary grounds will vanish with a better knowledge of the circumstances.

" The precipitation and disorder with which the exhibitors hurried off their products from the United States at the latest moment, their general, almost uniform neglect to furnish inventories of the contents of packages, and the arrival of every vessel but one in advance of the bill of lading and shipping documents such as they were, precipitated the business upon me in a condition which can only be appreciated by those who are familiar with the movements of commerce.

" For the most part I have had no means of furnishing the customs with the requisite inventories, nor of knowing the contents of packages, till they were opened and the inventories made, and many of them are not yet opened.

" The shipping lists have proved to be very inaccurate—several packages in them have not appeared, while many others not in them, nearly a hundred in all, have been delivered. Under these circumstances it is impossible to hold vessels to any strict account for delivery.

" The impossibility of making a correct catalogue under these circumstances is evident. I have made the best that was possible and it appears in the first edition of the imperial catalogue, but it is extremely imperfect.

" I have now in press a catalogue together with the statistics to accompany it; the catalogue is in three languages and the statistics in French. This will be more accurate, and will be out, I trust, by the 15th instant. But even this cannot be perfected before the second or third edition.

" The houses from Chicago have been a great embarrassment. The material for one of them was only got into the Park yesterday.

" The materials for the other arrived some weeks since, but instead of a house in sections ready to put up, it was lumber from the mills of which to build a house.

" Mr. Clark, the carpenter who came over to build the house, concluded that he could not do it either with French tools or French workmen. I sent him to England to procure carpenters and tools; he brought over fifteen workmen, and they are working on the first house, at heavy wages, and doing little, having evidently embraced the opportunity of coming to the Exposition rather than to work.

" I should not have felt justified in this course, but for the recent appropriation in Congress, which was telegraphed to me as intended for this purpose, and for the importance apparently attached to this exhibit by those who were interested in sending it, which seemed to leave me no choice, though so large an expenditure for this purpose is not in accordance with my own judgment.

" The pressure of work at this moment will be accepted as my apology, I hope, for so brief a report on the state of the work at the opening."

II.

THE PROGRESS AND CLOSE OF THE EXPOSITION.

SCIENTIFIC COMMISSION.

The importance of obtaining the assistance of professional and scientific persons to study the exhibition and aid in preparing suitable reports upon it, was pointed out by Mr. Beckwith in his letter of suggestions to Mr. Bigelow, April 3, 1865, printed upon p. 15. The department was also addressed by Mr. Bigelow in his dispatch from the United States legation at Paris, September 21, 1865, as follows :

" The circular, of which No. 1 is a translation, has been issued by the Commissioners of the Universal Exposition of 1867. It provides for the creation of an international scientific commission, whose duty it shall be to note the recent advances made in the sciences and arts, to contribute what they can to diffuse the knowledge of useful discoveries, to encourage international reforms, and, lastly, to point out, in special publications, the useful results to be derived from the Exposition.

" I invite your special attention to the provisions of this circular, and take the liberty of suggesting that our government can in no way turn this Exposition to better account than by sending a few of its cleverest men of science to make part of this commission. I say its cleverest, because it is not worth while to send men who would see nothing, and therefore describe nothing, which would not be seen, and as well or better described, by the French and other foreign exhibitors.

" The Exposition will be transitory, but the accounts that will be written about it have a chance of enduring. Europe will assign this duty to her choicest men. There is glory to be won in a successful competition with them. I think the opportunity should not be neglected."

The following is the translation of the circular referred to. Original was issued by the Imperial Commission, and signed by Rouher, the minister of state and vice-president of the Imperial Commission, September 20, 1865:

ORDER ESTABLISHING THE SCIENTIFIC COMMISSION.

" In accordance with the general regulations adopted by the Imperial Commission, 7th July, 1865, and approved by an imperial decree of the date of 12th July, 1865, which provides for the institution of a series of studies and experiments, under the direction of a scientific commission, and for the publication of results of general interest attained by these labors, (Article 63,) it is ordered:

"ARTICLE 1. There is established, in connection with the Imperial Commission, an international scientific commission, having for its object: 1st. To indicate the best means of representing, at the Exposition of 1867, the recent advances made in the sciences, in the liberal and industrial arts. 2d. To contribute to the extension of the employment of useful discoveries, and to encourage reforms of international interest, such as the adoption of uniform weights and measures, identical scientific unities, &c. 3d. To point out in special publications the results of general utility to be derived from the Exposition, and to undertake, if it be necessary, the researches required for their accomplishment.

" ARTICLE 2. The Scientific Commission is composed of Frenchmen, appointed directly by the Imperial Commission, and of foreigners appointed upon the nomination of their respective countries. These appointments will be made successively by special orders.

" ARTICLE 3. Scientific organizations, and, in general, persons interested in the progress of the sciences and the arts, are invited to submit to the Imperial Commission their opinions in regard to the researches to be undertaken, and the questions to be considered.

" ARTICLE 4. The members of the Scientific Commission will not be expected to hold stated meetings. They can labor separately upon the matters which are given them to treat; and can send, in their own names, the fruits of their labor to the Imperial Commission. It will also be permitted to them to meet with their colleagues of all countries."

" ARTICLE 5. The memoranda and reports will be submitted before the 1st July, 1867, to the Imperial Commission, and published, if necessary, under its direction. The whole will form the collection of the labors of the Scientific Commission.

" ARTICLE 6. The councillor of state, Commissioner General, is charged with the execution of these orders."

REPORTS UPON THE PROGRESS OF SCIENCE AND LETTERS IN FRANCE.

The following is a translation of a letter addressed, December 1, 1865, by M. Duruy, the minister of public instruction, to M. Le Play, the councillor of state and Commissioner General for the Universal Exhibition of 1867 :

"I have the honor to inform you that, in virtue of the approval given by the Emperor to my report of the 8th of November, the minister of public instruction will directly participate in the Universal Exposition of 1867, by producing there the works of diverse character which are comprised in the mission with which he is charged.

"He will at first present the best manner arising from a substantial rule which serves for the instruction of children and of adults in the primary public schools, and in order that we may be able to establish its value, he will make fully known also the results of the tuition. In addition, he will lay down a series of reports which will show, in the first part, the discoveries of scientific theories, from which emanate every industrial perfection, and on the other part, the moral ameliorations and administrative or economical reforms due to the influence of ideas that literature diffuses, that history verifies in the past, and of which political sciences provoke the application in the present.

"It is in the Classes 89 and 90 that the objects might be placed, which, by appealing to the eye, can allow it to appreciate the state of education.

"Among these objects will be found some works executed by the pupils themselves, such as drawings, modelings, &c., which it is usual to produce at every exposition, and of which the most meritorious have always gained some credit to the schools who have sent them.

"The most severe precautions will be taken by my administration in the public schools, in order that these objects may represent, with a scrupulous fidelity, the real labor of the pupils, without the assistance of teachers, and consequently what they will be truly in a condition to do upon the day when they will be left to themselves. It will be a true standard of primary education.

"The reports on the principal works produced by the French mind for the past twenty years in their intellectual order, and in their social order, will find, therefore, their natural place in the Class 90, which makes a part of Group X, where the Imperial Commission has united that which concerns the material and moral progress of populations.

"The reports will be made known as follows :

"1. The progress accomplished in France by the mathematical, physical, and natural sciences.

"2. The progress accomplished by the moral and political sciences in their applications to the wants of society.

"The character of French letters that they may study, at least with a view to their style or as a task of literary criticism, and in their effects upon the general education of the country.

"Some men, who are the light and honor of the Senate, of the Council of State, of the Institute, and of high education, have been willing to undertake to draw up these reports. Before speaking in the name of

French science, in presence of the wise men of the world, in an inclosure where every one will judge each other, they will study without troubling the serenity of the impartial historian; and in the same way, with a respect for their own labor, they will lay before their equals a testimony devoid of all personal interest.

"The ancients selected the sage to seek the beautiful, the true, and the perfect. The reports will tell whether the ancient formula is that of the modern sage, and whether French letters, faithful to the great traditions of Corneille and of Molière, seeking always the beautiful in order to diffuse the good, are still a school of manners, as the positive sciences and the moral sciences are a school of truth and justice.

"Before indicating the classifications of the matter comprised in the three divisions mentioned before, I believe it, Monsieur the Commissioner General, useful to communicate to you some explanations relative to the meaning and object of this work. It is of consequence to remark at first, that he is not to draw up at first an encyclopedial resumé of human knowledge. Proceeding in that way he would miss the mark by overshooting it. The interval which separates us from 1867 is not sufficient for the calculations of all the intellectual riches of humanity. It is already a sufficiently heavy task to measure their increase from the opening of the period which the contemporaneous generation completes by its labors; of that time even they will gather only the considerable facts and results well established. It is not the object in effect to write a complete history of each branch of human knowledge for twenty years. The vain efforts, the abortive experiments, the hypotheses not confirmed—all this scientific dross, which learning collects with curiosity, ought to be placed aside with the facts which may not have a useful character or a general interest.

"We do not purpose to burden ourselves with making for foreign countries a report of the things I have just-indicated, though they come within the limit of time prescribed. We will not be able, doubtlessly, in speaking of our progress, to abstain from touching upon that of neighboring nations.

"A joint responsibility closely unites to-day the scientific labors and moral preoccupations of the different nations. Sometimes the same idea spontaneously originates in several countries at once; sometimes an invention found on one side of the frontier has carried all the fruits which grace an accomplished perfection to the other side. Elsewhere, several peoples following, perhaps, our example, it is necessary to leave them the honor of pronouncing for themselves an authoritative and impartial judgment. France, in the reports which she undertakes, proposes exclusively to be occupied with herself, saving the exceptions which will be indispensable to place in the work a perspicuous and necessary justice; the minister of public instruction using the liberty left by the liberal programme of the Imperial Commission to all those who will wish, like herself, to exhibit in Class 90; and the classification which it presents ought to be considered as a simple memorandum, of which each will make such use as will be convenient to him.

"The programme of the subject to be treated in the report in question is determined principally in the following manner:

"1. PROGRESS ACCOMPLISHED BY THE MATHEMATICAL, PHYSICAL, AND NATURAL SCIENCES.—Geometry, analysis, mechanics, astronomy, geodesy. Physics, chemistry. Geology and paleontology, botany, zoology, anthropology, general physiology, medicine and surgery, hygiene, rural economy, and the veterinary art.

"2. PROGRESS ACCOMPLISHED BY THE MORAL AND POLITICAL SCIENCES IN THEIR AGREEMENT WITH THE WANTS OF SOCIETY.— Public right, administrative right, legislation—civil and penal, political economy, rights of nations.

"3. CHARACTER AND TENDENCY OF FRENCH LETTERS.—Literature— poetry, drama—philosophic doctrines, historical works, archæological discoveries.

"Around this collection of reports, and as an appendix in connection therewith, will be arranged some objects chosen for the purpose of indicating the most interesting results of scientific missions and archæological researches, accomplished in the same period under the auspices of the administration of puplic instruction.

"Accept, Mr. Commissioner General, the assurance of my most distinguished consideration."

Mr. Beckwith to Mr. Seward.

"PARIS, *December* 14, 1865.

"SIR: * * * I embrace this opportunity to allude to the subject of a scientific commission, for the purpose of studying and reporting upon the Exposition.

"The printed document hereto annexed, issued by the Imperial Commission, contains a decree forming a French scientific commission, and gives general directions for its guidance.

"The Scientific Commission is, first, to point out to the Imperial Commission itself the best means of exhibiting the progress recently made in the sciences and arts; secondly, to co-operate in propagating the adoption of useful discoveries and in promoting international reforms, such as the adoption of a uniform system of weights and measures; and thirdly, to indicate the useful results in general to be drawn from the Exposition, and to undertake, if there is occasion, the researches or experiments requisite to complete those useful results. Scientific bodies, and persons in general who interest themselves in the progress of sciences and arts, are invited also to express their views to the Imperial Commission on the researches which should be undertaken and the questions which should be examined by the Scientific Commission.

"The first part of the labor of the Scientific Commission, therefore,

6 P E

precedes the opening of the Exposition, and the results of it should reappear in the Exposition itself; the second part may commence at any time, and does not appear to be necessarily connected with the Exposition, but the third part relates more especially to the Exposition; and the reports of the commission, collective or individual, embodying the fruits of their researches, should be sent in to the Imperial Commission by the first of July, 1867, three months before the close of the Exposition, that they may be published.

" The members of the commission are Frenchmen, but foreigners may be added to it upon their nomination by the foreign commissioners, and acceptance by the Imperial Commission. And the members of the commission may unite in their labors and reports, or work separately, and make separate reports, if they prefer it.

" I do not perceive that any particular advantage would result from the addition of foreigners to this commission, as their reports are to be made to the Imperial Commission ; but foreign scientific commissioners might perhaps find it desirable at a later period to have one or more of their members on the French commission, as a channel of convenient mutual intercourse.

" In forming a scientific commission for the United States, and giving them instructions, the government will probably leave much to the judgment of the commission itself in choosing subjects of particular inquiry ; but the best results would probably be attained by limiting the range of inquiry, and making the study of fewer subjects more complete.

"Among the subjects to which attention might be particularly directed with advantage, I venture to suggest the following :

"1. A comparison of the most useful American products with similar European products, indicating the qualities and differences of each, whether of superiority or inferiority, pointing out in what these differences consist, and the causes of them.

" 2. The methods and processes by which these useful products and their various qualities are produced.

" The design of the Exposition is not limited to the display of products, but a prominent feature of its organization is the attempt to exhibit or disclose as far as possible methods and processes.

" Researches in this direction, which commence in the Exposition, must, in many instances, extend beyond those limits, and will be subject to the facilities for inquiry, greater or less, which may be afforded outside of the Exposition. But the utility of such researches cannot, I think, be doubted.

" Why is gas for lighting streets, houses, &c., so much dearer in America than in France ?

" The investigation of this subject would show, I think, that there is no great difference in the average cost of raw material in both countries; that apparatuses are as good in one as in the other; and that the difference in the price of manual labor is but a small element of the cost of gas in either case. But the methods adopted in France for utilizing secondary products, resulting from first processes, appear to have introduced economies which make gas in France cheaper than in America, and that these economies are applicable in America as well as in France.

" There will be in the Exposition specimens of rails composed of iron and Bessemer iron or steel. An inquiry into the method of making these rails would probably disclose several useful economies in the processes ; one of which results from laying an upper surface of steel on a body of inferior cheap iron, which combination gives at once solidity, weight, strength, and hardness of surface, producing a superior rail at a cost which admits of its introduction and use as an economy.

" On a recent visit to some of the great founderies in the north of France, I was informed that they were occupied with considerable orders for England, which orders they owed to the superior quality of their iron, the excellence of mechanical work, and moderate cost ; and I was surprised that such results could be attained in localities the most unpromising. Nearly all the raw material was brought from great distances, at great expense. Coal from England, Belgium, and distant mines in France ; iron ores from Spain, England, Belgium, and several French mines in different localities. I was informed, and investigation would probably prove, that under these great disadvantages, which are more than the equivalent to cheapness of manual làbor, good results are attributable, first, to the thoroughly scientific and careful analysis and mixtures of ores, by which superior metal is produced; second, to the excellent mechanical education of many of the workmen ; and third, to the economies introduced for utilizing secondary products of first processes, which secondary products are usually thrown away.

" It is observable that the genius of every country adapts itself to local circumstances, and takes its greatest development in the direction of its greatest wants.

" In America, where raw materials are abundant and cheap and manual labor is dear, mechanics and inventors and men of science and genius turn their attention with great success to the production of ' labor-saving' machines and methods, but exhibit at the same time comparative indifference or wastefulness in regard to raw materials.

" In Europe, where manual labor is cheap and materials are dear, the attention of the same leading class of minds is bent in the direction of economies in everything which relates to raw material, and in constant and successful efforts to utilize all secondary products, and in the steady improvement and perfection of processes by scientific means.

" Guided by those conditions, each country makes its own progress in its own way ; consequently there is something to give as well as to receive on all sides, and a universal exhibition should promote these exchanges.

"If it be true, in a general sense, that the agricultural, manufacturing, and industrial arts in America are, on the average, as fully developed as they were a few years since in Europe, it is equally true that Europe has advanced, and that the relative positions are not changed; and the proof and consequence of this is, that America continues to supply the raw materials and receive the manufactured products.

"This exchange is profitable to both sides; but as long as the skilled labor of one man exchanges for the unskilled labor of two men, the best of the bargain will be against us.

"The glory or vanity which each nation may derive or display in exhibiting its products will result in nothing valuable if not united with the serious studies of competent men. And however large the field of investigation which the government may prescribe to the commission, I hope their particular attention may be directed to the investigation of methods and processes, at once the most difficult and the most useful of researches."

Professor Joy to Mr. Derby.

"COLUMBIA COLLEGE, *New York, December* 6, 1865.

"DEAR SIR: In my letter of the 4th instant I spoke of the importance of the appointment by government of a scientific commission to report upon the Exposition of 1867, and I have since observed that Mr. Beckwith makes the same suggestion in his communication of April 3, 1865. I am glad, therefore, that the idea is likely to take root and come to proper development. 'The appointment of professional and scientific persons to study and aid in the preparation of a suitable report of the Exposition, to be subsequently published,' ought to be made as soon as Congress can act upon the matter.

"The scientific committee will need much time for the consultation of the reports of previous exhibitions. They will desire to carry on extensive private correspondence, first, in this country, for the purpose of obtaining the most recent information upon matters relating to the numerous subjects likely to be presented to them for study; second, with foreign scientific and practical men in order to learn the best sources of information. Without great previous study no person could prepare a clear and luminous report of any portion of the Exposition which would be of practical value. A report must not be a catalogue; it must sketch in a few words the history of the department under consideration, state its growth, point out its success, and give statistics and results in a way to enable any one, after reading the book, to invest money in new enterprises without the loss attendant upon a long series of experiments.

"These reports, in able hands, would become text-books for all branches of industry, and would tend to develop our resources as much as any papers Congress has as yet published.

"All parts of the country are equally interested in the publication and extensive circulation of such documents, and the wider this kind of knowledge is disseminated the better for the country.

"Let there be ten members of the scientific committee, corresponding to the ten groups of the Exposition, with power to appoint assistants where the amount of material is too great to be fully studied by one mind, viz: Committees on—

"1. Works of art.

"2. Materials and their applications in the liberal arts.

"3. Furniture and other objects used in dwellings.

"4. Garments, tissues for clothing, and other articles of wearing apparel.

"5. Products, wrought or unwrought, of extractive industries.

"6. Instruments and processes of common arts.

"7. Food, fresh and preserved, in various stages of preparation.

"8. Animals and specimens of agricultural establishments.

"9. Live products and specimens of horticultural establishments.

"10. Objects exhibited with a special view to the amelioration of the moral and physical condition of the population.

"It is obvious that ten men could not do justice to all these subjects, but it would probably be better to refer the matter to that number of persons to collate and prepare for publication the reports of the assistants they may select, the number and compensation of such assistants to be fixed by the Commissioner.

"By the early appointment of this committee of ten the Commissioner would have the advice and assistance of the ablest men in the country. He would be their presiding officer, if the committee were to be called together, and would have the right to call upon them for services at any time.

"I would suggest that the committee receive no compensation for their services further than a reimbursement of expenses actually incurred. For the purpose of control, let there be an amount fixed, beyond which expenses will not be paid.

"The committee not being business men, could not take charge of the collection and shipment of goods, but they could greatly assist the agents of each State in bringing out the most characteristic and representative articles. The literary work of the commission could be divided among them, and thus matters would be greatly facilitated.

"The members of the commission ought to be familiar with at least the French language. A knowledge of German would greatly aid in the preparation of a report, as the arts and manufactures of Germany as represented in the Exposition will, no doubt, equal in importance those of any other country. Immediately after Congress shall have made the necessary appropriations, the appointment of the scientific committee ought to be made by the Secretary of State, and the committee be accredited to the Imperial Commission in Paris, as the official scientific representatives of the government to the Exposition.

"This committee would in no way interfere with the commissioners appointed by the various State executives, as their duties are of a different character.

"I would confide to the committee a mission of a somewhat private character, viz: the duty of disseminating knowledge of our country for the purpose of encouraging emigration.

"They could accomplish an important work by making known the extent of unappropriated lands in this country, by editing short statements to be published in French, on sheets, and placed conveniently for every one to take a copy, and by writing articles for the newspapers.

"The magnitude of the work expands before me as one idea follows another, but I believe I have hit upon the principal points, and I shall be gratified if the views here expressed meet with your approbation."

Mr. Beckwith to Mr. Seward.

"PARIS, *December* 31, 1865.

"SIR: I had the honor to address you on the 14th instant, and took the liberty of making some suggestions on the subject of a scientific commission in connection with the Exposition of 1867.

"I omitted on that occasion to allude to a consideration which I think of importance, viz: that the members of the scientific commission be paid for their services.

There may be individuals who are competent and willing to serve without remuneration, but many of those who are best qualified by their attainments and studious habits for useful researches are actively employed.

"The interruption of their engagements for a twelvemonth would be a matter of moment to them, and they are not in general men of fortune who could afford so large a contribution of time and labor gratuitously.

"If therefore Congress omits to provide for their payment, the government will be restricted in its selection to those alone who may voluntarily offer their services.

"Under these circumstances I should much fear that it would be impossible to form a commission prepared to devote themselves to the continuous labor and serious studies which are indispensable to render their researches thorough and entitle them to appear as a national work.

"The labors of a competent commission could not fail to be of great value to the country, but an exhibition of products without a commission, or an insufficient one, would be a vain, if not a useless display, because it would fail or fall short in its educational effects, which are the proper object of an exhibition.

"Large sums are expended by civilized nations on voyages of geographical exploration and discovery in all parts of the world, and with beneficial results. But scientific and industrial explorations among each other would yield still better results.

"A commission of this kind from Europe to America at the present day could not fail to bring back a great deal of useful knowledge which does not now exist in Europe, and will be a long time in reaching here, but no such movement is contemplated.

"The self-complacency of nations is in proportion to their unacquaintance with each other, and the aversions which spring from this are a great obstacle to ameliorations.

"An evidence of this may be seen in the obstinate perpetuity of the cumbrous systems of weights, measures, and coins, the unification of which would promote the diffusion of statistical and economical knowledge, like a common language, simplify and facilitate exchanges and commerce, diminish their cost, and produce savings of great aggregate importance.

"But while the march of improvements in individual nations is constant and rapid, the transmission of those improvements from nation to nation is slow.

"New knowledge of many kinds is a long time in getting into books, after which it may become an article of merchandise, but much always remains, less attractive as an object of commercial speculation, but more useful, and is left unwritten to make its way by indirect channels, circulating with persons and with the general movement of commercial intercourse.

"The transmission might however be immediate and direct; nothing is more practicable, and if international exhibitions should give rise to international scientific explorations they will have accomplished their highest function.

"Scientific commissions may then supersede exhibitions, and divert the cost of them to less cumbrous and more effective methods, for these expensive collections of products, now so much in vogue and in fact so useful, are not indispensable to the investigations in question, but only necessary as leading to these researches which previously had no organized and established existence, and are still far from being perfected.

"It is thus evident, I think, that the real purpose and object of exhibitions cannot be attained without the aid of competent scientific commissioners, and I hope the government and Congress will incline to this view of the subject."

THE ORGANIZATION AND DUTIES OF A SCIENTIFIC COMMISSION.

Mr. Beckwith to Mr. Seward.

"PARIS, *January* 31, 1866.

"SIR: The commissions constituted by the principal nations of Europe, in connection with the Exposition, comprise fifty or sixty members each. They are divided into sub-committees, to which are attributed separate duties, such as the preliminary work of forming the exhibition, its subsequent installation and superintendence, serving on international juries, studying the Exposition in a scientific sense, and reporting upon it, &c.

"The labors of these large commissions will be facilitated by their proximity to the work; that, and inexpensive journeys, with frequent and brief visits to the Exposition, will enable them to make observations and memoranda which can be elaborated and perfected at home in their respective residences, surrounded by the conveniences of libraries, apparatus, and the aids pertaining to their habitual occupations. The large personal and office expenses necessary to the prolonged residence in Paris of a commission so numerous will be thus diminished. But this method is not fully applicable to the United States, and I have not thought it expedient on this occasion to ask for so large an appropriation as a continued residence of a commission so numerous would require. Very good results may be obtained at less cost.

"The resolutions presented to Congress on the 21st of December proposed appropriations for a scientific commission of ten members, corresponding to the ten groups of products. But this number, unassisted, will not be sufficient. It will devolve on them not only to make the requisite studies and reports, but to serve on international juries. The latter service, indeed, though requiring much time, will afford them the best opportunities for information resulting from the investigations, experiments, and discussions of the juries. But they will not be equal to the work without assistants, and they can be obtained at moderate cost.

"The services of scientific and professional assistants, draughtsmen, &c., can be engaged, whose special studies, colloquial knowledge of continental languages, familiarity with the continental nomenclature of the sciences and industrial arts, together with their personal acquaintances, access to sources of information, and works of authority and local knowledge in general, will render their services as assistants highly efficient. The Scientific Commission, thus supplemented, will be equal to the work required of it, and more useful labor can be accomplished at less cost in this way than in any other.

"I take the liberty, therefore, to recommend this subject to the consideration of the government, and to suggest that discretion be given to the Scientific Commission within such limits of expenditure as the government may determine. First, to accept of such professional assistants as may voluntarily offer without pay, for a long or short period of time; and secondly, to employ professional assistants, and pay them for their services.

"With regard to the management of the exhibition, it will be doubtless placed, in a general sense, in charge of the General Commission, which will comprise the Special Agent, the Commissioner General, and the Scientific Commission; and the work could be conveniently divided among them as follows:

"The minister of the United States, being the Special Agent, should preside at the opening ceremonies of the Exposition, and continue to be the channel of communication between the United States Commission and the French government.

"The Scientific Commission should undertake the scientific researches and reports; also the jury duties and the various experiments, essays, and trials of machinery, &c., which may be invited, or which they may institute, and, in addition, should be charged with advisory duties, which will be hereinafter mentioned.

"The Commissioner General should be charged with the general superintendence and care of the exhibition, and with the disbursement and accounts, and he will be the channel of communication between the United States Commission and the Imperial Commission, as provided by the imperial regulations.

"The numerous national exhibitions will all be conducted in conformity, first, with the imperial regulations; and, secondly, in conformity with their own regulations, respectively, which will be supplementary to the imperial regulations. The imperial regulations (for example) make a general provision for a police applicable to persons and property, and a guard for the protection of property; but much detail is left to the commissioners general, respectively, to provide for the daily sweeping, cleaning, and proper condition of their respective sections; in providing experts, linguists, &c., in case of need, to attend in the compartments for the safety of property, and to give such explanations and information respecting products as may be necessary or desirable. All details of this kind—and they are numerous—are left to the respective commissioners general, and both the work and employés required are under their orders, subject to the imperial regulations.

"Preparations should be made in advance by the Commissioner General for the reception and bonding of the products on their arrival, for their inland transport and installation in the Exposition, where they will remain in his charge during the Exposition, and finally at its close be returned by him to the United States, or delivered to owners who may apply for them here, they first paying the duties and charges, and releasing them from bond, at which point the control and responsibility of the United States government will cease.

"The suggestions above made in regard to employés to take care of, or to give explanations to visitors in regard to, any products in particular which may require it, are not intended to prevent exhibitors; but, on the contrary, exhibitors should be invited to be present at all times themselves, or to provide at their own expense proper persons to take care of, expose, and explain their own products, as their interest may require such service, being subject to the general regulations.

"The expenses of scientific assistants and other details, herein alluded to, will not, in my judgment, involve further appropriations of money than I have heretofore suggested, should the larger sums named be provided.

"The foregoing suggestions are made in view of the imperial regulations, and are so modified as to be in conformity with them, and at the same time to provide for the actual necessities of the Exposition, and for the execution of the work proposed to be done."

SCIENTIFIC COMMISSION UPON WEIGHTS, MEASURES, AND COINS.

Mr. Beckwith to Mr. Seward.

"PARIS, *June* 29, 1866.

"SIR: I had the honor to address you on the 14th December last in relation to the Scientific Commission formed by the French government and charged with various labors. The commission was directed in particular to promote international co-operation in the propagation of the use of new and important discoveries, and for the adoption of a uniform system of weights, measures, and coins.

"I beg now to advise you of the steps which have since been taken. At the instance of the Imperial Commission meetings have been held, composed of members of the Scientific Commission, the Imperial Commission, and the foreign commissioners, for the purpose of consultation regarding proper measures to be adopted in connection with the Exposition of 1867, for drawing public attention to the subject of uniformity in weights, measures, and coins.

"The following suggestions were made by the English scientific association and approved by the meetings:

"1st. To form a collection of the weights, measures, and coins of all nations, to be exhibited in the Palace of the Universal Exposition of 1867.

"2d. To organize an international committee charged especially with the formation and exhibition of this collection of weights, measures, and coins, and to devise the most efficacious methods of promoting uniformity.

"3d. In accordance with these views the Imperial Commission appropriated the space requisite for the exhibition of weights and coins in the Exposition palace, and formed a special committee connected with their Scientific Commission, which special committee is the commencement of the International committee alluded to to be charged with the subject.

"I annex hereto three documents, numbered 1, 2, and 3.

"No. 1 contains a brief report of the preliminary meetings before mentioned.

"No. 2 contains the approval of the proceedings of the minister of state, and a decree constituting a special committee, giving the names and professions of the persons appointed, which committee forms the nucleus of the International Committee on Weights, Coins, &c., to be constituted.

"No. 3 is a letter from this special committee asking my adhesion to the project in principle, and desiring me to take the further necessary proceedings.

" It will be observed that article five of the decree provides that additional members may be added to the International Committee by the foreign commissioners of those nations which take part in the exhibition of weights, measures, and coins.

" At the particular request of the Imperial Commission I now present the subject for the consideration of the government of the United States, and respectfully solicit their co-operation in the formation of the collection of national weights, measures, and coins to be exhibited, and in appointing or authorizing the appointment of commissioners to be added to the International Committee above named, and charged with the particular business herein described.

" I have read with great pleasure the recent proceedings in the House of Representatives relating to the introduction of the metrical decimal system into the United States, and I observe that those proceedings provide for a commission to be charged with the subject of a common unit of coin.

" If the general purposes and method of proceeding herein reported receive the approval of the government, I would venture to suggest that the commissioner to be appointed under the congressional authority alluded to be directed to prepare the proposed exhibition of weights, measures, and coins, and that he be nominated to the aforesaid International Committee.

" This arrangement will place the commissioner at once in direct relation with professional and learned persons occupied with coinage and analogous subjects, and best qualified to co-operate with him in the accomplishment of his particular object.

" The committee is now organized according to the usual forms on this side, to give additional weight to its proceedings, and it is probable that its numbers and nationalities will be increased to an extent that will comprise much ability and appropriate knowledge, and produce an influence favorable to the objects of its labor."

Mr. Beckwith to Mr. Seward.

" PARIS, *July* 17, 1866.

" SIR: I had the honor to address you on the 29th June, with documents relating to weights, measures, and coins.

" I beg now to wait on you with two legislative documents which are of interest.

" Document No. 216 contains the project of a law emanating from the council of state, and submitted for the consideration of the Corps Legislatif, which is designed to place the coinage of the empire in harmony with the recent monetary convention between France, Belgium, Italy, and Switzerland, and gives an exposition of the motives of the convention and the law.

" Document No. 282 contains the report of the committee of the Corps Legislatif on the proposed law, suggests amendment, and presents the law as finally adopted, on the 13th June, 1866, together with the monetary convention.

" It results from these proceedings that a uniform system of coinage is established in the four countries named—uniform as regards the unit, the metallic standards, and the value of the pieces to be coined. Each country retains the double standard of gold and silver, with the relative value of 1 to 15½.

" The composition of gold coin remains in the proportion of ($\frac{9}{10}$) nine parts of fine gold to one of alloy, and the coinage of gold is restricted to pieces of the value of one hundred francs, fifty francs, twenty francs, ten francs, and five francs.

"The composition of the silver five-franc pieces remains in the proportion of ($\frac{9}{10}$) nine parts fine silver to one of alloy; but the composition of silver coin of smaller values is reduced from $\frac{9}{10}$ to $\frac{835}{1000}$, (835 parts fine silver to 165 parts of alloy,) a reduction in value of about seven per cent·

" The coinage of this class is restricted to pieces of the value of two francs, one franc, fifty centimes, and twenty centimes, and limited in amount to six francs per head of population, which should give about 32,000,000 francs for Belgium, 239,000,000 francs for France, 141,000,000 francs for Italy, and 17,000,000 francs for Switzerland.

" This inferior money is a legal tender between individuals to the amount of fifty francs in a single payment, and receivable for dues to the government without limit. It follows from these measures that the unit of the monetary system (one franc of the standard of $\frac{9}{10}$) will cease to be coined; but it retains a nominal existence; it remains money of account and is still the unit of the monetary system, and the measure of all values, though it has no material existence except in its multiples, of which the quintuple or five-franc piece is the smallest coin.

" The reduction in the value of small silver coin brings the standard of this class in harmony, I believe, with the small silver coin of the United States, under the law of 1850. If this be so, the metallic standards both of the gold and silver coin of the United States are now in harmony with those of the four countries named, and the standards being in harmony and the system all decimalized, it only remains to harmonize the coin in order to produce reciprocal circulation. For this purpose a common unit does not appear to me to be necessary. However numerous the systems, if the standards are equal and the system decimalized, it is only necessary that the unit of each be capable of expression in the multiples or sub-multiples of the others to produce the uniformity of coinage requisite for reciprocal circulation.

" Coining a unit of either system will then be, of necessity, coining at the same time a unit multiple—a sub-multiple of all the systems—and these conditions are much easier of attainment than a common unit. Nor is it of moment what names may be given to coin in different countries, nor how numerous the varieties, or various their values; they will all be aliquot decimal parts of a common system, the coin of each referable to the unit of its own system, and referable with equal facility to the multiples or sub-multiples of the units of the other systems.

"To attain this result, no substitution of the unit of one country for that of another country is needed, because no common unit is required. That great difficulty may be obviated by very slight modifications of existing units, and modification is the easy way of all reforms.

"Our gold dollar, for example, is equal to 517 centimes; a reduction of seventeen centimes ($3\frac{1}{2}$ cents) would leave it an exact multiple of the French unit, or franc, and the equivalent of five francs. A reduction of twelve centimes ($2\frac{1}{2}$ cents) in the value of the British sovereign would leave it a multiple of the franc, and the equivalent of twenty-five francs, and consequently a multiple of the dollar, and equivalent to five dollars, nearly.

"In like manner, a small change in the standard of British gold—from $\frac{916}{1000}$ to $\frac{900}{1000}$—would complete the uniformity of the standard of gold coin; for nearly the whole civilized world, except England, has adopted the standard of $\frac{9}{10}$.

"Modifications of this kind are not difficult; they are common. They produce no inconvenience to the public; they do not disturb business, nor trench upon prejudices; they come in almost imperceptibly, and in this case would leave the unit of each national system, the great traditional measure of value, in effect undisturbed, with all their mottoes, emblems, and effigies, and with all the impregnable habits of mind, and even the superstitions, which cluster around them.

"The tenacity with which nations and peoples hold to their traditional measures of value is remarkable, and, whether it springs from a principle or a prejudice, it is a fact so firmly fixed that it is difficult to eradicate; nor is it worth the labor, if a common language of values can be otherwise attained."

Mr. Beckwith to Mr. Seward.

"PARIS, *September* 6, 1866.

"SIR: I have not been informed whether the government would consent to participate in the proceedings proposed in my letter of the 29th June, relative to weights, measures, and coins.

"I beg now to place before you a letter on the subject from M. Mathieu, president of the committee, together with a plan of the proposed structure for the international exposition of weights, measures, and coins, with explanations relating to the proposed collection.

"I took leave to suggest in my previous letter that the commissioner charged with the subject of a common unit of coin might advantageously be associated with this committee.

"I have not been able to learn whether such a commissioner was appointed by Congress; but I think it would be useful to take part in the proposed exhibition and in the proceedings of the international committee.

" The expense on this side, as set forth in the accompanying papers, is small, (about $300,) and I cannot doubt that some member of the commission for the Exposition already named would like to be charged with the exhibition of weights, measures, and coins, and be added to the international committee alluded to, if the government thinks proper to make this appointment.

" I trust your excellency will think favorably of this proposal, and enable me to inform M. Mathieu of the co-operation of the government of the United States in the useful aims of this commission."

M. Mathieu to Mr. Beckwith.[1]

" PARIS, *September* 4, 1866.

" SIR: In reference to our circular of the 25th June, relative to the international exhibition of measures, weights, and coins, I have the honor to send you, in the name of the committee, a project to fix the participation requested of each country, and respectfully ask you to submit it to the United States Commission, with the following explanations:

" This project was examined on the 25th instant at a conference in which the commissioners and delegates present in Paris took part, and the discussion showed the necessity of the concurrence of twenty states or groups of states, and the great probability that this participation would be obtained by reason of the importance of bringing together all the bases of exchange.

" The outside estimate is 30,000 francs, and the expense of installment is 1,500 francs for each state.

" The Imperial Commission gives this assessment for France, and, as the spaces for the different states are to be of the same dimensions, the committee proposes to guarantee the same quota of 1,500 francs to the United States Commission.

" The site for the special exhibition of measures, weights, and coins is isolated, (in the center of the grand entry,) and the Imperial Commission needs all information as soon as possible, to settle it definitely; and the construction of the tower of iron and glass must be executed with all the carefulness required by its destination and situation, so as to be one of the first objects that strikes the view of visitors.

" It is, therefore, highly necessary to arrive at a definite conclusion immediately, and we request the United States Commission to inform us of its decision as soon as possible.

" The inclosed photograph shows the nature of the building planned by the committee. The detailed programme of the series of measures, weights, and coins to be sent, will, of course, be considered hereafter by the full committee. In order to arrange this programme, as well as to fix the details of installment, it is desirable for the United States to appoint a person to represent them in the committee, and we request

---

[1] Translation.

that he be designated, and his name sent to us when the terms of our proposition are accepted.

" In regard to the series of weights and measures, in case the United States Commission will immediately collect a set, we must recommend the most common form, made, however, with the greatest possible care, so that the result of the comparison of the series in 1867, without pretending to the degree of exactness of scientific comparisons on legal standards, may offer that precision required in the practice of commercial exchanges.

"As to coins, it is only necessary to exhibit a double series of the current coin of the latest issue.

"INTERNATIONAL EXHIBITION OF MEASURES, WEIGHTS, AND COINS.— The exhibition is to take place in a glazed iron tower, accessible from the inside and outside. It is divided vertically into twenty equal sections, allotted to the different states or groups of states. The frieze and cornice have each section ornamented with the arms, emblems, and colors of the different countries. A double stairway will lead from the interior to the balconies of the first and second story. The ground-floor is for measures, weights, and coins. Measures of length are placed vertically in a circular rack near the partition. The weights are placed on pyramid stands. The coins, in a double series, for the face and reverse, are placed in frames against the wall, on a level with the eyes; and the shelf marked 4 displays the same coins of different countries on the same level. Measures of capacity occupy the basement. Road and field measures can be represented on the floor of the outside gallery in a uniform reduced scale.

" The first story is appropriated to paper money and documents. Paper money, postage stamps, and trade marks are placed on the walls. The inside of the glazed partitions has shelves or frames for pictures, manuals, and documents of all kinds relating to systems of measures, weights, and coins, and to methods of conversion.

" The second story is for measures of angles and time. Angular divisions of the circle and sphere are represented upon a globe, and the circles supporting it; and if any exhibitor will take the expense of making a clock, the globe turning on its axis may serve for it, with two hands and a dial at the poles indicating the hours from noon to sunrise. The glazed cage surmounted by the globe contains the calendars and almanacs. The portions of the world where the different systems of measures, weights, and coins are used are shown by different colors upon the globe."

Mr. Seward to Mr. Beckwith.

" DEPARTMENT OF STATE, *October* 4, 1866.

" SIR : I have to acknowledge the receipt of your communication of the 6th ultimo, relative to the proposed participation by the United States in the proceedings described in your letter to me of the 29th of June last, for the promotion of the adoption of a uniform system of weights, measures, and coins.

"I have to inform you in reply, that Samuel B. Ruggles, esq., one of the ten scientific commissioners, will be charged with the subject of a common unit of coin, and will be authorized to co-operate with the committee, of which M. Mathieu is president, in the manner which you suggest.

"Mr. Ruggles represented the United States at the late statistical congress at Berlin, and has already been in correspondence with the international committe, organized on that occasion, upon this important subject, to which he has devoted much study. A copy of your communication now under reply will be forwarded to him, and he will be requested to enter into correspondence with you. Any letter addressed to him, which may be sent to the care of Mr. Derby, the general agent at New York, will reach his hands."

Mr. Seward to Mr. Ruggles.

"DEPARTMENT OF STATE, *October* 9, 1866.

"SIR: You are hereby officially designated to take charge of that branch of the representation of the United States which relates to the question of the adoption of a uniform system of weights, measures, and coins, and you are authorized to co-operate with the committee, of which M. Mathieu is chairman, in the manner suggested in Mr. Beckwith's dispatch of the 6th ultimo. You will enter at once into correspondence with Mr. Beckwith on the subject."

Mr. Ruggles to Mr. Seward.

"NEW YORK, *December* 20, 1866.

"SIR: The Department of State having designated the undersigned, by written instructions of the 9th of October last, 'to take charge of that branch of the representation at the Universal Exposition at Paris, in 1867, which relates to the question of a uniform system of weights, measures, and coins,' and 'to co-operate with the special committee appointed by the government of France, of which M. Mathieu (member of the Bureau of Longitudes) is chairman,' the undersigned, on the 13th of October last, in entering on the discharge of the special duty thus committed to him, asked leave, in his communication of that date to the Department of State, to 'invoke the co-operation of any or all of the eminent individuals with whom he is associated in the commission authorized by Congress.' The Department having approved this request, the undersigned has invited the counsel and co-operation of Mr. James H. Alexander, of Maryland, he being one of the twenty commissioners appointed by the President of the United States. This selection was made in view of the pre-eminent qualifications of Mr. Alexander, in his well-known scientific attainments, and his long and careful study of the particular subject of uniform weights, measures, and coins, so fully manifested in the published works which he has already given to the world.

"It is gratifying to add that the association of Mr. Alexander in this important duty has met the special approbation of the honorable Mr. McCulloch, Secretary of the Treasury, who has given permission to carry to Paris, for exhibition, the standards of national weights and measures now in charge of the Superintendent of the Coast Survey, with such of the balances and other scientific apparatus as may be needed for their full illustration. A similar application will be made to the Director of the Mint for a series of all our national coins now in actual use, to which it is proposed by the undersigned to add, with the aid and co-operation of Mr. Alexander, a full collection of all our former coins, particularly including those which were issued by the separate States during the period in our national history preceding the adoption of the articles of confederation in 1781, and reaching back, as far as practicable, into our colonial era.

"Such an exhibition, it is believed, will conclusively show the value of our united government in unifying or extinguishing the multiform and widely varying measures of money, which otherwise would have overspread our continental republic from ocean to ocean. It will do more: it will demonstrate, by visible example, the transcendent import-ance of the higher problem of unifying the varying coins of the numer-ous nations of the civilized world, which is to be discussed in the coming concourse at Paris.

"In that discussion the salutary influence of our 'Old Confederation,' followed by the American Union, in simplifying the coins of our Western World, will doubtless be considered in comparison with the recent politi-cal consolidation of Northern Germany, under the guiding hand of Bis-marck, and its probable effects in reducing the heterogeneous coinage of that hitherto fragmentary portion of Europe to one common national system.

"Without entering prematurely into the broad field of scientific in-quiry embraced in the subject of uniform weights, measures, and coins, it will be evident that many questions must arise, not only scientific, but commercial and political in character. Especially must this be the case with respect to a uniform coin or unit of money, in the adjustment of which commercial habits and national prejudices must inevitably modify any decision on grounds purely scientific.

"The general proposition that the prosperity and civilization of man-kind would be greatly advanced by the use of a common coin of uniform fineness, no one will deny; but we shall be met at the threshold by the question, what shall be the weight of this unit of money; and especially by the inquiry, from which of the various systems of coinage now in use by the different nations shall that unit be selected? not to speak of a

7 P E

proposition which may be presented for a new unit wholly irrespective of any existing system.

"These questions were discussed at considerable length in the International Statistical Congress at Berlin, in 1863, but were found so difficult of solution that a 'special congress' was recommended by that assembly, at which they might be more carefully and deliberately examined. The disturbed state of the interior of Europe, with other causes, has hitherto prevented the convocation of this 'special congress;' but it is hoped and expected that the 'concourse of nations' at the coming Exposition at Paris will practically afford the opportunity for such a convention.

"From present indications it is fully evident that the projects to be proposed by scientific men from different portions of the world, profoundly involving not only the value but the accuracy of the present metre, will be numerous and conflicting; but it is probable, to say the least, that the discussion will at last be practically narrowed to the single inquiry: Shall the British pound sterling, or the French franc, or the German thaler, or the Russian ruble, or the American dollar, be selected as the common unit of money?

"In deciding such a question, necessarily involving so much of national feeling, we shall need at Paris something more than abstract science. We shall need the counsel and co-operation to their full extent of our most eminent bankers, ship-owners, merchants, and manufacturers. It is true that their opinions, with some of the results of their varied experience, might be partially collected by active correspondence; but all such opinions should be subjected to the legitimate and modifying influence of personal conference and mutual comparison in actual open convention.

"In many respects it is to be regretted that the joint resolution of Congress under which the United States Commission to Paris has been appointed should have expressly excluded members of Congress, and 'every person holding an appointment or office of honor or trust under the United States;' thereby depriving the country of the services of all the members of the national legislature, and of many others in public employment, civil and military, whose counsel and co-operation would have been in the highest degree valuable and important. Without any invidious distinction we may surely refer to the chairmen of the proper committees in the Senate and in the House of Representatives, whose experience and well-directed intelligence recently carried through the present Congress the important acts for the partial establishment in the United States of the 'metric system,' commencing, in truth, a new era in our commercial history.

"The undersigned would, therefore, venture respectfully to suggest, in behalf of himself and any other commissioner or commissioners associated or to be associated with him on this special committee, the expediency and propriety of their being allowed, under proper limitations, to invite the co-operation as associates in the committee of at least five of their fellow-countrymen of well-established ability, and not included in the present commission.

"It will be seen at page 56 of the third supplemental circular, published by the Department of State, that the special committee appointed by the imperial government embraces five distinguished individuals; four of whom are members of the Institute of France, and one an assayer in the mint. In justice to the special committee to represent the United States, it ought to embrace at least an equal number.

"A similar suggestion for increasing the numbers in all the branches of the national representation at Paris will be found in an able and interesting letter from Abram S. Hewitt, esq., of New York, one of the best informed and most intelligent manufacturers in our country, a copy of which is herewith furnished. Its clearly-conceived and well-expressed convictions, as well as the facts adduced in illustration, are well worthy of the attention of the commissioners, if not of the government. They are certainly in harmony with the general wish and expectation of the patriotic and far-seeing citizens of the United States that the opportunity so unusual, afforded by this Universal Exposition, the great occasion of the present century not only for exhibiting to the world the varied elements of our national strength, but for vastly increasing their value by careful study and full reports of the industries of other and older nations, may not be lost through any want of prompt and adequate support by the government of the United States."

Mr. Hewitt to Mr. Ruggles.

"NEW YORK, *December* 18, 1866.

"MY DEAR SIR: I have found no time to make a suitable reply to your request that I would give you my views as to the best mode of making the coming Exposition of Industry, at Paris, of real value to the people of the United States. I cannot, however, allow the opportunity to pass without submitting briefly a few suggestions, which your own experience will enforce in the proper quarters.

"These international exhibitions of industry are peculiar to the present age, and a practical recognition of the claims of labor on the attention of governments, as the real foundation of national greatness. To the people of the United States they have a special interest, as marking the steady progress in Europe of the cardinal ideas which underlie our political system.

"But in addition to this rather theoretical view of their significance, these exhibitions are of the greatest practical importance to us, if we avail ourselves of all the advantages which they offer.

"First an opportunity is offered to make known to the intelligence and capital of Europe the great natural advantages of this continent, for the cheap production of the great staples of human industry. The Exposition collects together, at one time and in one place, the most intelligent and enterprising men of all nations, who come either as commissioners or visitors. There is unusual freedom from restraint in personal intercourse, and more can be imparted and secured in a few mouths than could otherwise be possible with years of effort.

"However great may have been our need heretofore for capital and skilled labor to develop our resources, the late civil war has left one-half of the country in so devastated a state that our existing means are totally inadequate to the work of restoration. We want labor, skill, and capital from abroad, in order to restore the 'waste places' of the South at the earliest possible moment.

"The resources of the Southern States, especially in the way of coal and iron, are almost unknown in Europe, and but little understood even at home. There will not be another opportunity for ten years to bring this knowledge to the comprehension of Europe, and to lose ten years in the present age is to throw away an empire. It is, therefore, a matter of the highest national concern that the commissioners who represent this country shall be able to make known to the representatives of other nations the immense natural resources which in this country, and especially in the Southern States, offer such rich rewards to enterprise and capital.

"Secondly, the commissioners should be men of such technical knowledge in the various departments of industry that all improvements can be detected, described, and imparted by proper drawings and models to our own mechanics who cannot visit the Exposition. It is, doubtless, unreasonable to expect to secure such an official representation as would best accomplish this object; but power should be given to the board of commissioners appointed by the general government to add to their number the name of every American of special technical knowledge who either expects to visit the Exposition, or will agree to do so if appointed.

"This is not only just and proper, and in accordance with the spirit of our institutions, but in no other way can the results of the Exposition be made of permanent value to this country. Every such commissioner would, of course, be detailed to examine the special department with which he is conversant, and make a full and accurate report of the result of his observations. In this way it is reasonable to expect that the whole ground will be covered; and the reports, when collected and published, will convey to the nation a larger amount of technical knowledge than could be procured in any other way.

"It is idle to expect that mere visitors, as such, will either procure or make known this information. I was a visitor at the English exhibition of 1862. I was free to examine the results of labor, skill, and ingenuity which were there collected together. But to understand the processes by which these results were procured, it was necessary to visit the work-shops and the factories. To mere visitors those places were closed, but to commissioners every door was open. At Sheffield, for example, I had the greatest possible difficulty in getting access to the establishments where the new processes for making steel were carried on; and several times, when I was pleading my claims for admission, I saw foreign commissioners admitted, with smiling faces, on the mere production of their letters of appointment.

"Allow me to say that there is no branch of business, however trivial, which will not be greatly benefited by a knowledge of the methods employed abroad in the same department of industry. *The nation that possesses the best tools and the best processes will be the most powerful and the most highly civilized.* This is the true explanation of the present position of Great Britain. Her commercial supremacy rests on her steamships, and the superiority of these ships is due to her tools and her skill in using them. Great Britain builds the best engines, and hence her industry is more thoroughly and economically administered than that of any other country. We want the knowledge, and we want it diffused over all departments of business in which our people are employed. For want of this knowledge we are always years in arrear of our foreign competitors, even in the processes which affect the production of staple raw materials, such as iron, wool, silk, and even cotton.

"In Sweden the use of sulphury ores has been made practicable by a new process of roasting. Here they are totally rejected, and it is generally supposed that they cannot be used, and yet the Swedish process will greatly enlarge the business here, and reduce the cost of making iron. Again, the Bessemer process makes but slow progress here, from want of the proper and accurate knowledge of the results in Europe. We lose every day by the delay, and this lack of knowledge extends to nearly every branch of business.

"Hence I urge—and this is the main object of this hasty and imperfect letter—that you will press on Congress the propriety of giving the commissioners power to add to their number all such persons as may have special technical skill, who will agree to go to Paris at their own cost, and prepare for the commission a full and accurate report on the special department for which the appointment is made."

In compliance with a resolution of the Senate of the 19th December, 1866, the progress made in collecting the products, and the weights, measures, and coins of the United States, for exhibition, was reported by the Secretary of State, and transmitted in a message of the President, January 18, 1867, and was published.[1] A portion of the foregoing correspondence is reprinted from that document. Other letters from Mr. Ruggles will there be found, and a full report from him upon the proceedings of the committee on weights, measures and coins, and of the International Monetary Conference, was transmitted by the Department to the Senate, December 17, 1867, in compliance with the resolution of the 6th of December.[2] The subject has also been more fully reviewed and discussed in the Report upon the Precious Metals.

[1] Senate Ex. Doc. No. 5, 39th Congress, 2d session.
[2] Senate Ex. Doc. No. 14, 40th Congress, 2d session.

CATALOGUE AND STATISTICS.

Mr. Beckwith to Mr. Seward.

"PARIS, *July* 31, 1866.

"I have the honor to acknowledge the receipt of your letter of the 7th instant, with a copy of the resolution of Congress No. 52, a copy of the circular of the Department of State of the 6th July, addressed to the governors of States, and directions for me to make provision for the reception of such additional products as may be sent in to Mr. Derby hereafter.

"I have, in conformity, explained the situation to the Imperial Commission. They expressed the opinion that the latest date which could be safely given for closing the catalogue would be the 1st September, but they finally consented to the 15th September. I have, therefore, agreed to close the catalogue on the 15th September, and to deliver it to them on or before the 1st October, and have advised Mr. Derby of this agreement.

"It was feared that default would occur with Prussia and Austria, but the catalogue of the former was sent in a few days since, and the catalogue of Austria came in to-day, one hundred and seventy pages, well arranged and well printed."

PREPARATION OF A STATISTICAL SUMMARY.

Mr. Beckwith to Mr. Seward.

"PARIS, *September* 22, 1866.

"The annexed letter from M. Le Play, conseiller d'état, proposes the publication, in connection with the catalogues of the Exposition, of brief statistical information, which will be useful and interesting.

"The statements desired can probably be compiled, from materials in the Department of the Interior, in a short space of time and without great labor, as they will not require new researches. It is unnecessary to follow exactly the programme presented; statements which cannot be conveniently and readily made, such as the one which I have marked fourth, may be omitted; others thought useful may be added; and I would suggest, in addition, a synopsis of the present national banking system, which is a new institution worthy of notice.

"The opportunity thus presented of diffusing among the people statistical knowledge regarding the United States, authenticated by the government and sanctioned by imperial authority on this side, is rare, and I hope it will be considered, in your estimation, of sufficient importance to warrant the preparation of the requisite statements, and that they may be made as soon as possible, on account of the time required for translations and printing and the short period which remains for it."

M. Le Play to Mr. Beckwith.[1]

"PARIS, *September* 14, 1866.

"I had the honor to address you on the 24th August, with specimens of the general catalogue of the Exposition.

"The Imperial Commission now desires to add to the catalogue of each country statistical information relative to the territory, population, productive force, naval force, military force, and financial organization.

"In connection with the minister of agriculture, commerce, and public works, the Imperial Commission is now preparing for France a similar statement, which may serve as a programme for other countries, as follows:

"TERRITORY.—Geographical position.   Seas, mountains, and principal rivers.   Administrative divisions and principal cities.   Superficial area; development of coasts and frontiers; extent of principal basins or valleys.   Ways of communication—extent of roads; railroads and telegraphs; canals and navigable rivers; surface and circumference of lakes and inland seas.

"POPULATION.—Whole population; population by race, sex, and religion; by language, professions, and occupations; population of principal cities; average increase of population since 1855; emigration and immigration.

"PRODUCTIVE FORCE.—Quantity and value of the annual products of agriculture, of forests, of hunting and fishing, of mines and minerals, of textiles, and of manufactures of various kinds.

"COMMERCE.—Actual value of imports consumed in the country; value of exports being the produce of the country; aggregate imports and aggregate exports.

"INTERIOR TRANSPORTATION. — Transportation on roads, canals, rivers, interior seas, and lakes; coasting trade; railroad traffic; number of letters and number of telegraphic messages.

"MARITIME NAVIGATION.—Number and tonnage of vessels entered and cleared under the national flag; the same under foreign flags.

"ARMY.—Effective force of standing army and of militia.

"MARINE.—Personal force and material force of the navy and of the mercantile marine.

"FINANCE.—Receipts and expenditures forming the budget of the state; local receipts and expenditures of states, provinces, departments, and communes.

"This programme contains the elements necessary to a comparison, in an economical point of view, of the relative force and wealth of nations

Fourth.   "Nevertheless, to respond more fully to the idea of the Emperor, who has instituted a new Order of Recompenses for the Exposition of 1867, it will be useful to add to the above statistical information a succinct account of any special industrial institutions which have developed, in a remarkable manner, harmony among the different

---

[1] Translation.

classes of producers, and the material, moral, and intellectual well-being of the workmen.

"The Imperial Commission thinks, Monsieur le commissaire, that the United States will be willing to prepare, by the 1st of October, a statement on the basis analogous to that which I now have the honor to present to you, adopted by France.

"A collection of similar documents relative to each country will prove a useful complement to the catalogues, for the study of visitors to the Exposition.

"I have only to add that the intention is not a voluminous work, but a short notice, and the more summary, the better it will enter into our plans."

PRINTING AND PUBLICATION OF THE CATALOGUES.

Mr. Beckwith to Mr. Seward.

"PARIS, *January* 18, 1867.

"SIR : The publication of catalogues of the Exposition being a matter of importance, I submit a brief explanation of the existing circumstances relating to it:

"First. The general plan of the Imperial Commission proposed to publish an official catalogue in French, leaving other nations free to publish their own catalogues, each in its own and other languages except French, if they chose, and proposed also to sell the official catalogue at a low price, less than the cost, which they would be enabled to do by deriving a profit from advertisements permitted to be inserted in the catalogue, which propositions were, in general, considered satisfactory.

"Second. The Imperial Commission then contracted or sold a contract to a publisher conceding to him the right to print and sell the official catalogues at a fixed price. They also sold and conceded to him the right of advertisement in the catalogues, limiting the quantity of advertising space, but leaving the price of advertisements unlimited, and to be settled between the contractor and the advertisers.

"The right of advertisements was apportioned among nationalities, and these rights were resold by the contractor to sub-contractors or speculators, who offer them in turn to advertisers in different countries at prices which may be agreed upon between them.

"The original contracts of the Imperial Commission are construed by the holders as conceding to them exclusive rights in regard to the publication of catalogues, and in regard to advertisements, and that consequently the publication· and· sale of catalogues or advertisements by other nationalities is an infringement of their rights.

"Differences of opinion between the Imperial Commission and the contractor for the official catalogues have arisen, and are the subject of friendly adjudication in the courts, which differences have not yet been decided.

" I have not followed this litigation so closely as to be able to state, with legal accuracy, the exact points of difference between the Imperial Commission and the contractor, and may not have correctly indicated it, but accuracy in this is not essential to my present purpose.

" Third. The contracts made by the Imperial Commission were, however, of a nature, as construed by themselves, to trench upon the general plan suggested in my first paragraph, and were consequently the subject of reclamations by foreign commissioners.

" The Imperial Commission admitted that other nations had still the right to print and sell their own catalogues in their own language, but they thought that if the foreign commissions desired to advertise in their own catalogues, and to print them in other than their own language, that this would infringe upon the rights conceded to the aforesaid contractors, and that such publications should therefore be the subject of agreements between the foreign commissions and the said concessionaries.

" The foreign commissions, in general, were not of this opinion, but thought they were entitled to publish their own catalogues in any language they might choose, and to advertise in them for their own profit, as a method of defraying expenses, on the same principle as that adopted by the Imperial Commission.

" The questions thus raised are not yet definitively settled, but are the subject of frequent discussion, and I have no doubt will be soon definitively disposed of as between the Imperial Commission and foreign commissions, though I am not confident that they can now be so adjusted as to avoid all difficulties with the contractors.

" Fourth. My object in presenting these remarks is to bring to your notice the actual situation in regard to the catalogues. It is very desirable to print them in at least three languages, English, French, and German ; we are lamentably in arrears in preparing our catalogue, and my great anxiety is to perfect it in time to be published for the opening of the Exposition on the 1st of April, when the jury work will commence, which cannot be properly done without catalogue in hand.

" It should thus be published in the French language, but I have not any funds for its publication and *gratuitous distribution* in English or any other language.

" I have relied on the conviction that the right of sale and advertisement, which I consider included in the programme of the Imperial Commission of the 22d of August, would defray the expenses of publication in our own and other languages than French, which resource, however, was placed in question, and is still in suspense, as above stated.

" If, therefore, Congress should not appropriate money for the publication and gratuitous distribution of the catalogue, in our own and other languages than French, we shall not be provided with such catalogues unless by the means, before mentioned, of income from advertisements and sales; and believing that to be a proper method I am disposed to adopt it. But I think it would be going beyond my duty to do so without first presenting the situation and soliciting your directions, or, at least, the expression of your opinion for my guidance.

"My proposition is to print and sell the catalogue in English and in German, and to render accounts of the expenditures and of the receipts from sales and advertisements and to carry the balance of that account, be it a debit or credit, into the general account of the Exposition."

AGRICULTURAL MACHINES AT BILLANCOURT.

Mr. Beckwith to Mr. Seward.

"PARIS, *May* 17, 1867.

"SIR: In respect to the experiments to be made with agricultural implements at Billancourt, the trial of plows has already taken place.

"We have several varieties in the Exposition, and I regret to be obliged to state that none of our exhibitors gave any directions or made any preparations to put their plows to the trial, and consequently none of them took part in the competition.

"The competition of mowers will commence on the 23d instant, and that of reapers will take place in the last half of July.

"Several of our exhibitors will take part in these competitions. They are now training their horses, and will, I trust, be well prepared for the contest.

"I have authorized John P. Reynolds, esq., secretary of the State Agricultural Society of Illinois, and commissioner from that State, to superintend these trials, and have obtained from the Imperial Commission the appointment of Mr. Reynolds as a juror on these experiments, that the interests and machines of our exhibitors may be fairly appreciated."

Mr. Beckwith to Mr. Seward.

"PARIS, *August* 2, 1867.

"SIR: I have the honor to report the result of the final field trials of reaping machines and mowing machines, which commenced on the imperial farm of Fouilleuse, on the 26th July, but, being interrupted by bad weather, were terminated at Vincennes on the 31st ultimo.

"The number of mowers entered for competition was seventeen, and of reapers sixteen, being the products of several nationalities.

"The superiority of the American machines soon became apparent, and it was evident that the chief contest would be among themselves.

"The decision of the jury is as follows:

"Mowing—the 'W. A. Wood mower,' No. 1; 'I. G. Perry mower,' No. 2; 'C. H. McCormick mower,' No. 3.

"Reaping—'C. H. McCormick,' No. 1; 'Seymour, Morgan & Allen,' No. 2: 'W. A. Wood,' No. 3.

" These results will add to our successes probably two more gold, two silver, and two bronze medals. None of the other machines are rated higher than honorable mention or bronze."

INTERNATIONAL JURY AND THE AWARDS.

Mr. Beckwith to Mr. Seward.

" PARIS, *June* 24, 1866.

SIR : The document hereto annexed, marked No. 1, contains the project of the Imperial Commission for distributing 800,000 francs in prizes among exhibitors, for the organization of international juries to award the prizes, describes the nature and value of the awards, the manner of composing the juries, the regulations for their guidance, and the dates for the commencement and completion of their work.

." Document No. 2 contains the report of the minister of state upon document No. 1, and an imperial decree approving the report and legalizing the project of the Imperial Commission.

" The principal new feature of the plan is developed in that part of the report of the minister relating to section four of the project, which appropriates ten prizes of 10,000 francs each, and one capital prize of 100,000 francs, not for the excellence of material products exhibited, but for local establishments which are most successful by their organization and management in promoting the prosperity and harmony and the moral and intellectual well-being of the operatives."

ORGANIZATION OF THE INTERNATIONAL JURY.

The following is a translation of regulations fixing the kind of prizes, and organizing the juries for distributing them, discussed the 8th of June, 1866, and approved by imperial decree the 9th of June, 1866.

" TITLE FIRST.—GENERAL DISPOSITIONS.

." ARTICLE 1. The sum of 800,000 francs is appropriated for prizes to be awarded at the Universal Exposition for 1867.

"ART. 2. An International Jury is appointed to adjudge the premiums. The International Jury is composed of six hundred members, distributed among the different nations according to the space occupied by the products of each. The result of the distribution is laid down in tables A and B, annexed to the present regulation.

"ART. 3. The French members of the International Jury of Awards are appointed by the Imperial Commission. The foreign members are appointed respectively by the national commissions of each•country. All appointments must be made previous to the 1st of December, 1866. The Imperial Commission, after consultation with the foreign commissions, shall assign the members of the jury to classes.

"ART. 4. The International Jury must finish its labors between the 1st of April and the 14th of May, 1867. However, as regards Classes 52, 67 to 88, and 95, the duties of the jury shall continue during the Exposition.

"ART. 5. The formal distribution of the prizes is fixed for the 1st of July, 1867.

"TITLE SECOND.—GROUPS OF WORKS OF ART.

"ART. 6. The prizes at the disposal of the International Jury for works of art are arranged as follows: 17 grand prizes, each valued at 2,000 francs; 32 first prizes, each valued at 800 francs; 44 second prizes, each valued at 500 francs; 46 third prizes, each valued at 400 francs.

"ART. 7. The prizes in article six are distributed as follows, among the four sections of the fine arts arts, corresponding to the classes of Group I:

"First section, (Classes 1 and 2 united:) 8 grand prizes, 15 first prizes, 20 second prizes, and 24 third prizes.

"Second section, (Class 3:) 4 grand prizes, 8 first prizes, 12 second prizes, and 12 third prizes.

"Third section, (Class 4:) 3 grand prizes, 6 first prizes, 8 second prizes, and 6 third prizes.

"Fourth section, (Class 5:) 2 grand prizes, 3 first prizes, 4 second prizes, and 4 third prizes.

"ART. 8. The jury for the groups of works of art is composed of sixty-three members. The numeral proportion of the French and foreign members in each of the four sections is indicated by table A, annexed to the regulations. The French members of the four sections are appointed by the Imperial Commission among the members of the jury of admission. They will be chosen equally from the three lists forming the jury, instituted according to the decision of the 12th of May, 1866. The exhibitors who are members of the International Jury for works of art are not excluded from the competition for prizes. Each of the four sections is presided over by one of its members selected by the Imperial Commission. Two of the presidents are French.

"ART. 8. The four sections may assemble to propose modifications in the distribution of prizes, if necessary, as designated in article seven. The Imperial Commission selects one of its members to preside at the meeting of the four sections.

"TITLE THIRD.—THE NINE GROUPS OF AGRICULTURAL AND INDUSTRIAL PRODUCTS.

"ART. 10. The prizes at the disposal of the International Jury for the agricultural and industrial products are arranged as follows:

"Grand prizes and awards in silver to the total value of 250,000 francs.

"One hundred gold medals, worth 1,000 francs each.

"One thousand silver medals.

"Three thousand bronze medals.

"Five thousand honorable mentions, at least. All the medals are of the same form.

"ART. 11. The grand prizes are awards of merit for inventions or improvements in the quality of products and mode of manufacture.

"ART. 12. The prizes mentioned in article ten, for the nine groups of agriculture and industry, are adjudged by the class juries, the group juries, and the superior counsel.

"ART. 13. The numerical proportion of the French and foreign members in each of the class juries is fixed by table A, annexed to the present regulations.

"ART. 14. Each class jury meets from the 1st of April, 1867. At its first meeting it elects a president, a vice-president, and secretary. A secretary must be appointed before the 10th of April.

"ART. 15. Class juries may add members or select experts from the other classes of the International Jury, or outside; in the latter case the selection must be approved by the Imperial Commission.

"ART. 16. Exhibitors who have been made members of the International Jury cannot be competitors for prizes. Exhibitors in a class jury as members or experts are also excluded from competition in products of the class of which they are to be judges. However, the Imperial Commission may authorize certain exceptions to the exclusions mentioned in the preceding paragraphs.

"ART. 17. The foreign commissions are requested to select delegates for the class juries to furnish information to the jury about the exhibitors of their country. The place of residence of these delegates must be made known to the Imperial Commission before the 20th of March, 1867. The same duties are performed for the French Section in each class jury by the corresponding committee of admission.

"ART. 18. Between the 1st and 14th of April, each class jury of Groups II, III, IV, V, VI, and X shall examine the products, and class the exhibitors deserving prizes, without distinction of nationality. The class jury then makes out a list of exhibitors excluded by article sixteen, and proposes proper exceptions. It then classes the assistants, masters, and workmen, without distinction of nationality, to be recommended for services to agriculture or industry, or for their participation in the production of remarkable objects at the Exposition. The class lists, signed by the members who made them, shall be filed by the reporter with the Commissioner General, by the 14th of April at latest. Class juries of Classes 52 and 95 only furnish the information necessary to fix the number of awards for those classes, and propose associates to assist in the examination of the objects exhibited. If a class jury has not presented the lists by the 14th of April, the Imperial Commission shall attend to it.

"ART. 19. Between the 1st and 14th of April each class jury of Groups VII, VIII, and IX makes out a list of associates to assist in the examination of the products during the Exposition, and furnishes information to fix the number of awards.

"ART. 20. The presidents and reporters of the class juries are members of the group juries; in case of absence of the president, the vice-president shall take his place. One president and two vice-presidents are appointed outside of the members for each group jury. The arrangement of presidents and vice-presidents of group juries, among different nations, is designated by the table B, annexed to the present regulations. In accordance with article three, the French president and vice-presidents are appointed directly by the Imperial Commission ; the foreign, by the national foreign commissions. The secretary of each group jury is appointed by the Imperial Commission.

"ART. 21. Between the 15th and 28th of April each group jury of Groups II, III, IV, V, VI, and X examines the claims, fixes the lists of classification drawn up by the class juries, and writes the award after the proper name. For Classes 52 and 95, only the number of prizes is designated. It joins in turn each class jury in its deliberations, but only has a deliberative vote. These preliminary operations of the group juries must close and the result be communicated to the Commissioner General by the 28th of April, at the latest. If it is not done by that time the Imperial Commission will attend to it.

"ART. 22. Between the 15th and 22d of April each group jury of Groups VII, VIII, and IX makes out the lists of associates presented by the class juries and makes known the number of awards of each class to the Commissioner General.

"ART. 23. The president and vice-presidents of the group juries are called to constitute the superior council of the jury. One of the vice-presidents of the Imperial Commission shall be president of that council. The assistant secretary of the Imperial Commission shall aid the secretary of the council.

"ART. 24. Between the 29th of April and the 5th of May the superior council divides the total number of awards among the different groups. The council, if deemed necessary to increase the number of medals, may propose to the Imperial Commission to apply 50,000 francs for that purpose out of the sum for the grand prizes and silver sets. These duties of the superior council must be concluded by the 5th of May at the latest.

"ART. 25. A report on the exhibition of agricultural and industrial products shall be published under the direction and care of a committee proposed by the superior council and appointed by the Imperial Commission.

"ART. 26. Between the 6th and 12th of May each group jury mentioned in article twenty-one shall divide the awards fixed by the superior council among the proper classes. The result of this labor shall be sent to the Commissioner General by the 14th of May at the latest.

"ART. 27. During the Exposition the Imperial Commission will name temporary associates every fifteen days, to aid the class juries in their examinations of the products, processes or instruments of labor in Classes 67 to 88 presented for competition in the next two weeks. These associates are chosen from the lists made out by article twenty-two. On the second day of each fortnight each temporary committee will class the exhibitors, assistants, and workmen it thinks worthy of prizes, and arrange them in four lists, under the titles, first prize, second prize, third prize, honorable mention, of partial competition. That list may be made public at once.

"ART. 28. Between the 15th and 20th of October the group juries of Groups VII, VIII, and IX, after the arrangement of prizes and honorable mentions of the temporary committees, in conformity with the preceding article, shall make a total list of exhibitors for each class, as well as a list of assistants and laborers, and will award the prizes the superior council has assigned to them. The diploma has a list of the prizes and honorable mentions of the different temporary committees to the laureate during the Exposition.

"ART. 29. Class juries of Classes 52 and 95 shall present the proposals relative to awards which the group jury has reserved for them to the Imperial Commission by the 20th of October. The Imperial Commission confirms these proposals.

" TITLE FOURTH.—NEW ORDER OF AWARDS.

"ART. 30. A different order of awards is created in favor of persons, establishments, or localities, where, by special institutions, good harmony has been promoted among those who carry on the same labors, and the material, moral, and intellectual well-being has been secured among the operatives. These awards are: ten prizes of 100,000 francs and twenty honorable mentions. One grand prize of 100,000 francs may be given to the person, establishment, or locality, above all the rest in this respect.

"ART. 31. A special jury shall judge of the merits in this case, and determine the number and form of the prizes to be awarded. One of the vice-presidents of the Imperial Commission shall preside over it. The members shall be twenty-five, the president included. The distribution among different nations is fixed in table B. The functions of secretary shall be performed by the secretary of the Imperial Commission.

" ART. 32. In default of nominations before the 1st of December, 1866, as provided by article 3, the Imperial Commission will select foreign judges from among the persons accredited to it by the different governments.

" ART. 33. The number of members present necessary to make the jury decision valid is fixed at eighteen. The prizes and honorable mentions are decided by a majority vote. The grand prize must have a vote of two-thirds.

" ART. 34. The petitions and documents designating a person for the new order of awards, or an establishment or locality, must be addressed to the councillor of state, Commissioner General, before the 1st of December, 1866.

"ART. 35. The jury will hold its first session the 1st of December, 1866, to fix the rules for demands and to begin their examination.

"ART. 36. At a second and last session, between the 15th of April and the 14th of May, 1867, the jury will finally fix the distribution of the prizes. These prizes will be distributed at the same time as the other prizes—that is, on the 1st of July, 1867.

"Done and deliberated by the Imperial Commission on the 7th of June, 1866."

REPORT IN SUPPORT OF THE DECREE IN REGARD TO RECOMPENSES.

Monsieur E. Rouher, minister of state, vice-president of the Imperial Commission, made a report to the Emperor, attached to the preceding decree, of which the following is a translation:

"SIRE: In the name of the Imperial Commission I have the honor to submit to your Majesty a project for a regulation of the recompenses to be awarded, according to established tradition, at the Universal Exposition for 1867.

"This regulation fixes the nature, the number, and the mode of awarding the recompenses; it comprises four titles:

"The first title determines the general dispositions relative to the value of the recompenses and the organization of the jury.

"The second title answers to article 22 of the general regulations, and relates particularly to the group of the works of art. Its arrangement is similar to that used at the annual exhibition of the fine arts.

"Title three answers to article 62 of the general regulations, and relates especially to the nine groups of agriculture and industry. The dispositions of this title are similar to those of 1855, and tend to equalize the different degrees of recompense, to facilitate execution, and to get the jury's decision in six weeks. The limit of this delay is rigorously required by the date of the distribution of recompenses, fixed for the 1st of July, 1867. Special dispositions reserve for certain subdivisions of the jury the decision on products that are renewed often during the Exposition, but cannot be decided on till the close.

"These parts of the regulations for recompenses offer nothing new, and I confine myself to a summary. But the dispositions of title four constitute an important innovation, to which I must call your Majesty's attention.

"Preceding exhibitions have not shown all the merits that are due to agriculture and industry. The prosperity of those branches is not only shown by the good quality of the products and the improvements in their elaboration, but it also depends upon the comfortable circumstances of the producers and their friendly relations. These circumstances have been considered in preceding exhibitions, but the Imperial Commission has thought proper to create a new order of recompenses, as useful and in conformity with the principles inspiring so many acts of the government of the Emperor. These recompenses shall be awarded to persons, establishments, or localities, where, by an organization or special institution, good harmony has been promoted among those who carry on the same work, and the material, moral, and intellectual well-being of the operatives assured.

"This well-being and harmony of which we wish to find the best examples for your Majesty, is produced under very various forms. The local customs and secular traditions in certain countries maintain union among different kinds of producers; in other regions intelligent efforts remedy the spirit of antagonism that prevails. Here, workmen who have become masters in their trades find all the elements of success in themselves; in other countries, confined to large factories, they leave all care to their patrons. Producers sometimes apply themselves exclusively to agricultural labor or to manufactures: often the two kinds of industry are combined. But, in the midst of this diversity of conditions, well-being and harmony offer the same result; they assure public peace to the producers of all kinds, and to the locality enriched by their work. Apparent characteristics everywhere prove the existence of the two merits we propose to recompense. An inquiry of the prefects of the empire, made some years ago by your Majesty's orders, gave many examples that might be quoted at the competition instituted by title four to great advantage.

"The merits of competitors will be decided by a jury composed of eminent persons belonging to the different countries represented at the Exposition. The Imperial Commission thinks this jury should reject every previous system, and found its judgment entirely on facts.

"The value of the recompenses should be proportionate to the social importance of the competition.

"The commission, therefore, proposes to your Majesty to award ten prizes of the total value of 100,000 francs, with twenty honorable mentions.

"One grand indivisible prize of 100,000 francs might also be awarded to the person or locality distinguished by extraordinary superiority. This competition opens a new branch in exhibitions; it will cause a healthy rivalry between different nations, and will offer problems that have not yet been advantageously solved.

"If your Majesty designs to approve the considerations which form the object of this report, and which is a continuation of the deliberations of the Imperial Commission of the 7th of June, 1866, I beg you will sign the annexed decree."   *   *   *   *

THE NEW CLASS OF PRIZES.

Mr. Beckwith to Mr. Usher.

"PARIS, *December* 3, 1866.

"SIR: The documents issued by the Imperial Commission relating to the various classes of prizes were published by the government at Wash-

8 P E

ington in June last, (3d series, page 43,) but the new class of prizes does not seem to have attracted much attention in the United States.

"The new class is nevertheless most important of all. It is not composed of the material products of industry, but relates to the source of all industrial products, viz., the producers. It is composed of the persons, establishments, or localities which have developed in a remarkable manner good order and harmony, and the material, moral, and intellectual well-being of the workmen and laborers.

"The prizes are comparatively large—ten of 10,000 francs each, and one of 100,000 francs—which indicates the importance attached to the subject.

"Material improvement precedes all other improvements, and is the source of them.

"Associations and organizations which combine the sciences and industrial arts with labor and skill, augmenting the productiveness of both in an eminent degree, and at the same time improving the condition of the laborers, are among the most important institutions of the time, and they are without parallel in any past civilization.

"This new quality is that they not only make the rich richer, but they make the poor richer—a result which was never before studied nor attained; and in place of industrial populations devoted to immorality and poverty, we have now thriving communities acquiring property and education, and rising constantly in their material, moral, and intellectual condition.

"These organizations are not the result of accident or the mere growth of time, but they are the combined product of the ablest thinkers, the best judgments, highest motives, superior science, the practical skill, and the solid knowledge of the age.

"Each country makes its own development in conformity with its condition and its wants, and the attempt to bring together the most successful of these institutions, to be carefully studied and compared by competent men, cannot fail to be instructive and useful.

"We have many industrial establishments which, properly presented, might compete successfully for the prizes; but, whether successfully or not, they could not fail to improve by the competition.   ·

"The reports of these should contain their organization, administration, capital, number of hands of both sexes, their ages, wages, schools, libraries, churches, charities, publications, sanitary regulations, plans, marriages, births, deaths, savings and accumulations, investments, quantity and value of raw products, value and quantity of manufactured products, dividends, and, in fine, all the important facts relating to these small communities as if they were states.

"These should be tabulated and briefly and accurately stated, to facilitate the understanding of them and the comparisons.

"Similar reports will be sent in from all civilized countries, and will be submitted to the study of a special jury of twenty-five men, selected for the purpose from the different nations, and whose labor has no connection with class juries on products.

"This jury commenced their work on the 1st December instant, and will continue it through the winter, that they may complete their reports and render their verdict in May, in time for the general distribution of prizes on the 1st July next.

"Our representative on this special duty is Charles C. Perkins, esq., of Boston, a gentleman whose cultivation, general knowledge, studious habits, and acquaintance with continental languages, qualify him for the labor, and whose circumstances permit him to devote the requisite time to it.

"The reports, in conformity with the regulations, should all have been sent in by the 1st of December; but the jury resolved at their first meeting to remain open a short time longer, and will probably fix the last day for receiving reports at their next meeting, of which I will notify you. If this subject appears to you of sufficient interest to call for your exertions, be assured that any reports which you may send me, if received in time, will be clearly and amply developed to the jury by Mr. Perkins, and receive the consideration they may merit; but I need not dwell on the necessity of immediate action."

APPOINTMENT OF JURORS TO THE UNITED STATES.

Mr. Beckwith to Mr. Seward.

"PARIS, *April* 8, 1867.

"SIR : It was not in my power to report definitively on the nomination of jurors before the 26th March, when the list was completed and closed.

" The number of class jurors is six hundred, who are divided among nationalities on the basis of the relative space occupied by each nation in the Palace of the Exposition.

"The apportionment was made in June, 1866, (reserving fifteen jurors for subsequent distribution,) and the number which this method gave to the United States was eight.

"I applied for an increase of this number, and, of the fifteen reserved, four were subsequently added to our list, which gave us twelve class jurors.

"The juries are international (or mixed) and each jury comprises from four to fifteen members. There are ninety-four classes of products, and ninety-four separate juries, consequently there are eighty-two classes on which we have no juryman, our number being but twelve.

"Jurymen are presumed to be acquainted with the products of their own country in general, and to have a special knowledge of the products of the class on which they are placed, the chief service they can render being to make known to their associate jurymen the nature, quality, uses, and importance of these particular products, that they may be rightly appreciated and compared, and accurately judged.

"The most important products of nearly all competing nations fall into a few classes, and there is naturally great urgency among all to place their jurymen on these important classes, but this being impracticable there must be some disappointments.

"The Imperial Commission desired each foreign commission to send in a list indicating the classes on which they would prefer to place their jurymen, or else the order of preference, but stating at the same time that these indications could only be taken in a general sense as a guide in making allotments, and that it would doubtless be impossible to fully comply with them.

"Some of the important classes on which I desired to place jurors were conceded to me, some equally or more important were not conceded, and three or four classes were assigned to me which I did not ask for and which are of no particular value to us.

"A similar result occurred to others, and the distribution of classes on the whole produced among the foreign commissions a great deal of dissatisfaction, but I think I should add that in my opinion this could not have been altogether avoided.

"Mr. Derby was diligent in sending me early information in respect of the persons who were qualified to act as jurors, who were willing to do so, and who would be here in time.

"But unfortunately very few of the gentlemen thus indicated considered themselves particularly qualified for the classes which were finally assigned to us, and I have had much difficulty in making up a list of qualified jurors who could be relied on to be present in time for the work. Indeed, I have been obliged to nominate some for classes which they themselves did not wish to accept, but consented to do so because they saw that it was impossible for me to do better.

"The following is the definitive list:

"Group I, Classes 1 and 2 united: Paintings in oil, water colors, pastels, and designs—J. W. Hoppin, esq., New York.

"Group I, Class 3: Sculptures, bas reliefs, medallions, cameos, engraved stones, &c.—Hon. J. P. Kennedy, Baltimore, United States Commissioner.

"Group I, Class 4: Architecture and architectural designs.—R. M. Hunt, esq., New York, architect.

"Group I, Class 5: Engravings, polychromes, lithographs, crayons, &c.—Frank Leslie, esq., New York, United States Commissioner.

"Group II, Class 9: Proofs and materials for photography on paper, glass, wood, cloth, enamel, &c., instruments, materials, &c.—W. A. Adams, esq., Ohio, United States Commissioner.

"Group II, Class 11: Apparatus and instruments of the medical art, trusses, surgical instruments specially adapted to the army and the navy, &c.—J. R. Freese, esq., New Jersey, United States Commissioner.

"Group II, Class 12: F. A. P. Barnard, LL. D., New York, United States Commissioner.

"Group III, Class 20: Various kinds of fine cutlery—William Slade, esq., Ohio, United States Commissioner.

"Group VI, Class 51: Materials of chemistry, pharmacy, tanning, of glass, gas, &c.—Professor J. P. Lesley, Pennsylvania, United States Commissioner.

"Group VI, Class 59: Motors, generators, shafts, pulleys, steampipes, &c.—Ch. R. Goodwin, esq., Boston, United States Commissioner.

"Group VI, Class 54: Machine tools and mechanical apparatus in general; planing machines for iron and wood; elevators, pumps, machines, and tools of all sorts worked by steam, gas, water, &c.—J. E. Holmes, esq., New Hampshire, now residing in England; associate juror in this class with Mr. Holmes, Mr. Debauvais, machinist, of New York.

"Group VI, Class 57: Machines of all kinds for sewing, embroidery, &c., in cloth, stuffs, and leather; tacking, pegging, nailing machines, &c.—H. Q. d'Aligny, esq., Michigan, United States Commissioner.

"One special juror on the new order of prizes—Charles C. Perkins, esq., of Boston.

"One vice-president of Group V, comprising minerals and raw products of many kinds—Professor J. Lawrence Smith, Kentucky, United States Commissioner.

"Fourteen in all on the jury list.

"It will be observed that we have no class jurors on agricultural machines, nor on agricultural products, &c., where I was extremely anxious to have them. I should have been much gratified if I could have transferred three jurors from Group I to Groups V and VII. Failing to obtain from the Imperial Commission an allotment on these classes, I endeavored to effect an exchange by negotiation with other commissions, (which is permitted,) but found none who were not more tenacious of the classes I desired to obtain than desirous of the classes I wished to relinquish, and I could not effect the exchange.

"I repeat, therefore, that I am not satisfied with the distribution; but this feeling prevails in general, and I must acknowledge the evident desire of the Imperial Commission to be fair and impartial, and to give satisfaction as far as possible in a matter wherein the complete satisfaction of each commissioner, in conformity with his own judgment and wishes, was impossible.

"It is possible, also, that I exaggerate the importance which properly belongs to this subject, being influenced, perhaps, more by feeling than by judgment. I have witnessed the efforts of the foreign commissions to obtain the services of competent and skillful men for jurors, and most of them being near their own countries, with great facilities of communication, they have been able to assemble a body of able and experienced men for each class of products, whose investigations, it is impossible to doubt, will be thorough, and their verdict impartial and sound."

WORK OF CLASS JURIES.

Mr. Beckwith to Mr. Seward.

"PARIS, *May* 17, 1867.

"SIR: I had the honor to address you on the 3d of April, with a brief statement of the condition of our section of the Exposition at the opening, and suggested that fully a month would be required to complete it, which estimate was nearly correct.

"The work of the class juries commenced on the 2d of April, and was continued daily till completed.

"The necessity of getting the packages all opened, and the products in a condition to be examined by the juries, is obvious. For this purpose I employed a force as numerous as could work to advantage. The juries being also numerous, ninety-four in number, (counting six hundred members,) their visits were incessant in different classes and different parts of the section, and the necessity of attending to them delayed the work of placing the products, it being more important to·bring them under the inspection of the juries while it was possible, and to complete the final placing with more care afterward.

"The number of our exhibitors is from seven hundred to eight hundred, and it is remarkable that up to this date only sixty-four exhibitors have appeared, and ninety-six agents; and many of the agents being parties residing in Paris, take little interest in the business with which they are charged, and pay little or no attention to it. It is usual for exhibitors to be present with their products in person, or by agents who understand the products, and whose business and interest it is to make known their qualities to the juries, that they may receive proper appreciation. But the absence of exhibitors, and absence of substitutes provided by them, left the jury-work in a great measure to me; and considering it of great moment to the exhibitors, absent as well as present, I made the fullest provision for it in my power, and gave my whole attention to it while it lasted.

"Our products have therefore been well brought to the notice of the juries; their reports, when published, will show this, and show also, I think, that they have been fairly and intelligently appreciated.

"But exhibitors are anxious in presence of jurors; each is conscious of the value of his particular products, and naturally apprehensive that time enough will not be given to them. It seldom happens indeed that jurors spend as much time on a single exhibit as the exhibitor desires. The jurymen are chiefly experts, and do not require, or think they do not, as much time for accurate appreciations of products as the producer may think necessary. It also occurs frequently that the exhibitor or agent is not present, and cannot be found when the jury comes to his class, and the work goes on without him.

"This is unsatisfactory, and leads to numerous reclamations, to which the juries have been in general attentive and accommodating; but in most of the numerous cases of recall which I have had occasion to make, they have proved to be of small importance beyond satisfying the exhibitor, the juries having shown generally that their work had been properly done; and I conclude my remarks on this laborious part of the work with an expression of confidence that the result will prove satisfactory."

THE DISTRIBUTION OF PRIZES.

Mr. Beckwith to Mr. Seward.

"PARIS, *July* 2, 1867.

"The promulgation of the awards and distribution of medals, which is the principal event of the Exposition, took place yesterday in the Palais de l'Industrie, in the presence of an audience composed of all nations, and numbering about seventeen thousand persons.

"The awards to our exhibitors were mentioned in my letter of the 28th ultimo, and are now published in detail by the Imperial Commission, and copied generally in the daily journals.

"The proclamation of the awards was made the occasion of a brilliant and imposing pageant, but the event was suggestive of more grave and important considerations.

"Formerly the dignity, wealth, and fashion of surrounding nations met on the plains of France, 'in tents of cloth of gold,' to honor arms by mimic war and feats of personal prowess in the tournament; but at that period toil was remitted to serfs, and labor was degraded.

"To-day the civilized world assembles on the same ground to do honor to labor.

"We have been accustomed to read with admiration, real or romantic, the accounts of the fêtes of the Celestial Empire, where the ruler condescends annually to hold the plow in honor of husbandry. Yesterday an assembly of all nationalities, numbering seventeen thousand persons, including rulers, peasants, and every class, put their hand to the plow, and did homage to labor, skill, and science, which are the bases of civilization and progress.

"The participation of the people of the United States in this great competition, and the rank which the products of their industry hold, as determined by the deliberate judgment of ninety-four international juries who have decreed awards to more than one-half of our exhibitors, justify expressions of cordial congratulation and satisfaction."

HONORARY DISTINCTIONS.

Mr. Beckwith to Mr. Seward.

"PARIS, *July* 5, 1867.

"SIR: The volumes containing the official report of awards to exhibitors should have been published by the Imperial Commission on the 1st of July, but I am not yet able to send it to you, as it is still in the press.

"I intended also to allude more fully to the *grand prix* awarded to Mr. Hughes, to Mr. Cyrus W. Field, to the Sanitary Commission, and the *prix* (equivalent to a *grand prix*) to M. Chapin, in the new Order of Recompenses; but wishing to state, in the language of the report, the grounds of the awards, I am compelled to await its publication.

· "I have now to report the honorary distinctions, in connection with the Exposition, conferred by the government on the following persons:

Mr. ELIAS HOWE, JR.: - Inventor of the sewing machine, and founder of the great and important industry represented by that machine.

Mr. CHICKERING: - - A respectable house of long-standing and importance, continuing from father to son, which, by ability and great study to improve their work, have contributed to raise the standard of this class of products to a high level in the United States.

Mr. MULAT: - - - - Architect and engineer in the United States Section of the Exposition.

Mr. J. P. KENNEDY: - - United States Commissioner and member of the International Jury, Group I, Class 3.

Mr. J. LAWRENCE SMITH: United States Commissioner and vice-president of Jury for Group V.

Mr. S. B. RUGGLES: - - United States Commissioner and member of the committee on weights, coins, and measures.

Mr. C. C. PERKINS: - - Member of the special jury on the new Order of Recompenses.

Mr. C. R. GOODWIN: - - United States Commissioner, machinist, and member of the jury, Group VI, Class 52.

Mr. ROBERT BERNEY: - United States Commissioner.

"To the preceding, the honorary distinction of 'Chevalier de la légion d'honneur.' To:

Mr. N. M. BECKWITH, Commissioner General, the honorary distinction of 'Officier de la légion d'honneur.'

"The honorary distinctions above named have been received subject to the laws of the United States, which may be applicable to them and to the action of the government and Congress.

"A membership of the Legion of Honor is not a title of any kind, or a name by which men are called, but it confers the distinction which attaches to the membership of a numerous and respectable association, as a membership of the Institute or of the Royal Society.

"I am not aware, therefore, whether clause 7, section 9, of the Constitution, is applicable to it, or whether Congress has ever considered the subject and decided it.

"But I desire for my guidance (and in this I doubt not of the cordial concurrence of my associates) an expression of opinion from the government in regard to the propriety of accepting this distinction; and if not otherwise directed, I shall embrace an early opportunity of soliciting the action of Congress on the subject.

"The Exposition is international; its aims and effects, both national and moral, are international; its courtesies are international; and the services connected with it are necessarily in the same spirit.

"The awards, therefore, to be appropriate, should be of the same international character; and if the government and Congress approve the acceptance of these distinctions, they will thus impart to them the element which is requisite to complete their proper character and value."

EXHIBITION OF THE MEDALS AND DIPLOMAS AWARDED TO THE UNITED STATES EXHIBITORS.

Mr. Beckwith to Mr. Seward.

" PARIS, *November* 18, 1867.

"The great number of awards, and the preparation of the stamps for imprinting the name of each person on his medal as well as on his diploma, protracts the labor, and though the delivery of awards has commenced, it is not likely to be completed before February or March.

"Many of our successful exhibitors have not visited the Exposition, and of those who have done so most of them have now returned to the United States.

"It will be necessary, therefore, for me to transmit the awards, when received, to the government or its agents for right delivery, and in doing this I beg to suggest, for your consideration, the utility and expediency of an exhibition of the awards at the seat of government for a limited time before their final distribution.

"The collection of medals and diplomas thus exposed will make a visible display of some of the results of the Exposition, which, I think, cannot fail to be gratifying to the government and to Congress, and beneficial to the exhibitors and to the country.

"In making this suggestion I assume that the recipients of awards will be pleased with this course, which I infer from the nature of the proposal and from the uniform opinion of those with whom I have spoken on this subject.

"The additional expenditure required will be small, and will be kept within the limits of the appropriations already made for the exhibition.

"I venture to hope, therefore, that you may consider an exposition of the trophies a becoming and appropriate method of concluding a competitive international exhibition of the products of industry, and that you will be willing to assent to the proposal."

PRIZES FOR REAPING AND MOWING MACHINES.

Mr. Beckwith to Mr. Seward.

"PARIS, *January* 6, 1868.

"The adjustments of awards and delays in their distribution have prevented me from sending you a complete list showing the final result, and I am still unable to do so, the work being not yet completely finished.

"A distribution of awards was made by his Imperial Majesty yesterday, in the palace of the Tuileries, to the successful exhibitors in Groups VIII and IX, which, with previous additions and adjustments, will increase the number in the United States Section, reported in the letter which I had the honor of addressing to you on the 24th June last, from two hundred and sixty-two to two hundred and ninety-one.

"The superiority of the reaping machines of Mr. McCormick, and the mowing machines of Mr. Walter A. Wood, over all others exhibited, established by repeated experiments in the field during the season, together with the acknowledged importance and great utility of those machines, have secured for Mr. McCormick, from the International Jury, a *grand prix ;* to Mr. W. A. Wood a gold medal and an object of art; and from the Emperor, to each of those gentlemen, the cross of the Legion of Honor."

CONDITION OF THE INDUSTRIAL ARTS INDICATED BY THE AWARDS.

Mr. Beckwith to Mr. Seward.

"PARIS, *January* 21, 1868.

*   *   *   *   *   *   *   *

"I beg to solicit your attention to the tabular statement alluded to on page 6 of the preface,[1] as expressing briefly the opinion of six hundred international jurors, in a form that indicates the relative condition of the mechanical, manufacturing, and industrial arts and productive industries in the principal countries of the world, so far as that can be determined by an inquest of competent jurors. The table has been made with care and accuracy, and I believe is reliable.

"That the products of England should recede to the position of eighth on the list, is a conclusion that perhaps will not be readily accepted.

"But it is widely known that great changes have been going on since the first exhibition in 1851.

"England, it was admitted, excelled especially in machinery and in nearly every department of the working of metals.

"But her methods, her forms, and her models have been adopted and reproduced in many countries on the continent, and it is now contended that the better provision in the latter for industrial, mechanical, and scientific education supplies a greater number of superior producers,

---

[1] Vide preface to the " General Survey of the Exposition."

and that English models and methods have been perfected and carried beyond the originals.

"These observations are not limited to the metallic department, but are applied also to woolen fabrics and to nearly all the higher manufactures of cotton, while the supremacy in silks, linens, designs, dyes, and finish was already continental.

"It is also remarked that the principal progress in English products is in the department of pottery, in which she has adopted continental forms, colors, and designs, and improved her work, but content with her great aggregate production and great commerce, has generally remained stationary in regard to quality, while others have advanced; and that this change was strikingly visible in the exhibition of 1862, in which England was no longer in advance, and hardly maintained her level, having now distinctly fallen behind.

"It is also remarked, in addition, that while these relative changes and equalizations of the industrial arts have been going on, there has not been much invention, but that progress is chiefly noticeable in the perfecting of previous inventions, designs, and methods, and that for any new combinations which are creative and striking it is necessary to look to the other side of the Atlantic; that the great revolutions going on in war ships, guns, and small arms, and the remodeling of navies and armies are of American origin, and that the American Section of the Exposition was more fertile than any other in the original, the inventive, the peculiar, and the new.

"I need not express any opinion as to the accuracy or inaccuracy of these general views, but I reproduce them because I have heard them expressed during the Exposition by many Europeans connected with it and largely engaged themselves in manufactures and industry, and because these opinions appear to correspond in a general sense with the table of results to which I invited your attention.

"But admitting the accuracy of these indications, the value of them might be readily over-estimated.

"In an international competition many of the products exposed are made for show, and their qualities are in some degree meretricious. If the products of England, under these circumstances, take the position suggested by the awards, it does not follow that she has not preserved the medium of practical sense, and that her products are not best adapted to the wants, the means, and the consumption of the mass of mankind, on which her great commerce is based."

COMMISSION OF THE UNITED STATES.

JOINT RESOLUTION IN RELATION TO THE INDUSTRIAL EXPOSITION AT PARIS, FRANCE.

"Whereas the United States have been invited by the government of France to take part in a Universal Exposition of the productions of agriculture, manufactures, and the fine arts, to be held in Paris, France, in the year eighteen hundred and sixty-seven: Therefore,

"*Be it resolved by the Senate and House of Representatives of the United States of America in Congress assembled,* That said invitation is accepted.

"SEC. 2. *And be it further resolved,* That the proceedings heretofore adopted by the Secretary of State in relation to the said Exposition, as set forth in his report and accompanying documents concerning that subject, transmitted to both houses of Congress with the President's message of the eleventh instant, are approved.

"SEC. 3. *And be it further resolved,* That the general agent for the said Exposition at New York be authorized to employ such clerks as may be necessary to enable him to fulfill the requirements of the regulations of the Imperial Commission, not to exceed four in number, one of whom shall receive compensation at the rate of eighteen hundred dollars per annum, one at sixteen hundred dollars, and two at fourteen hundred dollars.

"SEC. 4. *And be it further resolved,* That the Secretary of State be, and is hereby, authorized and requested to prescribe such general regulations concerning the conduct of the business relating to the part to be taken by the United States in the Exposition as may be proper.

"Approved January 15, 1866."

The sixth portion of the first section of the joint resolution approved July 5, 1866, provided "for the traveling expenses of ten professional and scientific commissioners, to be appointed by the President, by and with the advice and consent of the Senate, at the rate of one thousand dollars each—ten thousand dollars—it being understood that the President may appoint additional commissioners, not exceeding twenty in number, whose expenses shall not be paid; but no person interested, directly or indirectly, in any article exhibited shall be a commissioner; nor shall any member of Congress, or any person holding an appointment or office of honor or trust under the United States be appointed a commissioner, agent, or officer under this resolution."

The following is that part of the supplementary joint resolution approved March 12, 1867, which provided for the appointment of twenty honorary commissioners:

"*Resolved by the Senate and House of Representatives of the United States of America in Congress assembled,*

"I. That the Commission of the United States at the Universal Exposition to be held at Paris in the year eighteen hundred and sixty-seven shall consist of the Commissioner General and honorary commissioner, whose appointment was approved by the joint resolution of January twenty-two, eighteen hundred and sixty-six; also of the thirty commissioners whose appointment was provided for by the joint resolution of July five, eighteen hundred and sixty-six; and of twenty commissioners, whose appointment is hereinafter provided for.

"II. That the Commissioner General shall be the president of the commission thus constituted, with a vote on all questions that may arise.

"III. That the commission shall meet at Paris as early as possible before the opening of the Exposition, upon the call of the Commissioner General, and, when properly organized, shall make such rules and regulations as may be necessary for efficient action, with power to elect a vice-president from their own number, who, in the absence of the Commissioner General, shall preside at all meetings of the commission, and to appoint committees and chairmen of groups.

"IV. That the commission may designate additional persons, not exceeding twenty in number, being citizens of the United States, known to be skilled in any branch of industry or art, who are hereby authorized to attend the Exposition in behalf of the United States as honorary commissioners without compensation.

"V. That the commission may employ a secretary and clerks for the commission, the necessary scientific assistants and draughtsmen, and may engage suitable rooms for the commission.

"VI. That no commissioner shall act as agent for the show or sale of any article at the Exposition, or be interested, directly or indirectly, in any profits from any such article."

LETTERS OF APPOINTMENT TO COMMISSIONERS.

Letters were addressed by the Secretary of State in the following form to the gentlemen appointed as commissioners, and to the ten professional and scientific commissioners:

To Commissioners without compensation.

"DEPARTMENT OF STATE,
"*Washington*, ———, 1866.

"SIR: The joint resolution approved on the 5th of July, 1866, a copy of which is annexed, authorizes the President to appoint twenty commissioners, who shall serve without compensation, to attend, on behalf of the United States, the Universal Exposition to be held at Paris, France, in the year 1867, commencing on the 1st of April, and closing on the 31st of October, in that year.

"These appointments are intended as honorary distinctions for gentlemen who are eminent for their patriotism and for scientific or professional attainments, or familiarity with some special branch of industry, and whose means enable them to serve gratuitously. Persons so appointed are required to aid the Commissioner General and the Scientific Reporting Commission in accumulating and arranging valuable information at the Exposition, in conformity with such regulations as may be prescribed by this Department.

"The President directs me to offer to you an appointment as one of the twenty commissioners thus authorized, provided you are not interested, directly or indirectly, in any article to be exhibited at the Exposition, and shall not, during your service as such commissioner, hold any other appointment or office of honor or trust under the United States.

"You are requested to reply to this letter at your earliest convenience, and, should you accept the appointment, to inform me of the place of your birth.

To scientific and professional commissioners.

"DEPARTMENT OF STATE,
Washington, ———, 1866.

"SIR: The President directs me to offer you an appointment as one of the ten scientific and professional commissioners of the United States to the Paris Universal Exposition, mentioned in sixth clause of the first section of the joint resolution of the 5th of July, 1866, a copy of which is annexed, provided that you are not interested, directly or indirectly, in any article to be exhibited at the Exposition, and that you shall hold no other appointment or office of honor or trust under the United States during your term of service as said commissioner.

"In the event of your appointment you will be required to attend the Exposition during the period for which it shall remain open, and to co-operate with the Commissioner General in the preparation of reports upon it, in conformity with such regulations as may be prescribed by this Department. You will be allowed for the payment of traveling expenses the sum of one thousand dollars, which is the only allowance or compensation provided by law.

"You are requested to reply to this letter at your earliest convenience, and, should you accept the appointment, to inform me of the place of your birth.

Instructions to commissioners.

"DEPARTMENT OF STATE,
"*Washington*, ———, 1867.

"SIR: I inclose, for your information, a copy of the several joint resolutions relating to the participation of the United States in the Paris Universal Exposition of 1867. Pursuant to the third clause of the first section of the resolution approved on the 12th instant, the commissioners are required to meet at Paris as early as possible before the opening of the Exposition, upon the call of the Commissioner General. As that resolution was not passed and approved until the 12th instant, sufficient time is not allowed for the transmission of a copy to the Commissioner General, and the issue of a call by him for the assembling of the commissioners at Paris, which would enable them to reach that capital before the opening of the Exposition, the date appointed for which is the 1st of April next.

"It is important that as full and as early a meeting as possible of the commission at Paris shall be held; and I therefore, on behalf of the

Commissioner General, request your attendance at that city at your earliest convenience, and in accordance with the regulations issued by this Department on the 20th of August last, under the authority of the joint resolution approved on the 15th of January, 1866, a copy of which regulations has already been forwarded to you.

REGULATIONS ISSUED BY THE DEPARTMENT.

The annexed is a copy of the regulations for the guidance of officers of the United States connected with the Paris Universal Exposition of 1867, issued by the Secretary of State:

"CHAPTER I.—The commissioners will make a report presenting a brief general survey of the Exposition, and a similar report upon the character and condition of the American exhibition.

" They will make special reports upon inventions, and upon the various products displayed which are most advanced in the sciences, in the arts, and in industry, giving a practical description of the methods and processes connected with such products.

" Similar reports will be made upon mineral and agricultural products of importance, and upon raw materials and manufactures of great general use, or displaying remarkable skill and merit; upon implements, machines, tools; on metallurgy, and the extractive arts in general; on the products of chemistry, and the preparations of food and clothing; and on any other subject connected with the Exposition, and relating to the material, moral, and intellectual well-being of the nation.

" To accomplish this work the commissioners will proceed as follows:

" First. On or before the opening of the Exposition, 1st April, 1867, they will assemble in Paris, and meet for the purpose of organization, at which meeting the Commissioner General will preside.

" Second. The commissioners will then constitute themselves a committee, and proceed to elect a presiding officer and secretary for committee meetings, which will be held as often as in their opinion the work may require.

" Third. At these meetings the committee will decide, from time to time, on the subject for special reports, and designate the persons who will undertake the investigation and studies each report will require.

" Fourth. Each report will be made on the responsibility of the person charged with it, and he will sign his name to it.

" Fifth. The commissioners who may be nominated on international juries will perform that service.

" Sixth. There will be regular meetings of the whole commission as often as once a month, at which meetings the Commissioner General will preside.

" Seventh. The committee will report at these meetings the progress made, and the reports which have been completed will be read to the meeting by their authors, and then delivered to the Commissioner General for transmission to the Department of State, (or disposed of as the Department may direct.)

"Eighth. The meeting shall not be competent to reject any report; but observations may be made on each report by any commissioner, and he may reduce his remarks to writing and send them with the report, if he so desire. It will be in order to decide upon subjects for special reports at the general meetings, as well as in meetings in committee.

"Ninth. The Commissioner General is entitled to call upon the members of the commission for their advice and assistance in his department. Their assistance so given will be voluntary, and their counsel advisory.

"Tenth. A brief record of the proceedings of the general meetings will be kept for transmission to this Department.

"CHAPTER II. The Commissioner General is charged with the supervision and management of the exhibition.

"He will receive the products on their arrival in France and place them for exhibition, and he will conduct the exhibition in conformity with the regulations of the Imperial Commission.

"The employés, whether paid by him or by exhibitors, and all persons engaged in explaining or exhibiting products, or occupied on the premises, will be under his direction; also the application of motive force to machinery.

"First. The 'over-head gear,' straps, pulleys, &c., for machinery in action, together with the materials and laborers for working the machinery and keeping it in order, will be at the expense of the owners of the machines.

"Second. The Commissioner General will provide laborers for keeping the apartments in order, linguists for explanations, and subsidiary police or guards for the protection of property and preservation of order; the number of persons to be so employed is left to his judgment, he having regard to the necessities and to the funds for this purpose placed at his disposal. Exhibitors may introduce employés in connection with their products, at their own expense, but they shall not so introduce them without the consent of the Commissioner General, and such employés will be under his direction.

"Third. The government will not be responsible for damages, thefts, or destruction of property, and insurance against all contingent risks to property, either in the Exposition or *in transitu*, is left to the care and cost of the owners respectively.

"Fourth. With the closing of the Exposition in conformity with the regulations of the Imperial Commission, the government control and charge of the property will cease, and the Commissioner General will deliver the products to the order of the respective owners, who will receipt for them; which delivery will be made at the place of exhibition. Property not applied for and removed by the owners within the period fixed by the regulations of the Imperial Commission for that purpose will be at the risk of the owners, and subject to the charges which may accrue upon it.

" Fifth. The Commissioner General will be responsible for the public money placed at his disposal, and will render the accounts of his expenditures to this Department, in conformity with the act of Congress."

MEETINGS OF THE COMMISSON.

Mr. Beckwith to Mr. Seward.

"PARIS *April* 12, 1867.

" I have the honor to report that a meeting of the United States Commissioners to the Paris Exposition was called by me at this office, in conformity with the act of Congress, on the 29th March, and that the following commissioners were present: Messrs. Ruggles, Evans, D'Aligny, F. Leslie, Slade, Seymour, Kennedy, Goodwin, J. P. Lesley, Berney, Norton, Smith, Valentine, Beckwith—14.

" The Commissioner General presided. William Slade, esq., was requested to act as secretary pro tempore, and business was commenced by the reading of the acts of Congress and the regulations of the Secretary of State constituting the commission and directing its organization and its labors.

" The meeting then proceeded to the election of a vice-president, which resulted in the unanimous choice of S. B. Ruggles, esq., to that office. Committees were appointed on by-laws and on the selection of suitable rooms for the regular meetings of the commission, whereupon the meeting adjourned to the following Tuesday, 2d of April, at 2 o'clock.

" Several meetings have since been held, which were occupied chiefly with discussions on the appointment of committees, the choice of new members, and other measures, which have not yet resulted in completing the organization of the commission for work.

" I append hereto a list of the commissioners who have arrived since the first meeting, and a list of those still absent :

" Since arrived.—Messrs. Barnard, Budd, Mudge, Stevens—4.

" Still absent.—Messrs. Bowen, Hewitt, Stewart, Freese, Adams, Jones, Archer, McIlvaine, Winchester, Leathermann, Garrison, Sweat, Leconte—13."

ATTENDANCE OF COMMISSIONERS.

Mr. Beckwith to Mr. Seward.

" PARIS, *December* 11, 1867.

" I have the honor to inclose herewith a list of the names of the commissioners appointed by the government, who have been present in Paris at any time during the Exposition; a list of the honorary commissioners elected by the previous commissioners, with the dates of their election, and explanatory remarks; and a tabular statement showing the number and dates of the sessions held by the commission, with a record of the attendance of each member.

" I transmit also to your address a volume containing the minutes of the proceedings of the commission, recorded by the secretary of the commission, Mr. L. F. Mellen.

" List of commissioners appointed by the government who have been present at any time during the Exposition.—Messrs. Slade, Evans, Norton, Stevens, Kennedy, Ruggles, Bowen, Berney, D'Aligny, Barnard, Seymour, Freese, Goodwin, Mudge, F. Leslie, Budd, Valentine, Smith, Hewitt, J. P. Lesley, Garrison, Stewart, Beckwith, Archer.

" List of honorary commissioners elected under the joint resolution of Congress of the 12th March, 1867.—Messrs. Elliot C. Cowdin, of New York, elected 29th April; George S. Hazard, of Buffalo, elected 29th April; W. S. Auchincloss, of New York, elected 6th May; William J. Flagg, of Ohio, elected 9th May; Alexander Thompson, of New York, elected 9th May; Professor William B. Rogers, of Boston, elected 9th May, (declined to accept;) Professor S. F. B. Morse, of New York, elected 9th May; Professor J. T. Frazer, of Philadelphia, elected 9th May; Messrs. B. F. Nourse, of Boston, elected 9th May, (resigned;) L. F. Mellen, of Alabama, elected 9th May; M. P. Wilder, of Boston, elected 9th May; J. P. Reynolds, of Illinois, elected 13th May; J. H. Chadwick, of Massachusetts, elected 26th May; Thomas McElrath, of New York, elected 8th June; Patrick Barry, of Rochester, elected 6th July; William E. Johnston, M. D., of Paris, elected 13th July; Professor J. W. Hoyt, of Wisconsin, elected 10th August."

Record of sessions and attendance of each member.

Names of commissioners.	March 29	April 2	April 3	April 10	April 15	April 22	April 29	May 6	May 9	May 13	May 20	May 26	June 8	July 6	July 13	August 10	September 5	October 4	October 7	October 8	October 10	October 14	October 18	October 21	October 23	October 25	October 28	October 29	October 30	October 31	Attendances of each commiss'r.
Beckwith	1	1	1		1	1	1	1	1	1	1	1	1	1	1	1		1	1		1	1	1	1	1	1	1	1	1	1	28
Slade	1	1	1	1	1	1			1	1		1		1	1	1			1	1		1	1	1	1	1	1	1	1		23
Evans	1	1	1	1	1	1	1					1			1	1			1	1	1	1	1	1	1	1	1	1	1	1	22
Mellen									1	1	1	1	1	1	1	1		1	1	1	1	1	1	1	1	1	1	1	1		21
Norton	1	1	1		1	1	1	1	1	1	1		1				1		1	1	1	1	1				1	1		1	21
Stevens				1	1	1	1	1	1	1	1	1	1	1	1	1		1	1	1				1		1	1	1			19
Kennedy		1	1	1	1	1	1		1	1	1	1		1					1		1		1			1			1	1	16
Ruggles	1	1	1	1	1	1	1	1	1		1		1						1	1	1	1									16
Bowen							1	1	1	1		1		1	1	1	1	1	1	1						1	1	1	1	1	16
Barnard		1		1		1	1		1		1		1	1		1	1	1	1			1					1			1	15
D'Aligny	1	1	1	1					1	1	1	1			1		1	1	1							1			1	1	15
Berney	1	1	1		1	1	1	1	1	1		1		1				1									1		1	1	14
Seymour	1	1	1						1	1		1			1	1	1									1	1	1			12
Morse									1	1		1	1		1					1			1	1	1	1	1	1	1		11
Freese												1	1	1	1		1			1		1	1			1	1	1	1	1	11
Goodwin	1	1	1		1	1						1						1									1	1	1		10
Mudge		1	1	1	1	1	1	1	1	1		1																			10
Thompson									1	1	1												1	1	1	1	1	1			9
Leslie, F	1	1	1	1	1					1	1	1																			8
Budd		1	1	1	1			1	1	1	1																				8
Johnson																1	1			1	1	1				1				1	7
Valentine	1	1			1			1			1			1		1										1					7
Smith	1	1	1					1					1			1															6
Flagg											1	1	1	1	1	1															6
Hewitt							1	1	1		1		1																	1	6
Lesley, J. P.	1	1		1					1			1																			5
Auchincloss									1	1	1	1	1																		5
Wilder										1	1	1	1																		4
Reynolds									1	1	1	1																			4
Hazard						1	1		1																						3
Stewart									1		1																				2
Cowdin									1		1																				2
Garrison									1																						1
Nourse								1																							1
Archer									1																						1
Chadwick																		1													1
Frazer																															
Barry																															
McElrath																															
Hoyt																															
Total *	13	17	14	12	13	11	10	13	15	18	18	13	19	22	11	12	6	11	11	6	7	8	7	9	7	9	14	14	10	16	

* The total number of meetings was thirty.

CLOSE OF THE EXPOSITION AND DELIVERY OF OBJECTS.

Mr. Beckwith to Mr. Seward.

"PARIS, *November 2, 1867.*

"The circumstances attending the close of the Exposition were as follows:

"On the 29th October the Imperial Commission published in the Moniteur the notice hereto annexed, to the effect that the rumors of a prolongation were unfounded, and that the Exposition would be definitively closed on the 31st October.

"On the same day letters were addressed by the Imperial Commission to the foreign commissioners, inviting them to a conference with the Imperial Commission on the 30th.

"The conference was presided over by the minister of state, assisted by the two vice-presidents, the minister of commerce and public works, and the minister of the Emperor's household.

"The minister of state informed the meeting that the applications for the prolongation of the Exposition for two weeks, from sources entitled to great respect, had become so numerous and pressing, that he thought it his duty to consult the foreign commissioners on the subject, hoping to find their views in accord with the public wishes, and suggesting that if the continuation of the Exposition for so long a period was not thought advisable, perhaps a shorter period might be agreed upon, and a portion of the proceeds during that time applied to public charities.

"A brief discussion followed, in which several members of the Imperial Commission and several of the foreign commissioners expressed their views. At this point the foreign commissioners desired permission, which was granted, to retire and deliberate by themselves and report the result of their deliberation to the Imperial Commission.

"The great majority of foreign commissioners, on coming together, appeared to be in favor of the prolongation; some were opposed to it; and of the latter the commissioners of Prussia, England, Russia, Austria, and the United States were in the outset of one opinion, viz:

"1. That they were without authority from their respective governments to assent to a prolongation.

"2. That the regulations confirmed by imperial decree of the 12th July, 1865, fixing the close of the Exposition on the 31st October, form the contract between the Imperial Commission and the exhibitors, and cannot be departed from without the consent of the exhibitors, which it is now too late to obtain. It is impossible, therefore, to assent to a prolongation; but the two first days of November being religious fêtes, and the third Sunday, not much work can be done; hence it is practicable to consider the Exposition closed on the 31st, in conformity with the contract, allow exhibitors to take possession of their products, commence the delivery of such as are sold, the taking down and packing of those not sold, and in general to begin the work; it is desirable at the same time to admit the public at the usual price during the three days of fête, as it will not materially embarrass the little work which can be done, and will create a considerable fund, to be given to the poor of Paris.

"These views, after considerable discussion, were agreed to by the foreign commissioners in general, reported to the Imperial Commission, and in substance accepted by them as developed in the annexed notice from the Moniteur, in which an appeal is also made by the Imperial Commission to the exhibitors to favor the charitable object of the three days as far as their convenience will permit.

"In conformity with this arrangement, the sale and delivery of products, the introduction of packing boxes and workmen, and the preparation for a general demolition of the Exposition are going on, in presence of visitors, whose admission will terminate on the 3d instant.

"To avoid any misapprehension, I beg to state distinctly that no disposition nor wish was shown to modify the implied contract resulting from the regulations, without the entire concurrence of all parties interested, and the conference was invited for the purpose of mutual consultation and harmonious action on that subject.

"I have also to report that the United States Commission, which for some time past has held daily sessions for the dispatch of the business intrusted to them, brought their meetings to a close with the close of the Exposition on the evening of the 31st October, and adjourned *sine die.* The proceedings were concluded by a unanimous vote of thanks to the Commissioner General, also to the commissioners who have acted successively as secretaries to the commission.

"The journal of the proceedings of the commission, and the reports which they have prepared, will be forwarded in due course to the Department of State."

CEREALS COLLECTED BY EXCHANGE.

Mr. Beckwith to Mr. Seward.

"PARIS, *December* 18, 1867.

"I have the honor to transmit herewith a catalogue of 1,442 specimens of cereals which I have collected, partly by exchanges, at the close of the Exposition.

"They are derived from each of the countries surrounding the Mediterranean, every country in Europe, and many localities of each country, and are consequently the product of great varieties of climate and soil, and being Exposition specimens are likely to be the result of the most approved methods of production, and free from noxious weeds and herbs.

"The collection comprises wheat, oats, beans, corn, millet, hemp, barley, buckwheat, linseed, rape, rye, beets, colza, mustard, carrots, clover, radish, canary, sorghum, sesame, peas, anise, timothy, &c.

"The quantity of each is unavoidably small, but probably sufficient for reproduction, and if distributed among many cultivators in suitable localities as regards climate and soil, may, perhaps, introduce some varieties which have been improved by laborious and intelligent care in the cultivation.

"The specimens are enveloped in paper sacks, each of which is numbered to correspond with the numbers and description in the catalogue, and are packed in a box addressed to the Agricultural Bureau, Washington, and forwarded by the ship Mercury, care of J. C. Derby, esq., agent, New York."

DISTRIBUTION OF MINERAL COLLECTIONS.

Mr. Beckwith to Mr. Seward.

"PARIS, *March* 27, 1868.

"SIR: Since the close of the Exposition I have delivered to the exhibitors in the mineral department all the specimens claimed by them, and taken receipts for them.

"But there remained after such deliveries a large quantity unclaimed, consisting principally of specimens, from numerous localities, of coal, iron, lead, copper, zinc, mica, franklinite, kaolin, hydraulic cement, building stones, marbles, slate, sulphates, carbonates, alum, salt, sulphur, pyrites, &c. These specimens were contributed in small quantities from many sources, and for the purpose of return to the United States they would only be worth the price of raw minerals. This would be covered by a small sum of money, not enough to defray the expenses of return or redistribution, even if ownership could be determined, which in most cases it could not be; these, and similar considerations, are, doubtless, the reason of their remaining unclaimed. In addition to the coarser minerals above named, Mr. Whitney, of Colorado, placed at my disposal a large and beautiful collection of the ores of that region, containing silver, gold, copper, lead, &c. But though the mineral value of these products for the purpose of returning them to the place of their origin is less than the expense, the question of value as specimens for analysis and study assumes a different aspect; the collection of minerals from remote regions involves considerable labor and cost, which are the proper measure of their value for the purposes alluded to, and I have thought the most useful disposition I could make of them would be to distribute them among the colleges, schools, museums, and other public institutions for the promotion of mineralogical studies, thus placing before the student classes the means of acquiring some knowledge of the quality and extent of the mineral products of the United States as a field for the young and enterprising.

"For this purpose the minerals were assorted, classified, labeled, divided, catalogued, and packed by Mr. D'Aligny, mining engineer, in separate parcels, and having obtained the requisite authorizations from the authorities, I have made, in the name of the contributor and of the government, the following gratuitous distribution of them:

	Boxes.	Specimens.
Association Polytechnique	10	70
Conservatoire Impérial des Arts et Metiers	5	68
École Centrale des Arts et Manufactures	7	162

	Boxes.	Speci-mens.
École Chrétienne de Passy	7	114
École d'Application du Génie Maritime	2	37
École Impériale de Grignan	4	43
École Impériale des Mines—department of geology, 4 boxes, 78 specimens; department of mineralogy, 2 boxes, 55 specimens	6	133
Muséum d'Histoire Naturelle	7	92
Musée de Toulouse	2	61
L'Union Centrale des Beaux-Arts, (building materials)	6	66
Collége Chaptal	3	32
Musée Royal de Bruxelles	1	71
Musée Industriel de Turin	4	56
Université de Christiania	2	48
École Polytechnique de Stockholm	2	53
Total	68	1,106

" With the minerals I distributed also a number of the General Land Office reports and maps of the mineral regions, which I had reserved for this purpose. I annex hereto a translation of the letter addressed to the minister of commerce, agriculture, and public works, which is similar to those addressed to each of the other departments; also copies of the several replies to these letters, with translations."

Mr. Beckwith to the minister of agriculture, commerce, and public works.

" PARIS, *March* 9, 1868.

"SIR : Several exhibitors in the mineralogical section of the United States exhibition having left their specimens at the disposal of the government, I have caused selections to be made of such as I deemed useful or interesting to some of the public institutions in the department of your excellency, and I hereby ask leave to present the collections thus made as follows :

"To the Central School of Arts and Manufactures, 7 boxes, containing 162 specimens of lead, coal, zinc, iron, silver, copper, emery, franklinite, &c.

" To the Couservatory of Arts and Trades, 6 boxes, containing 69 specimens of lead, coal, zinc, iron, silver, copper, emery, franklinite, &c.

" To the Imperial School of Mines, geological department, 6 boxes, containing 78 specimens of lead, coal, zinc, iron, silver, copper, emery, franklinite, &c.

"To the Imperial School of Mines, mineralogical department, 2 boxes, containing 55 specimens of lead, coal, zinc, iron, silver, copper, emery, franklinite, &c.

" To the Imperial School of Grignan, 4 boxes, containing 43 specimens of lead, coal, zinc, iron, silver, copper, emery, franklinite, &c.

" Detailed catalogues and envelopes containing duplicate catalogues addressed to the different institutions named above are herewith inclosed."

Similar letters were addressed to Messrs. Broström, consul general of Sweden and Norway, for Polytechnic School of Stockholm, Université de Christiania; Le Baron Haussmann, préfet de la Seine, Association Polytechnique, Collége Municipal Chaptal; Duruy, minister of public instruction, Muséum d'Histoire Naturelle; Minister of Marine, École Impériale des Génie Maritime; M. Guichard, president Association des Beaux-Arts appliqués a l'Industrie; Frère Philippe, École Chrétienne de Passy; S. Le Rayer, Muséum d'Histoire Naturelle de Toulouse; Minister of the Interior, Brussels, Musée Royal de Bruxelles; Le Chevalier Jervis, Muséum de Turin.

TRANSLATIONS OF REPLIES RECEIVED IN RESPONSE TO THE FOREGOING.

" PARIS, *March* 23, 1868.

" SIR : You inform me in your letter of the 9th March that several of the exhibitors in the mineralogical section of the United States of America have placed their specimens at the disposition of your government, and you have been so good as to express the desire to distribute among the establishments under the direction of this department a portion of the specimens named, such as you think of a nature to be useful or interesting to them, requesting the authorization to present the minerals in question to the institutions named in your letter, inclosing to me at the same time catalogues and letters addressed to the directors of those establishments, upon the receipt of which the packages will be delivered. I accept cordially this generous offer, for which I pray you to accept my thanks.

" I will transmit to the directors of those establishments the letters you have been pleased to address to them, and I beg you to receive the assurance of my distinguished consideration.

" *The Minister of Agriculture, Commerce, and Public Works,*
"DE FORCADE."

" PREFECTURE OF THE DEPARTMENT OF THE SEINE,
" *Paris, March* 26, 1868.

" SIR : You have been so good as to offer in the name of the exhibitors of the United States a collection of mineral specimens for the Municipal College of Chaptal, and for the Polytechnic Association, which institutions are under the patronage of the city of Paris.

" Those establishments will surely find these interesting collections useful auxiliaries in teaching, and I pray you to accept my cordial thanks for them, with assurances of my distinguished consideration.

" *Le Senateur, Préfet de la Seine,*
"HAUSSMANN."

"INSTITUTION DES FRÈRES DES ÉCOLES CHRÉTIENNES,
" *Paris, March* 14, 1868.

" SIR : I have received with your generous letter of the 9th instant the catalogue of seven cases of minerals, the produce of the United States of America, which you have been so good as to present to this institution, and have lost no time in taking possession of this rich and precious collection.

" A gift so gracious leaves us without words to express our thanks, but, penetrated with the liveliest sentiments, I offer, in the name of our institution, and in my own name, first, to you, and in your person to the great and generous government you so well represent, the tribute of our gratitude for the excellent gift we have received from your liberality.

" I pray you, sir, to accept this modest tribute, and permit me to add the homage of the respectful sentiments with which I have the honor to be
" Your obedient servant,
" *Le Supérieur Général,*
" FRÈRE PHILIPPE."

" MINISTÈRE DE L'INSTRUCTION PUBLIQUE,
" *Paris, March* 25, 1868.

" SIR : You have been pleased to express the desire, by your letter of the 9th instant, to offer to the museum, in the name of the exhibitors of the mineralogical section of the United States, an interesting collection of mineral products from that country. I pray you to accept my cordial thanks for this gift, which will add to the riches of the mineralogical department of the museum. I have requested the director of that establishment to take the necessary steps immediately to place the museum in possession of your contribution.

" Be pleased to receive the assurances of my distinguished consideration.
" *Le Ministre de l'Instruction Publique,*
DURUY.

" ÉCOLE IMPÉRIALE DE GÉNIE MARITIME,
" *Paris, March* 16, 1868.

" SIR : In reply to your letter of the 9th instant, I have the honor to inform you that I have received the specimens of coal from different localities presented to the School of Marine Engineering by the exhibitors of the United States of America.

" This collection possesses great interest for us. I hasten to express my gratitude for the destination you have given it, and pray you to convey these expressions to the exhibitors, whose names will be carefully inscribed on the specimens which they have given to the school.

" Be pleased to receive the assurance of my high consideration.
" *Le Directeur de l'École Impériale d'Application du Génie Maritime,*
" S. REECH."

"KINGDOM OF ITALY,
" *Royal Italian Industrial Museum at Turin.*

"The director of the museum returns thanks to the Commission of the United States at the Universal Exposition of 1867 for the objects named below, presented to the museum, and in consideration of their importance has directed that they form part of the collections intended to illustrate the latest progress of industry :

"Collection of mineral products from the United States of America, coming from the Universal Exposition of 1867, at Paris.

" *The Director,*
"DE VINCENZI."

"ÉCOLE IMPÉRIALE DES MINES,
" *Paris, March 25, 1868.*

" SIR : The minister of agriculture, commerce, and public works has transmitted to me the letter you did me the honor to address to me on the 9th instant, offering to the School of Mines six cases of minerals from the United States, which have been displayed at the Universal Exposition of 1867.

" I shall lose no time in taking possession of these cases, and I make haste to offer you my thanks for this interesting collection, which will be placed, with care, in the museum of the School of Mines, and a special mention will make known its origin and the names of the contributors.

" Be pleased to receive the renewed assurances of my distinguished consideration.

" *L'Inspecteur Général des Mines,*
Directeur de l'École Impériale des Mines,
"CH. COMBES."

"CONSULAT GÉNÉRAL DE SWEDE ET NORVÈGE,
" *Havre, March 11, 1868.*

" SIR : I had the honor to receive your letter of the 9th instant, announcing the generous offer which you were pleased to make, in the name of the mineralogical section of the United States of America, at the Universal Exposition of Paris, of four cases, containing samples of minerals, of which two are intended for the Polytechnic School at Stockholm, and two for the University of Christiania, in conformity with the two catalogues inclosed in your letter.

" I will immediately take possession of the cases, in conformity with your letter of address.

" Be pleased to accept the expression of my distinguished consideration.

" *Consul General for Sweden and Norway,*
"CH. BROSTRÖM."

" MINISTÈRE DE LA MARINE ET DES COLONIES,
" *Paris, March* 16, 1868.

" SIR : I have received the letter which you did me the honor to address to me on the 9th instant, proposing to present to the Imperial School of Marine Engineering two cases of mineral specimens from the Exposition of 1867.

" I hasten to inform you that I accept with lively pleasure this offer, and that I have transmitted to the director of that establishment the catalogue of samples which you have been so good as to present to my department.

" Be pleased to receive the assurance of my distinguished consideration.

" *L'Amiral Ministre Secretaire d'État de la Marine et des Colonies,*
"A. RIGAULT DE GENOUILLY."

III.

THE ACTION OF CONGRESS—ESTIMATES, APPROPRIATIONS, AND EXPENSES.

JOINT RESOLUTIONS PASSED BY CONGRESS—ESTIMATES BY THE COMMISSIONER GENERAL OF THE COST OF THE EXPOSITION—ESTIMATES, IN DETAIL, FOR TRANSPORTATION, UNPACKING, INSTALLATION, GUARDING, LINGUISTS—FOUNDATIONS AND FIXTURES FOR MACHINERY—DECORATIONS, CASES, STORAGE, LEGAL EXPENSES, ETC.—ESTIMATE OF EXPENSES OF SCIENTIFIC COMMISSION—DISCUSSION OF THE AMENDMENTS PROPOSING TO STRIKE OUT THE PROVISIONS FOR THE PAYMENT OF A PART OF THE APPROPRIATION IN COIN—REPORT OF THE ADVISORY COMMITTEE UPON THE NECESSITY FOR FURTHER APPROPRIATIONS—EXPENDITURES, REPORT FROM THE COMMISSIONER GENERAL—REPORT FROM THE AGENT IN NEW YORK.

JOINT RESOLUTIONS
IN RELATION TO THE INDUSTRIAL EXPOSITION AT PARIS, FRANCE.

" Whereas the United States have been invited by the government of France to take part in a Universal Exposition of the productions of agriculture, manufactures, and the fine arts, to be held in Paris, France, in the year eighteen hundred and sixty-seven : Therefore,

" *Be it resolved by the Senate and House of Representatives of the United States of America in Congress assembled,* That said invitation is accepted.

" SEC. 2. *And be it further resolved,* That the proceedings heretofore adopted by the Secretary of State in relation to the said Exposition, as set forth in his report and accompanying documents concerning that subject, transmitted to both houses of Congress with the President's message of the eleventh instant, are approved.

" SEC. 3. *And be it further resolved,* That the general agent for the said Exposition at New York be authorized to employ such clerks as may be necessary to enable him to fulfill the requirements of the regulations of the Imperial Commission, not to exceed four in number, one of whom shall receive compensation at the rate of eighteen hundred dollars per annum, one at sixteen hundred dollars, and two at fourteen hundred dollars.

" SEC. 4. *And be it further resolved,* That the Secretary of State be, and is hereby, authorized and requested to prescribe such general regulations concerning the conduct of the business relating to the part to be taken by the United States in the Exposition as may be proper.

"Approved January 15, 1866."

" JOINT RESOLUTION to enable the people of the United States to participate in the advantages of the Universal Exposition at Paris in eighteen hundred and sixty-seven.

" *Be it resolved by the Senate and House of Representatives of the United States of America in Congress assembled,* That, in order to enable the people of the United States to participate in the advantages of the Universal Exposition of the productions of agriculture, manufactures, and the fine arts, to be held at Paris in the year eighteen hundred and sixty-seven, the following sums, or so much thereof as may be necessary for the purposes severally specified, are hereby appropriated out of any money in the treasury not otherwise appropriated:

"*First.* To provide necessary furniture and fixtures for the proper exhibition of the productions of the United States, according to the plan of the Imperial Commissioners, in that part of the building exclusively assigned to the use of the United States, forty-eight thousand dollars.

"*Secondly.* To provide additional accommodations in the Park, twenty-five thousand dollars.

"*Thirdly.* For the compensation of the principal agent of the exhibition in the United States, at the rate of two thousand dollars a year—*Provided*, That the period of such service shall not extend beyond sixty days after the close of the exhibition—four thousand dollars, or so much thereof as may be found necessary.

"*Fourthly.* For office rent at New York; for fixtures, stationery, and advertising; for rent of storehouse for reception of articles and products; for expenses of shipping, including cartages, &c.; for freights on the articles to be exhibited, from New York to France; and for compensation of four clerks, in conformity with the joint resolution approved on the fifteenth of January, eighteen hundred and sixty-six, and for contingent expenses, the sum of thirty-three thousand seven hundred dollars, or so much thereof as may be found necessary.

"*Fifthly.* For expenses in receiving, bonding, storage, cartage, labor, and so forth, at Havre; for railway transportation from Havre to Paris; for labor in the Palace; for sweeping and sprinkling compartments for seven months; for guards and keepers for seven months; for linguists (eight men) for seven months; for storing, packing boxes, carting, and for material for repacking; for clerk-hire, stationery, rent, and contingent expenses, the sum of thirty-five thousand seven hundred and three dollars, or so much thereof as may be found necessary.

"*Sixthly.* For the traveling expenses of ten professional and scientific commissioners, to be appointed by the President, by and with the advice and consent of the Senate, at the rate of one thousand dollars each, ten thousand dollars, it being understood that the President may appoint additional commissioners, not exceeding twenty in number, whose expenses shall not be paid; but no person interested, directly or indirectly, in any article exhibited shall be a commissioner; nor shall any member of Congress, or any person holding an appointment or office of honor or trust under the United States, be appointed a commissioner, agent, or officer, under this resolution.

"SEC. 2. *And be it further resolved*, That the governors of the several States be, and they are hereby, requested to invite the patriotic people of their respective States to assist in the proper representation of the handiwork of our artisans, and the prolific sources of material wealth with which our land is blessed, and to take such further measures as may be necessary to diffuse a knowledge of the proposed exhibition, and to secure to their respective States the advantages which it promises.

"SEC. 3. *And be it further resolved,* That it shall be the duty of the said general agent at New York, and the said Commissioner General at Paris, to transmit to Congress, through the Department of State, a detailed statement of the manner in which such expenditures as are hereinbefore provided for are made by them respectively.

"Approved July 5, 1866."

"A RESOLUTION to provide for the exhibition of the cereal productions of the United States at the Paris Exposition in April next.

"*Resolved by the Senate and House of Representatives of the United States of America in Congress assembled,* That the Commissioner of Agriculture be, and he is hereby, instructed to collect and prepare, as far as practicable, and with as little delay as possible, suitable specimens of the cereal productions of the several States of the Union, for exhibition at the Paris Exposition, and forward the same in proper order and condition for shipment to J. C. Derby, agent of the United States government for the Paris Exposition, at New York: *Provided,* That it shall require no further appropriation from the public treasury.

"Approved January 11, 1867."

"A RESOLUTION supplementary to other joint resolutions to enable the people of the United States to participate in the advantages of the Universal Exposition at Paris in 1867.

"*Resolved by the Senate and House of Representatives of the United States of America in Congress assembled—*

"I. That the commission of the United States at the Universal Exposition to be held at Paris in the year eighteen hundred and sixty-seven shall consist of the Commissioner General and honorary commissioner, whose appointment was approved by the joint resolution of January twenty-two, eighteen hundred and sixty-six; also of the thirty commissioners, whose appointment was provided for by the joint resolution of July five, eighteen hundred and sixty-six, and of twenty commissioners, whose appointment is hereinafter provided for.

"II. That the Commissioner General shall be the president of the commission thus constituted, with a vote on all questions that may arise.

"III. That the commission shall meet at Paris as early as possible before the opening of the Exposition, upon the call of the Commissioner General, and when properly organized, shall make such rules and regulations as may be necessary for efficient action, with power to elect a vice-president from their own number, who, in the absence of the Commissioner General, shall preside at all meetings of the commission, and to appoint committees and chairmen of groups.

"IV. That the commission may designate additional persons, not exceeding twenty in number, being citizens of the United States, known to be skilled in any branch of industry or art, who are hereby authorized to attend the Exposition in behalf of the United States, as honorary commissioners without compensation.

"V. That the commission may employ a secretary and clerks for the commission, the necessary scientific assistants and draughtsmen, and may engage suitable rooms for the commission.

"VI. That no commissioner shall act as agent for the show or sale of any article at the Exposition, or be interested, directly or indirectly, in any profits from any such article.

"SEC. 2. *And be it further resolved*, That fifty thousand dollars, or so much thereof as may be necessary for the purposes severally specified, are hereby appropriated out of any money in the treasury not otherwise appropriated.

"For additional freights from New York to Havre.

"For transportation and freight from Havre to Paris.

"For return freight of articles owned by the United States or lent to the government by individuals.

"For marine and fire insurance on the articles thus lent.

"For additional steam-power at Paris, in the Palace and the Annex, or supplemental building, and in grounds adjacent.

"For the exhibition of machines, agricultural and other, and for the erection of buildings to illustrate the education and agriculture of the United States, and for the collection of specimens of agricultural productions under the joint resolution for that purpose.

"For the necessary expense of collecting, classifying, labeling, and packing mineralogical and metallurgical specimens to complete the exhibition of the mineral wealth of the United States.

"For the necessary expense of laborers and extra service in the offices at Paris and New York, and for the expenses of a secretary, clerks, scientific assistants, and draughtsmen, rooms, and other incidental expenses of the commission.

"SEC. 3. *And be it further resolved*, That it shall be the duty of the general agent at New York, and of the Commissioner General at Paris, to transmit to Congress, through the Department of State, a detailed statement of the manner in which the expenditures herein authorized are made by them respectively.

"Approved March 12, 1867."

ESTIMATES OF COST OF THE EXPOSITION.

Mr. Beckwith to Mr. Seward.

"PARIS, *January* 11, 1866.

"SIR: The joint resolutions on the subject of the Exposition, presented to Congress on the 21st December, provide for the expenses of a scientific commission, the freight of products to and from Europe, and the expenses of the agency in New York.

"I beg to suggest the expediency of introducing into these resolutions a similar distinct provision for the expenses on this side.

"The service here will continue longer than in New York, and will be more expensive, because it will require more employés.

"The principal items will be, rent and expenses of an office in which the business of the Exposition can be transacted, and in which the regular meetings and work of the scientific commission can be accommodated; the wages of clerks; the cost of stationery, printing, fuel, lights, &c.; and the wages of an engineer architect, for the constructions to be made.

"Notarial expenses: all the work should be put under notarial contracts in May or June next, at the latest. From the nature of the constructions and the distribution of the work, the contracts will be numerous; and unless put in notarial and legal form, so that the work can be pressed under penalties, it will not be done in time, and worse still, there will be numerous disputes and troublesome lawsuits about it, which should be avoided.

"The expenses of warehousing and labor at Havre and inland transportation, in and out, will be considerable.

"The reception of the products at Paris, and the unpacking and placing for exhibition, will take place in winter, when the days in this latitude are short, and the weather stormy and uncertain.

"The work can go on only by daylight; the distance from the entrepôts of the railway to the Champ de Mars is three miles, directly across the city; the expenses of cartage will not only be considerable, but the work must be carefully looked after throughout, or there will be much damage to property, and no redress.

"The item of cab hire will of necessity be considerable, and will be an economy as being less expensive than more clerks, which will otherwise be indispensable.

"The labor of placing machinery to be worked by motive force, or not worked, and the labor of unpacking and repacking, and of other products, and handling and placing of them for exhibition, must be done by a class of competent laborers, under the constant direction of the engineer architect before spoken of.

"It is impossible to compute in advance, with any useful accuracy, these incidental expenses.

"But the item of cost of installation, (fixtures, show-cases, &c.,) as stated in my letter of the 23d November, cannot, I think, be brought under $48,000, in Paris.

"The installations are the work alluded to, which must be put under contracts in May and June, and the contracts must be supervised in the making by the enginer architect, who alone is familiar with the technicalities requisite in such contracts.

10 P E

" My estimates for this item are based on a careful study of the details of the cost of similar work in London and in Paris, supplied to me by the Imperial Commission and used for their own estimates; and I may add, that the appropriations made in this department by Switzerland, as I am informed, amount to $80,000, to cover what I estimate at $48,000, it being my intention to dispense with the ornamental that is not useful and necessary.

" I have not thought it necessary to trouble you with even this much of detail, and my apology for doing so now is, that on reading the draught of resolutions referred to, it occurs to me that you may think it desirable to ask for the introduction in these resolutions of a more distinct provision for the expenditures on this side, which involve also exchanges, and for all which you will have occasion subsequently to authorize the disbursements, which must go on simultaneously with the work.

" I beg to repeat the opinion I have before expressed, that the exhibition cannot cost under $200,000, nor do I think it can exceed $400,000; probably it will not vary much from $300,000, and in my judgment it would be better not to undertake it than to do so on a less scale, which I am confident would result in disappointment and dissatisfaction.

" As soon as the final action of Congress is known, I will take the liberty of suggesting some regulations and instructions in that conformity, applicable chiefly to the work on this side, and to the disbursements and accounts, which suggestions, I trust, may recommend themselves to your consideration by their fitness in a manner to obtain your sanction."

Mr. Beckwith to Hon. N. P. Banks.

" PARIS, *February* 21, 1866.

" I have taken the liberty of addressing a letter to you as chairman of the Committee on Foreign Relations, presenting in considerable detail my estimates of the expenses of the exhibition.

" You will, I think, be convinced, if you had any doubt, that the appropriation proposed in Congress is inadequate to the occasion, and that the subject deserves reconsideration. Judging from the debates in the House which have reached me, the Exposition, in the estimation of that body, is not of much importance to the United States. I inclose herewith a list of the names of the English committee, present at their last meeting in London. Most of them will be familiar to you; they are those of men most prominent in politics, in industry, in the sciences and the arts.

" Similar organizations exist in most of the countries of Europe, and they indicate the importance which is on this side ascribed to the Exposition. These appreciations may be exaggerated, or they may not be, but there is no feature in the civil affairs of Europe so striking as the wonderful and steady growth of the commerce and wealth of the leading nations; and whether or not they overrate the value of the Exposition, I feel convinced that no country, not even France, can derive so much benefit from it as the United States, in every sense, scientific, industrial, commercial, and political.

"Under the circumstances which are now past, the government might perhaps have made or found reasons for declining to accept the invitation of the French government, though I think that would have been impolitic, and in the end unsatisfactory to the country. But Congress has accepted the invitation, and it now remains to consider the provisions suitable for it. The new feature of this Exposition is, that the producers of the different countries appear only through their governments.

"The government of the United States cannot come forward and present the products of the nation, scientific, industrial, mineral, and agricultural, in a manner satisfactory to itself and to the country, without the expenditure of a considerable sum of money; it is to be done not only in presence of the governments of Europe, but in competition with them, and they are doing their best in the same way for their own people.

"At a later period, when the entire press of Europe, able and powerful as it is in its influence on public opinion and affairs, becomes occupied with the Exposition, as it certainly will be, the importance of it in every point of view will become more apparent to those who at present have not the time nor occasion to reflect upon it.

"But in addressing you it would be superfluous, and in me presumptive, to dwell upon the numerous and interesting considerations which invite us to the Exposition."

Mr. Beckwith to Mr. Seward.

"PARIS, *February* 21, 1866.

"I have the honor to acknowledge receipt of your letter of the 29th January; also, the letter of the acting Secretary of State, of the 15th January.

"I have thought it might be useful to lay before the chairman of the Committee on Foreign Affairs, in the House of Representatives, estimates of the cost of the exhibition in much greater detail than those I have heretofore submitted, and I have accordingly addressed him on the subject.

"But it is impossible for me to know what the state of the business will be on the receipt of this communication. I therefore take the liberty to inclose it herewith, and leave it open that it may be read and sent to the address, or suppressed, as you may please to decide will be best."

Mr. Beckwith to Hon. N. P. Banks.

"PARIS, *February* 21, 1866.

"I take the liberty of submitting for your consideration the estimates n detail, which show the probable cost of the exhibition of 1867.

"They are based upon the results of previous exhibitions, upon local inquiries, and upon conjecture in regard to the quantity of products to be transported.

"The surface of ground within the Palace to be actually covered by products is about ten thousand square feet. If this area were covered to the height of four feet, the cubic contents would be forty thousand cubic feet, equal to one thousand tons of forty cubic feet each.

"I omit calculations relating to the groups in the Park, and assume one thousand tons as a convenient figure. The exact quantity is of small moment; the cost of transportation is but a small part of the whole cost; and the general expenses will be about the same, whether the quantity of freight be five hundred, or one thousand, or fifteen hundred tons.

"Transport on the railways is regulated by tariff, and it is impossible to know in advance what proportion will fall under the different rates. I therefore assume the medium rate from Havre to Paris, which is sixteen francs per ton:

1st. 1,000 inward, at sixteen francs, 16,000 francs, or at 5 francs	$3,250
Outward..	3,250
Landing expenses, bonding, storage, cartage, labor, and commissions for forwarding inward, 7½ francs................	1,500
Outward..	1,500
Paris, cartage a distance of about three miles from the depot to the Champ de Mars and labor, 6 francs................	1,200
Outward..	1,200
	11,900

"Unpacking in the Palace will commence on the 15th January, 1867, and end on the 30th March, 1867, a period of one and a half months. About the same time will be required for repacking and dispatch, making together three months. The number of laborers required for this work, which can only be done by daylight, will be increased by the shortness of the days in this latitude at that season of the year. The chef de service of the Imperial Commission, who superintends this work in the French section, and who has had great experience in the work both in London and Paris, informs me that I will require thirty laborers for six weeks inward and the same outward.

"The labor must be directed by an engineer architect capable of placing machinery in connection with motive force for action, and who is also familiar with the architectural plans and the arrangement of groups and classes for display. The greater part of the laborers must be those who are accustomed to work among machinery, &c., or in warehouses and shops, accustomed to packing and unpacking and the placing of varieties of fabrics and products for display. There will be a great demand for that kind of labor, and wages will probably average five francs per day :

2d. 90 days for inward and outward, at 5 francs = 450 × 30 = 13,500 francs...... .... $2,700

"During the Exposition, which will last seven months, the compartments must be sprinkled and dusted every morning and kept in proper condition :

3d. 214 days, at 10 francs, 2,140 francs..................... 428

"There must be a guard in each compartment during the visiting hours to prevent thefts and damages and report any misconduct of visitors. The French government provides police, &c., day and night, but does not respond for thefts and damages by visitors. This service will require seven men, which I estimate at 5 francs per day, and one man at 10 francs per day.

4th. 214 days × 7 = 1,498 days, at 5 francs, 7,490 ; 214 days, at 10 francs, 2,140—9,630 francs......................... 1,926

"It is usual for the Commissioner General to provide a number of linguists capable of giving explanations of machinery and other products to visitors. Without this many of the most curious and interesting objects cannot be understood ; important qualities are unobserved and the intended diffusion of information fails. I propose for this service two men in Group VI, and one man in each of the other groups, making—

5th. Eight men, 214 days = 1,912, at 10 francs per day, 19,120 francs.. 3,824

"This number will not be sufficient for the work, but I leave to exhibitors to give such explanations of their own works as they may be able to do, and to make further provision by interpreters if they think their interests require it ; also to provide the men for attending machines in action and the expenses of materials, &c., consumed in the working.

"The cost of foundations and fixtures for machinery, the cost of show cases, tables, and other fixtures or installations I take from the lowest average cost of similar work in the London and Paris exhibitions, excluding therefrom all decorations. The data for this have been supplied me by the Imperial Commission, and I cannot reduce the estimates I first reported :

6th.. 48,000

"Expensive decorations will be used in all other sections, and their absence from ours I do not expect will be wholly satisfactory to the Imperial Commission, to the public, nor to our own people.

"I could provide very respectable decorations of the kind most used for, I think, $10,000, but I have wholly omitted this item in my estimates.

"Packages (which are expensive here) will all require to be removed to a considerable distance to find storage for them for seven months and then returned to be used again.

7th. Cost of cartage, storage, recartage, and new materials for repacking... $1,000

"The contracts for the above works (installations) should be made in May and June, 1866, and should all be notarial and drawn by the aid of the engineer architect, who is familiar with the forms and nomenclature. Unless the contracts are so made and carry penalties which can be enforced, the work will not be ready in time; it will not conform to the agreements and the wants; and there will be disputes, references, and law-suits as well as disappointments.

"The notarials, &c., I hope to include in the above estimate, though it is not in the originals. The information relating to the preceding has been derived chiefly from the Imperial Commission and from the tariff of railways.

"The following estimate of office expenses is from Messrs. Munroe & Co., a respectable banking-house:

Office rent per annum...............................	$1,000
Fire and lights....................................	250
Porter...	500
Stationery...	300
Postages...	100
City taxes...	200
Office furniture[1]................................	600
Printing...	150
	3,100

8th. One year and nine months................................ 5,425

"The force required will be an accountant, a corresponding clerk, a copyist, and an out-door clerk:

9th. One at $1,600, one at $1,400, and two at $1,200, $2,400—$5,400, one year and nine months............................ 9,450

"An engineer architect familiar with French and English and acquainted with Paris, to superintend, the work as before stated, is indispensable. The four clerks must also be familiar with two languages at least, and acquainted with business in Paris. The cost of an engineer architect I estimate at $3,000 per annum.

10th. One year and nine months.............................. 5,250

11th. The extra expenses of a building in the Park, such as I have recommended, stands in my estimate at.............. 25,000

	114,903

[1] Probably can be sold for two-thirds of the cost.

" But I hope the report from Mr. Derby of the space required will enable me to reduce this estimate.

" These are the estimates of expenses on this side proper to the exhibition, and to this must be added exchange :

12th. Which with gold at 142 will be.................		$48,301
		163,204
A scientific commission of ten persons, and employed for one year, they paying their own traveling and all other expenses, I estimate at $6,000 each......	60,000	
They will require local professional assistants, as stated more fully in my letter on the subject to the Department of State, and I estimate for that purpose..........	$30,000	
	90,000	
Exchange as above...............	37,800	
13th...........		127,800
		$291,004

" I have not included herein sea freights, which I estimate at $7,000 each way in sailing vessels; nor the expenses in New York, to be determined by the length of time it will be requisite to keep that office open.

" Neither is anything here put down for contingent expenses, which are likely to arise from unforeseen causes.

" It is not probable the actual expenditures will run exactly with the estimates of items above; some of them will cost more and others less; but if the appropriations admit of transfers, as they should do, from one item to another, an appropriation of $300,000 I still hope will cover the cost of the work if carried out as projected; and I feel confident that if it is so carried out the cost cannot much exceed nor fall much short of that sum.

" But if reductions are to be made they will fall on the estimates for the scientific department. The estimates for the expenses on this side, which belong to the exhibition proper, cannot in my judgment fall below the sum above named, $163,000."

Mr. Beckwith to Mr. Seward.

" PARIS, *February* 24, 1866.

" The communication which I had the honor of addressing through the Department of State to the chairman of the Committee on Foreign Affairs in the House of Representatives, on the 21st instant, contained a statement in detail of the estimated expenses of the exhibition.

" To bring the subject before you in a convenient form I annex hereto, marked No. 1, a condensed statement of the items of expense enumerated in the estimates alluded to.

"For convenience of reference I also annex paper No. 2, which contains an extract from the letter which I had the honor of addressing to you on the 30th January, relating to the duties which will devolve on the Special Agent, on the Commissioner General, and on the Scientific Commission, separately and jointly.

"In framing the regulations for the exhibition and the instructions which you may please to send me, I solicit your consideration of the observations in paper No. 2, and would remark in continuation, that the Palace will be finished on the 1st December next, and ready to receive the fixtures and furniture, which must be previously constructed and prepared to be placed at that date.

"The great amount of work of this kind to be done for the different nations will create a large demand for that class of labor, and it will be necessary to make the contracts and commence the work early.

"If it be possible, as I trust it will be, for Mr. Derby to give me his definitive plans in time, I propose to make the contracts in May or June, and to begin the work; and in any event to do so at the earliest moment after the plans reach me."

No. 1.—Abstract of expenses.

Landing expenses at Havre, bonding, storage, cartage, labor, forwarding and commissions, 1,000 tons, inward $1,500, outward $1,500	$3,000
Railway transport to Paris, inward $3,250, outward $3,250	6,500
Paris, labor at depot and cartage three miles, inward $1,200, outward $1,200	2,400
1st	11,900
2d. Labor in the Palace, inward 1½ months, $1,350, outward 1½ month, $1,350	2,700
3d. Sprinkling and sweeping compartments, 7 months	428
4th. Guard and keepers, 8 men, 7 months	1,926
5th. Linguists, &c., 7 months (8 men)	3,824
6th. Show cases, tables, and other fixtures	48,000
7th. Storage of packing boxes, 7 months, cartage and material for repacking	1,000
8th. Office expenses per annum $3,100, 1 year 9 months	5,425
9th. Wages of clerks per annum $5,400, 1 year 9 months	9,450
10th. Engineer architect per annum $3,000, 1 year 9 months	5,250
11th. Estimate for Park building	25,000
12th. Scientific Commission, 10 men, 1 year	60,000
Assistants	30,000

Mr. Beckwith to Mr. Seward.

"PARIS, *March* 14, 1866.

"An attentive perusal of the resolution of Congress No. 52, appropriating money for the exhibition, suggests the following remarks:

"If the sum appropriated by the resolution were adequate to the wants, it would be necessary to ask at once for a modification of its provisions. Without this change the probable intentions of Congress could not be executed.

"The expenditures required by the exhibition will not agree with the definitive appropriations under each head named in the resolution, while many other expenses equally unavoidable are not named nor provided for, either specifically or in general, by a contingent fund.

"The business being new, and in many things without precedents for a guide in making the estimates, they are not likely to agree exactly in details with actual expenditures; many of the elements of cost are fluctuating as regards supply and variable in price, and although the total expenditure may not exceed the total estimate, if the appropriations are subdivided and restricted to the different heads which are based on such estimates in detail, they will not be found to correspond with actual wants in practice.

"But the most important feature of the resolution No. 52 is, that with the exception of the provision for clerks on this side, and a small contingent fund, no mention is made or any provision whatever for expenditures under different heads, proper to the exhibition itself, which will amount to not less than $160,000, United States currency.

"In the detailed estimates which I had the honor to transmit on the 21st February, the items of expense thus omitted to be provided for are enumerated, and it will be seen, I doubt not, that in principle they are unavoidable.

"It may hereafter appear that the estimates for some of these details are in excess, and that others are deficient, and again others may arise which have not been foreseen; but as they are in general based upon existing prices and upon the opinions of those who have had most experience in exhibitions, the total cost is not likely to vary much from the total of the estimates.

"If, therefore, Congress should make the necessary provision for the exhibition, they will greatly facilitate the work by providing at the same time for the transfer, in case of need, from the appropriations under such heads as may prove to be in excess, to those which may prove to be deficient, or to such as may have been unforeseen, provided that the total expenditure be kept within the total appropriation.

"Some discretion of this kind appears necessary in this case to avoid delays and embarrassments in conforming to the provisions of the appropriation bill, and to facilitate the work which Congress has in view."

Mr. Beckwith to Mr. Seward.

"PARIS, *July* 18, 1866.

" I learn by the mail just arrived that the report of the committee of conference on the appropriation for the exhibition has been accepted and passed by the House of Representatives, and is likely to be passed by the Senate.

"The amendments which strike out the provisions for the payment of a part of the appropriations in coin, and the high rate of exchange, make it necessary for me to solicit your consideration of the situation in which the exhibition is thus placed.

" The appropriations to be expended on this side are $48,000, $25,000, and $35,703; total, $108,703.

" The estimates for these expenditures were made upon a careful study of details, and in my judgment they do not admit of reduction.

" The product of $108,703 in coin, at the usual valuation at five francs per dollar, would be 543,515 francs. The same sum at the rate of exchange current at the last date, (3d July,) three francs, would produce 364,155 francs, which is a reduction of 179,360 francs, or $35,872—thirty-three per cent. on the gross sum, or forty-nine per cent. on the net sum.

" If the effect of this reduction were merely to diminish the proposed exhibition of products one-third, it would necessarily exclude so many important products that our exhibition would lose its character of universality and be no longer in any proper sense an exhibition of the products of the United States.

" But this would not be the whole effect of the reduction in the appropriations. The exclusion of one-third of the products would not produce an economy of one-third in the expenses. The organizations and preliminary expenses on both sides will remain unchanged, or nearly so, and the expenses of administration for seven months that the Exposition will remain open cannot be materially reduced. All the expenses will be unavoidably the same, or nearly the same, for two-thirds as for the whole of the products proposed to be exhibited.

" The only economy resulting from a reduction of the quantity of products will be in the cost of transport and of a part of the installations, and after examining this in detail I am satisfied that a reduction of nearly two-thirds in the quantity of products is required to produce the saving of $35,872, necessary to keep the expenditures within the appropriations.

" The space we have undertaken to fill in the Exposition Palace is small for the United States as compared with other nations, and a reduction of two-thirds, or even of one-half, will leave one-half that space vacant.

" An exhibition so limited in quantity, and so imperfect in its composition, will retain but small interest for the public, and smaller value for our producers; its cost will remain large in proportion to its importance, while the chief design of the undertaking and the hopes of a large portion of our producers will be nearly ruined.

"If, on the other hand, the work be continued as it has been begun, it will be a pretty fair representation of our products, and though still limited in quantity and variety it will be respectable and useful, but it will be barely installed and opened to the public when the appropriations will be exhausted, and to carry it on for seven months to its close without funds will only bring it to bankruptcy.

"Either of these results would be disagreeable and unsatisfactory to the country and to the government if not even embarrassing to the latter, and the probability of such a conclusion renders it incumbent on me to make these suggestions in advance, which I do with great reluctance.

"I am not informed whether any law or usage exists which carries exchange with the payments ordered by Congress and made by government in distant places or foreign countries; but if there is no method of covering the deficiency resulting from the reduction of the appropriations or the exchange, I respectfully suggest for your consideration whether the inadequate provision of Congress does not render the proposed exhibition in effect impossible.

"The deficiency, I believe, could be covered by an appeal to exhibitors and the public if I had time for it, but the work on this side is already so far in arrear that it will require my best efforts through the whole of the hot season to bring it up, which puts it out of my power to attempt that method of making up the deficiency. I shall therefore proceed with the preparations on their present footing, (in the absence of your instructions to the contrary,) but awaiting with solicitude the reply you may please to make to these suggestions."

Mr. Derby to Mr. Seward.

"NEW YORK, *January* 14, 1867.

"SIR : The report of my advisory committee, which was forwarded to you, calls for a further appropriation by Congress for freight, installations, insurance, additional steam power, and other necessary expenses; that is to say—

For additional installations of farm-house, school-houses, and western laborers' cottage, with all the appurtenances, including transportation; also for Palace and Annex	$25,000
Additional steam power	10,000
Additional freights from Havre to Paris	15,000
Return freights for articles owned by the government and individuals, including works of art loaned and not for sale, per steamer	20,000
Additional freights by steamer on products too late for sailing vessels, including the minerals and metals of nearly all the States and Territories	20,000
Necessary expenses for laborers and extra help in the Paris and New York offices	10,000
Say one hundred thousand dollars	100,000

"In addition to this sum a further appropriation for the United States Commissioners is necessary for the actual expenses incurred in preparing suitable reports of the results of the Exposition.

"The original minimum estimate of United States Commissioner General Beckwith, was $300,000 for the necessary expenses of the Paris Exposition. The additional amount now asked for approximates that amount, and is really necessary, or so much of it as is called for by legitimate expenditures on account of the Exposition."

REPORT OF EXPENDITURES.

Mr. Beckwith to Mr. Seward.

"PARIS, *April* 30, 1868.

"The several resolutions of Congress appropriating money for the Universal Exposition at Paris, 1867, require statements to be transmitted to that body, through the Department of State, of the manner in which the money has been expended.

"In conformity therewith I have transmitted to the Department quarterly statements of receipts and expenditures, with vouchers in detail for each disbursement.

"I have now the honor to inclose an account current which is the index and complement of the quarterly statements, and presents in a condensed form the receipts and expenditures under each of the twenty-one heads of account, for each quarter, commencing in 1866 and extending to the 30th April, 1868 ; on referring to the entries in this abstract, to the label, letter, and date corresponding to the entry, all the details and vouchers for that item will be found together.

	Francs.
The gross sum which I have received from the Department, being the proceeds of credits on London for £18,000, amounts to	452, 095. 00
From other sources, (enumerated in detail,)	4, 935. 05
Total	457, 030. 05
And the gross expenditures reach the sum of	453, 630. 68
Leaving a balance in my hands of	3, 399. 37

To be returned to the department minus some small items which remain to be settled.

The total receipts in round numbers, stated in dollars, at five francs, amount to	$91, 406
Expenditures	90, 726
Balance	680

"With regard to fixtures, furniture, materials, &c., for the Exposition, there were two methods of obtaining them : First. Upon plans furnished by commissioners to contractors, who undertook at prices agreed upon, in consideration of which the furniture became the property of the commissioners. Second. Upon plans furnished by commissioners to contractors who engaged at prices agreed upon, in consideration of which the furniture reverted at the close to the contractor.

"By the first method the risk and chances of resale were assumed by the commission, and in the second method by the contractor.

"Previous to the date when it became practicable to close our contracts, nearly all those of other nations had been closed and the works well advanced. Most of them being on the basis of the first method, made it for the interest of contractors, who would be buyers more than sellers, to combine at the close to put down instead of supporting prices. A knowledge of these circumstances, and an examination of the various bids of contractors to supply the work, with details of prices presenting the option, left no room for doubt that the second method of contract would be best, and it was adopted.

"There remained, consequently, at the close but a small quantity of tools and materials for sale which are accounted for in detail, the proceeds amounting to 370 francs."

Mr. Beckwith to Mr. Seward.

"PARIS, *June* 19, 1868.

SIR: I have the honor to transmit herewith my final account of expenditures, (with vouchers,) commencing with the balance of 3,399.37 francs, remaining in my hands on the 1st of May, showing a disbursement since of 961.15 francs, and a balance of 2,438.22 francs, which, to close the account, I have transmitted to the credit of the Department with Messrs. Baring Brothers & Co., in the sum of £96 15s. 1d., all which I trust will be found correct and satisfactory.

The proceeds of my drafts on Messrs. Baring amounted, as shown in my general account 1st May, with receipts from other sources therein enumerated, to francs		457, 030. 05
Disbursements to 1st May	453, 630. 68	
Present account	961. 15	
		454, 591. 83
Balance remitted to Messrs. B. B. & Co., francs		2, 438. 12

"Showing a total disbursement in gold, at five francs to the dollar, of $90, 918 33."

EXPENSES OF THE NEW YORK AGENCY.

The following is a statement of expenditures by J. C. Derby, agent of the United States government at New York, for the Paris Universal Exposition of 1867, transmitted April 1, 1868, to the Secretary of State, as directed by joint resolutions of Congress, approved July 5, 1866, and

May 12, 1867, vouchers in detail for which were forwarded to the Department of State :[1]

Salary of general agent, two years, at $2,000..	$4, 000 00	
Services previous to passing of bill, say from October 1, 1865, to April 1, 1866..............	965 00	
Salaries of clerks, not exceeding four in number, in 1865, 1866, and 1867...................	7, 514 17	
		$12, 479 17
Labor, storage, rent, cartage, fixtures, stationery, &c......		15, 267 00
Advertising.........!................................		3, 290 12
Safe....................................		250 00
Freight from New York to Havre........................		15, 726 27
Marine and fire insurance on works of art................		7, 460 77
Total.....		54, 473 33

[1] This statement, together with the financial reports of Mr. Beckwith, were transmitted to Congress July 13, 1868, and are printed in Ex. Doc. No. 334, 40th Cong., 2d session.

IV.

PUBLICATION OF THE REPORTS.

STATEMENT. OF THE AUTHORITY UNDER WHICH THE REPORTS HAVE BEEN PRINTED—
PUBLICATION IN A SEPARATE FORM, AND REASONS THEREFOR—GROUPING OF THE
REPORTS IN VOLUMES—LIST OF THE REPORTS BY TITLE, ARRANGED ACCORDING TO
SUBJECTS—ALPHABETICAL LIST OF THE AUTHORS OF REPORTS.

RESOLUTION AUTHORIZING THE PUBLICATION.

In the Senate of the United States, March 3, 1868, it was—

"*Resolved*, That the Secretary of State be, and he is hereby, authorized to have the reports of the Commissioners of the United States to the Paris Exposition printed at the Congressional Printing Office and laid before the Senate in a printed form; and that, in addition to the usual number, there shall be printed four thousand extra copies for the use of the Senate, and fifteen hundred copies for the use of the State Department, the reports to be bound separately or together as the Secretary of State may direct."

On the 8th of April this resolution was suspended, and the Committee on Printing was directed to inquire into the amount of material to be printed under the foregoing resolution, the cost of the same, and also whether such publication would involve the preparation and the printing of any maps, plates, or illustrations.

The Department having been called upon for this information, it was furnished in detail to the committee with estimates of the cost of engraving the illustrations, and it was ordered by the Senate that the printing should proceed under the original resolution.

The printing of the report presenting a "Brief General Survey of the Exposition," specially required by the instructions issued August 20, 1866, was commenced, and, inasmuch as at that time several of the most important of the reports were still in the hands of the authors undergoing revision, or not ready for publication, it was decided to print each report independently, and to issue a part of the whole number ordered in this separate form as soon as printed without waiting for the completion of the whole series, or of a sufficient number of reports to form a volume. Upon this plan each report has been printed with distinct paging and title, and one thousand copies of each in paper covers have been delivered to the Senate and five hundred copies of each, in the same form, to the Department of State. The remaining copies were reserved for the final grouping and binding in volumes. By this means the earliest possible publication was secured for each report, and their separate form permitted a wider and more economical distribution.

The following list gives the titles of the reports, alphabetically arranged according to the subjects. The reports all bear the imprint of the Government Printing Office and the year of publication. This imprint is omitted in the list, but the exact date of publication is supplied. The copies of reports not separately issued as above have been grouped together and bound in six volumes, under the general title of "Reports of the United States Commissioners to the Paris Universal Exposition of 1867; published under the direction of the Secretary of. State, by authority of the Senate of the United States."

A list of the reports, in the order in which they are grouped in volumes, will be found at the end of Volume I and of Volume VI.

LIST OF THE REPORTS, BY THEIR TITLES, ARRANGED ALPHABETICALLY ACCORDING TO THE SUBJECTS.

ARTS.—Machinery and processes of the industrial arts and apparatus of the exact sciences, by Frederick A. P. Barnard, LL. D., United States Commissioner.—pp. ix, 669. August 4, 1869. (In volume iii.)

ASPHALT AND BITUMEN.—Report on asphalt and bitumen, as applied to the construction of streets and sidewalks in Paris; also to terraces, roofs, &c., and to various products in the Exposition of 1867; with observations upon macadamized streets and roads, by Arthur Beckwith, Civil Engineer.—pp. 31. January 15, 1869. (In volume iv.)

BEET-SUGAR.—The manufacture of beet-sugar and alcohol and the cultivation of sugar-beet, by Henry F. Q. D'Aligny, United States Commissioner.—pp. 90. November 3, 1869. (In volume v.)

BÉTON-COIGNET.—Report on Béton-Coignet, its fabrication and uses—construction of sewers, water-pipes, tanks, foundations, walls, arches, buildings, floors, terraces; marine experiments, &c., by Leonard F. Beckwith, Civil Engineer.—pp. 21. January 15, 1869. (In volume iv.)

BIBLIOGRAPHY.—Bibliography of the Paris Universal Exposition of 1867, by William P. Blake, Commissioner of the State of California to the Paris Exposition of 1867. In press, April, 1870.

BUILDINGS.—Report upon buildings, building materials, and methods of building, by James H. Bowen, United States Commissioner.—pp. 96. September 28, 1869. (In volume iv.)

CEREALS.—Report on cereals: The quantities of cereals produced in different countries compared, by Samuel B. Ruggles, Vice-President of the United States Commission. The quality and characteristics of the cereals exhibited, by George S. Hazard, United States Commissioner.—pp. 26. September 28, 1869. (In volume v.)

CHEMISTRY.—The progress and condition of several departments of industrial chemistry, by J. Lawrence Smith, United States Commissioner.—pp. ix, 146. September 7, 1869. (In volume ii.)

CIVIL ENGINEERING.—Civil engineering and public works, by William P. Blake, Commissioner of the State of California.—pp. 49. March 5, 1870. (In volume iv.)

CLOTHING.—Report on clothing and woven fabrics; being classes twenty-seven to thirty-nine of group four. By Paran Stevens United States Commissioner. In press, April, 1870. (In volume vi.)

COAL.—Report on the manufacture of pressed or agglomerated coal, by Henry F. Q. D'Aligny, United States Commissioner.—pp. 19. October 8, 1869. (In volume v.)

COTTON.—Report upon cotton, by E. R. Mudge, United States Commissioner, with a supplemental report by B. F. Nourse, Honorary Commissioner.—pp. ii, 115. June 28, 1869. (In volume vi.)

EDUCATION.—Report on education, by J. W. Hoyt, United States Commissioner.—In press, April, 1870. (In volume vi.)

Report on school-houses and the means of promoting popular education, by J. R. Freese, United States Commissioner.—pp. 13. October 8, 1869. (In volume v.)

ENGINEERING.—Report upon steam-engineering, as illustrated by the Paris Universal Exposition, 1867, by William S. Auchincloss, Honorary Commissioner.—pp. 72. August 2, 1869. (In volume iv.)

FINE ARTS.—Report on the fine arts, by Frank Leslie, United States Commissioner.—pp. 43. February 6, 1869. (In volume i.)

The fine arts applied to the useful arts—report by the committee, Frank Leslie, S. F. B. Morse, Thomas W. Evans, United States Commissioners.—pp. 8, with 33 leaves of wood engravings. February 6, 1869. (In volume i.)

FOOD.—Report on the preparation of food, by W. E. Johnston, M. D., Honorary Commissioner.—pp. 19. October 8, 1869. (In volume v.)

GENERAL SURVEY.—General survey of the Exposition, with a report on the character and condition of the United States Section.—pp. 325. January 7, 1869. (In volume i.)

GOLD AND SILVER.—(See *Precious metals.*)

INTRODUCTION.—Introduction, with selections from the correspondence of United States Commissioner General Beckwith and others, showing the organization and administration of the United States Section. (In volume i.)

IRON AND STEEL.—The production of iron and steel, in its economic and social relations, by Abram S. Hewitt, United States Commissioner, 1868.—pp. 183. January 7, 1869. (In volume ii.)

MINING.—Report on mining and the mechanical preparation of ores, by Henry F. Q. D'Aligny, United States Commissioner, and Alfred Huet, F. Geyler, and C. Lepainteur, Civil and Mining Engineers, Paris, France. February 19, 1869. (In volume iv.)

MUNITIONS OF WAR.—Report on the munitions of war, by Charles B. Norton and W. J. Valentine, United States Commissioners.—pp. 213. January 7, 1869. (In volume v.)

MUSICAL INSTRUMENTS.—Report upon musical instruments, by Paran Stevens, United States Commissioner.—pp. 18. June 21, 1869. (In volume v.)

11 P E

ORES, MECHANICAL PREPARATION OF.—(See *Mining.*)

PHOTOGRAPHY.—Photographs and photographic apparatus, by Henry F. Q. D'Aligny, United States Commissioner.—pp. 19.   October 8, 1869.   (In volume v.)

PRECIOUS METALS.—Report upon the precious metals, being statistical notices of the principal gold and silver producing regions of the world represented at the Paris Universal Exposition, by William P. Blake, Commissioner of the State of California.—pp. viii, 369.   March 11, 1869.   (In volume ii.)

SCHOOL-HOUSES.—(See *Education.*)

SILK.—Report on silk and silk manufactures, by Elliot C. Cowdin, United States Commissioner.—pp. 51.   January 7, 1869.   (In volume vi.)

SURGERY.—Report on instruments and apparatus of medicine, surgery, and hygiene, surgical dentistry and the materials which it employs, anatomical preparations, ambulance tents and carriages, and military sanitary institutions in Europe, by Thomas W. Evans, M. D., United States Commissioner.—pp. 70.   January 28, 1869.   (In volume v.)

TELEGRAPHY.—Examination of the telegraphic apparatus and the processes in telegraphy, by Samuel F. B. Morse, LL. D., United States Commissioner.—pp. 166.   November 20, 1869.   (In volume iv.)

Outline of the history of the Atlantic cables, by H. F. Q. D'Aligny, United States Commissioner.—pp. 13.   'October 8, 1869.   (In volume v.)

UNITED STATES SECTION, REPORT ON.—(See *General survey*, &c.)

VINE.—Report upon the culture and products of the vine, by Marshall P. Wilder, Alexander Thompson, William J. Flagg, Patrick Barry, committee.—pp. 28.   October 8, 1869.   (In volume v.)

WOOL.—Report upon wool and manufactures of wool, by E. R. Mudge, United States Commissioner, assisted by John L. Hayes, Secretary of the National Association of Wool Manufacturers.—pp. 143.   January 7, 1869. · (In volume vi.)

WEIGHTS, MEASURES, AND COINS.—Extracts from the report of the International Committee on Weights, Measures, and Coins, with a notice of the introduction of the metrical system in the United States and its relations to other systems of weights and measures. (In volume vi.)

NAMES OF THE AUTHORS OF REPORTS, ARRANGED ALPHABETICALLY.

AUCHINCLOSS, WILLIAM S., Honorary Commissioner, Civil Engineer.

BARNARD, FREDERICK A. PORTER, S. T. D., LL. D., President of Columbia College, New York, United States Commissioner, member of the International Jury.

BECKWITH, NELSON MARVIN, United States Commissioner General, President of the Commission.

BECKWITH, ARTHUR, Civil Enginer.

BECKWITH, LEONARD FORBES, Civil Engineer.

BLAKE, WILLIAM PHIPPS, Commissioner of the State of California to the Paris Universal Exposition of 1867 and Delegate of the State Board of Agriculture.

BOWEN, JAMES H., United States Commissioner.

COWDIN, ELLIOT C., United States Commissioner.

D'ALIGNY, HENRY FERDINAND QUARRÉ, United States Commissioner, member of the International Jury, Mining Engineer.

EVANS, THOMAS W., M. D., United States Commissioner.

FLAGG, W. J., United States Commissioner, (honorary,) member of Reporting Committee No. 9. (Report upon the vine, &c.)

FREESE, JACOB R., United States Commissioner.

GEYLER, F., Civil Engineer, Paris, France.

HAYES, JOHN LORD, Secretary of the National Association of Wool Manufacturers. (The report on wool, etc.)

HAZARD, GEORGE S., United States Commissioner, (honorary.)

HEWITT, ABRAM S., United States Commissioner.

HOYT, JOHN W., M. D., United States Commissioner, (honorary.)

HUET, ALFRED, Civil Engineer, Paris, France.

JOHNSTON, W. E., M. D., United States Commissioner, (honorary.)

LEPAINTEUR, C., Engineer to the Syndicat of Class 47, Paris, France.

LESLIE, FRANK, United States Commissioner.

MORSE, SAMUEL F. BREESE, LL. D., United States Commissioner.

MUDGE, ENOCH R., United States Commissioner.

NORTON, CHARLES B., United States Commissioner.

NOURSE, B. F., United States Commissioner, (honorary.)

RUGGLES, SAMUEL B., Vice-President of the United States Commission.

SEWARD, Hon. WILLIAM HENRY, Secretary of State. (Introduction.)

SEYMOUR, CHARLES B., United States Commissioner.

SLADE, FREDERICK J., scientific assistant to Committee No. 6. (Report upon Bessemer steel, in the report upon iron and steel.)

SMITH, J. LAWRENCE, United States Commissioner, Vice-President of the International Jury upon Group V.

STEVENS, PARAN, United States Commissioner.

THOMPSON, ALEXANDER, M. D., United States Commissioner, (honorary,) member of Reporting Committee No. 9. (Report upon the vine, &c.)

VALENTINE, W. J., United States Commissioner.

WILDER, MARSHALL P., United States Commissioner.

V.

CLASSIFICATION OF THE OBJECTS EXHIBITED AND GENERAL INDEX.

THE CLASSIFICATION OF OBJECTS ADOPTED BY THE IMPERIAL COMMISSION—ITS COMPREHENSIVE AND EXACT CHARACTER—ITS VALUE AS AN INDEX TO THE EXPOSITION AND TO HUMAN INDUSTRY IN GENERAL—ENUMERATION OF OBJECTS IN EACH GROUP AND CLASS, AND REFERENCES TO THE REPORTS.

THE IMPORTANCE OF THE CLASSIFICATION.

The classification adopted by the Imperial Commission for the formation of the Exposition is the most comprehensive and exact ever made of the raw materials useful to man, and of the various products of industry and art. The Hon. N. P. Banks, in his speech delivered in the House of Representatives, March 14, 1868, says of this classification: "The plan of the Exposition of 1867 is the grandest classification of the products of human industry that the mind of man has ever conceived. There has never been presented, in the history of the world, such a comprehensive, systematic, and scientific grouping of the various branches of human industry as this plan unfolds. All the pursuits and products of its people are grouped in ten leading divisions, and are subdivided into ninety-five classes. * * * These ten groups embrace all the pursuits of man, all the products of industry; they represent the habits of life, and all the relations of men to each other, to society, and to progressive civilization."

This classification is now not merely an outline plan according to which the Exposition was formed, but, from the realization of that plan, it has become an index to the Exposition, and is not only valuable as such, but it has a permanent value as an index to the industrial arts, and may be used to facilitate reference to the reports on the Exposition and as a basis for future exhibitions. It has therefore been reproduced here, in connection with this introduction to the series of reports, and references to the reports have been added whenever the articles or subjects mentioned have been described or specially referred to in the series. It will thus, to a certain extent, serve as a general index to the reports. References are, in most cases, made to the volume in which the report will be found; and, as most of the principal reports are indexed, it will not be difficult to find the subject desired. As each report is separately paged, a general alphabetical index to the series could not be conveniently made.

CLASSIFICATION AND GENERAL INDEX.

FIRST GROUP.—WORKS OF ART.

CLASS 1—*Paintings in oil.*—Paintings on canvas, on panels, on glazing, and other surfaces. [Report on the Fine Arts, in volume i; also in General Survey, p. 19.]

CLASS 2—*Various paintings and designs.*—Miniatures, aquarelles, pastels, and designs of all kinds; paintings on enamel, on crockery, or porcelain; cartoons for frescoes and for glass windows. [Fine Art Report; also General Survey, p. 28.]

CLASS 3—*Sculptures and engravings on medals.*—Spherical, embossing, sculptured bas-reliefs, sculptures repoussées, pressed and chiseled medals, cameos, engraved stones, chemical engravings. [Fine Art Report; also General Survey, p. 32.]

CLASS 4—*Designs and models of architecture.*—Studies and fragments, representations and projects of edifices, restorations from ruins and from documents. [Report on Buildings, &c., (iv;) also General Survey, p. 34.]

CLASS 5—*Engravings and lithographs.*—Engravings (black) on copper, wood, stone, &c.; engravings in several colors; lithographs, in black, in crayon, in pencil, and in colors. [General Survey, p. 34; also Fine Art Report and Report on the Industrial Arts.]

SECOND GROUP.—MATERIALS AND THEIR APPLICATIONS IN THE LIBERAL ARTS.

CLASS 6—*Specimens of printing and publishing.*—Specimens of typography; proof-sheets of autography and lithography, in black and in colors; proof-sheets of engravings; new books and new editions of books already known; collections of works forming libraries on special subjects; periodical publications; designs; technical and school atlases and albums. [General Survey, pp. 35–39.]

CLASS 7—*Specimens of stationery, of book-binding, and of materials used in painting and designing.*— Papers, cards, pasteboards, inks, chalks, pencils, pastels, furniture of writing-desks, inkstands, letter-balances, copy-presses, &c.; registers, copy-books, albums, note-books, instrument-cases, bands, elastic bands; various articles for water-colors, aquarelles, colors in cakes, in bladders, in tubes, and in shells; instruments used by painters, designers, gravers, and modelers; specimens of paper-work, lamp-shades, lanterns, flower-pots, &c. [General Survey, p. 39.]

CLASS 8—*Specimens of design and plastic molding applied in the ordinary arts.*—Industrial designs; designs obtained, reproduced, or reduced by mechanical means; decorative paintings; industrial lithographs or engravings; models and rough sketches of figures, ornaments, &c.; sculptured work, cameos, lockets, and various objects ornamented by engraving; industrial medals molded by machines; reductions and photographs; sculptures; various objects molded. [General Survey, p. 44.]

CLASS 9—*Proofs and apparatus of photography.*—Photography on paper, glass, wood, stuffs, enamel; heliographic engravings, lithographic proofs, photographic stereotypes, stereoscopes and stereoscopic proofs; specimens obtained by amplification; instruments, tools, and materials for photography; materials and apparatus for photographic workshops. [Report on Photography; also General Survey, pp. 47, 260; also Report on Industrial Arts, in volume iii.]

CLASS 10—*Instruments of music.*—Wind instruments, not metallic, with simple openings, with windpipes, with reeds, with or without reservoirs of air; metallic wind instruments, simple, with extensions, slides, pistons, keys, key-boards; wind instruments with key-boards, organs, accordeons; instruments with cords for compression, or for the bow without key-boards; instruments with cords and key-boards, pianos, &c.; instruments for percussion or friction; automatic instruments, organs of Barbary, serinettes, &c.; detached pieces and apparatus for orchestras. [Report on Musical Instruments, in volume v; also General Survey, pp. 48, 261.]

CLASS 11—*Apparatus and instruments of the medical art.*—Materials and instruments for dressing wounds, sores, and for inferior surgery; instruments for medical explorations; materials and instruments for surgery; trusses and cases of instruments; cases of medicaments intended especially for army surgeons, navy surgeons, veterinary surgeons, dentists, oculists, &c.; apparatus for restoring sensation, general or local; apparatus (mechanical or plastic) of *prosthesis,* (the substitution of parts or members;) apparatus for deformities, ruptures, &c.; various apparatus for the sick, infirm, deranged; accessory objects used in the medical and surgical service, in pharmaceutics, and in hospitals and infirmaries. [Report on Instruments and Apparatus of Medicine, &c., in volume v; and General Survey, pp. 51, 262, 311.]

Materials for anatomical researches; apparatus for researches in medico-legal practice; special materials for veterinary medical fracture; apparatus for baths, medical baths, &c.; apparatus for the physical exercise of children, for healthful and for medical gymnastics, &c.; apparatus for aid to the wounded on the field of battle; ambulances, civil and military, for armies on land and at sea. [*Ibid.*]

Apparatus for aid to the drowning, suffocating, fainting, &c., and for electro-therapy. [Industrial Arts, in volume iii, p. 344.]

CLASS 12—*Instruments of precision, and apparatus for instruction in science.*—Instruments used in practical geometry, compasses, micrometers, levels, achromatic lenses, calculating machines, &c. [Industrial Arts, in volume iii, p. 613; General Survey, p. 53.]

Apparatus and instruments for surveying, for topography, for land measuring, for astronomy, &c.; apparatus for various observations; apparatus and instruments of the arts of precision, [*See* Industrial Arts;] weights and measures of different countries, moneys, medals, &c., [*See* Report on Weights, Measures, Coins, &c.; also Report on the Precious Metals, chapter x, volume ii; also Introduction, &c., volume i;] balances; instruments for physical observations, meteorology, &c.; optical instruments; apparatus for instruction in physical science, in elementary geometry, descriptive geometry, solids, and mechanics. [Industrial Arts, in volume iii.]

Models and instruments for instruction in the industrial arts in general; collections for instruction in natural sciences; figures and models for instruction in medical science, flexible anatomical models, &c. [Industrial Arts, in volume iii.]

CLASS 13—*Geography, cosmography, apparatus, maps, charts, &c.*—Maps and atlases, topographical, geographical, geological, hydrological, astronomical, &c.; marine charts, physical charts of all sorts, flat and in relief; celestial and terrestrial globes and spheres; apparatus for the study of cosmography. [General Survey, p. 54; also in Civil Engineering.]

Statistical works, tables, and ephemerides for astronomers and mariners.

THIRD GROUP.—FURNITURE AND OTHER OBJECTS USED IN DWELLINGS.

CLASS 14—*Rich furnishings.*—Sideboards, bookcases, tables, toilettes, beds, sofas, seats, billiards, &c. [Fine Arts Applied to Useful Arts, &c., in volume i; General Survey, pp. 59, 265.]

CLASS 15—*Upholstery and decorative work.*— Bedding, covered seats, canopies, curtains, hangings in tapestry and in stuffs; furniture and decorative objects in rich stone and other valuable materials; decorations molded in paste, in plaster, in pasteboard; decorative painting, frames, furniture; decorative ornaments for religious service. [General Survey, Group III, p. 59; Fine Arts, &c.]

CLASS 16—*Crystals, rich glassware, and glazing.*—Goblets in crystal, cutglass, double crystal, mounted crystal, &c.; glass for windows, furniture, and mirrors; glass figured, enameled, crackled, filigreed; optical crystals; ornamental glass-painted windows. [General Survey, pp. 61–65.]

CLASS 17—*Porcelain, faïence, and other potteries.*—Biscuit, hard, and tender porcelains; fine earthenware, glazed and colored; biscuit of *faïence, terre cuite*, enameled lavas. [General Survey, pp. 65–69; Building Report.]

CLASS 18—*Carpets, hangings, and other furniture tissues.*—Carpets, Wilton carpets, velvet tapestries; carpets of felt, of cloth, of clippings of wool, silk, or floss silk, of mat-weed, of India-rubber; furniture tissues of cotton, wool, silk, hair, vegetable leather, moleskin, leather hangings and coverings, oil-cloths, &c. [General Survey, p. 69.]

CLASS 19—*Painted paper.*—Papers printed on blocks with rollers, with machines; papers velveted, marbled, veined, &c.; pasteboards, book-covers, &c.; paper for artistic uses, spring blinds, &c., painted or printed. [General Survey, p. 72.]

CLASS 20—*Cutlery.*—Knives, penknives, razors, scissors, &c. [General Survey, p. 74.]

CLASS 21—*Goldwork.*—Goldwork for religious service, for table use and ornament, for toilettes, bureaus, &c. [General Survey, p. 76.]

CLASS 22 — *Bronzes, various artistic [castings, and repoussé works in metals.*—Statues and bass-reliefs in bronze, in cast-iron, in zinc; decorative and ornamental bronzes; imitations of bronze castings in zinc; castings coated with metallic coverings by the galvanic process; *repoussés* in lead, zinc, copper, &c. [General Survey, p. 79.]

CLASS 23—*Clocks and Clockwork.*—Separate pieces of clockwork; spring clocks, pendulum clocks, electrical clocks, watches, chronometers, regulators, second-counters, apparatus for measuring time, hour-glasses, sand-glasses, clepsydras, &c. [General Survey, p. 82; Industrial Arts.]

CLASS 24—*Apparatus and methods of warming and lighting.*—Fireplaces, chimneys, stoves, furnaces, accessory objects; apparatus for heating by gas, by hot water, by hot air; apparatus for ventilating and for drying; enameled lamps, blow-pipes, portable forges; lamps for oil—mineral, vegetable, or animal; other accessories of lighting; apparatus for lighting by gas; photo-electrical lamps; apparatus for lighting by magnetism. [General Survey, p. 86; also in Industrial Arts.]

CLASS 25—*Perfumery.*—Cosmetics and pomatums, perfumed oils, perfumed essences, liquid extracts, scents, aromatic vinegars, almond paste, powders, pastilles and perfumed sacks, combustible perfumes, toilet soaps. [General Survey, p. 87.]

CLASS 26—*Fancy articles, toys, basket-work.*—Small fancy articles of furniture, liquor-cases, glove-boxes, caskets, lacquer-work, dressing-cases, workboxes, screens, pocket-books, purses, portfolios, cigar-cases, memorandums; articles of checkwork; articles turned, sculptured, engraved, of wood, of ivory, in shell, snuff-boxes, pipes, combs, brushes, *corbeilles*, and fancy baskets; basket-work, grass-work. [General Survey, pp. 89–91.]

FOURTH GROUP.—GARMENTS, TISSUES FOR CLOTHING, AND OTHER ARTICLES OF WEARING APPAREL.

CLASS 27— *Yarn and tissues of cotton.*—Cotton, prepared and spun; tissues of cotton, plain and figured; tissues of cotton, mixed; cotton, velvets, tapes, &c. [Clothing Report, (vi;) General Survey, p. 93, (i;) Report on Cotton, (vi.)]

CLASS 28—*Yarn, and tissues of linen, hemp, &c.*—Flax, hemp, and other vegetable fibers, spun; linen and ticking; Baptiste tissues of thread, mixed with cotton and silk; tissues of vegetable fibers, equivalent to linen and hemp. [Clothing Report, (vi;) General Survey, pp. 95–98.]

CLASS 29—*Yarn and tissues of combed wool.*—Combed wools, tissues of combed wools, mousselines, merinoes, Scotch cashmeres, serges, &c.; galoons of wool, mixed with cotton, or thread, or silk, or floss; tissues of hair, plain and mixed. [Wool and Manufactures of Wool, (vi;) Clothing, (vi;) General Survey, pp. 98, 269.]

CLASS 30—*Yarns and tissues of carded wool.*—Carded wool and yarn of carded wool; cloths and other tissues of wool, carded and fulled; blankets, felts of wool or of hair, for carpets; hats, socks, tissues of wool carded and not fulled or slightly fulled; flannels, tartans, &c. [*Ibid.*]

CLASS 31—*Silk and tissues of silk.*—Silks raw or milled; silk or floss thread or yarn; tissues of silk, plain and figured; silk stuffs mixed with gold, silver, cotton, or wool; tissues of floss silk, pure or mixed; velvets, plushes, ribbons of silk, pure or mixed. [Silk and Silk Manufactures, (vi;) Clothing Report, (vi;) General Survey, p. 103.]

CLASS 32—*Shawls.*—Shawls of wool, pure or mixed; shawls of silk and of cashmere. [General Survey, p. 106; also Clothing and Silk Reports.]

CLASS 33—*Laces, embroideries, and trimmings for clothing, military clothing, furniture, carriages, harness, &c.*—Laces of thread or cotton, made with the lace spindle, needle, or machines; lace of silk, wool, or of goats' hair; gold or silver lace; tulle of silk or cotton, plain or figured; embroideries with the needle, the hook, &c.; embroideries in gold, in silver, in silk, in thread; tapestry embroideries and other hand-work; trimmings of silk, floss, wool, goats' hair, hair, thread, and cotton; laces, military trimmings, fine and coarse. [Clothing Report, (vi;) General Survey, p. 109.]

CLASS 34—*Hosiery, linen, and other articles of clothing.*—Stockings of cotton, thread, wool, cashmere, silk, and floss, pure or mixed; garments of linen for men, women, children; baby-linen; garments of flannel and other tissues of wool; corsets; cravats; gloves; gaiters; fans; screens; umbrellas; parasols; canes, &c. [General Survey, p. 115; Clothing, (vi.)]

CLASS 35—*Clothing for men, women, and children.*—Garments for men; garments for women; coiffures for men and women, wigs, and hair-work; boots and shoes; childrens' clothes; professional garments. [Report on clothing, (vi;) General Survey, Group III.]

CLASS 36—*Jewelry and precious ornaments.*—Ornaments of gold, platinum, silver, and aluminum, chiselled in filagree, or set with fine stones, &c. Diamonds; precious stones; pearls and imitations. [General Survey, p. 133.]

CLASS 37—*Portable armor.*—Defensive arms—bucklers, shields, cuirasses, casques; offensive arms—war clubs, maces, bludgeons, battle-axes, &c.; foils, swords, sabers, bayonets, lances, hatchets, hunting-knives, bows, cross-bows, slings.

Fire-arms — muskets, carbines, pistols, revolvers; accessory articles—powder-flasks, bullet-molds; projectiles, oblong, spherical, hollow, explosive; percussion caps, primings, cartridges. [Munitions of War, (v;) General Survey, pp. 138 and 270-273.]

CLASS 38—*Articles for traveling and for encampment.*—Trunks, valises, sacks, bags, &c.; dressing-cases, trusses, &c.; various articles, coverings, cushions, coiffures, costumes, shoes, walking sticks, parasols, &c. General Survey; p. 143.]

Portable, for traveling and scientific expeditions: photographic apparatus, instruments for meteorological and astronomical observations; necessaries for geologists, mineralogists, naturalists, settlers, and pioneers; tent and camp articles; military tent furniture—beds, hammocks, pliant seats, canteens, mills, ovens, &c. [Instruments and Apparatus of Medicine, &c., (v;) General Survey, pp. 143 and 273.]

CLASS 39—*Toys and gewgaws.*—Dolls and playthings; figures in wax; plays for children and for adults; instructive playthings. [General Survey, p. 145.]

FIFTH GROUP.—PRODUCTS, WROUGHT AND UNWROUGHT, OF EXTRACTIVE INDUSTRIES.

CLASS 40—*Products of mines and metallurgy.*—Collections and specimens of rocks, ores, and minerals; ornamental stones, marbles, serpentines, onyx, and other hard stones, [Building Report, &c., (iv;)] materials difficult of fusion; earths and clays; various mineral products, raw sulphur, [Industrial Chemistry, ii,] rock salt, salt from springs, bitumens, [Asphalt and Bitumen, iv,] and petroleums; samples of combustible, raw, and carbonized agglomerations of pit coal, [Pressed Coal, (v,)— See also Class 47;] raw metals, pig iron, iron, steel, [Iron and Steel, (ii,)] copper, lead, silver, zinc, &c.; metallic alloys; products of puddlers, (and cinders,) of refiners of precious metals, of gold beaters, &c. [General Survey, pp. 147, 273; Precious Metals, (ii.)]

Products of electro-metallurgy, objects coated with gold, silver, copper, steel, &c., by the galvanoplastic method.

Products of the elaboration of raw metals, molded castings, bells, iron of commerce, iron for special uses, sheet iron, tin, extra plates for constructions and for plating ships; sheet copper, lead, and zinc, [Building Report, (iv;)] wrought metals, forge work, heavy work for gates, fences, &c.; wheels, bandages, tubes without solder, chains, &c. [General Survey, 150.]

Products of wire-mills, needles, pins, trellis-work, metallic tissues, perforated plates; hardware; edge-tools; ironmongery; copper, brass, plate, and tin wares; wrought metal of various kinds.

CLASS 41—*Products of the forest.*—Specimens of different species of wood; wood for cabinet work and for building; fire-wood; wood for ship-work, for walking-sticks, for splintering; corks; textile barks; tanning, coloring, odoriferous, and resinous substances; products of forest industry; roasted and carbonized wood; crude potash; wood for cooperage, for basket-work, for sabots, for mat-work, &c. [General Survey, p. 151.]

CLASS 42—*Products of hunting and fisheries, and collections of natural growth.*—Collections and drawings of terrestrial and amphibious animals, of birds, of eggs, fish, cetacea, crustacea, mollusks. [General Survey, p. 157.]

Products of hunting—furs, peltries, hair, fine and coarse, feathers, down, horns, teeth, ivory, bones, shells, musk, castoreum, and similar products. [*Ibid.*]

Products of fisheries—whale oil, spermaceti, whalebone, ambergris, shells of mollusks, pearl, mother-of-pearl, corals, sponges, sepia, purple, &c. [*Ibid.*]

Collections from natural growth—champignons; truffles; wild fruits; lichens for dyeing, for food, and for fodder; fermented saps; Peruvian bark, useful barks, and filaments; wax; resinous gums; caoutchouc; gutta-percha, &c. [*Ibid.* Preparation of Food, in volume v.]

CLASS 43—*Agricultural products (not used for food) of easy preservation.*— Textile materials—raw cotton; linen and hemp, dressed and not dressed; vegetable textile fibers of all sorts; wool in fleece; cocoons of silk-worm. [Reports on Cotton and on Silk, in volume vi, and in General Survey.]

Products of agriculture used in manufactures, pharmacy, and domestic economy—oleaginous plants, oils, wax, resins, tobacco, tinder, substances for tanning and for tinting; fodder and provender preserved. [General Survey, p. 160.]

CLASS 44—*Chemical and pharmaceutical products.*—Acids, alkalies, salts of all kinds, marine salt, spring salt. [Industrial Chemistry, (ii;) General Survey, p. 164.]

Various chemical products—wax, soap, candles, matters for perfumery, resins, tar waters, essences, varnishes, coatings, waxings; manufactures of caoutchouc, of gutta-percha; substances for dyes and colors. [Industrial Chemistry, for candles, soap, and dyes.]

Natural and artificial mineral waters—gas waters, elementary pharmaceutic substances, simple and compound medicaments.

CLASS 45—*Specimens of the chemical methods of bleaching and dyeing, of stamping and preparations.*—Samples of yarn and tissues, dyed; samples of preparations for dyeing; linens, printed and dyed; tissues of printed cotton, pure and mixed; tissues of printed woolens, pure and mixed, combed or carded; tissues of printed silks, pure or mixed; printed carpets of felt or cloth; linens, painted or waxed. [Report on Wool and Manufactures of Wool, Clothing Report, General Survey, &c.]

CLASS 46—*Leather and skins.*—Elementary matters employed in the preparation of skins and leather; hides, green and salted; leather, tanned, curried, prepared, and dyed; varnished leather; morocco and sheep-skins; Hungary leather; chamois-skins, dressed with the hair or wool on; preparations and dyes; skins prepared for gloves; peltry and furs prepared and dyed; parchments. [General Survey, p. 165; Report on Clothing, (vi.)]

Articles of membrane work, cords for musical instruments, gold-beaters' skins, &c.

SIXTH GROUP.—INSTRUMENTS AND PROCESSES OF COMMON ARTS.

CLASS 47—*Apparatus and methods of mining and metallurgy.*—Apparatus for boring artesian wells and large wells; machines for drilling in mines, for digging coal, and for quarrying stone and breaking up rocks. [Mining Report, (iv;) Civil Engineering, (iv.)]

Models, plans, and views of works and labor in mines and quarries; ladders for mines worked by machines; machinery for lifting from mines; machines for exhausting and pumping; apparatus for airing, ventilators, safety-lamps, &c.; photo-electric lamps; apparatus for safety, parachutes; signals. [Mining Report, (iv.)]

Apparatus for the mechanical preparation of minerals; apparatus for the agglomeration of combustibles. [Mining Report, (iv;) Pressed Coal, &c., (v;) General Survey, p. 171.]

Apparatus for carbonizing combustibles; furnaces and hearths for metals; apparatus for consuming smoke; machines for metallic works; special apparatus for forges and founderies; electro-metallurgic apparatus; apparatus for the working of metals in all forms. [Iron and Steel, (ii.)]

CLASS 48—*Implements and processes of rural and forest work.*—Plans of cultivation; divisions by nature of the soil; requisite manures and successions of crops adapted to each; materials and methods of agricultural engineering; surface draining; under-draining; irrigation. [General Survey, p. 174.]

Plans and models of rural buildings; tools, implements, machines, and apparatus for preparing the ground for sowing, planting, and harvesting; for preserving and preparing the products of agriculture; carts, wagons, and apparatus for agricultural and rural transportation; for training and managing horses, &c.

Fertilizing substances, organic or mineral. [General Survey, pp. 175, 282, 283, 284.]

Apparatus for the chemical and physical study of soils.

Plans for replanting, cultivating, and managing forests; implements of forest work.

CLASS 49—*Apparatus and instruments for hunting, fishing, and for collecting natural products.*—Arms, traps, snares, machines, and equipments for hunting; fish-lines, fish-hooks, harpoons, nets, apparatus and bait for fishing; apparatus and instruments for gathering products obtained without cultivation. [General Survey, p. 176.]

CLASS 50—*Materials and methods of agricultural works and of alimentary industry.*—Apparatus for agricultural work, making manures, making pipes for drainage, dairies, corn and flour trade, disposal of fecula, making starch, oil, brewing, distilling, making sugar, refining sugar; works for preparing textile fibers, silk-worm nurseries, &c. [General Survey, p. 177; Beet-root Sugar.]

Apparatus for the preparation of food, bread-kneaders, and mechanical ovens for bakers; utensils for pastry and confectionery. [Preparation of Food, (v.)]

Apparatus for making dough, for sea-biscuit, for chocolate, for roasting coffee, for ices and sherbets, and for making ice. [Preparation of Food, (v;) Industrial Arts, p. 366, for ice manufacture.]

CLASS 51—*Chemical, pharmaceutical, and tanning apparatus.*—Apparatus and instruments for laboratories; apparatus and instruments for testing and experiments in industry and commerce.

Machines and utensils used in the manufacture of chemical products, soaps, candles, &c.; apparatus and processes for making essences, varnish, and objects of caoutchouc and gutta-percha. [Industrial Chemistry, (ii.)]

Machines and apparatus for gas-works; machines and methods for bleaching; machines and preparations of pharmaceutic products; machines and tools for workshops, for tanning and dressing leather. [Industrial Chemistry.]

Machines and apparatus for glass-works and potteries.

CLASS 52—*Motors, generators, and mechanical apparatus especially adapted to the uses of the Exposition.*—Boilers and steam generators, with safety apparatus; steam-pipes and accessory objects; shafts, fixed and movable; pulleys and belts; means of starting and stopping, shifting and regulating the movements of machinery; motors for furnishing water and the necessary motive power in the different parts of the Palace and Park. [Steam Engineering, &c., (iv;) Industrial Arts, (iii.)]

Cranes and all sorts of apparatus proposed for the handling of packages and objects in the Palace and grounds; rails and turn-tables proposed for use in the Palace and Park. [Steam Engineering.]

CLASS 53—*Machines and mechanical apparatus in general.*—Detached pieces of machinery, supports, rollers, slides, eccentrics, cog-wheels, connecting-rods, parallelograms, joints, belts, systems of ropes, &c.; mechanism for changing the gear of machinery, clicks, &c.; movement regulators and moderators; greasing apparatus. [Steam Engineering, (iv;) Industrial Arts, (iii;) Mining, (iv;) General Survey, pp. 286–290.]

Indicators and registers, dynamometers, manometers, weighing apparatus, gauges, and apparatus for gauging liquids and gases; machines for handling heavy objects; hydraulic elevators, pumps, water-wheels, rams, &c.; wheel and chain buckets for irrigation, reservoirs, wheels, wheels with vertical shaft, machines *à colonne d'eau;* steam machinery, boilers, generators, and accessory apparatus; condensers; machines moved by the vapor of ether, chloroform, ammonia, or by combined vapors. [Industrial Arts; Steam Engineering.]

Gas-engines, air-engines, compressed air-engines; electro-magnetic motors, wind-mills, &c.; aerostats. [Industrial Arts; Mining Report; General Survey, p. 286.]

CLASS 54—*Machine tools.*—Machine tools for preparatory wood-work; turning-lathes; planing and boring machines; mortising, piercing, and cutting machines; screw-cutting, nut-cutting, and riveting machines; various tools belonging to the yards of mechanical constructors. [General Survey, pp. 177 and 290.]

Tools, machines, and apparatus used in pressing, crushing, mixing, sawing, polishing, &c.; special machine tools for various uses. [General Survey, pp. 17–184.]

CLASS 55—*Apparatus and methods of spinning and rope-making.*—Apparatus for hand-spinning; detached parts of spinning machines; machines and apparatus for preparing and spinning textile matters. [General Survey, pp. 181 and 293.]

Apparatus and methods adapted to the complementary operations, such as drawing out, winding off, twisting, milling, &c.

Apparatus for classifying and determining the condition of the threads.

Apparatus of rope-yards, round, flat, and diminishing cables, rope and twine, wire cables, cables with metallic center, fuzes, quick-matches, &c.

CLASS 56—*Apparatus and methods of weaving.*—Preparatory apparatus for weaving; machinery for warping and for bobbins; glazing and smoothing; ordinary and power looms for plain tissues and for figured tissues; loom reeds; electrical looms; carpet and tapestry looms; mesh looms for hosiery and tulle; apparatus for making lace, for fringes, and for trimmings; looms for high warping and methods of shuttling; accessory apparatus, calenders, crimping, weaving, measuring, and folding machines, &c.

CLASS 57—*Apparatus and process of sewing and making clothes.*—Ordinary instruments for cutting and sewing and making; machines for sewing, quilting, and embroidering; tools for cutting up stuffs and leather for clothes, shoes, &c.; machines for screwing, nailing, and making shoes and boots. [General Survey, pp. 185 and 294.]

CLASS 58—*Apparatus and methods of making furniture and household objects.*—Machines for veneering; saws for cutting in profile, &c.; machines for moldings and frames, for ornamental floor-work and furniture-work, &c.; turning-lathes, and various apparatus for joiners' and cabinet-makers' shops; machines for pressing and stamping; machines and apparatus for working in stucco, in pasteboard, in ivory, in bone, in horn; machines for pointing, sculpturing, and reducing statues, and for engraving and chasing; machines for sawing and polishing hard stones, marble, &c. [General Survey, pp. 185 and 297.]

CLASS 59—*Apparatus and methods of paper-making, coloring, and stamping.*—Apparatus for stamping paper, colors, and tissues; machines for engraving cylinders; apparatus for bleaching, coloring, preparing paper and tissues; apparatus for making paper in vats and by machines; apparatus for crimping, ruling, glazing, and pressing paper; machines for cutting, paring, and stamping paper, &c.; apparatus and materials for letter-casting, stereotyping, &c. [General Survey, p. 187.]

Machines and apparatus employed in stereotyping, mezzotinting, autography, lithography, chalcography, paniconography, chromo-lithography, &c.; printing of postage-stamps; machines for composing and for classifying letters. [Industrial Arts.]

CLASS 60—*Machinery, instruments, and methods used in various works.*—Machinery for stamping money, for making buttons, pens, pins, envelopes, brushes, cards, capsules, for loading merchandise, and for corking and capping bottles.

Tools and methods of making lock-works, toys, ornamental boxes, baskets, &c.

CLASS 61—*Carriage and cart work.*—Separate pieces of carriage and cart work, wheels, bands, axles, wheel-boxes, tires, &c.; springs and various methods of suspension; systems of tackling and brakes; specimens of carts and vehicles for special uses, public carriages, private carriages, state carriages, hand carriages, litters, sleighs, and velocipedes. [General Survey, pp. 188 and 299.]

CLASS 62—*Harness-work and saddlery.*—Articles of harness-work, buckles, ornaments, &c.; saddles, donkey saddles, cacolet; harness and bridles for riding; harness for draught, stirrups, spurs, whips, &c. [General Survey, p. 190.]

CLASS 63—*Materials for railroads and cars.*—Separate pieces, springs, buffers, brakes, &c. [Steam Engineering, &c., in volume iv; also in General Survey, pp. 191–202 and 300.]

Fixed materials, rails, chairs, splices, switches, turn-tables, fenders, watering cranes, reservoirs, signals for sight and sound; rolling materials, wagons for earthwork, for merchandise, for cattle, for travelers. [*Ibid.*]

Locomotives, fenders, &c.; machinery and tools of workshops, for repairs and reconstructions. [*Ibid.*]

Material and machines for inclined planes and self-working inclines. [*Ibid.* Industrial Arts for "Mahovos."]

Material and machines for atmospheric railways; models of machinery; systems of traction, apparatus applicable to iron roads; models, plans, and drawings of termini, stations, sheds, and out-houses, necessary to railways. [Steam Engineering, iv.]

CLASS 64—*Apparatus and methods of telegraphing.*—Telegraphic apparatus, based on the transmission of light, sound, &c. [Report on the Telegraphic Apparatus, &c., in volume iv.]

Apparatus of the electrical telegraph, supports, conductors, tighteners, electrical batteries; apparatus for sending and receiving dispatches, bells, and electrical signals, accessory objects for the service; lightning-rods, commutators, prepared papers for printing, and autographic transmissions; special apparatus for submarine telegraphs. [*Ibid.* Industrial Arts, (iii;) General Survey, p. 301.]

CLASS 65—*Materials and methods adapted to civil engineering, public works, and architecture.*—Materials for building, wood, metals, ornamental stones, lime, mortar, cements, artificial stone, beton, tiles, brick, slate, pasteboard, and felt, for roofing. [Civil Engineering, &c., (iv;) Industrial Arts, (iii;) Buildings, (iv;) Beton, &c., and Asphalt and Bitumen, in volume iv; General Survey, p. 200.]

Materials and specimens of preserved wood; apparatus and methods of testing materials; materials of works for embankments, excavating machines; apparatus for stone-cutters' yards; tools and methods for draughtsmen, stone-cutters, masons, carpenters, roofers, tilers, slaters, locksmiths, joiners, glaziers, plumbers, house-painters, &c.

Ornamental iron-work, locks, padlocks, railings, balconies, balusters, &c.

Materials and machines for foundation work, pile-drivers, piles, screw-posts, pumps, pneumatic apparatus, dredging machines, &c.; machines for hydraulic work, seaports, canals, rivers, &c.; materials and apparatus used in water-works and gas-works; materials for repairing roads, plantations, and public works. [Civil Engineering, in volume iv.]

Models, plans, and drawings of public works, bridges, viaducts, aqueducts, sewers, canal-bridges, &c. [*Ibid.*]

Light-houses, public monuments for special purposes, private buildings, hotels, and houses to let, workmen's residences, &c. [Industrial Arts.]

CLASS 66—*Navigation and salvage.*—Drawings and models of ships, docks, floating docks, &c.

Drawings and models of all kinds of vessels for river and maritime navigation; types and models adopted by the navy; apparatus employed in navigation; boats and various craft; ship-chandlery; flags, signals, buoys, beacons, &c.; materials and apparatus for swimming exercises, for diving, and for salvage; floats, diving-bells, nautical impermeable clothing, submarine boats, apparatus for marine salvage, carrying hawsers, life-boats, &c. [General Survey, and the Report on the Industrial Arts.]

12 P E

SEVENTH GROUP.—FOOD, FRESH OR PRESERVED, IN VARIOUS STAGES OF PREPARATION.

CLASS 67—*Cereals and other farinaceous edibles, with their derivatives.*— Wheat, rye, barley, maize, rice, millet, and other cereals in grain or flour; hulled grain; meal.

Farina of potatoes, rice, lentils, &c.; glutens—tapioca, sago, arrowroot, cassava, and other fecula; specimens of mixed meals, &c.

Italian pastes, semouille, vermicelli, maccaroni; alimentary compositions as substitutes for bread, ribbon, vermicelli, pulp, domestic pastes, &c. [See, for this class, the Report on Cereals, the Report on Preparation of Food, and the General Survey, pp. 207 and 304.]

CLASS 68—*Baking and pastry cooking.*—Various kinds of bread, with or without yeast; fancy and figured bread; compressed bread, for traveling, campaigning, &c.; tea biscuits; specimens of pastry peculiar to every nation; gingerbread and dry cakes susceptible of preservation. [Preparation of Food, &c., in volume v.]

CLASS 69—*Fat alimentary substances, milk, eggs.*—Fats and edible oils, fresh and preserved milk, fresh and salt butter, cheese, various kinds of eggs. [*Ibid.*]

CLASS 70—*Meat and fish.*—Fresh and salt meat of various kinds; meat preserved by different methods; cakes of meat and portable soup; hams and preparations of meat; fowl and game; fresh and salt fish; barreled fish; cod-fish, herrings, &c. [General Survey.]

Fish preserved in oil; sardines, pickled tunny, &c.; crustacea and shells; lobsters, prawns, oysters, preserved oysters, anchovies, &c. [General survey.]

CLASS 71—*Vegetables and fruit.*—Tubers, potatoes, &c.; dry farinaceous vegetables, beans, lentils, &c.; green vegetables for cooking, cabbages, &c.; vegetable roots, carrots, turnips, &c.; spicy vegetables, onions, garlic, &c.

Salad, cucurbita, pumpkins, melons; vegetables preserved in salt, vinegar, or by acetic fermentation, sauerkraut, &c.; vegetables preserved by various methods; fresh fruits, dry and prepared fruits, plums, figs, grapes, &c; fruits preserved without the aid of sugar. [General Survey, p. 213.]

CLASS 72—*Condiments and stimulants, sugars and specimens of confectionery.*—Spices, pepper, cinnamon, pimento, &c.; table salt, vinegar, compound seasonings and stimulants, mustard, curry, English sauces, &c.; tea, coffee, and aromatic beverages; coffee of chiccory and sweet acorns; chocolate; sugar for domestic use, sugar of grapes, milk, &c. [General Survey, p. 215; Preparation of Food, in volume v.]

Various specimens of confectionery, comfits, sugar-plums, melting plums, nougats, angelicas, anise-seeds, &c.; sweetmeats and jellies, preserved fruits, citrons, cedras, oranges, apples, pine-apples; brandy fruit, sirups, and sugary liquids. [General Survey, pp. 215–219.]

CLASS 73—*Fermented drinks.*—Ordinary red and white wines, sweet and mulled wines, sparkling wines, cider, perry, and other drinks extracted from fruit. [General Survey, pp. 218–222; Report on the Culture and Products of the Vine, &c., (v.) Beet-root Sugar and Alcohol, in volume v.]

Beer and other drinks, drawn from cereals; fermented drinks, drawn from vegetable saps; milk and saccharine substances of all kinds; brandy and alcohol; spirituous drinks, gin, rum, tafia, kirschwasser, &c. [General Survey, p. 222.]

EIGHTH GROUP.—ANIMALS AND SPECIMENS OF AGRICULTURAL ESTABLISHMENTS.

CLASS 74—*Specimens of rural work and of agricultural establishments.*— Types of rural buildings of various countries; materials of stables, cow-houses, ox-stalls, kennels, &c.; apparatus for preparing food for animals; agricultural machinery in movement; steam-plows, reapers, mowers, haymakers, threshing machines, &c.

Types of agricultural manufactures, distilleries, sugar-mills, [see Report on the Manufacture of Beet-sugar and Alcohol,] refineries, breweries, flour-mills, fecula and starch manufactures, silkworm nurseries, &c.

Presses for wine, cider, oil, &c.

CLASS 75—*Horses, donkeys, mules, &c.*—Animals presented as characteristic of the art of breeding in all countries; specimens of stables.

CLASS 76—*Oxen, buffaloes, &c.*—Animals presented as specimens of the art of breeding in each country; specimens of cow-houses and ox-stables.

CLASS 77—*Sheep, goats.*—Animals presented as examples of the art of breeding in each country; types of sheepfolds, pens, and similar establishments.

CLASS 78—*Swine, rabbits, &c.*—Animals presented, &c.; types of hog-pens, and structures for raising animals of this class.

CLASS 79—*Poultry.*—Animals presented, &c.; types of hen-roosts, dove-cotes, pheasantries, &c.; apparatus for artificial hatching.

CLASS 80—*Hunting and watch dogs.*—Shepherds' dogs, hunting dogs, watch dogs; types of kennels and apparatus for training.

CLASS 81—*Useful insects.*—Bees, silk-worms, and various bombyxes, cochineal, insects for producing lac, &c.; apparatus for breeding silk-worms, bees, &c.

CLASS 82—*Fish, crustacea, mollusca.*—Living aquatic useful animals; aquariums; apparatus used in breeding fish, mollusca, and leeches.

NINTH GROUP.—LIVE PRODUCTS AND SPECIMENS OF HORTICULTURAL ESTABLISHMENTS.

CLASS 83—*Hot-houses and horticultural materials.*—Tools for gardeners, nurserymen, and horticulturists; apparatus for watering and for dressing grass-plots, &c.

Large hot-houses and their accessories; small green-houses for apartments and for windows; aquariums for aquatic plants; water jets and other apparatus for ornamenting gardens.

CLASS 84—*Flowers and ornamental plants.*—Species of plants and specimens of cultivation representing the characteristic types of garden and house plants of every country.

CLASS 85—*Kitchen-garden plants.*—Species of plants and specimens of cultivation representing the characteristic types of kitchen-gardens in all countries.

CLASS 86—*Fruit trees.*—Species of plants and specimens characteristic of the orchards in all countries; slips of forest species.

CLASS 87—*Seeds and useful forest plants.*—Species of plants and specimens of culture indicating the methods of replanting forests in different countries.

CLASS 88—*Hot-house plants.*—Specimens of the culture of various countries, with a view to utility and ornament.

TENTH GROUP.—OBJECTS EXHIBITED WITH A SPECIAL VIEW TO THE AMELIORATION OF THE MORAL AND PHYSICAL CONDITION OF THE POPULATION.

CLASS 89.—*Materials and methods for teaching children.*—Plans and models of school-houses, of school furniture, apparatus, instruments, models, wall-maps, &c., designed for facilitating the teaching of children; elementary collections suitable for teaching ordinary science; models of designs, tables, and apparatus suitable for teaching singing and music.

Apparatus and tables for instructing the deaf and dumb and the blind; school-books, atlases, maps, pictures, periodical publications, and journals for education.

Works of scholars of both sexes. [General Survey, pp. 229 and 308; and Report on School-houses, &c., (v.) Education, in volume vi.]

CLASS 90—*Libraries and materials for instruction of adults in the family, the workshop, the commercial and corporation schools.*—Works proper for family libraries, for the masters in workshops, cultivators, commercial teachers, mariners, traveling naturalists, &c.

Almanacs, memorandum-books, and other publications suitable for traveling venders.

Materials for school libraries, commercial libraries, &c.

Materials for the technical teaching necessary in certain manual pursuits. [*Ibid.*]

CLASS 91—*Furniture, clothing, and food, of all origins, distinguished for useful qualities, united with cheapness.*—Methodical collection of objects enumerated in the third, fourth, and seventh groups, supplied to commerce by large factories or by master-workmen, and specially recommended by their adaptation to good domestic economy.

CLASS 92—*Specimens of popular costumes of different countries.*—Methodical collection of costumes of both sexes, for all ages, and for pursuits the most characteristic of each country. [Clothing Report, in volume vi.]

CLASS 93—*Specimens of habitations, characterized by cheapness, uniting sanitary conditions and comfort.*—Types of habitations for families, suitable for various classes of laborers in each country. [Building Report, (iv.) General Survey, p. 310.]

Types of habitations proposed for workmen belonging to manufactories in the suburbs or in the country. [*Ibid.*]

CLASS 94—*Products of all sorts, made by master-workmen.*—Methodical collection of products enumerated in preceding groups, made by workmen who work on their own account, either alone or with their families, or as apprentices, for sale or for domestic use.

NOTE.—Such products only were admitted into this class as were distinguished for their own qualities, novelty, perfection of the method of work, or by the useful influence of this kind of work on the moral and physical condition of the people.

CLASS 95—*Instruments and methods of work peculiar to master-workmen.*—Instruments and processes (enumerated in sixth group) employed habitually by workmen working on their own account, or specially adapted to work done in the family or in the family circle.

Manual works which display in a striking manner dexterity, intelligence, or taste of the workman.

Manual works which, from various causes, have most successfully resisted the competition of machines.

LIST OF UNITED STATES COMMISSIONERS.[1]

N. M. BECKWITH,
Commissioner General and President of the Commission.

SAMUEL B. RUGGLES,
Vice-President of the Commission.

ALEXANDER T. STEWART, New York, New York.
JACOB R. FREESE, Trenton, New Jersey.
CHARLES B. NORTON, New York, New York.
W. J. VALENTINE, Massachusetts.
THOMAS W. EVANS, Paris, France.
FRANK LESLIE, New York, New York.
JAMES ARCHER, Missouri.
ENOCH R. MUDGE, Boston, Massachusetts.
WILLIAM A. B. BUDD, New York, New York.
CHARLES B. SEYMOUR, New York, New York.
CHARLES R. GOODWIN, Paris, France.
C. K. GARRISON, New York, New York.

PAID COMMISSIONERS.

F. A. P. BARNARD, LL. D., President of Columbia College, New York.
J. LAWRENCE SMITH, M. D., professor, &c., Louisville, Kentucky.
J. P. LESLEY, professor, &c., Philadelphia, Pennsylvania.
ABRAM S. HEWITT, Esq., New York, New York.
SAMUEL B. RUGGLES, LL. D., New York, New York.
Hon. JOHN P. KENNEDY, Baltimore, Maryland.
WILLIAM SLADE, Esq., Ohio.
HENRY F. Q. D'ALIGNY, Esq., New York, New York.
JAMES H. BOWEN, Esq., Chicago, Illinois.
PARAN STEVENS, Esq., New York, New York.

HONORARY COMMISSIONERS, WITHOUT PAY, DESIGNATED BY THE COMMISSION IN PARIS.

ELLIOT C. COWDIN, Esq., New York, New York.
GEORGE S. HAZARD, Buffalo, New York.
WILLIAM S. AUCHINCLOSS, New York, New York.
JOHN P. REYNOLDS, Esq., Springfield, Illinois.
WILLIAM J. FLAGG, Ohio.
ALEXANDER THOMPSON, New York, New York.
SAMUEL F. B. MORSE, LL. D., New York, New York.
JAMES T. FRAZER, professor, &c., Philadelphia, Pennsylvania.
B. F. NOURSE, Boston, Massachusetts.

---

[1] The names of those who did not serve are omitted.

L. F. MELLEN, Esq., Alabama.
MARSHALL P. WILDER, Boston, Massachusetts.
JOHN P. REYNOLDS, Esq., Springfield, Illinois.
J. H. CHADWICK, Esq., Boston, Massachusetts.
THOMAS McELRATH, Esq., New York, New York.
PATRICK BARRY, Esq., Rochester, New York.
WILLIAM E. JOHNSTON, M. D., Paris, France.
JOHN W. HOYT, M. D., professor, Madison, Wisconsin.

EXECUTIVE.

N. M. BECKWITH,
Commissioner General and President of the Commission.

W. C. GUNNELL,
A. P. MULAT,
Engineers and Architects.

J. N. PROESCHEL,
Secretary.

J. C. DERBY,
United States Agent, New York.

LIST OF THE MEMBERS OF THE INTERNATIONAL JURY ALLOTTED TO THE UNITED STATES.

CHARLES C. PERKINS, New Order of Awards.
W. T. HOPPIN, Group I, Classes 1 and 2.
J. P. KENNEDY, Group I, Class 3.
R. M. HUNT, Architect, Group I, Class 4.
FRANK LESLIE, (supplemented by Dr. T. W. EVANS,) Group I, Class 5·
W. A. ADAMS, (supplemented by W. T. HOPPIN and Dr. T. W. EVANS,) Group II, Class 9.
J. R. FREESE, (supplemented by Dr. T. W. EVANS,) Group II, Class 11.
F. A. P. BARNARD, President Columbia College, Group II, Class 12.
WILLIAM SLADE, Group III, Class 20.
J. LAWRENCE SMITH, professor, Vice-President of Jury, Group V.
J. P. LESLEY, professor, (supplemented by Professor T. S. HUNT,) Group VI, Class 51.
C. R. GOODWIN, Engineer, Group VI, Class 52.
J. E. HOLMES, Engineer, Group VI, Class 54.
H. F. Q. D'ALIGNY, Engineer, Group VI, Class 57.
J. DEBEAUVAIS, Engineer, Associate Juror in Group VI, Class 54.
J. P. REYNOLDS, Juror on Agricultural Trials at Billancourt.

PARIS UNIVERSAL EXPOSITION, 1867.
REPORTS OF THE UNITED STATES COMMISSIONERS.

GENERAL SURVEY OF THE EXHIBITION;

WITH A REPORT ON THE

CHARACTER AND CONDITION

OF THE

UNITED STATES SECTION.

WASHINGTON:
GOVERNMENT PRINTING OFFICE.
1868.

ERRATA.

Page 24, for Troyon, read Troyon.

Page 59, *et infra*, for furniture and other objects for the use of dwellings, read furniture and other objects for use in dwellings.

Page 64, 8th line from the bottom, insert comma after " this."

Page 80, for Montague, read Montague.

Page 101, for Vienna, read Vienne.

Page 102, for Vanquelin, read Vauquelin.

Page 102, for National Association of Wool Growers, read National Association of Wool Manufacturers.

Page 103, for Oiseet Ere, read Oise-et-Eure.

Page 105, for fuchshine, read fuchsine.

Page 133 to 146, the head lines should be changed to conform to the classes.

Page 149, 5th line, for *pounds*, read *poods*.

Page 153, for Lannet, read Lannes.

Page 154, for usages, read uses.

Page 265, for Madona, read Madrona.

Page 287, for steam, read steam pump.

Page 315, at the foot of the page, the title of the catalogue should be corrected to read as follows: Catalogue Officiel des Exposants Récompensés par le Jury International. Paris, E. Dentu, Libraire-Éditeur de la Commission Impériale.

PREFACE.

THE examination of products and making awards was committed to international juries, numbering in all six hundred members.

The number of jurors taken from each nation was in proportion to the ground occupied by each in the Exhibition, and the general commissioner of each nation nominated the jurors allowed to his national section.

The organization comprised one special jury, ninety-four juries of classes, ten juries of groups, and a superior council.

The work was divided and distributed among them as follows:

I. The subjects which were presented for the new order of recompenses, intended for persons, establishments, or localities, which, by organization or special institutions, have developed harmony among co-operators and produced in an eminent degree the material, moral, and intellectual well-being of the workmen, were submitted to a special jury of twenty-five members, whose decision was final.

II. The examination of Group No. 1, comprising the five classes of fine arts, was committed to four separate juries, whose reports were subject to revision and adjustment by a group jury formed by the four class juries united, numbering sixty-four members, whose decision was final.

III. The remaining ninety classes of products were submitted to the inspection of the corresponding ninety class juries, whose work was subject to revision by the group juries and superior council.

Each class jury elected from its own body a president, vice-president, and reporter.

The nine group juries were composed of the presidents and reporters of the ninety class juries, with the addition of a president and two vice-presidents to each group jury, not taken from the class juries, but specially appointed by the respective general commissioners of the national sections to which these appointments were allotted. The secretary for each group was appointed by the imperial commission.

The superior council was formed of the presidents and vice-presidents of the nine group juries, presided by one of the vice-presidents of the imperial commission.

IV. The organization thus comprised—

	Members.
One special jury on new order	25
One class and group jury on fine arts	64
Ninety class juries, numbering in all	483
Nine group juries, numbering—	
Presidents and vice-presidents of classes	180
Added, nine presidents and eighteen vice-presidents	27 27
	207

		Members.
One superior council—		
Presidents and vice-presidents of groups	27	
One presiding officer added	1	1
		28
Total		600

V. The duties of the class juries were to examine the products in detail in their respective classes, and make lists of the exhibitors whose products they considered deserving of awards, naming the award they proposed for each, and the reason of it, which completed their work.

The reports on products and exhibitors thus drawn up were passed to the group juries, whose duty it was to revise them, concurring in the recommendations of the class jurors as far as approved, modifying the parts not approved, and sending them in this form to the superior council.

The duty of the superior council was to decide upon the whole number of awards to be made, and the number of each grade of awards, for which purposes they had a limited authority to add to the whole number which had been recommended, and power to diminish the whole number called for by the juries. Having determined the whole number and the grades, they apportioned the numbers and grades to each group for distribution, and in this form returned the work to the respective group juries, whose remaining duty it was to adjust the awards made to the numbers and grades thus placed at their disposal, retrenching the names, if any in excess of their means; and this adjustment was final.

The classification of products adopted by the imperial commission having been made known two years in advance, and the national allotments of jurors made public at an early period, ample time had been given for the selection of jurors qualified to appreciate the particular class of products on which each was to be placed.

A more highly competent body of experts in the products of every industrial art and science was probably never assembled for a similar purpose. The rapidity of their appreciations, in many cases, was not in conformity with the views of exhibitors, who thought more time and explanation would have made their products better understood. But men devoted to special studies, familiar with first principles, and acquainted with their application, modified by human skill, in almost every form, seldom meet with a product in their line so entirely new in principle, so ingenious in design, or so complicated in structure, as to make it difficult for them to arrive at a correct opinion upon its general merits in a short space of time. Exceptions occur, but the inventive skill of producers rarely exceeds the comprehension of experts, and the general accuracy of the conclusions of the juries will, without doubt, be proved by experience and largely confirmed by public opinion.

In the ceaseless struggle to gratify human wants, scientific, mechanical, and industrial progress are developed unequally in different countries and in different localities of the same country. Bringing together the best fruits of industry and skill from all regions facilitates the exchange and diffusion of the arts and methods of production, and equalizes the common stock of intelligence. All are gainers in the highly civilized commerce which consists in the gratuitous exchange of useful ideas and practical knowledge, together with the methods of their application in every form to ameliorate the material and moral condition of mankind.

The united verdict of the international jury, composed in great part of professional men of known skill and established reputations, is the ablest and soundest judgment that will be pronounced on the relative condition of the arts of industry at the present time, as displayed in the products of all countries.

Ninety-five juries, working simultaneously and independently, and rendering in every department separate reports, produce, when collated, revised, and confirmed, an aggregate verdict of reliable value.

The relative condition of national industries thus indicated will be most easily and readily understood by a tabular statement, divested of the embarrassment of superfluous figures and variable numbers, showing merely the percentage of awards to exhibitors.

Percentage was not the object, but is the inevitable result, of awards, and it is the most unquestionable expression, in a concentrated and reliable form, of the united opinion of the whole body of jurors, the importance of which is not diminished by its being unforeseen and unpremeditated.

The table which follows shows in the first line the percentage of awards of each grade, and the total average percentage. The percentage of awards in each grade results from a comparison of the whole number of awards in each grade with the whole number of exhibitors in the Exhibition; and the total average percentage results from a comparison of the whole number of awards with the whole number of exhibitors; this total average results equally from the sum of the averages of the grades.

The subsequent lines show in like manner the percentage applicable to each country. In these the percentages of awards in each grade result from the whole number of awards in each grade, made *to the country named*, compared with the whole number of exhibitors *from that country;* and the total average percentage of each country results from a comparison of the total number of awards and total number of exhibitors pertaining to the country named, or equally from the sum of the preceding percentages.

The lines read horizontally show, therefore, the percentage of grades and awards to each country, and the columns read vertically present the relative grades and awards of each country compared with the other countries.

The percentage of awards to the exhibitors of the remaining twenty-five countries falls below the succeeding.

PERCENTAGE OF AWARDS TO EXHIBITORS.

Name of country.	Percentage of grand prizes.	Percentage of gold medals.	Percentage of silver medals.	Percentage of bronze medals.	Percentage of honorable mentions.	General average percentage.
General average percentage of awards to exhibitors	0. 00175	0. 02221	0. 08113	0. 12759	0. 11265	34. 53
Special average :						
France	0. 00306	0. 04272	0. 14742	0. 20086	0. 16166	55. 57
United States	0. 00932	0. 03171	0. 13432	0. 17910	0. 17350	52. 79
Austria...............................	0. 00095	0. 02722	0. 12273	0. 18194	0. 14326	47. 60
Prussia and North Germany.............	0. 00226	0. 02890	0. 10760	0. 18497	0. 15028	47. 40
Belgium...............................	0. 00161	0. 01834	0. 10318	0. 15428	0. 15326	43. 26
Russia	0. 00073	0. 01538	0. 06593	0. 14945	0. 10915	34. 06
Switzerland	0. 00092	0. 01944	0. 07500	0. 11388	0. 10926	31. 85
Great Britain and colonies..............	0. 00178	0. 01829	0. 06217	0. 09531	0. 08338	26. 10
Italy	0. 00122	0. 00589	0. 02826	0. 06311	0. 09338	19. 18
Spain	0. 00000	0. 00794	0. 02950	0. 07630	0. 07333	18. 70

The ardor of competition in a great international assembly, with the eagerness and suspense which precede the declaration of awards after that event, display the reaction common to all excitements. The awards of the successful, so desirable by anticipation, diminish in importance by possession, and seldom give satisfaction; while the unsuccessful, with more courage or more philosophy, find little difficulty in adopting the conclusion of their friends who have succeeded, that the whole affair has been greatly overrated.

Neither of these impressions is probably very accurate. Experience on former occasions has in the main justified the awards of the juries, and they have served not only to confirm established reputations, but to bring into more prominent notice the excellent products of thousands of skilful and worthy producers, who labored previously in comparative obscurity, and whose improved fortunes date from those periods. But the benefits resulting from this are not limited to the successful exhibitors. They are naturally stimulated to renewed efforts to maintain their new positions, which quickens their invention, improves their products, and raises their own standards, whilst their rivals and competitors, who, if equally skilful, are less lucky, are thereby compelled to work up to this higher level. A new spirit is thus breathed into every department of industry, and the benefits of increased production, improved qualities and varieties, and diminished cost become universal.

The influence of exhibitions in producing the remarkable rise and equalization of the industrial arts over a large portion of the civilized world, increasing useful products and augmenting the growth of commerce, is conspicuous everywhere and obvious to every intelligent mind which has been turned to the subject under circumstances favorable to observation.

Their effects also in a scientific, economic, and political sense are subjects of great interest, but may be with more propriety separately considered.

The high position conceded by the verdict of the juries to American industrial products is not due in general to graceful design, fertile combinations of pleasing colors, elegant forms, elaborate finish, or any of the artistic qualities which cultivate the taste and refine the feelings by awakening in the mind a higher sense of beauty, but it is owing to their skilful, direct, and admirable adaptation to the great wants they are intended to supply, and to the originality and fertility of invention which converts the elements and natural forces to the commonest uses, multiplying results and diminishing toil.

The peculiar and valuable qualities of our products will be adopted and reproduced in all parts of Europe, improving the mechanical and industrial arts, and it is reasonable to expect and gratifying to believe that the benefits will be reciprocal, that our products will in time acquire those tasteful and pleasing qualities which command more admiration and find a quicker and better market than the barely useful.

The reports of the United States commissioners upon the important subjects selected by them will undoubtedly command attention.

For a general survey of the Exhibition I refer with confidence to the able sketches of Commissioner Seymour, written with clearness and freedom, in a flowing and agreeable style, free from the stiffness of technical language; and to the observations on the American section, which will convey to those interested, especially in that department, correct general information on the products of our own country.

I refer with equal confidence to the special reports of a more practical character, on subjects of particular importance to the great industries of the country. Several of these reports are from professional men whose established reputations guarantee the thoroughness of their studies and the accuracy of their work, whilst the authors who have not yet acquired this authority may reasonably expect to obtain it from the just appreciation of the public. In this connection I cannot deny myself the pleasure of alluding to the assiduity, the ability, the zeal, and the excellent spirit which have animated the commissioners in devoting so long a period to labors adapted to promote the common welfare and prosperity of the country.

N. M. BECKWITH,
United States Commissioner General.

PARIS, *January* 17, 1868.

GENERAL SURVEY

OF THE

PARIS UNIVERSAL EXPOSITION OF 1867.

INTRODUCTION.

THE following report has been prepared in conformity with the instructions from the Department of State, August 20, 1866, which require the Commission to make a "report presenting a brief general survey of the Exhibition, and upon the character and condition of the American department." The committee formed for the purpose consisted of three members Messrs. Seymour, Evans, and Auchincloss.

It has been attempted in the following pages to present to the reader a sketch of one of the most important events of the nineteenth century, and to describe certain objects of general interest in a rapid and, it is hoped, popular way. There were 95 classes in the Exposition, many of them subdivided into other classes, and all worthy of deep consideration. To obtain information, and to collate and compile it, were matters of difficulty, and hence absolute brevity, although it has been attempted, could not, in the nature of things, be attained. But details have been avoided; they belong properly to the special reports. Nevertheless, it has been thought desirable to reproduce in English, from the French official catalogue, some of the introductions to the principal classes. They have been prepared with the greatest care, contain many interesting particulars, and offer the latest data on the subjects treated.

Before concluding their labors the committee think it proper in this place to acknowledge the valuable assistance of the Commissioner General, Mr. N. M. Beckwith, in the preparation of these reports. Involved in duties which were alike arduous and ungrateful, because seldom properly appreciated, he was able, by unflagging attention to the interests of the commission, by great executive ability, unyielding integrity of purpose, and inflexible resolution, to render great assistance to exhibitors and to all who sought his knowledge and advice.

To Professor W. P. Blake, of California, the committee is indebted also for much useful matter.

CHAS. B. SEYMOUR,
Chairman of Committee.

ORGANIZATION AND LOCALITY.

The Exposition of 1867 takes its origin from the imperial decrees of the 22d June, 1865, and subsequent dates, instituting an International Exposition, to be opened at Paris on the 1st April, 1867, and placing it under the direction of an imperial commission of 60 members, of which the Prince Imperial was named president; M. Rouher, minister of state, M. Forcade de La Roquette, minister of commerce and public works, and Marshal Vaillant, minister of the imperial household, vice-presidents; and M. Leplay, councillor of state, commissioner general.

The locality selected for the Exhibition was the Champ de Mars, the great military parade ground, extending from the military school to the Seine, and from the avenue Labourdonnaye to the avenue Suffren, forming a rectangle of 48 hectares, or 119 acres. To this was annexed the island of Billancourt, giving an additional area of 21 hectares, or 52 acres; making a total of 171 acres appropriated to the Exposition. Although somewhat removed from the most attractive parts of the city, it was easy of access; and being also the property of the government, and without any constructions which needed to be removed, it was suitable for the intended edifice, and was free from expense on the score of rent.

The ground was given up by the government on the 28th of September, 1865, and the first iron pillar of the building was raised on the 3d of April, 1866. At the end of the year the structure was comparatively ready for the exhibitors.

It is proper to use the word "comparatively," for there was delay and backwardness on many sides; and the opening, although it took place on the day and hour announced, was a regulation rather than a necessity. A few only of the groups were in a condition to be fairly presented to the public, and still less to the jurors whose work was to commence and terminate within the first 14 days of the opening month. Thanks, however, to the efforts of the respective commissions, and the hearty good-will of the exhibitors, those who had seen with dismay the condition of the building on the day when the Emperor and Empress dedicated it to its beneficent and instructive purposes, were certainly the most gladdened and surprised to find a fortnight later that order had sprung from chaos, and that the vast idea of this colossal undertaking had crystallized into an object of beauty.

As the season progressed the enclosure known as the Park advanced in clearness and interest. Structures that ranged from the nomadic hut of an Esquimaux to the gilded palace of a sultan, sprang up on every side. These buildings, being constructed by the various governments represented, were eminently national, and, in many instances, were faithful reproductions of edifices that are of world-wide fame. They were rendered additionally interesting from the fact that, to whatever use they were devoted, the attendants, either as workmen or servitors, were almost invariably national. It was thus possible in many ways to visit the habitations and witness the customs of the most remote as of the most intimate nations of the earth—a study which can hardly be considered inferior to any other that was afforded on this occasion.   It may be mentioned in this place that, amidst all the allurements of strange designs and blazing decorations, the simple structures contrived for cheapness, and intended for working-men and their families, attracted not only the attention of the public, but won the highest prizes of the juries.   It may surely be added as a matter of congratulation that the Emperor Napoleon, who planned this immense and splendid show, was himself a competitor in the simple walks of useful ingenuity.   He gave the world a palace of unequalled splendor, and contributed himself a design for small dwellings, suitable for the commonest order of laborers.   The latter was so excellent that it received the principal prize awarded in such competition.

Thus in a short time the appearance of the Champ de Mars was totally changed.   It was no longer an arid, gravelly surface without vegetation or adornment.   It became a place where the palace and the cottage, often together by accident, were purposely put side by side for examination; where the traditions of generations could be contrasted with the latest discoveries and experiences of to-day.   The vast elliptical building in the centre occupied 190,000 yards, or 39 acres.   The circumference was 1,600 yards, or nearly a mile.   Externally the effect was heavy, and by no means imposing; but it speedily became apparent that it was admirably adapted to the purposes for which it was intended.   The entire length between the Quai d'Orsay and the military school was 1,125 yards, and the width between the two avenues De Labourdonnaye and De Suffren 515 yards.

The Exposition was divided into three portions; the first, called the Park, comprising the palace and structures before referred to, and the banks of the Seine; the second, called the Reserved Garden, containing the botanical, horticultural, and piscicultural collections; the third, called Billancourt, from the name of an island in the Seine, where the agricultural implements were exhibited.   To facilitate the practical trials of the latter, the Emperor was also good enough to give up to the competitors all the land and crops they required.   Thus the mowing machines were tried at the Emperor's farm at Fouilleuse, near St. Cloud, and the reapers at the imperial establishment at Vincennes.

PREVIOUS EXHIBITIONS.

The Champ de Mars was the site of the first French industrial exhibition, held in the year 1798. This had 110 exhibitors and lasted three days. It was succeeded by other exhibitions with a constantly increasing interest and number of exhibitors, as will be seen from the annexed table.   In 1820 Belgium and Holland united in an exhibition at Ghent.   Prussia held an exhibition at Berlin in 1844, and Austria at Vienna in 1846. But the first exhibition which it was proposed to make *universal* was opened in London in 1851 at the Crystal Palace, constructed for the purpose. It was followed by an exhibition in New York in 1853, and by the Universal Exhibition at Paris in 1855, held in the Palais de l'Industrie, also specially constructed for the purpose, and which was the scene of the distribution of the prizes by the Emperor on the 1st of July, 1867.

The second international exhibition in England was opened in 1862 and covered about 17 acres, exclusive of annexes, and had over 26,000 exhibitors.

The relative importance of these different exhibitions, the space covered by each, and number of exhibitors and visitors as far as ascertained, is given in the following table:

Year.	Name of country.	Space in sq. meters.	Number of exhibitors.	Number of awards.
1798	France	23	110	23
1801	France		220	80
1802	France		540	254
1806	France, (empire)		1,422	610
1819	France, (restoration)		1,662	869
1820	Belgium and Holland			
1823	France		1,642	1,091
1827	France		1,695	1,254
1834	France, (Louis Philippe)		2,447	1,785
1839	France		3,281	2,305
1844	France		3,960	3,253
1844	Prussia, (Berlin)			
1846	Austria, (Vienna)			
1849	France, (republic)		5,494	4,000
1851	London, (great exhibition of all nations)	88,027	13,937	5,248
1853	New York, (world's fair)			
1854	Germany, (Munich)			
1855	France, (Paris universal exhibition)	118,786	23,954	10,811
1862	London, (international)	119,994	28,653	
1867	France, (exposition universelle)	694,153	50,226	

THE BUILDING.

The buildings erected for previous great exhibitions are generally known as *palaces*, but the structure on the Champ de Mars had nothing in its appearance, as our previous remarks have hinted, suggestive of the name. In its plan and construction architectural effects were subordinated to the great end in view—the exhibition of the objects of all nations in such a manner as to invite and facilitate comparison and study. This end was attained by the classification of the objects in groups, and their arrangement in a corresponding number of galleries disposed side by side concentrically. As three out of the ten groups—such as the agricultural exhibitions, live produce, &c.—could not be properly placed in the building, only seven galleries were required and constructed. These galleries, ellipsoidal in form and one story in height, composed the building.

The ground plan was not exactly an ellipse, it was rather a rectangle with rounded ends and the sides running parallel with the adjoining avenues. The exterior lines of the two sides ran straight for a space of 120 yards, one facing the quarter of the Gros-Caillou, the other the quarter of Grenelle, and were united by two demi-circumferences of equal diameter, with one side of the rectangle facing the bridge of Jena, and the other the military school. An open space in the centre, prettily ornamented with flowers, statues, and fountains, served as the point of radiation for the seven enclosing galleries. It was also the site of a central pavilion which contained the exhibition of the weights, measures, and moneys of all countries.

In the construction of this building upwards of 370,000 cubic metres of soil had to be removed to make room for foundations, drains, air passages, and water-pipes. The outer circle was excavated so as to give a succession of vaulted cellars built of stone and concrete and lime with cement. The two interior galleries of the building were built of stone and the seven others of iron.

The outer circle, devoted to the engines and machinery, was the highest and the broadest of all. Its width was 114 feet, and its height, to the top of the nave, 81 feet. The roof was formed of corrugated iron and supported by 176 iron pillars (each weighing 24,000 pounds) upon which the arches or ribs were placed. Along the centre of the whole length of this great machinery gallery or arcade an elevated platform was supported upon iron columns, and afforded a safe and convenient promenade and point of view for the machinery below. It appeared to support the line of shafting by which motion was communicated to the various machines, but this shafting was sustained by a separate frame.

The supply of water for this enormous structure, and for the Park and its various buildings and fountains, was obtained from the Seine, and was raised by powerful steam pumps to a reservoir placed upon the high ground on the opposite bank. This reservoir had a capacity of over 4,000 cubic yards of water, and was made water-tight by a lining of concrete. The main conduit leading from this reservoir crossed the Seine by the bridge of Jena, and traversed the whole length of the Champ de Mars. Complete details of the hydraulic service and of the ventilation and mechanical appliances generally will be found in the subsequent part of this report under class 52, Group VI.

AVENUES OF COMMUNICATION AND CLASSIFICATION OF OBJECTS.

The avenues of communication within the buildings and in the Park may be best understood by reference to the map. Both the Park and the building were bisected through the entire length by one straight avenue leading from the grand entrance opposite the bridge of Jena to the front of the military school at the opposite extremity of the Champ de Mars. This was crossed at right angles by three other broad avenues leading to the side entrances upon the public streets. These principal avenues, together with several others at each end, radiating from the central garden to the outer circle, intersected each gallery at right angles, and divided the whole building into 16 sectors of nearly equal area.

The objects exhibited by France and its colonies occupied seven of these sectors; England filled two and a half, and the United States one-third of one, exclusive of the displays in the buildings outside.

It will be seen that the form and arrangement of the building and the disposition of its contents was in harmony with the classification and grouping adopted by the imperial commission.

This classification included 10 groups, subdivided into 95 classes, as follows:

Group I.—Works of art, classes 1 to 5.

Group II.—Apparatus and applications of the liberal arts, classes 6 to 13.

Group III.—Furniture and other objects for the use of dwellings, classes 14 to 26.

Group IV.—Clothing, including fabrics, and other objects worn upon the person, classes 27 to 39.

Group V.—Products, raw and manufactured, of mining industry, forestry, &c., classes 40 to 46.

Group VI.—Apparatus and process used in the common arts, classes 47 to 66.

Group VII.—Food, fresh or preserved, in various states of preparation, classes 67 to 73.

Group VIII.—Live stock and specimens of agricultural buildings, classes 74 to 82.

Group IX.—Live produce and specimens of horticultural works, classes 83 to 88.

Group X.—Articles exhibited with the special object of improving the physical and moral condition of the people, classes 89 to 95.

To each of the first seven of these groups a gallery of the building was assigned. Thus Group I, works of art, occupied the inner circle or gallery 1, and so on to Group VII, which occupied the outer circle.

By following one of these galleries the observer passed in succession among the productions similar in kind of different countries. By following the avenues he passed successively through the different productions of the same country. The student therefore could investigate the condition of any particular art or industry as manifested by different nations, or he could pursue his studies geographically and note the characteristic productions of each country, and compare them as a whole with those of other countries. The arrangement facilitated exhibition, prompted study and comparison, and in these respects fully realized the intentions of its authors.

After the adoption of this classification, it was decided to devote a portion of the inner gallery, next to the central garden, to antiquities, so as to give a history of human labor.

The order in which the various countries were ranged in the building, the space occupied, and the number of exhibitors from each country, are shown in the following table:

Name of country.	Space occupied in square metres.				No. of exhibitors.
	In the Palace.	In the Park.	On the shore.	Total.	
France	65, 228. 84	88, 507. 00	2, 756. 52	156, 492. 36	15, 025
Republic of Andorra	2. 00	....	....	2. 00	1
Holland	1, 995. 84	4, 764. 50	....	6, 760. 34	538
Luxembourg	6. 60	....	....	6. 60	6
Belgium	7, 325. 60	9, 273. 90	....	16, 599. 50	1, 853
Prussia and North Germany	12, 365. 31	9, 408. 14	....	21, 773. 45	2, 249
Hesse					242
Baden	4, 396. 26	2, 553. 75	....	6, 950. 01	203
Wurtemburg					259
Bavaria					414
Austria	8, 381. 25	9, 820. 60	....	18, 201. 85	2, 094
Switzerland	2, 855. 37	3, 819. 28	....	6, 674. 62	1, 080
Spain	1, 771. 88	1, 574. 00	....	3, 345. 88	2, 644
Portugal	759. 38	1, 530. 00	....	2, 289. 38	1, 648
Greece	759. 37	....	....	759. 37	480
Denmark	1, 012. 50	453. 00	....	1, 465. 50	283
Sweden	1, 940. 62	3, 008. 00	....	4, 948. 62	605
Norway					411
Russia	3, 037. 50	3, 146. 40	....	6, 183. 90	1, 365
Italy	3, 459. 37	3, 035. 28	....	6, 494. 62	4, 069
Rome	709. 38	410. 00	....	1, 119. 38	172
Roumania	663. 02	1, 767. 00	....	2, 430. 02	1, 056
Turkey	1, 187. 53	2, 889. 00	....	4, 076. 53	4, 817
Egypt	587. 55	6, 005. 00	....	6, 592. 55	14
China					80
Japan	1, 784. 18	4, 075. 37	....	5, 859. 55	139
Persia					1
Siam					1
Tunis	890. 22	3, 498. 00	....	4, 388. 22	1
Morocco					13
United States	3, 576. 95	5, 183. 60	....	8, 760. 55	703
Brazil					1, 138
Republics of Cent'l and South America	1, 387. 82	815. 20	....	2, 203. 02	394
Hawaii					52
Great Britain	23, 033. 42	12, 137. 20	1, 175. 04	36, 345. 66	6, 176
Interior promenades	3, 472. 47	....	....	3, 742. 47	....
Central garden	5, 882. 65	....	....	5, 882. 65	....
Reserved garden	....	48, 350. 00	....	48, 350. 00	....
Vestibules	....	77, 792. 96	....	77, 792. 96	....
Restaurants	....	....	1, 053. 00	1, 053. 00	....
Roads and warehouses	....	....	10, 308. 44	10, 308. 44	....
Floating exposition	....	....	6, 300. 00	6, 300. 00	....
Total	158, 742. 88	303. 817. 12	21, 593. 00	484, 153. 00	50, 226

GBS_FOLDOUT

GBS_FOLDOUT

GALLERY OF THE HISTORY OF LABOR.

It was a happy thought to so arrange the antiquities as to give a connected view of the progressive development of the arts and form a fitting introduction to their present advanced condition. Even the pre-historic relics of the human race were displayed there to complete the series. The Exposition was thus not only of the present, but of the past. It gave the history of human labor in various countries from the earliest periods, and became to a great degree an exposition of the mental development of the human race. It was impossible to pass successively from the inspection of the implements of stone, bronze, iron, and finally of steel, without recognizing a progressive development of humanity. The galleries of antiquities made the Exhibition an unwritten history of civilization which every one could read, of whatever nation or language. It attracted the peasant and the scholar, and taught history and philosophy by the contrast of the productions of human labor of all periods and countries.

The French exhibit was the most complete as a whole, and was divided by partitions into a series of halls or apartments, so as to more distinctly mark the different periods.

The pre-historic period was brought boldly forward by the extensive collections which have been made in various parts of Europe during the past ten years—such as implements of stone from the bone caverns, peat bogs, and from the lake dwellings of Switzerland.

The cases were filled with enormous spear heads of flints, hatchets and other rudely-made implements formed by chipping and without polish. These occur in association with the bones and teeth of the extinct cave bear, the elephant, and the mastodon, and specimens of these were displayed in the same cases. These rudely-made implements are supposed to belong to the first or earliest stone period. A second or later period of the stone age is indicated by implements of a superior finish; such as were ground down to smooth surfaces, and in some instances polished.

All these objects of the pre-historic period were classified and displayed under the direction of a commission with Mr. Edward Lartet at the head. The interest attached to the exhibit was greatly enhanced by the meeting during the progress of the Exposition of the " *Congrès International d'Anthropologie et d'Archéologie Préhistoriques,*" the members of which were enabled to make studies and comparisons of the various collections.

The next hall contained instruments of the bronze period, extending to the Gallo-Roman. The objects consisted chiefly of cutting instruments, agricultural implements, lamps, and objects of ornament, such as bracelets of bronze and of gold, rings and pins. Of the latter a large collection contained pins with a shield for the points, and a spiral spring at

2 U E

the back almost identical in form with some of the patent pins of the present day.

The next hall was devoted to the Celtic and Gallic relics, and contained the remarkable golden necklaces from the museum of Toulouse. The representation of the work of the middle ages was characterized by a variety of church ornaments and relics, such as oak chests, seals, caskets, croziers, bronzes set with masses of rock crystal, like those of China and Japan; ivory carvings, illuminated missals of vellum, swords, and chain armor.

The fifth hall contained objects of the sixteenth century, or the Renaissance period. Here were found curiously fashioned iron locks and keys, cutting instruments, jewels, and a few nearly spherical watches. The enamels of Limoges occupied a large space, and came in great part from the collection of Baron Rothschild.

At the entrance of the sixth hall, representing the arts of the seventeenth and eighteenth centuries, a curious collection of high-heeled boots and shoes attracted considerable attention. Here, also, were seen the faïences of Rouen, and the productions of the renowned Palissy, old furniture, mirrors, inlaid cabinets, black letter books, and specimens of bookbinding. The collections of this period were continued in the halls beyond, and contained the porcelains of Sevres, richly wrought table services of silver, tapestries, miniatures, snuff-boxes, thread lace, and elaborately decorated fans.

Among the curious relics from other countries the most noteworthy were the cradle of Charles XII of Sweden, the elaborately fashioned trappings of the horse that Mahommed rode in 1331 at the siege of the town of Castro el Rio, and a variety of specimens of ancient arms and armors. The richest collection of ancient arms was sent from the Imperial Museum of Austria, and contained a number of guns with ivory stocks, richly inlaid with metal, and steel bows, also mounted in ivory.

Among the ancient ornamental works and jewels of Austria of the sixteenth and seventeenth centuries, there was a remarkable display of tankards, vases, and goblets of rock crystal, of great size, and showing a high degree of taste and skill in the art of the lapidary at that time.

There were several interesting relics and works of ancient art in the English section, among them a table covered with silver in *répoussé*, or beaten work, belonging to her Majesty Queen Victoria, and another table made in 1700. A selection of old armor from the Tower of London occupied one of the cases, and in another were various specimens of silver and gold plate, and tablets of Wedgewood's porcelain.

Although the collection of antiquities as a whole was very large and interesting, it could not be regarded as a complete exhibit of the progress of human labor up to the present time. The wonderful advances made in the mechanical arts of the present century, and the various applications of science to the arts, were not historically shown. The collection was also deficient in representations of the ancient arts and civilization of China, Japan, of Egypt, Mexico, Central America, and Peru.

GROUP I.

WORKS OF ART.

Class 1. Paintings in Oil.—Class 2. Other Paintings and Drawings.—Class 3. Sculpture, Die-sinking, Stone, and Cameo Engraving.—Class 4. Architectural Designs and Models.—Class 5. Engraving and Lithography.

CLASS 1.—PAINTINGS IN OIL.

The interior circle of the Exposition was, as already indicated, devoted to works of art. Thus, by an arrangement which, if accidental, was, at all events, poetic, we passed from the gross necessities of life such as the cereals, the wines, &c., to the machinery which represents industrial force; to the manufactures which conduce to individual comfort; to the instruments which add to the intelligent perception of all natural phenomena, and so to that last and refining phase where the immagination excites its most powerful and refining influence.

The fine arts naturally involve certain cognate professions. Group I was therefore made up of five classes, thus tabulated: 1. Paintings in oil; 2. Other paintings and drawings; 3. Sculpture, die-sinking, stone and cameo engraving; 4. Architectural designs and models; 5. Engraving and lithography. The various articles exhibited in these classes occupied a considerable but broken space in the Exposition. Several nations feeling that the space allotted to them for pictures in the first gallery, which, in accordance with the original plan, was subdivided into fourteen compartments, was inadequate, declined to avail themselves of it. They found it preferable to erect structures of their own in the Park. The statuary was more houseless than the pictures, and was scattered, not always disadvantageously for effect, through the entire surface of the Champ de Mars. The theory on which the central gallery was devoted to the fine arts was, perhaps, good, but practically it was open to serious objection. The rapidly closing concentric lines had the effect of presenting many of the best works at inconvenient angles. This was particularly the case in the small portions devoted to foreign countries, which, being in the elbows of the building, were exposed to many cross lights. Probably no two sections were more unfortunate in this respect than the American and English. They occupied the same gallery and worthily. But the United States, with nothing to complain of in their portion of the gallery itself, were unhappily compelled by the number of their contributions to take refuge for the surplus in the adjacent passage, called by a ludicrous accident of neighborhood "the street of Africa." In no respect of light or atmosphere could this be considered a favorable location; but it had its advantages in point of popularity. A large proportion of those who even transiently visited the Exposition passed through this artery, and, it may be presumed from their expressions, were gratified with and interested in the display which was provided for their examination.

One-half the entire space—and the best half because the lateral half—was occupied by works of art contributed by French artists. It does not fall within the province of a brief review like the present to discuss the merit of individual pictures, or to contest the claims of the French school of art, which most assuredly is capable of taking care of itself, and which, without question, was nobly and amply represented. It is agreeable to the writers of this report to state this at once, for, from some discussions in the preliminary committees appointed on the subject of the fine arts, it was understood that the collection, although admirable, did not by any means represent the full vigor of the nation. Owing to this cause it has been stated by writers of eminence that the display was not equal to that made at the Palais de l'Industrie in 1855.

But it was rich in the French masters who are most known and admired in America, many of whom indeed were on the jury and received the highest honors that were awarded to the class. Gérome was represented by a large and valuable collection of singularly accurate and impressive scenes, depicting for the most part the savage side of eastern life or the similar episodes of Roman history. There was nothing from this artist, however, that was unknown to Americans. The majority are familiar in a photographic form, and several are owned by our private collectors who loaned them for the present occasion. The same remark applies to the productions of Meissonier, whose minute masterpieces, difficult to obtain and highly prized by their fortunate possessors, are great favorites in America. Each of these masters contributed more than a dozen works—children of studios that had been scattered for years, but had been brought together by the interest of the Exposition and the worthy pride of their creators who gathered them together for this solemnity. Gérome and Meissonier represent the most popular form of French art, or rather that phase of it which, requiring the greatest accuracy of detail and closeness of study, produces its results at long intervals, in small forms, and with extreme concentration of thought and action. The canvases are of the most modest cabinet dimensions, and protest with singular emphasis against the vastness which vulgarizes the many battle-pieces of the larger national picture galleries. Nothing could be more dramatic or free from the clap-trap of commonplace than Meissonier's picture of "Napoleon I in Russia." The tone of the work, expressing a disaster without depicting it; its fulness of detail and clear faithfulness of particulars, cannot be sufficiently praised. A work of almost equal importance represents "Napoleon III at Solferino." Both indicate a larger scope in composition than we are apt to expect in this fine colorist and genial but microscopic artist, who usually is content with one or two figures. Gérome has been accused of hardness in the matter of drawing, and a selection of subjects which are ordinarily painful, or, at all events, repulsive. Conscious of this reproach he exhibited a painting called "Louis XIV and Molière," in which the monarch and the poet are exhibited, greatly to the advantage of the latter in point of condescension. The courtiers express their amazement and contempt at the easy ways of the writer, but the King is obviously overwhelmed. Such a subject naturally affords an opportunity for the contrast of many physiognomies, and for the display of much variety in the matter of color and costume. The success of Gérome in this new field has not been pronounced as positive. Most impartial spectators regarded the stern, nay, dismal tragedy of the "Duel after the Masquerade" with more interest than the insipid smiles and supercilious sneers of the big-wigged actors who make up the tableau of "Louis XIV and Molière." Thus it would seem, so far as the Exposition of 1867 permits us to judge, that Meissonier can step more easily and successfully out of his ordinary sphere of action than Gérome. Both, be it added, are great and strong, and the deviations noticed are a matter of curiosity rather than of criticism.

Very different from these pillars in art is Corot, a painter whose every work is extolled to the skies or condemned to pitiless ridicule by his countrymen. So far, no other people has put itself to the trouble of going to either of these extremes. In New York, Corot's pictures were exhibited without producing even a pecuniary result. They were returned with promptness to the country of their birth, and many visitors from the other side of the Atlantic were surprised to find them turn up again in the Exposition. The artist has touched the whole range of art, and his knowledge is as undisputed as his eccentricity. He has a style of his own, inasmuch as no one has ever thought of imitating it; nevertheless it has many admirers. It is characterized by a singular vaporiness of color, and a consequent faintness of outline which suggests haste, but is the result of an elaborate effort to be dreamy. To live in a constant atmosphere of fog, surrounded by objects of ghostly aspect, is not agreeable to most spectators; but such as are predisposed this way will find congenial feeling in the canvases of the eccentric Corot.

Classical art was represented by Cabanel, who had six pictures—three of the number being on epic subjects, and the other three portraits. The largest of the former was from "Milton's Paradise Lost," and represented the Deity surrounded by his heavenly ministers—an effort which is seldom successful, and was not rendered so on this occasion. This and its companions, however, displayed great academic skill and the influence of a school which makes the study of the form the first necessity of its existence, and which has recently lost its greatest exponents in the lamented Ingres and Flandrin.

In a semi-classical vein, but with a quaint infusion of sentiment and allegory, were many works, mostly by artists who owe their education to

the liberality of the government.   The productions of Hamon, Bougereau, and others, are of this agreeable class.   These gentlemen have
each, at various times, taken the *prix de Rome*.   This is more than a
recompense: it is like a presentation to a college, and means a classical
education.   Those who are fortunate enough at the academy competition
to gain it are, for five years, nursed and cherished as men of superior
ability and trained in a settled and severe way.   They are sent to Rome,
and during each year of their sojourn in the Imperial City they are
expected to send specimens of their progress to the powers of the
academy in Paris.   These specimens are preserved with national care,
being placed in a building where they are at all proper times exposed to
the view of the public.   The Palais des Beaux Arts and of the Luxem-
bourg are, to a great extent, representations of the art progress of the
country, and the pupils who each year contribute to their treasures
remember that they have vast reputations to contend with.   They are
encouraged, too, with the reflection that these reputations were no greater
than their own when their fortunate possessors sent their first contributions to the academy.   It may be interesting to state that the earliest of
these canvases was sent by Sarrabat, and bears the date of 1688.   The
school, which also comprises an academy of architecture, was established in 1648.—the architectural section being founded in 1671, and the
pupils being sent to Greece instead of Italy.   It includes also three
studios for sculpture, one for copperplate engraving, and one for engraving on medals and fine stones.   A competition for the *Grand prix de
Rome* takes place every year for painters, sculptors, and architects;
every two years for engravers; every three years for engravers on
medals and fine stones.   After remaining two years at Rome the young
students are permitted to travel.   Engravers on medals and fine stones
have only three years' provision made for them, and must remain two
years in Rome.   The governor of the establishment sends official reports
every six months of the progress and pursuits of the pupils.

So far as painting is concerned strict attention to design is of the
highest importance.   But as the manner of the age drifts slowly from
the stern manner of the ancients we find, as in the case of the artists
just named, a tendency to fanciful subjects, with just sufficient of the
classic element to remind the spectator of good training and of the intellectual restraint of other and older schools where inanimate art was con-
ventionalized by uniformity and straightness in such things as foliage,
and animate art was confined almost exclusively to the exhibition of the
nude figure.

Hamon's pictures are familiar everywhere.   They have been reproduced
by the process of the engraver equally with that of the photographer.
Every one has seen in some way a reproduction of his "Aurora," where
the goddess of morning sips from the lips of the cup the first liba-
tion of day.   The pose of the figure is charming, and whilst showing
the coquettish knowledge of the female form which French artists pos-

sess and display with a gracefulness all their own, it seems also to answer the purposes of the life study which those who win the *prix de Rome* are expected to pursue. Bougereau is perhaps less known in America. He is more severe than Hamon, and his sense of color is more positive. The object of referring to these artists is not so much to explain what they have done, and still less to tell the American people how they have done it, which indeed would be a difficult task. But, to add a further statement, they occupy a very prominent position before the most intelligent community in the world. Whatever comes from their easel is in demand—great demand—a demand which can scarcely be supplied. It is pleasant to know, therefore, that a portion of the time so much occupied is devoted to other purposes. It is to the co-operation of such thoroughly informed artists that the government manufac. tories of France owe their unquestionable pre-eminence.

The government of France indeed exercises a direct and practical influence on art which cannot be overestimated. It is paternal in the means it affords to its youth to avail itself of the opportunity to study, and it is liberal in purchasing what has been done. Out of the 625 numbers mentioned in the catalogue as appertaining to France, no fewer than 252 are contributed by the government. Many of the others, as we have before hinted, were loaned for the special occasion of the Exposition, being traced by their painters to their distant homes in the Old and New Worlds.

The dramatic phase of historic art—that in which an action is expressed to the eye—was very largely represented. From the soldier who wraps his wounded leg in his pocket-handkerchief, to the tyrant who lays his head prayerfully on the block, it is in this department the same thing, namely, a matter of what can be remembered or felt, and mainly in French art, of what can be remembered. The innumerable, colossal, and tedious battle pieces which prevail in every museum of France are an evidence of this. Versailles tells the history of France with the coarse, smoky gusto of a dragoon. Throughout the pitiless range of chambers there is not a scene which recalls a pleasing incident of battle, of triumph, or of defeat. The battle pieces at the Exposition were almost entirely of this character. They displayed an idea of action, a thorough sense of what is called situation, and an utterly faithful amount of details, topographical, military, and otherwise. To the eye not inanately tutored to the beauties of red, there seemed too much in these productions, but the uniform which offends the foreign eye from its brilliancy is naturally the recognition point of Frenchmen, and appealing with earnestness to the recollection, recalls the liveliest interest. The government, of course, was the principal exhibitor in this department. The pictures were the product of commissions given to various artists and intended for, or borrowed from, various museums of the country. History and poetry alike delight to record the triumphs of valor, but it is only of late years that painting has attempted to do so. The attempts have nowhere been so successful as in France. It may be questioned if any one will desire to essay more than Yvon has accomplished, an artist of splendid abilities, whose two pictures of the "Taking of the Malakoff," and the "Struggle in the Gorge of Malakoff," are perfect, but it may be asked if such gigantic productions are desirable even as records of patriotism. As works of art they excite the regret that such splendid ability should be thrown away on a scene which could be rendered with greater effect, and precisely the same color, at a minor theatre of the city.

Nearly five per cent. of all the pictures exhibited in the French department were battle pieces. The three which from their real sentiment and vigor of drawing attracted the most attention were by Protais: "The Morning before the Attack," the "Evening after the Combat," and the "Return to Camp," a work of very singular vigor, although windy, and which was contributed by the celebrated Bellanger. It depicts the episode of Waterloo, described in every French history, but which Victor Hugo has put himself to the trouble of refuting, namely, that the Old Guard was prepared to die, but not to surrender.

In animal paintings the French department was represented by Rosa Bonheur, Fromentin, and Troyon, deceased. Animal paintings, or, to speak more closely, the desire for animal paintings, is the fancy of a day. Judging from the productions of the artists named, it would seem that the fancy is somewhat out of fashion. Rosa Bonheur's powers were finely represented, but recent productions of the lady do not maintain her very high reputation.

Of that large class of subjects which are called "genre," and which relate to little episodes of life or peculiarities of costume, there was an endless variety. Among the most prominent of French artists in this respect, may be mentioned Plassan, Fichel, Poulmouche, and Wetter, who each exhibited a number of interesting figure subjects charmingly suggestive and exquisitely painted. Of the painters of rustic life, Breton and Millet preserved their well-known pre-eminence.

In the way of landscape artists, the most agreeable and well known were Theodore Rousseau, Lambinet, Daubigny, Cabat, and Dupré; the most singular was Corot.

The French collection, as before remarked, consists of no fewer than 625 pictures, of which many were the personal property of the Emperor or the nation. It was said by French critics that the display did not indicate any progress, and contained very little that was new. With few exceptions all the important pictures had been exhibited elsewhere. This remark, however, applies with equal force to every other nation. The fine arts department of the great undertaking was intended as an exposition, not as a competition. Otherwise it would have been unfair to have given such marked preference to reputations. As an exposition it was exceptional excellence, and represented very forcibly the prominent position occupied by several artists of France.

There were four nations who, not finding themselves sufficiently provided with space in the interior, obtained permission to build, and thereupon erected galleries of their own in the Park. These were Belgium, Switzerland, Holland, and Bavaria. Of these outside collections the most important was that made by the government of Belgium, it consisting of 186 pictures, and, as in the case of France, it was more a display of individual and well-established reputations than a competition of numbers. Of the 186 frames no fewer than 52 were contributed by five artists only. These were Leys, Stevens, Willems, Verlat, and Clays, (marine.) The names suggest almost everything that is vital in the Belgian school. Of the five, the least known in America is Alfred Stevens. This artist has no fewer than 18 pictures, all of them of cabinet size, and having for subjects familiar episodes of life, many of them touching and simple, and all of them interesting to the eye. Thus the picture called "Tous les Bonheurs," representing the serene content and bliss of a young mother nursing her infant, may be cited as a happy illustration of the artist's powers. Stevens paints with great boldness, and his coloring from its brilliancy is occasionally offensive to the eye, but his power is unquestionable. In his selection of subjects, however, he sometimes borders on the "demi-monde." This is a fault which cannot be charged against his colleague Willems, whose extreme delicacy of fancy is apt to invade the realms of the insipid. No one ever understood the swirl of a lady's satin dress better than Willems, whose knowledge of this texture is singularly exact. Indeed, the details of all his work are remarkable for their truth and delicacy. They are never in the way, and interest the mind only as a part of the recollection of a very charming impression. The subjects selected by Willems are of the simplest character, and neither suggest invention nor any other form of intellectual activity. But as they invariably represent a lady of refined appearance and elegant costume, with hair and eyes of exquisite hues, they never fail to be interesting. As specimens of faithful and conscientious work they are unequalled. The most important work exhibited by this renowned artist, and one which marks an ambitious step in the way of composition, was "L'accouchée," a quiet interior which two visitors are entering on tip-toe. A young wife sleeps peacefully on a bed, and not far from her is the nurse holding in her arms the first offspring of a happy house. The tone of the picture and the treatment are in every way admirable. The subject too is clearly expressed; a soft and tranquil stillness, not of death, but of exhaustion, hangs about the apartment like a spell. It would be a sin to disturb that fair young mother. "Two lovers exchanging a ring," another large picture—if the term can be used of this artist, whose canvases are always of the smallest—displays the indications of a new style, bolder in color and in treatment than that with which heretofore we have been familiar. There is no artist, possibly with the exception of Coomans, who understands so thoroughly how to harmonize the most delicate tints.

A thorough contrast in this respect is found in the 12 works exhibited by the Baron Leys, the pre-Raphaelite prophet of the Netherlands. The characteristics of this singular mediæval style are too pronounced to escape notice. The prevailing color is dead red or brick color. Bricks indeed of every color are favorite objects with the baron, as also are the cobble-stones which line his thoroughfares with painful distinctness. All the figures stand with their legs astride, a position more comfortable than graceful. All the legs are in red stockings, which, added to the cobble-stones and the bricks, contribute to a massive monotony of tone which, no doubt, is highly characteristic of the period and might serve as a warning to the present generation. In the faces there is invariably a painful expression, as if the toothache were a mediæval invention that had recently been discovered. It is impossible to resist the laughable side of this school. But it has another and a serious significance. These lurching and lugubrious figures that seem to be falling out of the frames are at least correctly garbed. Every detail of dress or habitation or decoration is the result of learned investigation and study. The details of Leys's pictures are revelations of archæological lore. To a certain class of minds, too, this seeming antiquity is irresistible. Leys's pictures were of all shapes and sizes. The subjects were taken for the most part from the stirring period of the great struggle with Spain for religious and civil liberty in the sixteenth century.

Verlat's tendencies are more classic. He exhibited a very beautiful "Virgin and Child," a work quite exceptional in its excellences. Also a "Dead Christ at the foot of the Cross."

It would be superfluous to speak further of the pictures in this excellent collection. The tendency of the Belgian school is ambitiously French, except in the case of Leys, who is individual and pre-Raphaelistic.

The government of Holland exhibits 170 pictures, among which are many works of unquestionable excellence. Israels is the head of this school, and is distinguished by delicacy of sentiment and simplicity of statement. He had five frames, all of which were worthy of attention. But it is evident that this artist and nearly all the others in the gallery attach more importance to the teachings of the French school than to the traditions of the Dutch. Bles, Alma, Tadema, Bukkerhorff, Schendel, Scheltema, and Verveer, contributed acceptably to the display.

The Swiss collection was composed of 112 pictures, most of them of local interest. Where indeed could a Swiss artist find grander scenes for study than those of his own country?

Bavaria, as we have before mentioned, had, like the three preceding countries, her own building in the Park for the display of her art treasures. Her principal artists were Piloty, Horschelt, Adam, Schnets, Schwind, and Lizzenmayer, in figure subjects; Woltz, in cattle pieces; Lier, in landscapes; and Lenbach, in portraits. The number of oil paintings contributed by Bavaria was 211. A large proportion were, avowedly, sent for sale, and hence the display was neither so national nor so good as in other countries.

Prussia, for reasons of various kinds, did not do justice to herself. Many of her best artists were unrepresented. The number of works in all was but 98, and a large proportion was the property of the artists. Nevertheless there were several works of interest, such as Knaus's "Saltimbanque," well known by the engraving, and others equally familiar to the frequenters of our print-shops. Knaus's style is genial and earnest, and he possesses the power of concentration in an eminent degree.

Austria contributed 89 pictures, the most important of which was the "Diet of Warsaw, 1773," by Matejik, a very bold and well-distributed composition, laid on in heavy but effective masses of color.

Spain was represented by 42, Portugal by 23, Greece by 4, Denmark by 29, Sweden by 54, and Norway by 45 oil paintings.

Among the 63 contributions of Russia were several that attracted attention. The subjects were mostly original, but the treatment had no distinctive national characteristic. It was, however, good, and worthy of comparison with the best in the gallery. Such comparison would be out of place here. The principal contributions were Gué, sacred subject; Simmler, history; Peroff, Rizzoni, and Popoff, genre; Kotzebue, battles; and Clodt, landscapes.

Italy, the mother of arts, contributed 51 oil paintings, none of which were distinctive, and but few of which were above mediocrity. The Papal states sent 25, Turkey 7 paintings.

Next in the order of the catalogue—which we have followed, except when speaking of the establishments in the Park—came the limited space allotted to the United States of America. In another portion of this report, devoted to the special consideration of objects exhibited in the American department, will be found a description of the 75 works there put on view. The collection was in every way a creditable one. The foundry scene of Weir was the best work of the kind in the Exposition; indeed, it was entitled to even greater consideration, for it was the only work of its kind. The landscapes of Church, Kensett, and Bierstadt were also eminently national, and the productions of Boughton, Huntingdon, Hart, Johnson, Healy, Hunt, Whistler, &c., drew the attention of connoiseurs who knew nothing of their origin. For, be it remembered, most of these paintings occupied the extreme end of the English gallery, and it was natural to suppose that they formed a portion of it. This in itself was no advantage. Nothing can convince a continental critic that art is either known or practiced in the British isles; and, owing to this cause, the stranger paid but passing heed to what was there displayed.

In the schools we have so far hastily glanced at there has been a certain uniformity of effort, which we have explained by describing the mode of study practiced by France. French influence in art at this moment extends to every continental country. The distinctiveness of the Dusseldorff school is rapidly disappearing. That, too, it is evident, will become French. It is useless to look elsewhere. But if we cross the channel we shall find a totally different state of affairs.   Instead of 5,000 men who paint precisely alike, and differ from each other only in the order of their intellectual, emotional, or mechanical force, we shall not find five who have agreed on any settled plan or style.   The lack of regular methodical instruction, combined with a total, or almost total, deficiency of government support, throws the art student entirely on his own resources.   He is compelled to seek the manner which is readiest to him, and select the subjects which are more congenial to private taste. The government will neither show him the way which is best, nor reward his efforts for pursuing it successfully.   A certain number of picture galleries, to be sure, are provided, and the student may do as he likes about following the style of any master there exposed.   No direct influence controls his studies, and he consequently wanders.   There is something to be regretted in this, but a great deal, also, to be commended.   Self-help is tedious and slow in its results, but it has often proved that it is the best kind of help, and certainly in art, as in everything else, it has shown on many occasions that it is better than blind subjection to established rule.   There is character in the English exhibition, as there is in the American—so much character, so much contrast, so much individual effort, that the dilletant who is familiar only with the smooth competition of the schools is bewildered, and condemns where, perhaps, it might be better to investigate.   Certain it is that the French critics have been unusually severe on the English exhibition, and also on the pictures exhibited in the American section.   The remarks we have made may seem an easy way of accounting for this severity.   They have, at all events, their value with unprejudiced persons.

CLASS 2.—OTHER PAINTINGS AND DRAWINGS.

Under this general head were comprised miniatures, aquarelles, pastels and drawings of all kinds; paintings on enamel, on porcelain, on crockery; cartoons for frescoes and for glass windows; mosaics.

Water-color drawing (aquarelle) or painting is, comparatively speaking, a new art.   It has been brought to its greatest perfection in England, where Turner is still regarded as its best exponent.   On the continent it has attracted some attention, but it is regarded with distrust. Water-color drawing differs from oil-color painting in many mechanical matters of detail.   The separate names of these two arts suggest the most important of these differences; the one is wrought in oil and the other in water.   But beyond this there is a general distinction, which is often overlooked: In a water-color drawing all the colors are transparent; the "lights" are obtained from the original surface on which the drawing is made.   In oil color, all the lights are superimposed on the canvas, and the original surface is of no value at all.   Some of the finest artists that England has possessed have devoted attention to this very pleasing branch of art; among others, may be mentioned Turner, Cox, Dewint, Hunt, Copley, Fielding, and Stanfield.

The only important collection was in the English gallery, where the pictures, glazed and framed, occupied swinging panels in the centre of the apartment. Other nations, in their respective departments, contributed a few specimens; but the whole, put together, were greatly inferior in number and quality to the English. It was intended in this, as in the case of the oil painting, to illustrate the past ten years, not to assert positively what had been done from the very recent date of its birth. The drawings were, of course, entirely supplied from private sources, the government having no museums from whence to draw a supply. Of late years these private sources have been called upon very often to give up their treasures. Local art exhibitions have been rife in England, Ireland, and other parts of the British isles. Pictures have been borrowed, and, after due exhibition, returned to their owners in an injured condition. It has been affirmed that, owing to these causes, the owners of valuable works declined to run the risk of sending them across the channel, and that, in consequence, the collection, good as it was, could scarcely be said to represent satisfactorily the present condition of the art in England. Nevertheless, there were many works of sterling value, and nearly all were worthy of examination. It would be useless to describe the excellencies of particular frames, but it may be serviceable to refer to the comments of an admirable artist, who seems to think that the art has taken a downward tendency. He bases this opinion on the ground that in nearly every picture exhibited opaque colors were used. By this expression he meant little masses of mineral substance placed in prominent places, and heightening, by a sort of embossed brilliancy, the effects of the lower tones. It is affirmed by the best critics that water-color drawing should be entirely transparent, and that this tendency to overlay the natural source of the light is meretricious. Moreover, it is known to be detrimental to the permanent value of the drawing. The imposed substance drops off, from climatic causes, and is especially effected by the glass covering which gives protection to the other parts of the picture. This point is of importance to purchasers of water-color drawings, and of interest to artists who may not themselves be familiar with a fact which, while increasing their present popularity, endangers their permanent fame. Mr. Horsley speaks feelingly on the subject. He says:

"A water-color draughtsman who cherishes the beautiful ground he works upon for his lights, or, if he has lost this, scrapes or washes them out, has a far harder and more anxious time of it than he who, by the aid of opaque mixtures, dabs them on in a moment and renews them at pleasure. It may, however, readily be conceded that another and worthier reason for the use of opaque color is the yearning of the artist to have substance and solidity in his material; but when he feels this, and that he is flagging in devotion to those qualities of art which water-color, and water-color alone, can produce, he should become an oil painter, and cease to be a water-color draughtsman."

There were scanty displays of water-color drawings from France, from Austria, from the Pontifical States, from Greece, from Sweden, and from Russia. The latter were by far the best. China, too, exhibited a distemper painting of almond-eyed beauties, with skins that seemed to have been wound up tight by means of the hair-dresser, and their under lips painted green.

Of pastels and drawings of all kinds there was no end. Every design, indeed, could be brought under one of these two heads, and almost every country contributed to the store. The word "pastel" in these days means anything from chalk up to body color. The French department offered fine specimens of the various processes. Bavaria presented a remarkable display of drawings and models, showing the various stages of study from the cast and from life.

The subjects of "painting on enamel, earthenware, and on china," do not greatly interest the American community, except in their practical bearing on housekeeping; but in Europe they engage the attention of the better classes, and give occupation to the highest kind of skilled labor. Thus, while it happens that beauty and permanency are often attained, it is often the case that the local fame of an artist and his tedious patience take the prize which the former should have commanded. Mr. Horsley, writing on this subject, says:

"It seems necessary to bear in mind not only the principles of art that should be applied to these various branches of industry, but also to suggest that peculiar abstraction of mind is in some instances requisite in order to appreciate the results, as far as the arts of pictorial designs and execution are concerned. Take, for instance, what it is presumed would be considered the highest class of enamelled works in the Exhibition—those of Lepec and Rudolphi, who exhibit enamels on gold and other metals. The pictorial art exhibited in those works is both puerile and bad, as, for example, the 'Angelique and Roger,' by Lepec, which is placed among the French miniatures. Nothing can be less worthy of regard, in an artistic point of view, and his portrait is little better. Lepec has also a case of enamelled vases, executed with the rarest skill and ability, with fabulous prices attached to and given for them; yet the painting which is intended to ornament these *objets de luxe* is quite beneath notice. Again, look at the series of elaborate enamels in porcelain in the Bavarian *annexe*, by Wimmer, of Munich, and other German artists, after well-known pictures. What are these but wicked copies of immortal works?—so bad as to be irritating to the artist who looks at them; copies, which, if made on canvas or paper, would not fetch as many pence as the pounds which are now given for them. Then, what quality is it that makes these productions so readily marketable? It can be only that of permanency—a quality appealing to minds so constituted as to derive satisfaction in the possession of 'Angelique and Roger,' of Lepec, or one of Wimmer's travesties of Raphael and Rubens, simply because they are works which will never tone with age or fade with time. Great as may be the charms to some minds of the sense of permanency, it must be permitted to those of more artistic sensitiveness to assert that this quality does not compensate for other wants."

The same able critic also makes the following remarks, which are in every way worthy of attention:

"To come to what may be termed painting proper upon porcelain, *i. e.*, the decoration of vessels of various forms for ornament and use, it may be submitted that the general principle to be observed in applying art to such work is that it should harmonize in every way with the forms receiving it. As these forms are of a well-defined and architechtonic form, so the pictural adjunct should, as far as possible, partake of the same qualities. Thus, speaking broadly, all landscape subjects and those requiring picturesque treatment are undesirable and incongruous for the object in view. Occasionally in the present exhibition you will come upon a vase on which a landscape is painted, which commencing on the body of the vessel, is made to meander (trees, sky, buildings, and all) over the concave and convex forms to be found at its neck. Can there be a more absurd departure from true taste in ornamentation than such an example as this?"

There were many cartoons for stained glass and fresco, but they were of interest chiefly to artists who work in this extensive way. It is hardly desirable to refer to productions which may never come before the public again. In America everything that is painted on a ceiling or a wall is called a fresco. Such work is ordinarily executed in distemper, in wax, water-glass, or oil. True fresco has a peculiar quality of its own which eminently distinguishes it from all other methods of painting. It is this: that a fresco is a non-absorbent of light. The fresco ground is composed of certain proportions of lime (from which the heating element has to a great extent been washed out) and sand, and this mixture is used by the painter in its moist state. The wet lime, absorbing carbonic acid from the atmosphere, becomes carbonate of lime, and in combination with the sand produces an impermeable cement which is formed over the surface of the ground during the day's labor, and in which the color used is incorporated and fixed. This cemented surface has been stated to be sufficiently crystalline to reflect light; but whether this be so or not, its non-absorbency of light is unquestionable. Thus, where an oil painting would be invisible a fresco is clearly seen.

The Russian mosaic work was by far the finest in the exhibition and deservedly attracted much attention. It came from the atelier of Michel Chmielevski, of St. Petersburgh, and was designed by Professor Noff. The subject was a group of ecclesiastics in their vestments, and the object the decoration of a Greek church. The Roman mosaics were far inferior.

CLASS 3.—SCULPTURE, DIE-SINKING, STONE AND CAMEO ENGRAVING.

It would be impossible in the space devoted to this report to do justice, even cursorily, to the many specimens of sculpture exhibited in the various sections of the Exposition, and it may be added, too, that it would be entirely uninteresting to do so. To the majority of people, statuary, at best, is a sealed book. It creates no sensation when it is visibly before them, and it would certainly create less, if it were possible, when simply described by the feeble power of a reporter. Nevertheless, it is the grandest, most ancient, and most durable of the arts. The works which delight the critic of to-day and are believed to mark the golden age of statuary, date their origin many centuries before the Christian era. The full beauty of the human form has never been so accurately described as by the Greek sculptors. The mythology of the country gave to their efforts an elevation and purity of thought which in these days cannot be conveyed to similar subjects however skilfully manipulated. Hence the tendency of sculpture has been to moderate the severity of the ancient school and to create another in which clothes should not be wholly disregarded. The toga imposed itself on the thoughts and consciences of artists. Were it a booted warrior with a cocked hat that had to be depicted he was found clad in the garb of a Roman senator. An absurdity so conspicuous could not long continue. A new school sprang up. Its aim was to call a spade a spade. If top boots and a cocked hat were wanted the disciples of that school were ready to supply them. Nay, if Achilles, himself, in addition to his one natural defect, had also had a pimple on the top of his nose, they would have alighted upon it with enthusiasm. Excess of any kind naturally leads to reaction, and a reaction took place. But the various theories still remain. The purists and the realists contend for their separate ideas, and the able men on either side prove how easy it is for both to be right.

There never was a better battle-field than the Champ de Mars, where statuary of colossal proportions contended with the humbler but equally interesting productions of our own Roger, whose small domestic groups for the mantelpiece are well known to loyal people. Nothing could be more realistic than these touching incidents of the late war. While thus bending, as all young nations will, to the ideas which are newest, it happened curiously enough that the gem of the classical school was also of American origin. The composition referred to was by Miss Hosmer, and was called the "Sleeping Faun." The attitude of the principal figure is graceful and natural, the expression of the face thoroughly winning. A mischievous child faun is most happily introduced in the group. He is partly hidden behind the trunk of the tree beneath which the elder faun is reposing, and amuses himself by knotting the tail of the latter into the tail of a lion's skin upon which the elder faun reposes.

The French statuary, by its numbers and the variety of its styles and subjects, was considered the best. The Italians also exhibited much that was very marked in character, and sufficient to show that in this respect Roman art yet maintains her own. One of the most striking statues in the Italian vestibule was "The Last Days of Napoleon I." This was another realistic work, and, so far as execution went, its details were worked out with a skill and power of execution that was not to be found elsewhere. But its subject was painful. It may be questioned whether any amount of skill justifies an artist in exhibiting a hero in so decrepid, diminute, and hopeless a condition. Sculpture has nothing whatever to do with decrepitude. Its office is to ennoble and idealize the grandest types of humanity. Napoleon seated in his arm-chair, with his head drooping forward, his eyes heavy and sad, and the hour of dissolution visibly upon him, is a spectacle which robs history of a hero. The French, however, were satisfied with the work, and a gold prize was awarded to the artist. It may be added here that there was a very curious and interesting collection of busts of Napoleon I. They were six in number; but only three or four of the six were derived from authentic sources. The authority for the last, "Napoleon at St. Helena," may be disputed, and the first, representing him as a child, has no other authority than an apocryphal sketch in pencil which may be seen yet at the Louvre. Taking them, however, as real presentments of the boy and the man, they are in the highest degree interesting and valuable.

In the Belgian department were exhibited some small terra cotta models belonging to the familiar picture sculpture school and representing scenes from domestic life and from Shakespeare and Moliere. Their merit consisted in their broad humor and true expression, to which may be added great care and ability shown in the modelling.

The sculptor Westmacott, in concluding his official report on the statuary of the Exposition, says: "The impression left by a careful examination of the works in sculpture of different nations is on the whole of a favorable character. That there is much that challenges criticism must be admitted; but the general practice of the art affords satisfactory evidence that while its employment is very greatly extended there is also manifest improvement in sculptors, in knowledge of form and in a feeling for the beautiful, showing the value of close observation of nature regulated by the discipline derived from a careful study of the best ancient examples. There is also considerable technical power shown in execution, in carving, modelling, casting, and chasing, proving beyond question that in the material exercise of the art there is good ground for congratulation."

French artists have long been eminent for their attention to and skill in medal engraving and die-sinking. It has always been the practice of France, from a very early date, to encourage these arts, and the sculptors have worthily responded to the patronage and protection thus accorded. Some of these works in the present Exposition were of large size, consisting of groups and compositions admirably treated. Others displaying beautiful workmanship, although merely portraits, were, in fact, gems of art.

3 U E

CLASS 4.—ARCHITECTURAL DESIGNS AND MODELS.

The display of architectural designs and models was ample. The latter especially exhibited remarkable skill of production and elaborateness of detail. Both pertain to subjects that do not come within the range of this report, which is not technical but general. Among professional men it was thought that a better show might have been made, particularly in the case of works that are now actually progressing. The most perfect exhibition was made by the Suez Canal Company, which, topographically, architecturally, and otherwise, exhibited the difficulties which beset that great undertaking, the way they have been overcome, and what yet remains to be accomplished. These details occupied an entire building in the Park, and formed a special attraction of themselves.

CLASS 5.—ENGRAVING AND LITHOGRAPHY.

The subjects in this group appeal in a thousand ways to every taste, and are especially valuable alike for amusement as for instruction. There is hardly a work of any importance in the scientific world that does not in some way appeal to or depend upon one or other of these sister arts. The larger and more important part of all engravings are transcripts from paintings, and this mode of reproduction has of late become so popular that the number of those who pursue the profession, which was declining, has greatly increased. Of the innumerable body of engravers on wood it is impossible to speak. A fair exposition of their products would have filled half the building. There has been no marked improvement either in engraving or lithography during the past decade, save what could be traced to increased skill on the part of those who exercise these professions.

GROUP II.

APPARATUS AND APPLICATION OF THE LIBERAL ARTS.

CLASS 6. PRINTING AND BOOKS.—CLASS 7. PAPER, STATIONERY, BINDING, PAINTING, AND DRAWING MATERIALS.—CLASS 8. APPLICATION OF DRAWING AND MODELLING TO THE COMMON ARTS.—CLASS 9. PHOTOGRAPHIC PROOFS AND APPARATUS.—CLASS 10. MUSICAL INSTRUMENTS.—CLASS 11. MEDICAL AND SURGICAL INSTRUMENTS AND APPARATUS.—CLASS 12. MATHEMATICAL INSTRUMENTS AND APPARATUS FOR TEACHING SCIENCE.—CLASS 13. MAPS AND GEOGRAPHICAL AND COSMOGRAPHICAL APPARATUS.

CLASS 6.—PRINTING AND BOOKS.

The principal contributions in this class were from France, Austria, England, and the United States. The following extracts from the introduction by E. Dentu, to the catalogue of the exhibitors in the French section, present a condensed view of the condition of the publishing trade in France, and some general observations upon the present state of the typographic art:[1]

"The productions comprised in Class 6 may be divided into eight sections: I. Specimens of typography. II. Autographic proofs. III. Lithography in black and colors. IV. Engravings. V. New books and new editions of various works. VI. Collection of works forming special libraries. VII. Periodical publications. VIII. Drawings, atlases and albums, technical or educational. This class includes 144 exhibitors from seventeen departments of France. Paper and ink, and in a less degree vellum, and objects in paper and pasteboard, are the raw materials of printing and the library. These articles make part of class 7. Good quality of the raw material, and perfection in the manufacture, are the essential requisites for paper, which, in the form of books, lithographs, or engravings, is destined to bear the test of time. The facilities afforded for the export of rags from France have not yet been counterbalanced by the employment of substitutes so eagerly sought in the manufacture of printing-paper. Periodical publications, produced in large numbers and of ephemeral interest, alone, employ paper containing ligneous or other substances mixed with waste textile materials. Parchment and vellum are only used for a few special matters; such, for instance, as patents and diplomas. The imitations of vellum in paper, having the strength and surface of the skin, are more generally employed in choice editions. The quality of the ink has a great effect on printing and on the beauty of the work produced; its price varies according to the degree of fineness.

---

[1] This and the subsequent extracts from the Official Catalogue have been taken from the English version, published under the authority of the Imperial Commission by J. M. Johnson & Sons, London.

It should dry rapidly, give clear lines, and reproduce the finest strokes. The manufacture of colored printing-inks has been much improved, and they are now applied in many ways in printing. The series of colors and tints is very varied; some are remarkable for tone and brilliancy; but, unfortunately, their price is relatively high, especially in the case of those which include the aniline colors in their composition. Since the day when Guttenberg conceived the idea of producing the characters of the text accompanying engravings in movable types, to the commencement of the present century, the improvements introduced in the art of printing were but few. Sixty years ago hand-presses were still in use, with the vertical pressure which had replaced the originally lever arrangement; the ink was still ground by hand with a muller, and the ball still inked the type or engraving in relief. The impression was still taken from the forms composed of movable characters. The progress of modern society soon rendered these primitive means insufficient. The problem to be solved was, how to arrive at the most rapid and most economic production. This was resolved by the invention of stereotyping, or method of converting into single plates the pages composed in separate types. The galvano-plastic process afterwards enabled the stereotyped plates to be formed with increased rapidity, and, moreover, assured their preservation. The transformation was completed by the invention of cylinder machine. Chromo-lithography, or lithographic printing in several colors, in consequence of improvements in the methods of registering, and in the facilities of multiplying without great cost the number of stones necessary for the printing in various colors, has assumed enormous importance. It has thus been made applicable to the demands of trade, especially in the production of decorated tickets and show-cards. One of the happiest applications of chromo-lithography is the reproduction of the miniatures and stained glass of the middle ages, and the publication of *fac simile* copies of ancient manuscripts and illuminated missals. Independently of designs executed directly on the stone, lithography is applied to the printing of maps, engraved drawings of machinery, to writing transferred to stone by means of autographic paper, to copperplate and wood engravings, and to typographical printing. Photo. lithography, which has for its object the obtaining of photographic pictures on stones, and the production of printed impressions, begins to yield some practical results. Copperplate printing, which consists in inking a copper, steel, or pewter plate by the ball or by the hand, is still executed by hand-presses; the mechanical processes attempted have yet yielded but small results. Engraving and ornamental printing has been greatly aided by the galvano-plastic process, which supplies stereotype plates as perfect as the plates or block cut by the engraver, and which thus allow an unlimited number of impressions to be taken without affecting the original. The plates furnished by this process for chromo-typography, or typographic color printing, possess an exactness which it has been found impossible to obtain by other means. They enable the printer to produce for a few halfpence excellent impressions worked from fifteen to twenty plates, in register, each with a different color or shade.   The numerous and persevering attempts made to reproduce in relief the original designs of the artist, and to convert drawing and writing on stone into typographic stereotype plates, have yielded, if not perfect results, at any rate sufficient proofs that the problem is in reality solved.   Paticonography, a chemical process which produces blocks in relief from the hollows of engraved plates, is now employed in the illustration of many important publications.   It is used with success for printing maps, *fac similes*, and music.

The publisher is, at the present day, a real producer; carrying on, not a house of business, but a sort of collective workshop, in which the designer, the engraver, the printer, the paper-maker, &c., work together under his guidance with a fixed object.   He has also another claim to the title of producer.   He not only issues new or old works in choice or popular editions, but he creates collections of works with special objects, periodical or encyclopedical publications, and supplies subjects for treatment.   It is by such combinations that the greater part of the extensive publications now issued are brought to light.   The extension of the home trade in books would be considerably increased if the law of colportage, (hawking and sale at stalls,) and the limitation of printers' licenses, did not diminish the means of action.   Working printers are divided into two classes : those who work by the task and those who are paid by the day.   Compositors employed at task work receive for a thousand letters [ens?] 55 centimes to 1 franc 40 c., according to the type employed, and the language in which the copy is written.   Those who work by the day are paid according to a tariff arranged by the employers and workmen in common, and of which the lowest rate is 5 francs 50 centimes per day for ten working hours.   The pressmen stand in the same condition, and their wages are as high as those of the compositors.   The workmen who attend the machines only earn 4 francs a day, and the children employed as assistants receive from 1 franc to 1 franc 50 a day.   Wages in the provinces are about 30 per cent. lower than in Paris.   The employment of women in printing establishments, after having encountered great opposition, has at length been carried out, and gives very satisfactory results.   The wages which they receive are very nearly the same as those of the men.   The great printers have established relief funds; but only one in Paris, equally prominent for the importance of his business, and his personal character, has admitted his workmen to a participation of profits.   The principal centres of the business are: Paris, Tours, Rouen, Lille, Lyon, Limoges, Rennes, and Epinal.   Strasburg stands in the second line; and afterwards come Bordeaux, Marseilles, Grenoble, Caen, and Chatillon.   The printers are divided into typographical printers, who number about 900 in France; and lithographic printers, amounting to 800, of whom 391 are in Paris.   As to the copperplate printers, Paris possesses about 138.   There are but very few in the provinces.

The number of works printed in the year 1866, including new books as well as reprints of all works, amounted to 13,883. Of this number the "Belles Lettres" and novels form the greater portion. Political and religious works amounted to nearly 2,000; history, geography, voyages, and travels to nearly 1,500; scientific works, 1,900; works on commerce and agriculture to nearly 1,000. The production of engravings, lithographs, photographs, plans, maps, charts and drawings of all kinds, amount to about 30,000; to which must be added 9,000 publications of vocal and instrumental music. These productions represent on an average 20,000,000 of francs in the total exportation of France, and employ 2,500 tons of paper. There are also printed in France 1,771 periodical publications, of which 336 are political journals, and the remaining 1,435 literary, scientific, and miscellanous. Among the improvements introduced into the printing and bookselling trades since 1855, the following may be pointed out: 1. The variety and clearness of the types produced in the foundries, and the better choice of types employed in the printing of books, as regards the subject and the object of the publication. 2. The progress made in chromo-lithography and chromo-typography. 3. The improvement made in stereotyping, both as regards rapidity and perfection; the development of stereotyping by the galvanoplastic process, and the employment of paniconographic stereotype plates. 4. The improvement and cheapness of the impressions obtained by the excellent method of cutting employed in engraving, and the general introduction of improved printing presses driven by steam; the satisfactory result obtained by the application of these presses to lithography and chromo-lithography; the skill exhibited in the composition of tabular matter; and, above all, the increasing number of printing establishments capable of executing difficult work with great perfection."

The exhibition from the United States was by no means as complete as it should have been. Only two or three of the prominent publishers were represented by their publications. D. Appleton & Co. sent a bound copy of the New American Encyclopedia; Merriam & Co., of Springfield, sent specimens of their printing, and Brewer & Tileston sent a copy of Worcester's Dictionary. The choice and beautifully printed works from the presses of Cambridge, New York, and Philadelphia, were not to be found. The books and apparatus for the use of the blind attracted much notice.

The very interesting display made by the American Bible Society should be noted here as one of the most remarkable of the typographical and publishing exhibits of the Exposition. This society, organized in 1816, has issued 22,118,475 copies of the Holy Scriptures, in about 50 different languages, at home and abroad; such as English, German, French, Spanish, Italian, Portuguese, Welsh, Irish, Gaelic, Dutch, Danish, Swedish, Latin, Greek, Hebrew, Polish, Russian, Esthoman, Hungarian, Finnish, Syriac, Arabic, Armenian, Hebrew-Spanish, Armino-Turkish, Arabo-Turkish, Mpongwe, Zulu, Arrawack, Grebo, Benga, Choctaw, Chickasaw, Ojibwa, Dakota, Mohawk, Delaware, Creolese, Hawaiian, Micronesian in several dialects, Chinese in several dialects, Siamese, Hindu, and Urdu.

A very interesting and valuable series of publications upon science, art, medicine, and morals was sent by the Viceroy of Egypt, as specimens of typography from the government establishment, Boulac, Cairo. The government of Hawaii sent various specimens of native publications in English and the Hawaiian language. The latter works were curiosities, simply showing the mechanical march of letters into regions where education had scarcely penetrated. They had no claims to typographical merit. The perfection of a printed page is to look clear. It must never look crowded, whatever be the type in which it is printed. The proportion of each letter must be mathematically correct. The capitals must bear a true relation to the small letters, and neither escape the attention nor attract the eye too much. The spaces—or intervals between the letters and words—must be well determined, not capricious, for in the latter case the effect would be spotty. In this art, modern printers may yet learn much from their predecessors. The regularity of black letters was favorable to uniformity, and the contrast of black and white was more positive from the heaviness of the characters used. In the earliest books, the capital letters were left to be illuminated by hand, but very soon wood engravings were used both for the capitals and as borders for the last. Later, the borders were abolished and large ornamental capitals cast in type metal were used for the capitals of each chapter. These were succeeded by engravings on copper with head and tail pieces, many of which were the works of the first artists of their time. The process was a slow one, inasmuch as it involved two distinct modes of printing. It was in due time abandoned, and the fashion has now returned to borders cut in wood, or types, and to illumination, a new process involving lithography as well as common printing, expensive but very beautiful.

There were admirable specimens of books in the Oriental languages. The Hebrew types are the clearest and most elegant that exist. They have long had this renown, and the Arabic, although stiffer, are still more elegant than any other type cast in Europe. The charm probably lies in the respective alphabets.

CLASS 7.—PAPER, STATIONERY, BINDING, PAINTING AND DRAWING MATERIALS.

The following statistical data are extracted from the report of Messrs. Haro and Roulhac, members of the committee of admission of class 7 in the French department. The facts relate chiefly to France, but are of general interest.

The articles exhibited in class 7 comprehended stationery proper, bookbinding, the various objects comprised under the title of office requisites, and artistic materials.

STATIONERY, AND PAPERS.

There are few departments which do not possess several paper-mills. Angoumois, Ardèche, Vosges, Isère, and the basin of the Loire are the most important as regards the number of the mills. The rags employed in the manufacture are nearly all purchased in France. Since the treaty of commerce, these materials, of which the export was previously prohibited, may be exported on the payment of a small duty, which is gradually being reduced to extinction. The importation of cotton and linen rags and old cordage amounted, during the first nine months of 1866, to 2,830 tons. The importation of foreign rags, including cotton, linen, and old cordage, during the same period, amounting to 7,914 tons. The number of vats for hand-made paper in France is said to be 140; that of great machines for making white or colored paper, sized or unsized, 270; and of machines for making wrapping papers, 230. These vats and machines occupy about 34,000 persons, of whom 11,000 are women, and produce more than 129,000 pounds of paper. The annual consumption of the rags may be estimated at 115,000 tons. The average price of hand-made paper does not amount to more than two francs the kilogram; that of printing and writing papers is about one franc ten centimes the kilogram; that of packing and wrapping papers, forty centimes the kilogram. The greater part of the paper manufactured in France is consumed in the country. Exportation, however, tends to develop itself; it has considerably augmented since 1865. During the first nine months of 1866, it rose to 7,578 tons. As to the importation of foreign papers it is unimportant; the amount, during the same period, did not exceed 100 tons. The committee of admission of class 7 points out, among the improvements carried out in the paper manufacture: 1. The use of motive power, which during the last few years has increased at least 10 per cent.; 2. The gradual and intelligent application of substitutes for rags in those places where the latter are wanted or are dear; 3. A positive amelioration in the general economy of the manufacture, which has surmounted all difficulties by reducing the price, in spite of the constantly increasing cost of the raw material and of everything which contributes to the production of paper.

PASTEBOARD.

Pasteboard is divided into three sorts: 1. Pasteboard in sheets, which is obtained by uniting sheets of paper one upon the other by means of pulp paste; 2. Pulp pasteboard, which is made in the frame with waste paper, old paper collected, paper cuttings, and often with the aid of a mixture of straw and other materials; 3. Machine-made cardboard, which is nothing more than cardboard made by machines similar to those employed in making paper. This mode of manufacture only dates from 1846. Among the pasteboard which is employed in a special manner must be cited bitumenized pasteboard, the pasteboard which serves for the Jacquard loom; the pasteboard of which railway tickets are made; and especially the glazed pasteboard used in the dressing of shawls, stuffs, and papers. This last manufacture is developing very considerably, and there is no country that can equal France in this kind of product. The manufacturers of ordinary cardboard are to be found in all the districts of France. They have little connection with foreign countries; they exported, however, during the first nine months of 1866, 211 tons to various countries. Paris employs in this branch of trade more than 500 work people, and the annual amount of business exceeds £120,000.

PLAYING CARDS.

The manufacture of playing cards comprises the making of the card; the impression of the design; the coloring of the engraved figures; the glazing. The French cards, that is to say, those of which the designs and the ace of spades are furnished by the government, are divided into fine cards, demi-fine, and common. The fancy cards, of which the price is higher, are charged with a tax of 50 centimes. Foreign cards, intended for exportation, pay no duty. The home consumption of this article is increasing, but the exportation is not extending. A large number of playing cards is exported to Mexico, to Hayti, to Peru, and South America generally.

FANCY PAPERS.

This name is given to all papers gilt, silvered, colored, printed, embossed, pierced, &c., which are used in making objects in paper for bookbinding, confectionary, pharmacy, drugs, and laces. Among these papers, some, such as marbled papers, are made entirely by hand; others, printed, watered, and shagreened, are machine-made. All these articles are manufactured with white French paper, more or less fine. This trade exports little, in spite of the incontestable superiority which an immense assortment and excellent taste confer upon it. The manufacture, in France, of these fancy papers amounts to nearly £280,000. Paris is the centre of this interesting specialty, which employs more than 1,200 work-people.

OBJECTS MADE OF PAPER AND PASTEBOARD.

This class includes a multitude of articles small and large, rich and common, for offices, warehouses, travelling necessaries, packing, and the makers of fancy articles. This trade is essentially Parisian, and is continually on the increase. There are nearly 400 makers in the two branches of the trade above indicated; they employ more than 2,500 work-people, and the total amount of business may be safely estimated at £400,000.

OFFICE STATIONERY, ETC.

This term includes account-books, pocket-books, ink-stands, sealing-wax, wafers, pen-holders, pencils, and miscellaneous articles. This trade is essentially Parisian. Its various branches include 309 makers, who employ 1,436 work-people, and do business to the amount of 9,220,860 francs, (£368,834.) The article of account-books is the most important; it is treated with great care and superiority in all parts of France, but particularly in Paris, where 130 manufacturers and stationers do business to the extent of not less than £252,880 in this one article. The invention of artificial lead for pencils has given rise to an industry which is essentially French. The sealing-wax manufacture is interesting from the progress which it has made since the treaty of commerce. The custom which prevails of gumming envelopes interferes seriously with the fabrication of sealing-wax and wafers.

ENVELOPES.

There are few trades which exhibit a development equal to that of envelope making. This specialty dates from 1838, but only began to grow into importance in 1851. All the envelope makers are found in Paris, and they do not produce less than 2,500,000 a day. Nearly all the operations are performed by mechanical means: folding and gumming are done by machines; even the boxes in which the envelopes are sold are produced mechanically. The annual product of this article exceeds £80,000.

ARTISTS' MATERIALS.

The number of painters, professors of drawing, of water-color and miniature painting, pastel drawing, of engravers, wood and lithographic draughtsmen, &c., amounts to more than 6,000. These 6,000 artists—all of whose names are not, doubtless, celebrated, but at least obtain a living by their pencil, chisel, or burin—employ more than £240,000 worth per annum of fine colors, canvas, panels, brushes, varnish, &c. To the cost of materials to these artists must be added the still larger sum expended by their pupils and by amateurs every year. It is quite safe, therefore, to estimate the total amount of this industry at £800,000. Machinery plays a certain part in the preparation of colors, trituration, grinding, and washing, but it is not universally employed. Each establishment has still the aspect of those of the alchemists of the middle ages, and works without publishing its processes, its secrets—in a word, that which constitutes its specialty. It is admitted that France makes the best of oil colors, pastels, and canvas; the last are superior, as regards finish and dimensions, to those made in other countries. The proofs lie in the orders received from foreign artists, and even foreign governments. It must be admitted, however, that with respect to water colors the French makers have serious competition to contend with, as regards quality, especially in the case of England; but some French houses have made great efforts to rival the quality of the English colors, while at the same time selling them at a lower price. The instruments and apparatus employed by painters, engravers, lithographers, architects, engineers, and sculptors, present an immense variety.   Pencils and brushes occupy in their production more than 2,000 men and women.   French brushes are greatly preferred by foreigners to those of their own make, and amount in value to several millions of francs.   Drawing-boards, T—squares, &c., used especially by architects and engineers, form a remarkable branch of industry, and the same may be said of Indian ink, printing ink, chromo-lithographic colors, and engravers' and lithographers' materials.   The making of lay figures for painting draperies calls for serious study of anatomy and mechanism.   It requires encouragement, as it does not supply sufficient remuneration to the persons engaged in it.   Nevertheless, by perseverance, several manufacturers have achieved results which deserve to be noticed.   The same remark will apply to easels, color-boxes, and, above all, to the metal tubes which enclose color ground in oil.   The transfer from their canvas, the remounting and the reparation of pictures—in short, the means used for preserving works of art, form a branch of art to which too much attention cannot be invited.   As an industry it is equally useful and interesting, and it may be affirmed that the best results and the greatest study have been made in France in connection with it, and that it is still the object of highly praiseworthy efforts.

There were but two exhibitors of paper from the United States.   Jessup & Moore, of Philadelphia, sent specimens of paper made from wood, straw, and hemp.   The other display consisted of white and straw papers, of excellent qualities, from the San Lorenzo mills, Santa Cruz county, California.   This establishment has been in operation about six years, and now produces annually about 31,000 reams of straw paper and 7,000 reams of white newspaper; the total production is valued at over $100,000.

In the Wurtemberg section a machine for making paper pulp or paste out of wood was shown in operation.   Logs of wood at one end of the machine were cut into billets a foot long by a circular saw.   These billets were then subjected to the action of the machine, and were delivered at the other end in the form of a white paste or pulp, which is used to mix with rag pulp to the extent of from 25 to 60 per cent.   This invention is claimed by the firm of H. Wolker & Sons, at Heidenheim.

There are now 20 paper establishments in operation at Wurtemberg, having 28 machines and 237 rag-mills, and 29 establishments where hand labor alone is employed.   The total production of paper is about 15,800,000 pounds, representing a value of £265,708, most of which is exported. The principal localities of the manufactures are Dettingin, Faurndeau, Göppingen, Heidenheim, Helbronn, and Pfullingen.

In addition to paper made from wood and straw, there was exhibited in the French section paper made of " esparto," (the Spanish rush,) the fibres of the palm tree, the aloe, the Indian fig or cactus, and from sea-weed.   Excepting the last, these are all fibrous plants, possessing in some instances a length of fibre sufficient even for other manufacturing purposes.   The sea-weed, in addition to its known tenacity, possesses a sort of glue, which, it is claimed, renders it valuable as a mixture with other substances.

France excels in many varieties of paper, especially those used for printing and fancy purposes. England manufactures most of the finest qualities, and enjoys almost a monopoly for certain kinds used in the arts. Holland was once famous for its paper. It had but two exhibitors. The paper of Venice, inferior to that of Holland, enjoyed a great reputation 200 years ago, and up to a late period the letter paper of Naples was considered the best in the world. No one would have that opinion now. Spanish paper has also had its vogue; but the only branch of the manufacture in which Spain now excels is in the paper for cigarettes. Linen is still the ordinary wear of the peasantry in Spain; linen rags are there more easily obtained than in other countries, and from these a thin and admirably tough paper is conscientiously made.

CLASS 8.—APPLICATIONS OF DRAWING AND MODELLING TO THE COMMON ARTS.

Class 8 comprises artistic productions applicable as models and ornaments for industrial purposes. They are: 1. Designs for printing—Dresses, fancy silks, foulards, ribbons, muslins, cotton fabrics, woollen goods, chintzes, &c. 2. Designs for weaving—Shawls, carpets, hangings, &c. 3. Designs for embroidery, lace, &c. 4. Designs for furnishing—Paper hangings, furniture, pottery, &c. 5. Designs for ornamentation, models, &c.—for jewelry, plate, fine iron and lock works, cameos, engravings, wood, copper, ivory, bronze, and other metals, stained glass, &c. 6. Designs and objects of industrial modellings, obtained by mechanical means, (reductions, enlargements, and photo-sculpture.)

It will thus be seen that the number and variety of objects exhibited in this class was very great, comprising not only drawing upon paper for tissues, but models for carvings in wood, ivory, metal, glass, and stone.

There were but two exhibitors in the United States section—one of embossed locket and miniature frames; the other, J. Rogers, of New York, three groups of statuettes.

The Science and Art Department of the South Kensington museum, London, sent a series of illustrations of the course of drawing, painting, and modelling, and studies for the improvement of manufactures pursued in that institution, and also a collection of reproductions of works of art, for the use of museums or similar schools of art.

Inasmuch as these articles form an entirely new branch of commerce as well as of useful instruction, and have for their end the instruction of labor where skill is required, it is thought desirable to give a full description of what they consist. They are commercial to the extent that any museum or school can procure exact copies of them, and thus be on a satisfactory level at once with the material of a good art school. A few particulars will explain the value of this fact. In all countries examples of more or less excellence for the use of art schools have been prepared. They are easily obtained. But, besides the production of work to be used as a course of study for training the hand and the eye, the culture of taste and of sound principles of art have to be promoted by placing before the decorative artist the purest specimens of ancient and modern production, wherein handicraft skill has realized beautiful design. For this end all countries have gradually awakened to the necessity of founding museums and collections of rare and beautiful objects for use and reference. Such works, however, were difficult to obtain, and as museums multiplied the difficulty naturally increased. It then became absolutely necessary to discover a means of reproduction that was at once faithful and cheap. The various processes of the electrotype, of photography, of chromo-lithography, of gelatine and gutta-percha moulding, &c., were called into play. The English government, in the interest of their own schools of industrial art, left no means untried, and at length succeeded. After the Exposition of 1855, the French Emperor responded to its request to allow the most valuable jewels, crystals, enamels, &c., in the Louvre to be photographed, and he placed at the disposal of the English government the means of carefully coloring those photographs after the originals. On a subsequent occasion he added permission to mould, for electrotyping, the finest pieces of armor in the Musée d'Artillerie, and allowed repetitions to be made from the casts prepared for France from the Trajan column. Other countries have since permitted similar reproductions, so that now almost any remarkable object, exactly reproduced in size, color, and present appearance, can be obtained. The boon is of inestimable value. It places within the reach of small associate bodies of students the power of studying the finest specimens of art from all quarters of the earth, to visit which, apart from the matter of expense, would be the work of a lifetime. An idea of the material may be gathered from the fact that in the British section were shown plaster casts from the pulpits of Giovanni and Nicolo Pisano; of part of the door of Santiago de Compostella in Spain; electrotypes from the gates of the cathedral of Pisa; from the bases of the standards on the piazza at Venice; electrotypes of armor in the Musée d'Artillerie; of the coronation plate in the Tower of London; of rare objects in the South Kensington Museum, and of colored imitative drawings, photographs, etchings, and chromo-lithographs of the choicest works of Europe.

There were 22 exhibitors in this class from England, 36 from Italy, and 41 from Switzerland. In the French section there were 240 exhibitors, mostly of designs and engravings. The following observations upon the relations of the French school of design to the manufactures of the empire are translated from the introduction to the Class in the catalogue:

"Schools of design, established in most of the great manufacturing centres, have contributed to disseminate in France the most elevated notions of industrial art. Paris is the centre par excellence from which radiate to the varied branches of our national industry the highest inspirations of taste, elegance and novelty. The most distinguished pupils of the provinces come to Paris to perfect themselves in design, and many establish themselves advantageously there. It is in Paris, then, that we must seek the source of the great artistic current. In certain industries many large manufacturers who formerly had designers attached to their establishments now prefer to apply to Parisian artists for designs more novel in themselves and more adapted to the various demands of the consumer. Some artists work alone, or assisted by a small number of pupils; but all those who have made themselves a name have created *ateliers*, where young men come to perfect themselves in their art. Some of these workshops confine themselves to one specialty; others, veritable sources of industrial information, combine several branches of design. The raw material holds an insignificant place. The intrinsic value of the drawings and models is merely nominal. Their importance and merit are due to the artistic inspiration alone. The methods employed are extremely simple; in fact, it may be said that there is no manufacture, properly so-called, because the mechanical processes merely serve to carry out or to produce the designs or the models, which are the personal work of the artists. The manufacture only commences with the industrial execution, that is to say, with the manufactured product; the design itself, whatever may be the material to which it is applied, has few essential differences. As already remarked, the establishment of ateliers is on the increase. In such cases the artist selects his assistants and portions out the work according to circumstances. He remunerates his employés by the day or by task work; sometimes, even, by annual salaries, according to their merit or to the value of their work. From the first idea placed upon paper or plaster to the finished design or model which is to serve for the manufacture, each sketch passes through a long series of artistic elaborations. The master-artist finds in the co-operation of others acting under his orders at once economy as regards time and greater perfection of execution. It is almost impossible to supply any exact information relative to the value of works of industrial art, because the cost is included and mixed up with the price of the manufactured objects. The price of this artistic contribution varies with the products. It is higher or lower in proportion to the demand for the objects themselves."

The fullest exhibition of the works of pupils in art schools was made by Wurtemberg. The students, as in England, seem to be taught practical geometry, perspective and mechanical drawing, of which good examples were exhibited; the course of orthographic projection being very full. In freehand drawing, a clear and precise system of outline seems to be sought after, and the early training of the hand and eye to correctness carefully attended to. The shading from the casts was more with the point than with the stump, the object of the schools apparently being to form good draughtsmen and modellers—intelligent artisans skilled to handle the pencil and the modelling tool, and able thoroughly to comprehend working drawings rather than to instruct designers for manufacture or to instil the principles of decorative art. Italy, Bavaria, and Austria also exhibited specimens of their schools of art. They were similar to those from Wurtemberg without being better.

CLASS 9.—PHOTOGRAPHIC PROOFS AND APPARATUS.

Class 9 includes: 1st. Photographs on paper or on glass; 2d. Photographic enamels; 3d. Photographs obtained in printing ink by the various processes of heliographic engraving, or of photo-lithography; 4th. Photographs obtained on metal or on paper, with the colors of nature; 5th. Specimens of the various applications of photography; 6th. Apparatus and wood-work for photography, chemical, and all other accessories.

The Exposition was exceedingly rich in the number and variety of photographs exhibited, but the specimens were in general widely separated and not displayed to advantage. If all could have been assembled in a special gallery the interest in them would have been greatly increased and there would have been an opportunity for direct comparisons. France had 165 exhibitors, Great Britain 105, Austria 58, Prussia 52, Italy 42, and the United States 17.

There does not appear to have been any recent marked advance in the art. The progress has been chiefly in the direction of production of photographs in enamel and upon porcelain and glass, and in the heliographic process, by which the pictures are engraved upon copper or steel, so that they may be multiplied by printing. There are several exhibitors of such plates and of photographs engraved upon lithographic stones. Lackerbauer, of Paris, exhibited lithographic engravings of objects and microscopic preparations magnified from 5 to 2,500 times. No satisfactory results in the attempts to produce colored pictures appear to have been obtained.

In the English section there was a very interesting series of views of the ancient architecture of India, as shown in the temples and palaces of the interior of that country.

The most notable display from the United States was made by Mr. C. E. Watkins, of San Francisco, who sent a series of 30 views of the Yo Semite valley of California, and views of the great trees. These photographs were not only interesting as pictures but as splendid specimens of the art. The jury awarded a bronze medal. A similar series was sent by the firm of Lawrence & Houseworth, of the same city, with the addition of a great number of stereoscopic views of the interior mining regions of California, showing in a very distinct manner the various processes in use there for the extraction of gold from the soil.

The contributions in this department of Mr. L. M. Rutherford, of New York, are to be particularly noted for their high scientific value as well as peculiar excellence as photographs, and for the subjects represented. One is a large photograph of the moon, representing its pitted surface as seen through a powerful telescope; and the other is a photograph of the solar spectrum, two feet long, showing the almost infinite number of dark lines. These two photographs, although scarcely noticed by the multitude, excited great attention and interest among the savans, and received a silver medal from the jury.

The exhibition of photographic apparatus and chemicals was very large. It is to be noted that the photographic art has exerted a very marked influence upon various branches of manufacture, particularly of chemicals, and that it has given great impulse to industry and commerce in these directions. The demand for photographic apparatus and materials is so large as to require many considerable establishments devoted exclusively to their production.

Certain substances, such as hyposulphite of soda, which formerly were rarely employed and therefore rather expensive, have been so much used in photographic operations as to cause them to be made on a large scale, and thus to reduce their prices to half or one-third, or even one-sixth, of their former value. We may mention also the sulphocyanides of potassa, and ammonia, which were only used before in the chemist's laboratory, but are now manufactured extensively at gas works, where large quantities can be obtained from the distillation of coal. Photography in France has given rise to considerable trade with foreign countries. Not only are apparatus, paper, and chemicals largely exported, but also stereoscopic views on paper and other materials.

CLASS 10.—MUSICAL INSTRUMENTS.

The following information upon the variety of the objects exhibited in the French section, and upon the condition of the French manufactures of musical instruments and materials, is extracted from the translation of the report of Mr. Wolf, of the committee of admission:

"The products exhibited in Class 10 include eight principal series, viz: 1st. Church organs; 2d. Harmoniums; 3d. Pianos; 4th. Stringed instruments; 5th. Wind instruments; 6th. Percussion instruments; 7th. Accessories for the manufacture; 8th. Editions of musical works."

"Paris is the only important manufacturing place for organs, pianos, and harmoniums. Then follows, according to importance, Marseilles, Lyons, Nancy, Toulouse, and Bordeaux, where pianos are chiefly manufactured. Stringed instruments are made principally at Mirecourt; wind instruments, in wood—such as flutes, clarionnets, hautbois—are more specially manufactured at Lacounture, (Eure.) All kinds of instruments are also made in Paris. Chateau-Thierry has, likewise, no specialty; nearly all kinds are manufactured there.

"The woods for musical instruments are produced from France, Russia, Norway, Brazil, St. Domingo, and Isle Bourbon. The native woods most frequently employed are oak, fir, lime, beach, maple, box and pear. These vary in price from 55 to 200 francs the cubic metre. Box is sold from 50 to 60 francs the 100 kilograms. The exotic woods most used are rosewood, mahogany, cedar, and cedrine, ebony and grenadille, which cost from 15 to 150 francs the 50 kilograms. Those more generally used are oak, fir, and beech for the heavy parts of pianos, organs and harmoniums; cedar, lime, maple, and pear-tree for the mechanical parts; rosewood and mahogany for veneering and ornamentation; box, ebony, and grenadille for wind instruments. Beech and mahogany are chiefly in use for bassoons. Ivory for piano keys is sold from 22 to 45 francs the set (50 keys.) The felt, woollen stuffs, skins, and glue for pianos are manufactured in France. Part of the felt comes from England. There is in France no manufacture of metallic cords. Those in steel are imported from England and Germany, and are worth about 8 francs per kilogram. The copper covering for strings is worth from $5\frac{1}{4}$ francs to $7\frac{1}{2}$ francs per kilogram. The metals most in use are iron, lead, copper, for wind instruments; tin for organ pipes. The gut cords are manufactured in France.

"The tools employed for working the wood are the ordinary tools of the joiner and cabinetmaker. However, we must notice the profile machine for making panels, which is only an improvement of the parquetry machine; and also the special steel perforators for wooden wind instruments. The only special tools in use for working metals are mandrils, employed in the manufacture of wind instruments. We must mention also, wheels for covering cords. All these tools were unknown in 1855, or rather have been very much improved since then.

"In Paris and all the large towns the men employed in the manufacture of musical instruments work together in the workshops; scarcely any work at home. At Mirecourt, on the contrary, the men, about 250 in number, all work at home. Half the Paris workmen work by the piece; the other half by the day. The salary varies from 3 francs 25 centimes for common workmen, and from 5 to 11 francs for the superior artisan. The musical instrument trade employs few women and children.

"Part of these articles are sold in France, and part to commission merchants, who buy for exportation; a third, perhaps the most considerable, is exported direct, to order, to all parts of the world. The small instruments are worth from 50 to 200 francs; harmoniums from 100 to 1,500 francs; violins and violoncellos from 200 to 500 francs; copper instruments, 80 to 400 francs; wind instruments, in wood, 80 to 300 francs pianos, 500 to 4,000 francs; church organs, from 2,500 to 100,000 francs. The profits of the manufacturers vary from 12 to 18 per cent. The manufacture of musical instruments represents a sum of twenty or twenty-three millions of francs per year. Raw materials are imported into France to the value of five or six millions. About half the produce goes to foreign countries, and is exported to all parts of the world, but particularly to America, and chiefly to South America. The importation is next to nothing.

"The committee of admission for class 10 points out among the improvements made during the last ten years, in the manufacture of

4 U E

musical instruments: 1st. The considerable extension given to mechanical processes, and the general use of steam machinery; 2d. The application, as far as possible, of the principles of the division of labor; 3d. Piece-work substituted in most cases for work by day."

There was no class in the Exhibition more thoroughly and completely represented than this. Every nation contributed its quota to the huge aggregate. That the art of music "hath charms to soothe the savage breast" was amply demonstrated. The wildest and strangest countries contributed their eccentric contrivances of bamboo and hide—instruments that were dulcet to native ears, but hideous to the average tympanum of civilized Europe. There were large and small drums, in wood and clay, used by the Tinkaonis; rude violins covered with gazelle skin and ornamented with horns and men's heads; trumpets made of antelope horn and elephant's tusks, of which the sound is heard at the distance of a league; and perhaps worse than this, the bagpipes of the Arab tribes used in the region of Cordovan. Many of these instruments were of the greatest antiquity, and were played upon by "professors" in the various departments precisely in the same way as when they were invented. A few steps sufficed to take the spectators to an adjacent section where the latest improvements of Europe were standing side by side—improvements which require the greatest technical skill to appreciate or use.

The United States had nine exhibitors. Numerically considered, the display was insignificant, but the objects comprised in it were of the highest excellence. The piano-fortes contributed by the New York firm of Steinway & Sons and the Boston house of Chickering & Sons were considered the best in the entire Exposition. Each was awarded a gold medal. The latest improvements are to be found in these instruments, which are almost wholly constructed on original plans and produce results of a very satisfactory character. For length of tone, brilliancy, sympathetic quality, and magnificence of power they are unrivalled. The broad merits of both pianos were found to be so superior that the jury, having but four gold medals to award, unanimously voted two to America—an honor which cannot be overstated, for it was remarkable enough that pianos should be sent at all from America to Paris, and still more singular that they should there be regarded as the best.

The harmoniums and cabinet organs of Messrs. Mason & Hamlin were also objects of much interest, and gained the award of a silver medal. They were, like the pianos, admired for their workmanship and for the singularly pure tone which they possessed. The mode of producing this tone was the subject of much curiosity, inasmuch as it differs essentially from the European plan, and in America has entirely superseded it.

The wind instruments of the Schreiber Cornet Manufacturing Company and the string instruments of Gemünder, both of New York, also obtained prizes. The brass instruments of the former were regarded as excellent specimens of manufacture. The violins, &c., of Gemünder were greatly admired for their forms and for certain improvements which that maker has introduced into the construction of the instrument.

The general display of piano-fortes was unusually large. All the European centres of the trade sent their best specimens. These were of the ordinary forms—grand, upright, oblique, square, and cycloid. (There was a specimen of the latter in the American department and several specimens elsewhere.) France was represented by Erard, Pleyel & Wolff, and Henry Herz; England by Broadwood and Kirkman; Prussia by Blüthner and Bechstein; Wurtemberg by Schiedmayer; Austria by Streicher, &c. Of these Broadwood took the prize for England and Streicher for Vienna. The pianos of the latter house are made on the plan of Messrs. Steinway & Sons.

Of the infinite variety of wind instruments it is impossible to speak. It will suffice to say that the efforts of all modern makers is to introduce a homogeneous quality into the separate families, namely, that all those composed of wood should sound like each other; that all those formed of brass should bear a respectable and not overbearing relation to the rest of the family. Many curious instruments have been invented for military bands by which orchestral effects can be better imitated.

There were innumerable specimens of stringed instruments, but with this it is the singular fashion to go backwards, and progress therefore had to be looked for in an inverse ratio. The ancient model of Stradivarius seemed to be the model most in favor.

There were but few organs in the Exhibition, and the best, on account of the size, had to be accommodated in the machinery department. It was of French make and is intended for the new church at Nancy.

CLASS 11.—SURGICAL INSTRUMENTS.

In this class France had 101 exhibitors, Prussia 18, Austria 19, Italy 38, United States 22, Great Britain 31.

"The articles exhibited in this class were very numerous and varied; they related to the practice of medicine, surgery, and hygiene, and included—1. Surgical instruments used in operations, such as cutting instruments, forceps, tenaculæ, suture needles, instruments employed in amputations, setting broken bones, &c. 2. Special instruments used in certain operations connected with diseases of the organs of sense, the respiratory passages, and the male and female genito-urinary organs. 3. Instruments or apparatus intended to cure natural or accidental deformity, such as orthopœdic instruments, bandages, belts, and elastic stockings. 4. Articles relating to dental art which have greatly improved of late years. 5. Instruments usually employed in determining the diagnosis of diseases of the heart, lungs, eyes, &c., and those which are used in experimental physiology. 6. Apparatus used in public or private hygiene, such as bathing and hydropathic appliances, instruments employed in friction, the two systems of application, electricity, artificial lactation, and various appliances used in domestic gymnastics.

In France the principal centre of the manufacture of surgical instruments and apparatus is Paris, after which rank those large towns which possess a faculty or a secondary school of medicines, such as Strasburg, Montpellier, Lyons, Toulouse, &c. In the different manufactures connected with class 11, the principal materials employed are the metals, such as steel, iron, gold, platinum, silver, and German silver, so that, from one point of view, this class is closely connected with metallurgy.

Besides the metals, many products of the mineral and vegetable kingdoms are used; such as ivory, horn, skins, gum, and, above all, India-rubber. The articles exhibited in class 11 partake of the nature of cutlery and of mathematical and philosophical instruments, but this class of manufactures possess comparatively limited resources, and the trade is necessarily spread over a large area and a more numerous constituency.

" The delicacy of manipulation and intelligence necessary in making surgical, hygienic, and orthopœdic instruments requires the greatest attention and care on the part of the workmen. A certain number make only special articles, either at their own homes or in their employers' workshops. Women are employed in a large proportion. It is calculated that the manufacture of surgical instruments and orthopœdic apparatus in Paris and the provinces, gives employment to from 3,500 to 4,000 workers, male and female. A large number of these articles are sent into the country or abroad. Foreign manufacturers also have closely copied our own inventions, and have, nearly everywhere, endeavored to reproduce French instruments and apparatus. It is difficult to estimate the value of the articles produced in this complicated trade. Simply taking into consideration surgical instruments, bandages, and orthopœdic apparatus made in France, the productions may be valued at 13,000,000 to 14,000,000 francs. These figures would be largely increased if hygienic, hydropathic, and other apparatus were included in the estimate. This committee has but few changes to point out since the exhibition of 1855, either in the instruments themselves or in the mode of manufacture, but the improvements in the instruments have been very numerous; such as the extended application of certain products, caoutchouc, for instance, and the progress made in the management of baths and thermal establishments."—(Extract from the report of the members of the committee of admission of class 11.)

Among the many interesting objects from the United States in this class, the exhibition made by the Surgeon General was particularly complete and worthy of attention. It included ambulances, medicine wagons, army field hospitals, and litters which were used throughout the war, and the best artificial limbs which have been invented.

The display of American artificial teeth and of dental instruments and apparatus was very creditable to this branch of the healing art.

Pertaining to this class and described elsewhere was the exhibition of the societies for aiding wounded soldiers and sailors.

CLASS 12.—MATHEMATICAL INSTRUMENTS AND APPARATUS FOR TEACHING SCIENCE.

The French exhibition in this department is decidedly the largest and most interesting. The manufacturers of instruments of precision have fully maintained their high reputation for the accuracy, elegance, and cheapness of their productions. Of the 107 exhibitors in this class, we may note the following as prominent in their respective specialties: Ruhmkorff exhibits a variety of electrical, magnetic, and electro-magnetic instruments, and particularly several enormous coils bearing his name, but known in the United States as the Ritchie coil. Auchet & Son exhibit a great variety of excellent microscopes, arranged for one or more observers at the same time, and an apparatus for microscopic projection and reproduction. Soleil exhibited several optical instruments of interest to mineralogists and chemists, particularly a polarizing microscope upon the pattern of M. Descloizeaux, together with numerous crystals cut and polished so as to show their optical characters. Deleuil exhibited philosophical and assay balances, photometers, machine for the solidification of gas, pneumatic machines, force pumps, Foucault's pendulum for demonstration. M. Deschanel, member of the committee of admission, subdivides class 12 as follows, and adds some interesting general observations:

"1. Instruments intended for scientific research and education. 2. Special optical instruments, microscopes, telescopes, and field-glasses. 3. Mathematical instruments, graduated rules and compasses, levels of all kinds, and geodetical circles, whether for the use of the marine or of engineers. 4. Barometers and thermometers, of which glass forms the principal element of manufacture. 5. Apparatus intended to carry a new idea into effect, or to execute a known operation by a new process, and special apparatus, which, without being new, have a special object, and consequently do not enter into the preceding series. Lastly, collections of natural or artificial preparations intended to illustrate the three great natural kingdoms. This series of the naturalist's preparations, logically connected as much with that of medical art as with natural philosophy, also forms part of another class. The production of philosophical instruments is confined almost exclusively to Paris. In some of the ports, however, there are special makers of mariners' compasses for ordinary navigation. In the Jura and in Picardy are to be found some manufactories of optical glasses, intended for common instruments, and which draw their materials from the works of St. Gobain. For carefully constructed instruments, glass of a special kind is produced in Paris itself. For other kinds of instruments ordinary glass is employed. According to the statistics collected by order of the Chamber of Commerce in the year 1860, the value of this manufacture in Paris amounted to 15,861,720 francs. Since the exhibition of 1855 the progress in the construction of scientific instruments has followed that of science itself.

Among the articles which exhibit a decidedly enhanced importance, we may mention telescopes with silvered reflectors, apparatus for the production of inductive currents, electro-magnetic machines, the regulators for the electric light, and optical indicators of the vibrations of sonorous bodies."

In the American section the instruments sent by the United States Coast Survey comprised some of the instruments of navigation, and a thermometer for measuring the temperature of the sea at great depths. The beautiful graduated rules, squares, and gauging instruments made by Darling, Bangor, Maine, attracted much attention from those interested in such objects.

CLASS 13.—MAPS AND GEOGRAPHICAL AND COSMOGRAPHICAL APPARATUS.

The following complete and instructive *aperçu* of the exhibition made by France in this class, is a translation of the introduction to the class by the committee of admission:

"The objects of class 13, which figure for the first time in a universal exhibition, may be divided into four series: 1. Maps, whether separate or forming atlases, including geographical, cosmographical, astronomical, marine, hydrographical, topographical, geological, agronomical, historical, itinerary, meteorological, or other maps. 2. Terrestrial or celestial globes, uranographic apparatus, &c. 3. Maps and charts in relief. 4. Works, tables, pictures, and other accessories of geography and cosmography."

"Paris is the only centre of production of these works, even of those which are edited or published in the provinces. Among the material employed, the copper and steel are prepared in Paris by the planers and polishers, who supply the engravers. The lithographic stones come principally from Bavaria, (near Munich,) but during the last few years they have been obtained also at Vigan, (Gard,) nearly of equal value and at a much lower price. Quarries have also recently been discovered in Isère, (Cerin, Crey.) The sized paper, almost the only kind employed, is produced in the Vosges, Isère, (Rivès,) and Angoulême; the unsized paper is obtained from Nièvre (Clemency) and the Marais, (Seine and Marne.) The processes of the manufacture are: Engraving on copper, steel, and stone; engraving on wood or metal; lithography or drawing on stone; autography; the transfer on to stone of engraved work, and chromo-lithography. The laying down of the maps on cloth constitutes a supplementary operation frequently called into use. The persons engaged in this branch of industry include designers or draughtsmen, map and writing engravers, colorists, mounters and binders, globe and sphere makers, lithographers and few photographers, and lastly, copperplate and lithographic printers. The last named are generally engaged in large establishments, in which the printing of maps is but a secondary matter; the rest work either at home or in small work-shops, where not more than 25 or 30 persons at the utmost are engaged.

"There are in Paris about a dozen employers and 150 artists and workmen engaged in the specialty of geographical industry. The productions of French cartographic establishments are principally destined for the institutions of public instruction in France; the remainder is sent to South America, Russia, Germany, North America, Italy, Spain, and England. Egypt and Algeria also take a certain quantity. The depot of the war department and that of the marine assist greatly in this exportation, either through the mediums of booksellers or by their depots abroad. The maps and globes imported come principally from England, and the total value is between 40,000 and 45,000 francs. The export is estimated at about 150,000 francs, out of a total production of the value of 400,000 francs. The military and naval establishments furnish about one-quarter, not including the large number of maps which they supply to various public departments, and especially to the naval service belonging to the state."

Among the improvements which have been made in this business during the last 12 years, the following may be pointed out: "1. Imparting a steel-like hardness to copper plates, with the view of making them serve for a longer period. 2. The employment of photography as an expeditious method of multiplying, enlarging, and reducing drawings. 3. The heliographic method of engraving on stone or copper. 4. Chromo-lithography and chromo-printing. Many colors are now applied to maps, particularly for special maps. 5. The extended use of relief maps, made to the same scale, as regards the vertical and horizontal measurements. 6. Printing on curved surfaces for globes. 7. The increased use of various methods of projection. The above information applies to the drawing out of maps. As to the original operations of surveying and projection, they are centralized, as far as regards terrestrial geography, at the war office, and for navigation at the admiralty. The publishers generally avail themselves of the productions of these two departments for the preparation of maps for the trade, making use, however, also of official documents furnished by foreign governments, as well as of the works of travellers and literarti of all countries."

MAPS FROM VARIOUS COUNTRIES.

Among the numerous, large, and interesting maps from various countries, one from Russia, a relief map of Caucasus, is worthy of particular notice. It was upon a large scale, so that the highest mountains rose fully six inches above the sea level, and every detail of the topography and the distribution of forests appeared to be accurately given. In the English section a relief map of India on a large scale showed the chain of the Himalaya, the high plains of Asia, the valley of the Ganges, and the lines of railway, in a most striking manner.

The Geological Survey of the United Kingdom sent a full series of its geological maps and sections, all elegantly bound, and specimen sheets of the ordnance maps were to be found in the same section.

In the United States section the principal contributions were from the Coast Survey office, Washington, of a full series of the published maps of the survey, printed on large, thick paper.

In the preparation of the British ordnance maps, resource has been had to processes comparatively unknown, and the facilities thus discovered have led to interesting results. These were exhibited in the same department, and may be briefly described. They consisted of two very remarkable publications, namely, a photozincographic *fac simile* of "Doomsday Book" and the "Ordnance Survey of Jerusalem." Every boy who has studied history has heard of "Doomsday Book." It was made by order of William the Conqueror, in 1086. It contains a description of the owners and inhabitants of every manor, hundred, village, &c., in England, with the exception of Northumberland, Cumberland, Westmoreland, and Durham, and an estimate of the area of the lands and their cultivation. Besides its value as an ancient historical document, it is of great interest as showing the ownership of the country at that remote period. A copy of "Doomsday Book" was published at the end of the last century at great cost, but the type was accidentally destroyed by fire, and copies are now rare and expensive. By the art of photozincography an exact *fac simile* has been reproduced at a comparatively insignificant cost. A county, for instance, costs from $2 to $6 currency. The *modus operandi* by which the result is obtained is thus described: A photographic negative of the map or drawing to be reproduced is first made. A positive print is then taken on paper prepared with a solution of gelatine and bichromate of potash, mixed up with lithographic ink. The effect of the light on this solution is to render it insoluble; therefore, after the print has been taken, those portions which were protected from the light can be washed away, leaving intact the remaining insoluble portions which correspond to the lines of the map or drawing. This can then be transferred at once to zinc or stone, and printed in the same way as if the plan had been drawn on lithographic transfer paper. Thus a plan which it would take weeks or months to copy by hand for lithography, can, by this process, be executed in a few hours, and with a fidelity which no copyist could hope to rival.

The survey of Jerusalem was made, in 1864–'65, by a party of sappers who were detached for that purpose from the ordnance survey of England. The funds were provided by Miss Burdett Coutts, and others, the special object being to obtain a basis on which to work for improving the sanitary condition of the city, especially as regards drainage and water supply. Thus for the first time a map of the Holy City has been produced that can be relied on with certainty in discussing the localities and events connected with its history, which are of such deep interest to the whole civilized world. The map is accompanied by photographs of different parts of the city and neighborhood, which were taken while the survey was in progress, and also by photozincographs. During the course of the survey the ancient aqueduct from Solomon's pools, which supplied the city with water, was traced. It is a work of the highest engineering skill, and in so good a state of preservation that, at very little cost, it has been put into such a state of repair that water has actually been again conveyed through it for the use of the city. In connection with this survey a line of levels was also run between the Mediterranean and Dead seas, in order to settle accurately the much vexed question of the amount of depression of the latter below the former. The result showed it to be 1,290 feet.

Austria and Switzerland exhibited beautiful maps, showing and *not* showing the hill features of the respective countries. The object of the latter is to secure greater clearness. Austria exhibits a map of Bohemia, the most ridgy and mountainous country in her possession, without the indication of an elevation. The well-known map of Switzerland, by General Dufour, is still regarded as the best work of its kind in existence. It is a complete and picturesque representation of the most romantic country in Europe.

Elsewhere in this report mention has been made of a well-executed model of the Isthmus of Suez. In the French court was also exhibited an interesting map of the region by Goujon, of Paris, showing the course of the proposed canal. The undertaking being of general interest, a few particulars here of its progress may not be out of place. The map and model show the works not as they are, but as they will be when completed. These works consist of two distinct portions, a fresh-water and a maritime canal; the former is about six feet deep, and 50 or 60 feet broad at the top, with shelving sides. It starts from the Nile at Cairo, runs in an easterly direction as far as Lake Tinisah, which forms a portion of the maritime canal, and then, bending to the south, terminates at Suez. This canal is completed. Its object is to supply with fresh water the laborers and machines employed in the maritime canal, and also to afford to the town of Suez a plentiful supply of fresh water, which was much needed.

The maritime canal is still in a very miniature state. It starts from Port Saïd, on the Mediterranean, and is to traverse the isthmus to Suez, with sufficient works at its extremities to afford good harbors in both seas. Its breadth is to be 100 metres (rather more than 100 yards) at the level of the water, and its depth about 26 feet. The portion between Port Saïd and Lake Tinisah is so far advanced that barges, towed by steam tugs, can traverse it to the latter point, where, by means of locks, they are placed on the fresh-water canal, by which they can reach Suez. Thus, there is already water communication from sea to sea; but, before the fleets of the world can be transported across the isthmus, much time must elapse, and a vast amount of money must be expended. The cost of the works up to this time is said to be $45,000,000 specie.

In all the countries of Europe the necessity has been recognized of having a detailed and accurate map which should be available for military and other purposes. The scales on which these surveys have been published have varied from about three inches to a mile to about three miles to an inch. First in point of scale comes the little electorate of Hesse, now politically extinct, the excellent map of which is published on a scale of 1 in 25,000, or about three inches to a mile. Belgium comes next, 1 in 40,000; then Baden, Bavaria, Sardinia, Holland, Wurtemberg, Oldenberg, Grand Duchy of Hesse, 1 in 50,000; Saxony, 1 in 57,600; Great Britain, 1 in 63,360, (afterwards increased to 1 in 10,560, and again to 1 in 2,500;) Denmark, France, and part of Prussia, 1 in 80,000; Lombardy, Venetia, States of the Church, Tuscany, Parma, Placentia, Guastella, 1 in 86,400; Hanover, Portugal, Prussia, Sweden, Switzerland, 1 in 100,000; Schleswig-Holstein, 1 in 120,000; Russia in Europe, 1 in 126,000; Austria, 1 in 144,000; Norway and Greece, 1 in 200,000. The survey of Great Britain commenced in 1784; that of France (the new map) in 1818. Belgium has been late in the field. Only a few sheets are published. The last country in Europe to recognize the necessity of a government survey (except Turkey) was Spain. Within the last year or two steps have been taken to remedy the defect. In Turkey no government survey exists.

It will be seen by the above particulars that the only important step in the way of topographical precision has been taken by the British government. It was found that the scale of one inch to a mile was not sufficiently large to make the maps available for many important operations in which maps are specially necessary—such as the apportionment of taxation, the registry and transfer of property, &c. It was therefore determined to increase the scale to six inches to a mile, and subsequently to 25 inches to a mile, on which scale the ordnance survey of England and Scotland is now being carried on. Specimens of the English maps on all three scales were exhibited, and also of town surveys on the extraordinary scales of five and ten feet to a mile.

GROUP III.

FURNITURE AND OTHER OBJECTS FOR THE USE OF DWELLINGS.

CLASS 14. FURNITURE.—CLASS 15. UPHOLSTERY AND DECORATIVE WORK.—CLASS 16. FLINT AND OTHER GLASS; STAINED GLASS.—CLASS 17. PORCELAIN, EARTHENWARE, AND OTHER FANCY POTTERY.—CLASS 18. CARPETS, TAPESTRY, AND FURNITURE STUFFS.—CLASS 19. PAPER-HANGING.—CLASS 20. CUTLERY.—CLASS 21. GOLD AND SILVER PLATE.—CLASS 22. BRONZES AND OTHER ARTISTIC CASTINGS, AND RÉPOUSSÉ WORK.—CLASS 23. CLOCK AND WATCH-WORK.—CLASS 24. APPARATUS AND PROCESSES FOR HEATING AND LIGHTING.—CLASS 25. PERFUMERY.—CLASS 26. LEATHER WORK, FANCY ARTICLES, AND BASKET WORK.

CLASS 14.—FURNITURE; AND CLASS 15.—UPHOLSTERY AND DECORATIVE WORK.

Class 14 included furniture, such as sideboards, bookcases, tables, bedsteads, chairs, billiard tables, &c.; and class 15 comprehended upholstery, bed furniture, coverings, curtains, hangings, articles of ornament, and ecclesiastic as well as domestic furniture.

The principal displays of fine furniture were to be found in the French, English, and the Italian sections. In the former there were 220 exhibitors, in the next 41, and in the latter 66. The United States numbered only nine. It cannot be said that there were any prevailing styles. The principal objects might be referred to the Gothic, Renaissance, Egyptian, Etruscan, and Pompeian.

The observer from the United States, accustomed only to the furniture there, could not fail to be impressed with the general use of ebony as the material for ornamental furniture, and with the richness of the decorations of ivory, porcelain tablets, and enamels, and with metallic bas-reliefs, medallions, and figures. It is evident that the use of these decorations is largely increasing, and that a great impulse is thereby to be given to the reproduction of suitable ornaments and to new designs. The metallic ornaments consist chiefly of the choicest productions of the galvano-plastic art, and they are left either in their usual bronzed condition, or are silvered, and in some cases thickly gilded by the same process.

The finer and highest colored stones, such as lapis-lazuli, malachite, and the choice marbles, are now freely used in the fronts and sides of fine cabinets, sideboards, and similar pieces. Some of the richest examples of such inlaying were found in the Italian and Russian sections. In the former lapis-lazuli tablets and columns ornamenting ebony cabinets were abundantly displayed, together with choice Florentine mosaics and a series of splendid inlaid tables.

The Russian cabinets in the style of Louis XIV were remarkable for their beauty and value, being made of ebony and inlaid with broad tablets of the finest colored lapis-lazuli, and adorned with bunches of fruits and flowers, carved with wonderful fidelity to nature, out of precious stones. These cabinets were made at the imperial establishment of Peterhoff, and the finest was valued at 27,418 roubles.

In the English section the displays made by Trollope & Sons, Holland & Sons, Gillow & Co., Wright & Mansfield, were particularly noticeable for their elegance and excellence. The last mentioned firm received the gold medal for their display, which included fine specimens of inlaid maple, ornamented with porcelain tablets.

In the Prussian display in this class the most notable feature was an alcove filled with carved walnut furniture in the Renaissance style. The Wurtemberg section was characterized by the beauty of the samples of inlaid floors sent by Wirth & Sons, Stuttgart. Other fine exhibitions of parquetry were noted from Bembe and from Knussman, of Mayence, Hesse.

The exhibition from the United States did not in any degree represent the actual condition of the manufacture of either common or fine furniture. A few folding steamer, or camp chairs, and rocking-chairs from Massachusetts, with an inlaid table from Wisconsin, (honorable mention,) and an ornamented door from San Francisco, composed the exhibition. The door from California was a beautiful specimen of the laurel wood of that State, and of excellent workmanship—superior, decidedly, to anything of the kind in the Exhibition.

Denmark has an interesting and curious exhibition, contributed by a society for the encouragement of art workmen. There was a cabinet on legs of ebony, lightly carved in parts and inlaid with red and green tortoise shell. The green color, like the red, is given by painting the ground on which the shell is laid. This is applied in large medallions, each surrounded by a line of brass. Several other objects were exhibited, all of them conveying a high idea of the solid good taste of the country that sent them.

Italy made a great show of artistic cabinet work. The trade appears to be reviving in the land which gave it birth. The Italian models of the sixteenth and seventeenth centuries are still regarded as the most perfect in existence. They show what is not always remembered in the present day—the proper way of treating ivory in combination with ebony. Not only must the ebony be almost covered with delicate traceries of ivory, but in the parts where the ivory forms masses its whiteness must be corrected by engravings filled in with black. In this way all violent contrast is avoided, and the decorator, with only two elements to work upon, obtains a third means of effect from the power of modifying to any extent the tone of the ivory. For instance, where a plate of ivory intended to be engraved with a subject is inserted on a flat surface of ebony, a close hatching gives a border which forms an easy transition from the black to the white, and, as we have said, the pure white is only used in very fine or closely interlaced lines in direct contact with the ebony. This work was frequent in France about the year 1550.

The following general observations upon the manufacture of furniture in France are extracted from the official catalogue:

"All the principal furniture makers who have given real importance to their trade have experienced considerable advantage by adding to it the sale of everything connected with decoration and ornament, and with very few exceptions their establishments undertake upholstery as well. On the other hand, the best upholsterers manufacture, or commission the manufacturers to make for them in their name, all kinds of elegant furniture and cabinet work. It is the same in the case of beds and bedding, now made by manufacturers of furniture as well as by upholsterers.

"A few years ago the manufacture of elegant furniture in France was almost exclusively confined to Paris; but of late some important firms have arisen at Bordeaux, Lyons, Nantes, and in several other towns, such as Troyes and St. Quentin. These, however, are not numerous, and the Paris trade has much extended since the last universal exhibition, and become more important than ever, on account of the increasing demands caused by the greater comfort and elegance of the new habitations. The reports on the international exhibition of 1862 showed, in relation to all the trades connected with furniture and decoration, the valuable assistance obtained by great establishments from artists of approved merit, and the great improvement thus produced, both as regards good taste and practical fitness. The manufacturers have understood the advantage to be derived from art, together with that technical ability that French industry possesses in so high a degree, and have boldly entered into the new path, which has already in some cases led to the most brilliant successes. The most important improvements to be noted during the last twelve years are these: Considerable increase of production; the introduction, in the case of ordinary articles, of the use of cutting machines and mechanical processes, often producing the cheapest possible results; and the employment, in all the trades connected with furniture and decoration, of distinguished artists, whose co-operation has introduced art and good taste into the manufacture.

Class 15, being intimately allied with the preceding, was included in it by the jury. Both were fused in one.

CLASS 16.—FLINT AND OTHER GLASS; STAINED GLASS.

The articles included in this class were divided into eight sections, and involved eight separate processes of manufacture: 1. Crystal glass, with basis of lead, for table services, lustres, candelabra, ornamental and fancy crystal glass, cut and plain, white and colored, threaded, gilt and painted. 2. Fine and common table glass; articles for restaurants and cafés;

mineral water bottles or syphons; retorts and other chemical apparatus. 3. Glass for mirrors and windows; moulded glasses for light-houses and paving; rough glass, channelled and plain, for glazing conservatories. 4. Window glass, plain and colored; cylinders, globes and shades of various shapes; glass tiles. 5. Bottles for wine and mineral waters; bell glasses for gardeners, &c. 6. Flint and crown glass for optical purposes. 7. Enamel in block and in tubes, for jewellers, enamellers, &c. 8. Stained glass.

The finest exhibitions of plates and mirror glass in the Exposition were from the establishment of St. Gobain & Chauny. Enormous plates standing in the outer circle of the building on each side of the main entrance were unobserved by many, for their great perfection of surface and transparency permitted objects beyond to be seen as if nothing were interposed. These plates were nearly 18 feet high and 12 feet broad; another plate measured 5.94 metres in height by 3.65 or 21.68 square metres of surface. Among the silvered plates there were two of 18 square metres and 20 square metres respectively. These firms made exhibitions not only in the French but in the Prussian and Baden sections. The products comprised mirror plates, glass for flooring, roofing, light-houses, and for telescope reflectors.

The saloons in the French section devoted to flint glass in its divers forms, for table services, decanters, pitchers, chandeliers, &c., &c., were exceedingly brilliant. In one of the saloons, the most striking of the large objects were the grand candelabra at each end, rising some 20 feet above the floor, with an enormous chandelier between them. These splendid objects were displayed by the Joint Stock Company of St. Louis, Moselle. The long ranges of tables and supports around these chandeliers were covered with other splendid productions of the works, such as urns, vases, and table services of various patterns.

The next saloon contained another magnificent display around a colossal fountain, made entirely of flint glass and rising some 25 feet above the floor, with the lower basin in massive crystal 10 feet or more in diameter. Around this remarkable object were displayed the most exquisite productions in the art of painting, enamelling, and engraving upon glass.

There was also a fine exhibition of flint glass in the English section, from London and Birmingham, particularly of finely engraved glass for the table from the firms of Millar & Co., Edinburg, and of Dobron & Green, London. These specimens of engraving were recognized by the French as even superior to their own.

There was an exhibition of moulded flint glass in the American department, which, although not attractive in point of quality or color, was remarkable as demonstrating the success with which large vessels can be moulded in a single piece without showing any trace of the mould.

In the process of manufacturing glass, the most important change that has taken place of late years is the employment (for melting the materials of which glass is composed) of Mr. Siemen's regenerative gas-furnace instead of the ordinary furnace heated by coal.

From a report presented by Mr. H. Chance to the British Association in 1865, it appears that the weekly produce of plate glass in Great Britain is about 100,000 feet. There were seven manufacturers of crown and sheet glass, three of whom made 75 per cent. of the whole quantity produced. The number of workmen engaged in these works was stated to be 2,500, and the quantity of glass produced 17,000 tons. The annual produce of flint glass in the Tyne and Wear district *only*, was estimated at £10,000,000. Birmingham produces about £5,000,000, and Stourbridge £3,500,000 annually. The make of glass bottles in the Tyne and Wear district in the year 1862 is stated by Mr. Swinburne to have been about 4,230,000 dozen.

GLASS MANUFACTURE IN FRANCE.

The introduction to this class by the French committee of admission gives the following datas upon the glass manufacture of France:

"The products of this class are chiefly manufactured in the departments of the Nord, the Aisne and the Seine, the Meurthe and the Moselle, the Rhone, the Loire and the Allier. The raw materials of the glass manufacture principally comprise silica, which, in the shape of sand, forms one-half the bulk of flint glass and three-fifths of other kinds of glass; oxide of lead, which forms one-third part of the composition of crystal glass; carbonate of lime, which represents one-fifth of the composition of common glass; and sulphate and carbonate of soda, which also form a fifth of the composition. With the exception of the lead, these materials are all of home produce; the lead is derived from Belgium, England, and Spain.

The fusion is performed in crucibles, heated by coal or wood; but the substitution of the former for the latter fuel is becoming universal. The glass manufacture depends principally on the skill of the workmen; machinery plays but a secondary part. It is only in the case of plate glass that machinery is indispensable. Glass making is carried on in houses provided with furnaces, glass-cutting, dressing and polishing shops. The workmen generally work by the piece, and there are no middlemen employed; the glass houses employ few women, but the number of children employed about the furnaces is nearly equal to that of the men. Paris is the chief market for flint, table and plate glass, as well for home consumption as for exportation. Window glass is sold to wholesale dealers, who retail it to the glaziers. The bottle makers sell to the wine producers, bottlers of mineral waters and wholesale dealers. The annual production of flint glass has risen since 1862 from 9,000,000 to about 11,000,000 or 12,000,000 francs. Ordinary table glass is extensively manufactured in France, and the importance of this trade is at least equal to that of flint glass. The production of plate glass is estimated at 350,000 to 400,000 square metres per annum, and the trade at 12,000,000 francs or 13,000,000 francs. The quantity of window glass produced may be set down as 5,000,000 to 6,000,000 square metres, of the value of 12,000,000 to 15,000,000 francs. The number of bottles is estimated at 100,000,000 to 115,000,000, of the value of 18,000,000 to 20,000,000 francs. The glass trade is increasing in all parts of the empire; and it is probable that the glass stainers here will speedily rival the skill of the old masters. Finally, the value of the whole industry reaches about 75,000,000 francs, one-third of which represents the salaries of 35,000 men, women and children. Among the improvements introduced into the glass trade since the last Exhibition must be mentioned a new method of fusing glass, by means of a combination of combustible gases, derived from coal, wood or peat, with the aid of special apparatus. This transformation, which promises important results, is the most remarkable fact in the glass trade."

The art of glass painting, says Mr. Gambier Parry in his admirable resumé of the specimens exhibited at the Exposition, published by order of the board of the council, can rarely receive justice in a general exhibition. Its diminished light is injurious to most other objects. It is as exclusive in an exhibition as a beech tree in a forest, under which nothing else will grow. Manufacturers, conscious of this fact, were careful not to undergo an ordeal which exposed them to danger. The well-known names of Bertini in Italy, Capronnier in Belgium, Aismüller in Germany, Gerente at Paris, and Clayton and Bell in England, did not appear in the catalogue. The art therefore was facilely represented. But there were, notwithstanding, many interesting specimens of excellent work. No comparison could be made between the respective merits of the various countries which exhibited, inasmuch as the specimens were scattered, and while on one side of the building the light was good, on the other it was necessarily bad.

A few general remarks on the subject of painted glass will suffice for the purposes of this article, and we shall borrow their substance from the report already referred to.

In France the system now generally prevails of giving a semi-opaque solidity to the glass, by the use of various enamels. If the light be strong outside, this dimness gives clearness to the design and makes the subject more important than the material, which theoretically is correct, but which practically, in the case of stained glass, is open to objection. In the earlier styles of the 13th and early 14th centuries the "dim religious light," of the cathedral or church was produced only by the quality of the glass. The taste of the present day is for pictorial effect only, and to produce this opacity is more or less necessary; at all events it facilitates the operations of the mere designer. A lustrous reflective glass is always preferable for the mellowness which much semi-opaque enamel would mar. The genuine gothic feeling and drawing, both in figures and ornaments, are much better represented by the English painters on glass than by their continental competitors, but their knowledge of drawing is infinitely less, and is sometimes awkward to absurdity. There is no more fatal mistake than that any one can draw well enough for a gothic window. The continental artists are educated to their profession; elsewhere it is considered an easy thing to construct a window on the old plan. The wretched result of consigning this art to inferior hands and minds is to discard so many styles and modes of expression, and to bring all work to one level of tameness and insipidity.

The principal exhibitions, in this department were from France, Belgium, Prussia, and England. France bore off the palm. The glass pictures of Marechal were art gems. Produced by the combination of opaque and transparent enamels, they seemed to exhaust the resources of the art.

CLASS 17.—PORCELAIN, EARTHENWARE, AND OTHER FANCY POTTERY.

"The productions exhibited in class 17 and designated by the title of ceramic, were for domestic use or decoration. They may be divided into four sub-classes: Terra cotta, earthenware, faïence, and porcelain. 1. Terra cotta includes all plastic objects, which, by the application of fire, are rendered fit for decoration. 2. Earthenware is hard unalterable pottery, employed to satisfy the artistic taste of the day and for the manufacture of chemical products. 3. Fine and common faïence are both used equally for domestic purposes and for decoration. Tin-glazed faïence supplies the decorative arts with indispensable elements. The ground lost in the case of common faïence has been gained by the finer sorts, which now answer perfectly the demands of the public, both as regards perfection of form and decorative appearance. 4. Hard porcelain, characterized by its whiteness, is the pottery "*par excellence*" for the service of the table and for domestic use. It is, also, advantageously applied in many cases in indoor decoration. Fine porcelain, on account of the brilliant colors which it is capable of receiving, is exclusively reserved for ornamental purposes. It is not adapted for domestic use on account of its fragility. Terra cotta is made almost everywhere. Earthenware is made in Paris, at Beauvais, and in some parts of Normandy. Fine faïence is made chiefly at Creil, Montereau, Sarreguemines, Choisy le Roi, Gien, and Bordeaux. Artistic faïence has its centre in Paris and its environs.

Steam power tends to replace hand labor to a certain extent in the making of faïence. The introduction of the methods employed in England have transformed this branch of the manufacture. As regards porcelain, the softening caused by the high temperature required for the baking deforms pieces made in any other way than by hand; and to the present time no mechanical assistance has been found available. However, there is good reason to hope that in the shaping and preparation of the material mechanical art may eventually lend its aid. The workmen are almost always paid by the piece. In consequence of the tenderness of the production, especially before baking, the men can only work in factories. As regards the decorative portions of the work, even when the artistic element predominates, the workman is compelled to manipulate the pot-

5 U E

tery.   No less than 1,362 men and 458 women are employed in Paris in the decoration of china alone.   The greater portion of the potteries have agencies or depots at Paris or send their wares to the wholesale dealers there for sale.   The latter often take the decorative portion into their own hands and equalize the productions of the various provinces by making one supply the deficiency of another.   Paris is, therefore, the grand centre of the porcelain and faïence trade.   Limoges, which comes next, sends its ware to all parts of the empire by the aid of travellers and agents.   Artistic ware, however, finds an almost exclusive market in Paris.   The treaty of commerce has made little change in the importations.   The home production has greatly augmented; the annual value of fine faïence is estimated at 10,000,000 francs, and that of porcelain at 20,000,000 francs.

The improvements realized during the last 12 years are as follows: 1. The increasing use of terra cotta in the decoration of public and private edifices.   2. The almost complete renewing of the plant of the faïence potteries, so that good organization, from being an exception, has become the rule.   3. The substitution of coal for wood in the baking of porcelain and the consequent reduction in the cost of the process.   4. The improvement introduced in the art of decoration through the chromo-lithographic process."   *Extract from the translation of the report of Messrs. Salvetat and Dommartin, members of the committee of admission of class 17.*

The word "faïence" is of recent origin, and its employment indicates an elegant extension of the business which was formerly carried on under the vulgar name of pottery.   A fine pot is no longer a piece of delf, but a specimen of faïence.   No business has grown more rapidly and satisfactorily than that represented in class 17.   Beyond all doubts, pottery is the most ancient of arts.   Drinking cups, hardened in the sun of the tropics, were, perhaps, the first utensils fashioned by man.   From this first step, long since forgotten in the series of uncounted ages, the art of the potter has maintained its ground as the most important in the series of human economy.

Pottery is the most fragile and at the same time, from its very nature, the most durable of the works of man.   In the term are included all kinds of earthenware from the rude jar and brick of the Sakkara pyramid to the porcelain of China, and the "*pâte tendre*" of Sevres.   It is, however, in the modern application of earthenware that the present age excels.   This divides itself into two important groups.   The first comprises all pottery composed of a non-vitrified body, such as terra cotta and fine and coarse earthenware.   This is the lower order.   To the second belongs all pottery composed of a vitrified body, such as stoneware and porcelain.   This is the higher order.   Each of these groups is capable of being subdivided into a very great number of different kinds.   Proceeding, then, from the simple to the more complex, we find, first, terra cotta, which is intended to serve as a substitute for stone in architectural decoration.   When an ornament has to be repeated many times, terra cotta has the advantage of cheapness over stone, and, if well prepared, possesses greater durability.   Stone that can be easily worked by the chisel and at the same time resist, for centuries, the changes of climate, is rare and difficult to obtain.   The resistance of terra cotta, on the contrary, is well known.   A glance at that of the ancients is sufficient to prove that after several thousand years it remains unchanged.   In Greece the use of terra cotta was general.   The Romans employed it in great profusion, and it descended naturally to the Italians.   The employment of terra cotta in England has revealed a remarkable fact.   "It does not," says Mr. Arnona, "blacken in the atmosphere as readily as stone and can be much more easily cleaned."   It is in Italy, France, Prussia, and Belgium that it is most generally used.   The best example in the Exposition was the fragment of a façade exhibited in the garden near the Italian section by Mr. Boni, of the national manufactory at Milan.   This specimen, very elaborate, in which the artist endeavored to show all the resources that could be made available in this material for external decoration, was in the form of a gateway, the framework of which was ornamented and decorated with figures and the panels elaborated in the style of the Renaissance.   It was regarded as the finest piece in the collection, although the Prussian and English work in terra cotta was very admirable. For practical purposes the latter specimens were, perhaps, the best.   We noticed, particularly, magnificent specimens of glazed drain-pipes, ranging from ten to thirty inches in diameter and sounding to the touch like a tube of metal.   They were made of coarse stoneware, a material harder than common earthenware and glazed.   There are three kinds of glazes commonly used, the bases of which are chloride of sodium (common salt) and salts of lead and tin; the last for majolica and other light ware in various proportions and with various adjuncts.   Salt only is used in glazing the ordinary drain-pipes.

This hard brown stoneware is also used in forming the vessels used in many of the arts.   Among other things were a large distilling retort with a well constructed worm, a barrel the size of a half-hogshead, and some gigantic jars, besides retorts and filters of all shapes and sizes. Machinery is now largely used in the preparation of this description of pottery.   It kneads the clay and moulds it into shape, and it is, thanks to the facilities thus obtained, that the economic application of hard earthenware has received such extension.

Bricks and terra cotta are the same form of pottery, differing only in treatment.   Numerous specimens of the hollow or perforated bricks were exhibited.   These, although invented twenty years ago, have only recently come into general use.   They have many advantages over the solid brick; not the least of which is their lightness.   They hold the mortar with great tenacity, and, when properly used, make drier, warmer, and healthier houses.

In a higher order of faïence the specimens were innumerable, and here it may be proper to give the latest definition of the word faïence. Any clay, which after having passed through the fire, preserves a certain amount of porosity, and which is then covered with a glaze, takes the name of faïence. When composed of a common body and covered with transparent and colored glazes, it is a faïence of the same description as that of Palisse. If it is made of common clay, but coated with an opaque enamel, it is the Italian, the delft, or the old French faïence, according to the degree of opacity in the enamel. Again, if clays of different colors are worked some upon the other, or some into the other, it becomes similar to the old ware of Perugia or that of Voiron, known as Henry II ware. If the clay contains sands and is covered with a transparent and uncolored glaze, it is the style known as the Persian ware; then, again, if the clay or the body is of a fine description, white and covered with a transparent uncolored glaze, it is the cream-colored ware or the ordinary earthenware. Those processes are often combined together, sometimes on the same piece.

Pottery that is not porous is of a vitreous texture—that is to say, porcelain biscuit and porcelain itself. There is no natural clay or mixture of clays which, being submitted to the action of fire, does not lose its porous nature, and acquire a degree of vitrification, which for the same clay will be in proportion to the heat applied. All clays have in them a natural flux. In the inferior sort this is lime and the metallic oxides—oxide of iron, chiefly; in the superior sort, which is the clay arising from the decomposition of feldspar and granite, it is a very small amount of potash or soda. The vitrification known can be arrested by mechanical means. In this process, requiring great experience and skill, the English houses excel.

Porcelain itself is the perfection of the potter's art. There are two kinds: hard—the true porcelain—the eldest, which is that of the Chinese and the Japanese, of very simple composition; and soft porcelain, an invention of the last century, in which transparency—the characteristic of porcelain—is obtained by artificial means.

The principal centres where hard porcelain is manufactured are China, Japan, Germany, and France. The manufacture of soft porcelain is even more limited, for it is the most difficult to produce of all pottery. For many years it was confined almost exclusively to the imperial factory at Sèvres. In 1804, however, the manufacture of soft paste ceased there. Investigations into the nature of hard porcelain had never been discontinued. The secret of the manufacture was known at the royal manufactory of Saxony, but every precaution was taken against its being divulged. The royal manufactory of France, founded at Vincennes in 1745, and removed to Sèvres in 1753, had always pursued its researches, and the first success dates in 1768. From 1753 to 1768, therefore—a period of 15 years—soft porcelain was exclusively produced at Sèvres. Starting from 1762, the two were produced together, with a gentle preference to one or the other, according to the taste of the directors for the time being. The French Revolution and the emigration of the nobility, which followed, struck a severe blow at a manufacture which was conducted only for the most delicate tastes. It ceased to be appreciated, and by degrees attention was diverted from a complicated and costly manufacture to one comparatively easy, capable of producing larger prices, and which offered to the artist painter the advantage of being able to estimate, during the execution of his work, the real value of the tints he employed. Hard porcelain thus became the national pottery—a source of wealth to many departments, and an important article of exportation. Since 1847, however, the old *pâte tendre* has been again produced at Sèvres.

Specimens of every kind of porcelain were exhibited in the Sèvres court. The display in every respect was superb, and worthy of a government which, without regard to cost, has established a school of pottery entirely without any equal.

There were two exhibitors in this class from the United States.

CLASS 18.—CARPETS, TAPESTRY, AND FURNITURE STUFFS.

The productions included in this class were: 1. Silk and satin damask; 2. Reps and table-covers; 3. Velvet, in goat's hair, wool, and cotton; 4. Woollen, damask, poplin, Algerian stuffs, and horse-hair fabrics; 5. Chintz, cretonne, textile fabric, and printed cloth; 6. Carpets and tapestry; 7. Embroidered and figured muslin; 8. Tick for furniture, blinds, and bedding.

In this class there were about 60 exhibitors in the French section; in the Prussian, 28; Great Britain, 39; United States, 2; but in the section of Turkey there were no less than 260, nearly all, however, of carpets; Algeria also contributed a great number of carpets and mats, but generally of small size.

The display was also exceedingly good in the Persian and Russian sections. The coarser description of Persian carpets were hung side by side with the ribbed rug woven in Koordish tents, and there were beautiful specimens of both. In color, precision of outline, and beauty of texture, some of the Persian specimens seemed more like shawls than carpets. Such work is generally intended for the mosques; men never tread on them but barefooted.

European carpets are sufficiently well known. Among the French, those of the *savonnerie*, as a short-pile carpet, are still unequalled, and in the furniture department were some very creditable imitations of this manufacture, which has been abandoned. Of long-piled carpets there was a large display, but none to equal those of the imperial factory of the Gobelins. In the French *moquettes* (velvet pile) there were many beautiful imitations of Smyrna and other ornamental carpets.

Carpets are comparatively of modern introduction in private houses in France; they were reserved for the mansions of the wealthy, small

rugs or mats laid before the seats being the only provision made to preserve the feet from the cold of the waxed oak or brick floors. Carpets
are now in very general use, but they are much more expensive than in
England.

In the English section there was a fair show of Brussels and velvet
piles, chiefly from Kidderminster.

There was a very remarkable carpet in the Austrian section. It was
designed for gas-light, being intended for the saloon of the Emperor's
box at the new opera-house. The peculiarity of coloring was not seen
to advantage under the softer influence of daylight. The pile was rather
long, like that of Smyrna carpets, but the texture was close, and parts
of the design were exceedingly delicate.

PRODUCTION IN FRANCE.

"In France the principal centres of production are: For group 1, Lyons
and Tours; 2, Paris and Nimes; 3, Amiens; 4, Roubaix, Courcoing,
Mulhouse, and Paris; 5, Mulhouse, Rouen, Claye, (Seine and Marne,) and
Paris; 6, Aubusson, Amiens, Abbeville, Beauvais, Nimes, and Courcoing;
7, Tarare and St. Quentin; 8, Lille and Flers. The imperial manufactories of Gobelins and Beauvais produce the beautiful tapestry which is
only used for the imperial palaces. That which is sold in the trade is
made at Aubusson.

The raw materials used in the manufacture of fabrics for upholstery
are very numerous. The organzines of France and Piedmont, the wefts
of China and Japan, are used in the manufactures of the silk fabrics.
The price of these materials has much increased during the last few
years; it is now at 120 francs to 130 francs for the warp, and 110 francs
to 120 francs for the weft. The French silk is the dearest and the most
esteemed. The manufacture of reps and table-cloths is composed of
French wool, valued at 10 to 15 francs the kilogram, and floss silk,
worth 40 to 60 francs, which is chiefly derived from Switzerland. Utrecht
velvet is made of goat's hair, spun in England, and sold at from 9 to 30
francs the kilogram, according to its purity. Horse-hair fabrics are woven
of materials of French origin; that which comes from Buenos Ayres is
much more expensive, costing from 16 to 30 francs. Woollen damasks
are woven with wool coming from the north of France; the weft is worth
from 7 to 8 francs the kilogram, the warp from 9 to 10 francs. For the
mixed silk fabrics they use warp at a price of 50 to 60 francs the kilogram. The Algerian fabrics are composed of cotton warps and woollen
wefts, worth 5 to 6 francs the kilogram. The price of cotton fabrics,
such as calico and cretonne, used for making prints and chintzes, is
from 50 centimes to 150 centimes per metre. These fabrics are woven
in Alsace and Rouen. The cloth used in upholstery is manufactured at
Mouy; the widest, used for table-covers, is worth, in its rough state, 3
francs the metre, and that used for covering furniture about 8 francs the
metre. The printing of the calico, cretonne, and textile fabric is per

formed principally at Mulhouse, Rouen, and Claye; the cloth is printed in Paris. The carpet manufacture employs English and French wool; the minimum price for the ordinary qualities is 8 francs the kilogram. Tapestry is made of unmixed English wool, which costs, without dyeing, from 12 to 15 francs the kilogram. The embroidered cotton fabrics come from Tarare and its neighborhood, the figured muslins from St. Quentin. The flax yarn for tick is spun at Lille. The figured fabrics used in upholstery are woven in the Jacquard machine; the plain fabrics are partly woven in power looms; the embroidery and tapestry is produced by hand, but they are beginning now to manufacture carpets by machinery; the printing is accomplished by cylinders or plates.

The cost of manufacture amounts to 10 or 15 per cent. of the value of the common article, to 20 or 25 per cent. in that of the better fabrics, and to 30 or 40 per cent. of that of the most expensive articles. The average amount of general expenses is 10 per cent. of the value of the production, without counting the cost of the designs and the inventions, which is often very considerable. Plain fabrics, at least those which are worked by hand, are manufactured in the homes of the workmen, in the neighborhood of the principal manufacturing centres; for instance, Utrecht velvets are woven in the environs of Amiens, by workmen who also cultivate the ground. Figured and fancy fabrics are usually manufactured in large workshops. In the upholstery trade only about 30 per cent. of the hands employed are women.

Paris is the principal market for all kinds of fabrics for upholstery; those manufacturers who have no depots in Paris have always an agent of some kind. Many manufacturers only work for one or two Parisian wholesale houses, and refuse all other business; and this association between the manufacturer and the Parisian salesman results from the absolute necessity of dividing, and thereby diminishing, the risks of manufacture (often considerable) in the production of those fancy articles, of which the consumption is relatively small and variable. The manufacturers of hand-made tapestry only work to order, for a new pattern has to be made for almost every buyer. Those who make carpets by machinery prepare their designs beforehand of the different sizes accepted in the trade, so as to always have a large assortment on hand. The manufacture of fabrics for upholstering is one of those for which France is most justly celebrated; the tapestry of the imperial manufactories of Gobelins and Beauvais is without a rival. The production of these fabrics is estimated at about 60,000,000. The exportation of carpets and tapestry is now very large. French woollen manufactures bear comparison with those of the best foreign markets, and their silk fabrics are unrivalled.

The committee of admission points out, among the principal improvements introduced since 1855: "firstly, the great extension of steam machinery; secondly, the introduction of a machine with eight and ten rollers, printing fabrics with that exquisite perfection of coloring which formerly could only be produced by hand."—*From the introduction to class* 18 *in the official catalogue.*

CLASS 19.—PAPER-HANGINGS.

The products exhibited in class 19 comprise: 1. Paper-hanging; 2. Painted or printed blinds.

Paper-hangings are principally made in Paris, and particularly in the faubourg St. Antoine, where are collected about 130 large factories, in which are employed 4,500 workmen, and whose produce per annum amounts to about 18,000,000 francs in value. There are also some works at Bixheim, Lyons, Metz, Caen, Toulouse, Epinal, and Mans. The raw materials employed in the production of paper-hangings, that is to say, the papers, the colors, the gelatine, &c., are now all of French origin. The designs for the decorations are always produced by French artists; the cuttings of the blocks and rollers have the same origin; and the machinery is constructed in our workshops. The materials, in the production of which foreigners for a long time held the monopoly—ultramarine and German gold, for example—are now all made in the French manufactories. The introduction of machinery into the French paper-hanging trade does not date more than 30 years back. Limited first to the production of striped papers of a single color, machine working was rapidly improved as to enable it to produce designs in many colors. In the year 1851, the number of machines employed by the paper-hanging manufacturers of France scarcely amounted to 20; they number more than 100 at the present moment. Each machine produces, on an average, 25 times as much as a hand printer; still, the introduction of mechanical means has not had the effect of diminishing the number of workmen in the same proportion. The number of slabs for hand printing has only fallen from 900 to 700 since 1851. It is to the increase in the trade itself that this result is to be attributed. If the statistics of the last 15 years be compared, we find that the amount of trade in 1865 was about 20,000,000 francs, or double that of 1850. The workmen nearly all work by the piece, and in shops where rarely less than 10 or more than 100 persons are employed, one-third of these consist of boys of less than 16 years of age; there are few women employed in the trade. There is no special market for paper-hanging, the trade being always carried on by means of commercial travellers and samples. The exportation of French paper-hangings, after having increased rapidly until the year 1860, was suddenly arrested at the time of the treaty of commerce; but it has risen again to the amount of the best years. In 1855 it was 4,074,916 francs; in 1857 it had risen to 5,948,331 francs; and finally, having fallen to 3,407,675 francs in 1861, it rose again to 5,085,000 francs, or nearly to the level of 1857; but the average price of paper-hanging fell in the same period from two francs sixty centimes to two francs twenty centimes the kilogram. From 1863 to the present day the imports have remained steadily at about 450,000; they are almost exclusively from England. The committee of admission may point out, among the improvements which have taken place in the trade: 1. The development of machine printing, and the daily improvements of the process. At first restricted to papers with two or three colors, it is now applied to the production of papers and borders with from 15 to 20 colors. 2. The recent introduction and immediate adoption of machines for strengthening or deepening colors. 3. The invention of some special kinds of paper-hangings, such as stamped, velvet, and gilt imitation of leather, of silk damask, &c.; the application of some new colors, such as the anilines and Guignet-green, in place of arsenical green, &c.

PAINTED OR PRINTED BLINDS.

The use of painted or printed blinds is much less general in France than in some other countries, Sweden and Germany for instance. They are often produced by artists working on their own account, and seldom attain the position of manufacturer. Nor is their production confined to any particular district or locality; they are produced in small workshops in nearly all great towns. There are about 30 of these in Paris, employing from 100 to 150 artists and workmen, and doing business to the extent of about 700,000 francs annually. One-fifth of the production is for exportation. The blinds made in France are for the most part painted by hand; block-printing is sometimes used, but in general only when the design employed is regular and geometrical. There has been no progress in the trade worth mentioning during the last 12 years; the processes remain the same, but the quantity produced is notably increased. In place of the unsightly blinds, overcharged with pretentious designs, and loaded with a mass of heavy opaque colors, we now see elegant compositions, produced in fine transparent colors, and worthy to take part in house decoration.

Paper-hangings had their origin in the desire to produce a material for the decoration of walls which should be less expensive than tapestry. At their first invention they were so expensive that they were literally hung on the walls, not pasted on them. They were carefully treasured, and were moved from house to house like other goods of the proprietor. They were first made at Rouen, in the early part of the sixteenth century. The earliest specimens resemble the flock-paper of the present day. They were imitations of tapestry, made by painting a pattern in adhesive oils and powdering it over with the colored wool obtained from the dressing of cloths. The next step seems to have been in marbled papers, in many of which gold and silver were introduced. It was not till the latter part of the eighteenth century that the use of chintzes suggested the application of printing to this manufacture. Of course, like the original chintz-printing, as it is still practiced in India and Persia, the design was produced by a number of engraved blocks, each charged with one color. Mechanism has long since abolished this tedious process. The paper, no longer in small sheets, but in an endless roll, passes under a succession of engraved cylinders so accurately combined that, when it issues from the press on the other side, it is completely and accurately colored. France, England, and Germany contend for the superiority in this branch of manufacture, and it is not certain that either nation maintains the superiority, although each has its own peculiar character to maintain. The French common papers surpass those of any other country in elegance, and, perhaps, also in cheapness. There were some papers with a white pattern on a gray ground, which could be bought at the rate of eight yards and a half for three cents; brown, blue, and white upon grey, a combination of three colors, cost four cents. These were for modest purposes, but the papers in imitation of Cordovan leather, gorgeously colored and gilt, were costly objects of luxury. Several of the French flock papers were excellent imitations of velvet, cloth, and reps.

Many of the English hand-made papers were admirable in execution and exceedingly rich, although generally inferior in design. Mr. Owen Jones, the famous mediæval decorator, exhibited a curious design in the way of paper-hanging. It looked like the border of one of his favorite mediæval manuscripts surrounding a page of the blue sky, powdered with a microscopic gold pattern.

The German display was not regarded as a satisfactory one.

CLASS 20.—CUTLERY.

Cutlery, properly so called, exhibited in this class may be divided into several descriptions: 1. Table cutlery, which includes knives and forks, with blades of the precious metals, commonly known by the appellation of small table plate; 2. Pocket cutlery, including spring knives of all sorts, certain huntsmen's knives and penknives; 3. Cutlery with fixed blades, such as hunting knives, poignards, and cutting tools of various kinds for business purposes; 4. Scissors and shears of all kinds, including gardener's shears, &c.; 5. Razors of every kind.

The familiar articles in this class were represented by 60 French and 94 foreign exhibitors. Of late years the French have made vast progress in the difficult art of manufacturing cutlery. England has enjoyed for years the reputation of excelling in this industry, and it is recorded that at the World's Fair in London, 1851, there was but a single exhibitor of cutlery from France. Of late years a vast stride has been taken in this business, not only by France, but by the German States and Belgium. A large proportion of the cheap cutlery which is called English has never crossed the channel, but is native and of continental production.

The English cutlery exhibition was a fine one. Sheffield was still able to maintain her own. The forms of table cutlery have gained in lightness and elegance; the razors are as keen as ever, and the scissors were very pretty, without having yet attained to the quaint ornithological forms of the French. Among the curiosities was a knife containing 28 blades, from Solingen, and any quantity of travelling knives from all parts of the world. In the six exhibitions of England were to be found clasp-knives, in the handles of which means had been found to place a spoon, a fork, a corkscrew, a pair of scissors, a saw, a file for the nails, a gimlet, a bodkin, a cutting punch, and four or five other objects, the use of which it is difficult to devise. It must be dreadful to own such a knife.

FRENCH CUTLERY.

The following, from the translation of the official catalogue, gives some interesting facts regarding the manufacture in France:

"There are four principal centres of cutlery manufacture in France: 1. The Puits de Dôme, represented by the town of Thiers, which is by far the most important as regards the amount of business. The number of pieces of cutlery produced annually in the factories of the Puits de Dôme amount to 48,000,000. This enormous production consists exclusively of cheap articles. 2. The Haute Marne, represented by the town of Nogent, produces cutlery of all qualities. It is from Nogent especially that the Paris cutlers obtain the blades for their table knives. The trade is considerable. 3. Paris, whose manufacture of articles of cutlery is far more interesting as regards the quality than the quantity of its productions; fine articles of all kinds are made at Paris, but principally table knives and razors. Lastly, Châtellerault, in the department of Vienna, which produces principally table knives and ordinary razors. The raw materials used in the cutlery trade are numerous; as iron, steel, gold and silver, employed for the blades and the ornaments; ivory, mother-of-pearl, ebony, bone, and many sorts of hard wood and horn, are used for the handles. English cast steel forms about one-half of the material for the manufacture of Parisian cutlery, and the cast steel of St. Etienne supplies the remainder. For ordinary cutlery, the cast steel of St. Etienne, the ordinary steel of Rives, (Isère), and the iron of Berry, are all employed. The principle of the division of labor is generally carried out in the cutlery trade in a very complete manner. Still manual labor predominates in this branch of industry, as the workmen only make one sort of article, and that always the same. He buys his raw material and finishes the article himself. There are, however, some important manufactories, where a certain number of mechanical tools are employed, such as stamping and cutting presses. In the centres of the great cutlery districts the workmen work at home, with apprentices, living in the surrounding villages. In Paris, however, and in a limited number of large establishments, the men work at the shops. There are but few women employed in the cutlery trade. The great centre of sale is in Paris; the manufacturers of Thiers, Nogent, and Châtellerault have depots in Paris and many other towns. Middlemen, who travel through the provinces to supply the retail houses, obtain their goods at these depots. The depots also supply the merchants for the export trade and the Paris cutlers. The value of the French cutlery trade amounts to about 20,000,000 francs; and by far the larger part of the productions of this trade is for home consumption. Thiers and its environs produced about 12,000,000 francs' worth of cutlery; the department of the Haute Marne about 4,000,000 francs' worth; Paris, 2,000,000, and Châtellerault about 1,000,000 francs' worth. A certain description of knives, called "Eustache," which formerly were in very great demand on account of their low price, are made at St. Etienne and at Nontron, in the Dordogne. The exports amount to about one-quarter of the whole production. These knives are sold at present at from 35 cents to 85 cents (3½d. to 8½d.) per dozen. A certain amount of progress has been made in the cutlery trade since the year 1855; there has been a constant improvement in the machine tools which have been applied to the production of very many kinds of articles; and in spite of the very decided increase in the price of most of the raw materials employed, and also of the advance of wages, the amount of the production has undergone little change.

CLASS 21.—GOLD AND SILVER PLATE.

This class comprises: 1. Artistic goldsmiths' work; 2. The major part of small table plate in gold, silver, and in alloyed metals, silvered or gilt by electro-chemical process; 3. Bronze ornaments for the tables and dessert services; 4. Plated ware; 5. Gold, silver and church plate; 6. Gold, silver and copper enameled ware. The goldsmith's trade is almost entirely concentrated in Paris, but there are some makers of church plate at Lyons. Fine silver is worth on an average 220 francs the kilogram. The law allows the employment of two different standards of alloy for solid plate, but the first of these is almost exclusively employed. This is worth 212 francs 62 centimes, while the second is worth only 180 francs the kilogram. Silver and gold are applied by the electro-chemical process upon articles made either of brass or of white metal, (maillechort,) which is brass, with the addition of nickel. The prices of the metals which enter into the manufacture of these alloys are as follows: Copper, 200 to 300 francs the 100 kilograms; zinc, 75 to 80 francs; nickel, 12 francs to 13 francs. The manufacture of plated ware is rapidly disappearing. The operations which contribute to the production of goldsmiths' work are very numerous. The metallic alloys are melted in crucibles; they are afterwards cast in moulds of beaten earth and sand. When taken from the mould the articles pass into the hands of the chaser. The chaser's work is, however, economically replaced in the case of stamped work by presses and steel dies. By means of these processes are produced table ornaments, certain objects of art, and various pieces of goldsmith's work, which are also made by means of the latter, the hammer and stamping. Mounting consists in uniting the various parts of a work together. This is done by means of soldering, and also of screws and nuts. Spoons and forks are made by means of rollers, on which the forms of the articles are engraved. The other processes are hand engraving and biting in with acid, enamelling, engine turning and polishing with special lathes; and, lastly, finishing, which includes rouge polishing and burnishing with steel, agate, and other tools. Goldsmiths' work is done almost exclusively either in large shops or at the houses of master workmen, employing a certain number of assistants and apprentices; very few work entirely alone. The proportion of men to women employed in the business is four to one. The number of females engaged has, however, increased since the introduction of electro-plated work, the polishing of which is entirely performed by them. The average rate of wages in Paris is 5 francs a day for men and 2 francs (40 cents) for women. The manufacturers generally sell their productions either to retail dealers or to merchants and agents for exportation. The annual value of productions, including plated ware, is 43,000,000 francs, of which only about 4,000,000 francs' worth are exported.—*Translation of the introduction of Paul Christofle, member of the committee of admission of class* 21.

The oldest establishment in France, the well-known house of Odiot, made a large display. There was nothing, however, that claimed the merit of novelty, unless it were the three massive pieces of plate which were intended in some way to celebrate the fame of the Creusot Iron Works. These were remarkable for the introduction of figures in the ordinary artisan's costume of the day, smiths resting from their toils with their implements in their hands, and cog-wheels, piston rods, and cranks filling up the details of the foreground. The idea was an innovation, and the difficulties to be overcome were no doubt great. But in these matters the effect is all that need be judged, and this did not give general satisfaction.

The collection exhibited by the brothers Fannière, besides its high order of artistic merit, had the extremely rare peculiarity of being the work of the hands of the exhibitors themselves. The brothers Fannière, pupils of Vechte, from being art workmen in the employ of others, have risen by their talent and industry to an independent commercial establishment, and in this exhibition carried off the first gold medal awarded to silver plate. Their *specialité* is a very high perfection of *repoussé* sculpture. Two shields, one in iron and the other in steel, were the most remarkable of their productions. The amount of relief was considered greater than had ever before been attained in the material, and as steel is not a tractable metal, it was deserving of attention, not only for its great artistic merit, but as defining the limit within which bold embossing, almost amounting to alto-relievo, retains its genuinely metallic character. With silver it is different. If it be burst by forcing it into a relief beyond its powers of expansion, it may be patched up by soldering in new pieces neatly enough to escape observation, unless the back be carefully examined, and even the back may be so cleaned up by files and other implements as to show no seam.

The largest collection was by Mr. Christofle, whose inumerable stores all over Paris are easily recognized by the invariable sign of windows filled with table spoons tossed into confusion with a prodigal hand. The house is one of the largest in the world, employs an enormous number of workmen, and manufactures everything, from the commonest articles of plated ware to the most expensive art productions for the table. The mass of material put on show was of a very heterogeneous character.

A collection of great artistic value and beauty was also exhibited by Mr. Le Pec, whose specialty is enamelling on a solid gold ground—gold being the only metal that can withstand the firing necessary for the superimposed work which Mr. Le Pec employs. When a vase has been thoroughly finished by this elaborate process it looks more like the production of the potter than the goldsmith.

The German collection by Wagner, the court silversmith of Berlin, was well worthy of examination. He exhibited two important works—bucklers—one given to the Prince Royal on his marriage, the other to Francis the Second of Naples, in 1864, in memory of the siege of Gaeta. Both were examples of the art skill for which the house is renowned.

Russia had a superb collection of thoroughly characteristic silver ware, mixed with occasional imitations of Arabic and Persian art. The Muscovite style is a combination of the various contrasts of whitened silver, oxydized silver, both obtained by the aid of acids, and gilding. The designs are striking, and, in not a few, inscriptions in the Russian alphabet, either pierced or engraved, are used with quaint effect. The hammered and chased silver work was regarded as the best in the class.

In the English section there were three names that challenged attention, Hancock, Hunt & Roskell, and Elkington, the English Christofle, but Elkington only exhibited silver ware. The collection was exceedingly fine. A silver swan exhibited in one of these cases occasioned a good deal of amusement, and was certainly one of the most ingenious pieces of mechanism in the building. It was of life size, and was gracefully poised on a basin of artificial water represented by revolving spirals of crystal. In this water a shoal of artificial fish were seen swimming. The swan moves the feathers of its neck gracefully, takes a proud and dignified survey of the situation, perceives the fish, seizes one in its bill, and then raises its neck and straightens it so that the fish disappears. Satisfied with this frugal but somewhat indigestible repast, the automaton curls its neck under its wings and goes to sleep. The whole is effected by means of clockwork machinery, which is said to be old, the present exhibitor only having refitted it.

In this class there were but two American exhibitors. A small collection of chased silver ware was forwarded by Messrs. Tiffany & Co., of New York, which was good enough of its kind, but inadequate to the occasion. Two pretty models of steamboats in precious metals were much admired. They were from the same house. A collection of Connecticut table ware was shown and used in the American restaurant.

CLASS 22.—BRONZES AND OTHER ARTISTIC CASTINGS AND REPOUSSÉ WORK.

The alloy forming what is called imitation bronze consists of tin, regulus of antimony, and lead. The productions of this alloy are remarkable for sharpness; but it is dear and almost always wanting in solidity. At the present day it may be almost absolutely declared that the manufacturers have given up the use of this alloy in favor of pure zinc, and particularly that prepared sort known as the Vieille Montagne zinc. Zinc, then, remains nearly the only metal in use, and when covered with a coating of copper by the electro process produces a good imitation of bronze. This galvano plating, however, entails considerable expense; and in order to produce very cheap articles certain establishments use a mere varnish, either of the color of bronze or gold. In some shops steam power is employed, but this cannot in any case supersede manual labor. All that it does is to aid the workman by saving him a considerable amount of fatigue, especially in the turning shop. The apparatus included in class 22 employ about 11,000 workmen, some of whom are paid by the day and others by the piece; the wages of the former range from four francs fifty centimes to eight francs a day. There are, however, many instances of men earning much higher wages. Piece work is of course affected by the laws of supply and demand, and is a matter of special arrangement. About 4,000 men work at home or with the designers, the others are employed in large shops, the day's work consisting of 10 hours.

The annual value of the productions of the trade reaches about 70,000,000 francs, nearly £3,000,000 sterling. In 1863 the export amounted to 44,000,000 francs, but it fell to 40,000,000 francs in 1864, and 34,000,000 in 1865. The returns for 1866 are not yet made, and we cannot, therefore, give exact figures, but, in all probability, there is still a further falling off. This diminution in the exports is attributable to the efforts made in England, Belgium, Germany, and even Russia, to establish works for the production of bronze, zinc, and iron castings. These nations are making great efforts to develop these valuable manufactures in order to compete with French producers, not only in the markets of these nations themselves, but also in the general trade.

The importation of manufactured articles is valued, at the above-named periods, at 480,000 francs, 545,000 francs, and 495,000 francs, divided between England, Belgium, and, since 1864, Germany.

The improvements to be noted are those which have arisen out of elevation of taste and knowledge of art, which are progressing daily, rather than to any improvements in the tools, &c., which have remained unchanged for a long time. There remains, however, a great deal to be done. The study of drawing and modelling becomes more and more indispensable every day, in order to enable the workmen to maintain our productions in the high esteem which they have hitherto enjoyed. Like that of bronze, the zinc and iron casting manufactures are greatly improved. They produce in the present day works which formerly belonged exclusively to art, and which, on account of their cost, were rendered almost unavailable for the decoration of private dwellings. Finally, all these industries are closely connected, and each of them, in various degrees, has undergone a perfect revolution during the last 20 years, with the aid of the fine arts. Our best artists have readily met the demands of the bronze manufacturers, and the production of a host of articles for various usages within doors bears witness to the increasing alliance of art with industry.

The articles exhibited in this class form six principal groups : 1. Artistic bronzes and ornamental bronzes, including statues, statuettes, clocks, vases, tazza, decorative candelabra, &c. 2. Iron castings, comprising figures, vases, tazzi, fountains, candelabras, railings, balconies, crosses, and miscellaneous articles. 3. Imitation bronze, (composition,) including compositions for clock cases, tazzi, vases, candlesticks, &c. 4. Repoussé work, including figures, vases, ornaments, &c. 5. Galvanized cast-iron. 6. Zinc figures and ornaments, statues, statuettes, clocks, vases, &c. The bronze, as well as the imitation bronze and zinc trade, is essentially Parisian. The art, taste, and fancy which preside over these productions have given them a special character, which, to the present moment, has kept them above rivalry. The same may be said of repoussé work, adding, however, that this industry, which is in its youth, or rather renaissance, may be expected to assume great development. The galvanization of metals, as regards France, is concentrated in Paris, but it is practiced in all parts of Europe. The application of cast-iron to ornamentation is comparatively of recent date ; its progress has been marked at each of the great exhibitions of 1851, 1855, and 1862 ; the low price of the raw material allowing of its application to monumental works, and therefore to contribute to the adornment of large public places, and edifices of all kinds, parks, gardens, &c. Iron foundries exist in almost every part of France, but there are but few that produce artistic work. For these, as for bronzes, the study and the production of the models are made in Paris. Paris is also the principal market for the disposal of these productions. The principal metals employed in the manufacture of bronze are : The copper of Chili, Russia, New Zealand, Minnesota, or Lake Superior, but the greatest portion is from Chili ; zinc, from Silesia and the Vieille Montagne ; tin, from Banca, Sumatra, and Cornwall. In this branch of manufacture the metal represents two-ninths of the value of the production, the rest being divided between the moulder, the founder, the chaser, the mounter, the turner, &c.

The principal exhibition of bronzes was from the establishments of France. The above general description of the bronze trade and manufacture is from the translation of the introduction to the class by Barbedienne, member of the committee of admission of class 22, and one of the largest producers of artistic bronzes.

Although the actual business of France in the articles of bronzes does not seem to be on the increase, her supremacy in the manufacture is unquestionable. At all periods bronze has been a favorite material for art. The small bronzes of antiquity, occasionally found in Greece and Egypt, and of which a vast collection has been exhumed from Herculaneum and Pompeii, prove that the ancients employed this material preferentially for the decoration of their houses, as well as for celebrating the virtue and valor of their heroes. At a later period bronze was used for ecclesiastical purposes, and it is for the most part the objects in this comparatively valueless metal which have been preserved as specimens of the church art of the past. The rapacity of enemies, and the impecuniosity of religious bodies, have consumed almost all the works in the nobler metals which the church had accumulated. The Renaissance was not slow to adopt this material, and in Italy schools of bronze workers have flourished from the beginning of the sixteenth century.

In France the art has passed through many trying ordeals, and survived a variety of styles. The French section of the Exhibition contained three large compartments exclusively filled with bronzes. The majority of these were in zinc bronzed, real bronze itself being vastly more expensive. Thus the Buveuse of Moreau, a crouching girl or nymph drinking out of a shell, could be produced in zinc for 550 francs, while in bronze it would cost 2,000. For such purposes, however, many prefer cast-iron, of which work fine specimens were to be seen. Some of the castings exhibited in the state in which they had left the mould were exceedingly beautiful, testifying to the great perfection to which the French have brought this art. Statues made of this material must be painted or bronzed, or covered with copper by the galvano plastic method, as much to prevent rusting as to hide the unpleasant color of the metal, which, in its natural state, is as dull and ugly as anything can well be. Iron is unquestionably the metal of the present day, and if protected by a proper coating of copper, it is not only as lasting as bronze, but much less exposed to the cupidity of revolutionists. The metal is almost worthless. No Jew, says an amusing writer on this subject, will buy it by the ton, like him who loaded so many camels with the Colossus of Rhodes; it would hardly pay the carriage. No revolution will coin it into pence, unless, indeed, posterity returns to the manners of Sparta, which is by no means the direction in which the world seems moving.

There were, of course, many fine specimens of real bronze. The reductions of famous statuary, by Barbedienne, in this substance, deserved the most unlimited praise. He is regarded as the best in the *specialité*, but there are many other names of almost equal renown.

The only American exhibitors in this class were Messrs. H. Tucker & Co., of New York, who brought a good collection of iron ornaments bronzed by a new process of their own, which is claimed to be better than the French method, and practicable at one-fourth its cost. The objects here shown were of general interest, and engaged the particular attention of all who were in the business. Cheapness, durability, and sharpness of outline are the characteristics of iron when wrought successfully. The Tucker company have made considerable progress in these directions—apart from any consideration of the special merit of their invention for bronzing—but the models and forms of their goods can very easily be improved.

CLASS 23.—CLOCK AND WATCH WORK.

The exhibits of this class were divided into three series: 1. Clocks for public buildings and their parts, such as winding apparatus, escapements, chimes, hands, illuminating apparatus, &c. French monumental clock-work is an entirely national and superior industry, taken altogether, as compared with that of foreign countries, and the value of the manufactures, principally confined to Paris, may be estimated at about 2,000,000 francs per annum. 2. The ordinary watch and clock work of commerce, which includes the making of the rough parts of both, pendulums included; dials and time-pieces for apartments, portable time-pieces, common silver watches, and watches of higher finish, whether in silver or gold cases. 3. Astronomic regulators, and marine and pocket chronometers. This branch of trade only occupies a secondary rank, but it holds the first place for its scientific importance and the beauty of its products. 4. The accessories of horology, including the manufacture of main and balance springs, the working of precious stones and machine tools. 5. Wooden clocks, the use of which is so general in villages and country places. The total value of the productions of the trade in France is estimated at 35,000,000 francs.

The centres of manufactures in France are, for the finishing of clocks, Paris; for finishing of watches, Besançon, Doubs; for the movements of watches, Beaucourt, Haute Rhin, the districts of Montbéliard and Cluses, upper Savoy; for the wheels and parts of turret and portable clocks, St. Nicolas, D'Aliermont, Seine Inferieure, Beaucourt, and Montbéliard; lastly, Morez, Jura,.for large iron clocks and those called *de Comté*, principally used in workshops and large factories. The productions of the last-named places form a considerable portion of this national industry, and it is valued at more than 4,000,000 francs. All these factories feed the French markets, and their manufactures are also exported to a considerable extent.

The number of persons engaged in horology at Besançon is about 15,000, men, women, and children. It is about one-seventh of the whole population of the arrondissement. There are 110 watchmaking shops, 20 engravers, and two large establishments which refine and prepare gold and silver for the trade. One hundred and fifty licensed manufacturers supply work to a number of isolated workmen, or to families of three or four persons, men and women, working together. These work-people are divided into classes which correspond with the various parts of the watch. Thus, there are separate workmen for the dial, the hands, the springs, the pendant, winders, &c. The shops that feed Besançon extend at present all along the Swiss frontier in the arrondissements of Moreau and Pontarlier, in the district of Montbéliard, and the mountains of the Doubs. The last two centres, represented by large factories, only make the rough pieces and detached parts, such as wheels, pinions, balance wheels, cylinders, &c. The produce of Besançon amounts to 300,000 gold and silver watches per annum, of the value of about 10,000,000 francs. In this amount labor is represented by about two-thirds, and material by the remainder. At the present moment the watch manufacture of Besançon represents four-fifths of the entire consumption of France. Its progress is very rapid, as the following figures will show: In 1845, the total production was 54,192 watches; in 1855, 141,943; and in 1865, 296,012. Within the same period importation has considerably fallen off. It diminished from 200,000 watches in 1855 to 45,454 in 1865. There exist many mutual aid societies in Besançon, and a school of horology, towards which the municipal authorities have voted a grant of 20,000 francs per annum. There are turned out annually, in addition to a large number of alarms, musical boxes, &c., more than 200,000 clock movements from Beaucourt, Badevel, and the district of Montbéliard. The town of Cluses, upper Savoy, also possesses a school for young watchmakers. The boys are employed for making rough movements and detached pieces, especially pinions, which are sent to Besançon or to Geneva. The manufactures of St. Nicolas d'Aliermont, although far from equalling that of Franche Comté in importance, still furnishes a considerable share to the horological trade of France. Out of a population of 2,500 inhabitants, about 1,000 are employed in the watch trade. Chronometers and astronomical regulators are produced there, the prices of which range between 600 and 1,200 francs, besides a large quantity of wheels for clocks, alarums, and electrical apparatus. The produce amounts annually to 144,000 pieces, the value of which is estimated at more than 1,000,000 francs. As at Besançon, numbers of workmen live in their own homes, and work, with their families, around the manufactories. Women are employed in preference to men for polishing, pivoting, and mounting the wheels. The weight of the raw material employed is 50 tons per annum, copper forming nearly the entire bulk. The articles manufactured at St. Nicolas d'Aliermont are sent principally to Paris and London.—*Translation of the introduction to Class 23, by Làngiere. (Official catalogue.)*

At the London Exhibition of 1862 there were, of all nations, 300 exhibitors in this class, of which 54 were French and 97 English. In the Exposition of 1867, the number had increased to 535, France being represented by 223 exhibitors, and England by 29.

The manufacture has made more progress in France than elsewhere, but for scientific and the higher purposes of horology the English makers still occupy the first rank. From the period when timekeepers, in the form of the quaint "Nuremberg egg," were invented, it has been the constant effort of horologists to improve the construction of horological instruments; and the efforts in this direction have been so successful that ships in the middle of vast oceans are enabled, by means of chronometers, to ascertain their position with extraordinary precision; and parties in dense forests provided with these instruments cut paths through them with unerring accuracy. To the marvellous precision of chronometers the laying of submarine telegraph cables is, in a great measure, due, and without their aid the picking up of the lost Atlantic cable—one of the most astounding feats of the century—could not have been effected.

This perfection has been attained after incessant thought, experiment, and trial. The principal difficulty that had to be contended with, and which even now has only been relatively overcome, was that of compensation. Metals, however carefully prepared, expand and contract with the atmosphere, and these variations naturally interfered with rate of speed. The errors were of vast importance to the navigator, and admonished him that he should be very careful that his chronometers were adjusted for high and low temperatures in the ice-chambers and gas-stoves of their makers. Bad oil was another cause of imperfect working, but to correct the temperature error was the chief aim of the makers of these sensitive and valuable pieces of mechanism. Vast progress has been made in this direction. The faults of the chronometer have been brought down to a matter of statistics, like the rising and setting of the sun, so that every deviation is regular and anticipated. The Arnold-Earnshaw compensation balance, composed of brass and steel laminæ, corrects every temperature error to a daily rate of four seconds, which may be regarded as pretty nearly uniform in all temperatures between 30° and 90°. Mr. Charles Frodsham exhibited some curious compensation balances, involving various new constructions; also a micrometric balance affording a simple means of adjusting chronometers without removing the balance or disturbing the mean time.

English chronometers are, in general, constructed to go two days, or 54 hours, and to be wound up daily. A considerable number, however, are constructed to go eight days, and are to be wound up every seventh day. The same gentleman exhibited an astronomical regulator combining every accumulated improvement, including new brass tubular mercury compensation pendulum and connecting galvanic apparatus for recording the time of observations. This clock was especially interesting to Americans, inasmuch as it was made for Cambridge University, Massachusetts. It was regarded by experts as the most perfect instrument of its kind in the Exposition. It is a model of the celebrated clock made by Mr. Frodsham for the Melbourne Observatory. The results of the performance of this clock during three years were submitted to the jury and pronounced to be the most remarkable for accuracy on record.

Mr. Frodsham attributed its wonderful precision not only to mechanical excellence, but also to the discovery that few pendulum-rods are ever so perfectly homogeneous as to lengthen directly by heat and shorten directly by cold. On the contrary, experiments show that they often expand into a bow form. In submitting six rods to a temperature of 600° only one of the rods remained perfectly straight, and the others bowed and warped into such shapes as to be entirely useless until they were reannealed; and what was even more surprising was the fact that the flat rods not only warped more than the round ones, but also warped edgeways. The pendulum rods used in the clock for the United States were submitted to this test of 600°.

The French collection was admirable not only in fashionable and other kinds of watches, but also in instruments of precision for astronomical and marine purposes. Gaurdin exhibited a turret clock built for the new cathedral at Buffalo, United States, and containing chimes of 43 bells, with machinery by which the airs may be varied. The bells are sweet enough, but it is to be presumed that the airs *will* be varied, for they are of a singularly trashy character, and entirely unsuited to the purposes for which they are intended.

A few electrical clocks were exhibited in the French department, and also some specimens of clocks made by machinery at Dieppe. But in the latter art the French have not yet approached the precision of American manufacturers.

Very ingeniously constructed, small, portable alarm clocks were exhibited by Phillipe. They strike an alarm and light a candle at any desired hour.

Among the revolutions attempted to be effected by the French makers is the ten hours' movement. They wish to introduce the decimal system of time in watches, dividing the day into ten hours and the minutes into 100 seconds.

The watch manufacture of Switzerland was represented by 163 exhibitors, 67 of whom were from the Bernese Jura. Watches were there to be seen ranging in price from eight francs to 1,250 francs. Among the cheap watches were some curious specimens constructed for exportation to China. A school for teaching watchmaking, founded in Geneva in 1824, turned out some extremely fine work. Pupils are admitted at the age of 14, and may remain in the establishment for four years and a half, during which time they are taught all horological processes. The terms are, for natives of Switzerland, five francs a month, and for those of other countries, 20 francs. Natives of Switzerland also enjoy the advantage of being provided gratuitously with all necessary watchmaking tools. During the winter months the pupils have the privilege of attending free courses of lectures, given in the evening, on geometry, mechanics, and linear drawing. There are also four other schools in Switzerland with professors at their heads.

Watches that are wound up with the pendant, or, as they are popularly called, the keyless watch, were very general. The fashion is convenient and advantageous, inasmuch as the watch need never be opened, and is therefore kept free from dust and moisture. The invention, however, is by no means so novel as is generally supposed. It was first introduced, says Mr. Weld, by John Arnold, in 1823, for the convenience of a naval officer who had lost his right arm.

There were two exhibitors in the American department. The workmanship of Fournier's turret clock was regarded as extremely good. It was, in every respect, a carefully constructed instrument. The contributions of the New Haven Clock Company were remarkable mainly for the processes by which they were made.

CLASS 24.—APPARATUS AND PROCESS FOR HEATING AND LIGHTING.

In this very extensive class were included the following subjects: Fireplaces, chimneys, stoves, furnaces, calorifiers, accessory objects; apparatus for heating by gas, by hot water, by hot air; apparatus for ventilating and for drying stoves; enamelled lamps, blowpipes, portable forges; lamps for oil—mineral, vegetable, or animal; other accessories of lighting; apparatus for lighting by gas; photo-electrical lamps; apparatus for lighting by magnetism.

There were fourteen exhibitors in the American department. The processes employed did not vary materially from the most advanced principles of Europe, and both in warming and lighting it may be claimed that the United States are ahead of other nations. European makers address themselves mainly to the utilization of fuel, and where they attempt warming a building they contrive to throw a small stream of heat into many apartments without interfering with the boiling of the pot in the kitchen.

The uses of gas are as yet imperfectly understood in Europe, owing to the fact that there is still a wide-spread prejudice against its use. Most of the contrivances were for regulating the supply, and measuring it with extreme accuracy.

An ingenious contrivance was shown in the English cottage. It was for lighting grate fires without the troublesome use of wood, paper, and other combustibles. Two small tubes containing burners similar to those used in gas stoves are placed besides the chimney jambs. They are on moveable joints, and can be turned to any bar. The grate is filled with coal and these tubes are lighted. They blow a blue flame into the grate, and rapidly ignite the coal.

In the French department was exhibited a plan for heating the new Grand Opera.

CLASS 25.—PERFUMERY.

This class comprehends, under the head of perfumery, all the numerous articles of the toilet. A great number of perfumery establishments exist in Paris, and there are also very important ones in Nantes, and nearly all over the south of France, particularly at Grasse, Marseilles, and Nice. The raw materials employed are oils and greases, impregnated with perfumes of flowers; distilled waters, with and without alcohol; cinnamon, cloves, &c.; odoriferous chemical essences, in their natural state and prepared, are also used. Algeria and south of France supply the flowers for perfumery at a price relatively low. Those who produce special articles, such as soap, the preparation of which involves complicated operations, employ in their workshops machines of all sorts, the use of which is becoming general everywhere. One of the Paris exhibitors who produces the raw material is a soap and perfume maker, and retails his own manufactures. A large proportion of the work people are women; they are employed both in the preparation and the making up of the perfumes. Children could also be employed, if required. The ordinary journeyman perfumers take very little time to learn the trade. They are divided into producers of raw materials, the purifiers of fatty substances, and the perfumers, who select the perfumes, incorporate them in certain substances, and sell them made up in forms more or less elegant, according to their qualities. The products of perfumeries, which attain a large total, are delivered for home consumption and to agents for exportation. The exports reach the sum of 15,000,000 francs, while the imports do not exceed 1,000,000 francs, including a certain quantity of raw materials. The exports from France are made to all parts of the world; the excellent preparation of the ingredients, the care with which they are made up for sale, and their incontestible quality, cause them to be in great demand, and daily increase their value and importance. It is to be regretted that the numerous counterfeit imitations from abroad tend, every now and then, to interfere with the impulse acquired by this branch of industry. We must signalize, however, the considerable and interesting progress which has been made in perfumery during the last few years. The methods of working have been improved, as much in regard to the processes as in an economical point of view. The plant and utensils employed in the production of toilet soap have undergone a complete transformation. The use of certain machines has become general in the greater number of workshops. Finally, in spite of the duties which weigh upon some of the raw materials, we can safely assert that the trade of perfumery has not attained its greatest development, and that the formation of the syndicate will open up a new outlet, which will tend to maintain it in the high rank it now occupies among the great French industries.—(*From the Official Catalogue.*)

Perfumery, in the present sense of the word, owes its origin to the Egyptians. The process of embalming involved the use of scented substances of all kinds, and for toilet purposes aromatic preparations were used in great profusion. The unguents used by the priests were compounded with such skill that a specimen in the museum of Alnwick castle was found, a few years ago, to have retained its scent after the lapse of 3,000 or 4,000 years. The Jews, after the Israelite captivity in Egypt, possessed themselves of all the secrets of the Egyptians, and improved upon them. They became the greatest experts of the ancient world in preparing odors of all kinds. All the Asiatic nations exhibited an intense love of perfumes. The Greeks were addicted to fine scents, and the wise Solon enacted sumptuary laws on the subject. The Romans brought many Greek customs from parts of southern Italy which had been settled by the Hellenes, and among others that of perfuming the body. Julius Cæsar issued a mandate like Solon against the importation of these dangerous articles, but without success. Caligula the Gross constantly bathed in perfumed waters, and in Nero's golden palace the drinking tables were made with concealed silver pipes, which cast on the guests a spray of essences. The unctuarium of a Roman bath contained innumerable preparations for the hair, the beard, and body. The boudoir of a Roman beauty was a complicated laboratory, where nature's idea of beauty was corrected according to the latest code of fashion, even to the particular of changing the obstinate color of the fair one's hair, which then, as now, was considered beautiful if auburn, light brown, or golden. The dye used consisted of a soap from Germany made of goat's fat and ashes, no doubt containing some very powerful alkali.

Arabia discovered the secret of extracting perfumes from flowers by the process of distillation, and the first flower to surrender its sweets was the rose. Hence the earliest commercial perfume was, and still is, known by the name of "rose water." This must not be confused with "otto of rose," which is an Indian preparation of singular potency and great price. The story of its discovery is related by Mr. Rimmel and other writers on this very interesting topic. A fair princess, while walking in her garden, through which meandered a gentle stream of rose water, observed certain oily particles floating on the surface, and this turned out to be the veritable "otto." In the present day the essence is, of course, procured by means of distillation.

Musk, although known to many nations of antiquity, seems to have been the special favorite of the Chinese, owing, perhaps, to the fact that many of the northern provinces of China are the "habitat" of the musk deer, a little animal about the size of a greyhound, from whence the perfume is obtained. When once musk has been used, its obliteration from the sense of smell is almost impossible, as an instance of which it is stated by Dr. Piesse that the walls of Malmaison, inhabited more than forty years ago by the Empress Josephine, though since then repeatedly rubbed and painted, and even washed with aquafortis, still retain the odor of this imperishable scent, of which, it is needless to add, the empress was inordinately fond.

Soap, it may here be added, whether perfumed or otherwise, was known to many savage nations long before it was discovered by Europeans.

These historical particulars, and the precise statistics of the French department, will suffice for a rapid glance at a class which cannot be made interesting by description, save by him who can paint the lily and perfume the rose. The French display was fine, not only in the manner in which these delicacies were "put up" for the market, but especially fine in the exhibit of essences and materials employed by perfumers of all countries in the fabrication of their goods. There were sixty-two exhibitors.

After France, England, except in the article of eau-de-Cologne, in which Prussia, of course, bore off the palm, ranked next. She had fifteen exhibitors. There were two contributions from America.

The contribution from Egypt was made by his Highness the Viceroy, and consisted of "galena" in powder, called "lohle," used for darkening the eyebrows and eyelids; henna powder, "lansonia alba," used for the toilet of Arab women; soap made at Cairo, small caskets; scented wood, used for perfuming rooms; "dilka" (cosmetic) and ostrich grease, used by the women of Nubia and the Soudan; wooden bottles, covered with embroidered tissues, containing bladders of crocodile musk and various perfumes used in the "Sennar;" wooden bottle and pencil used for the coloring of the eyebrows and eyelids; ivory horns used for perfumery by the nomade Arab tribes; wigs worn by the negroes of Niams-Niams on fête days.

His Highness the Bey of Tunis sent: Metikaux, essences of roses, cassia, behar, cloves, amarante, double jasmine, aloes, ambergris, sfax, jasmine, and mixed perfumes; ambergris pastilles, zebed pomade, chenouda, and oil of jasmine; "sousse" soaps, with and without scent; orange flower, "nesri," jasmine, rose, and other waters.

CLASS 26.—MOROCCO WORK, FANCY ARTICLES, AND BASKET WORK.

The articles exhibited in class 26 represented several trades which are closely connected; we may say in a general way that they belong to that kind known under the name of "articles de Paris." There are three principal series: 1. Articles in Morocco leather, and other small fancy articles; 2. Articles in fancy wood; 3. Basket work.

MOROCCO WORK AND OTHER SMALL FANCY ARTICLES.

The small fancy articles included under this head are pocket-books, dressing and travelling-cases, purses, cigar-cases, &c. The manufacture of articles in morocco leather is chiefly confined to Paris, and particularly to the third arrondissement. For these manufactures a great variety of materials are used, of which the principals are sheep, goat, boar, and other skins, specially prepared; paper, silk, velvet; rosewood, mahogany, oak, and other woods derived from Algeria; bone, horn, ivory, tortoise-shell, gold, silver, and veneers are also employed, besides iron, steel, copper, white metal, and sometimes aluminium is used.

A great number of instruments and tools are used to work the different materials; turning-lathes, presses, stamping and drawing machines, dies to cut out stuffs, frames, &c.; paring, piercing, and hinge-making machines; sewing and stitching machines; polishing and nail-making and tempering machines. The last-named description are moved by steam, the former by hand. The great variety of articles in morocco makes it difficult to reckon the value of the materials used; we can say, however, that it is of no great importance, when the articles are plain, and require no ornamentation in gold or silver. Most of the manufacturers have no working establishments, and do not employ any men in their workshops; they resort to cabinet-makers, jewellers, and others, who work by the piece. One-third of those employed are women; they almost all work for employers. The salaries vary in Paris from five francs to six francs for men, and from two francs fifty centimes to three francs for women. The articles are delivered direct to the retail venders, and to the agents for exportation. Two-thirds or so are sold in France; the remaining third is exported, principally to America, England, Germany, Spain, Russia, and several other countries. The production of articles in morocco, including small fancy articles, dressing and other cases, represents more than 12,000,000 francs. The manufacture of these articles has been much improved since 1855, and is constantly on the increase; and, at the present time, the articles are remarkable for great finish, good taste, and variety of shape.

ARTICLES IN FANCY WOOD, BASKET-WORK, ETC.

These include small articles in ivory, tortoise-shell, mother-of-pearl, shell, horn, bone, cocoa, hard wood, &c., such as ivory statuettes, billiard balls, combs, snuff-boxes, brush mountings, fans, screens, chessmen, dominoes, draughts, tric-trac counters, parasol and umbrella handles, and quantities of other articles in general use. The small lacquer-boxes belong to the same class. Their manufacture is carried on chiefly in Paris, Dieppe, St. Cloud, (Jura,) Beauvais, and in the cantons of Meru and Noailles (Oise,) Beaumont, (Seine and Oise,) in the arrondissement of Eureux, (Eure,) and in the departments of the Aisne, Marne, and Loire, Moselle, and Vosges. The articles exhibited in class 26 belong almost exclusively to the Paris trade. The materials employed are of great variety, both as to price and origin. The following are the most generally used: gold, silver, tortoise-shell, mother-of-pearl, ivory, horn, cocoa-nut wood, pasteboard, waxed leather, &c.; for the manufacture of pipes, meerschaum, brier-root, common and yellow amber, horn, ivory, bone, all the white woods, colonial woods, cherry, ebony, &c.; for combs, tortoise-shell, ivory, common horn, Irish horn, and buffalo horn, wood, hardened India-rubber, and, in some cases, metals.

The mode of manufacture of these articles is extremely varied; it changes with the articles produced.   The work is usually done by hand; nevertheless, the comb-makers have used machinery to cut out the plates of horn and tortoise-shell.   The daily wages are five or six francs for men, and two francs fifty centimes or three francs for women.   Two-thirds of the workmen work by the piece, and about two-thirds of the workwomen are employed in workshops.   The trade includes many specialties.   The principal men employed are sculptors, engravers, painters, lacquerers, horn-flatteners, workers in bronze, pasteboard cutters, decorators, filers, inlayers, moulders, polishers, turners, &c.; for women, pasteboard shapers, polishers, and piercers, *(reperceuses.)*   Most of the tradesmen employ workmen at home, and have no workshops; a certain number of workmen work on their own account, and sell their articles to the special houses in Paris, or the provinces, and to commission merchants, for exportation.   The amount of production of these small fancy articles represents as much as 50,000,000 francs.   Paris alone, whose products are almost exclusively shown in class 26, makes 11,000,000 francs.

The greater portion of the products are sent to America, England, Russia, Spain, and Germany.   During the last 10 years, the manufacture of fancy articles has become very important; brush-making particularly has made great progress.   We may note, in the first place, an important decrease in the price of almost all the products, and we can add that the Paris workmen are particularly skilful in the manufacture of fancy boxes.

BASKET-MAKING.

Basket-making has but a small space in class 26; however, a few fancy articles, which are only manufactured in Paris, may be seen there.   These are baskets and flower-stands in osier, painted, varnished, bronzed, gilt, and remarkable by the variety of their ornaments.   Few common baskets are made in Paris.   It has become a most active branch of industry in several departments, and chiefly in the Aisne, at Brigny-en-Vierache, near Vervins.—(*Extracted from the translation of the Official Catalogue.*)

The articles embraced in this class were so numerous that it would be easier to describe one-half of the fancy stores of Paris, and two-thirds of those of Vienna, than to give an idea of their infinite variety and extent. Nevertheless, they were divided into three families, called, in French, *maroquinerie, tabletterie, et vannerie.*   Each of these families was numerous enough, and distant offsprings were to be found in every part of the building.   *Maroquinerie* proper relates to large objects, such as travelling bags, &c.; *la petite maroquinerie,* to small articles, as purses, &c.   They are, as the name implies, made from morocco leather, or imitations thereof. *Tabletterie* comprises all articles turned in ivory and wood; *vannerie,* everything that is wrought by the basket-worker.

The French had 93 exhibitors in class 26, and for ingenuity, elegance, and beauty combined, were incontestably ahead of any other nation. The English excelled in leather articles, where substantiality (as in dressing-cases) was the desideratum.   The Austrians were formidable rivals to both nations.

A small amount of usefulness and a large proportion of style are the characteristics of all the well-known objects of class 26.   Most of the novelties were consequently dependent on the latter quality, no new material having been lately introduced into the manufacture of these charming objects.

Mr. Latry (France) exhibited several articles in hard wood, which were not exactly what they pretended to be.   They were, in reality, composed of fine wood-dust, mixed with the blood of animals.   This curious process is new and a trade secret.   The intensely black appearance given to the articles is ascribed to the carbonization of blood, caused by the action of heat—boiling or baking.

Another exhibitor displayed a slab of ivory of unusual proportions and vastly larger than could be obtained from the diameter of the elephant's tusk.   It was obtained by sawing spirally, in concentring rings, a longitudinal portion of the solid ivory and then opening the coils into one sheet by means of steam or some other softening process.   The specimen was 1½ foot long by 1 foot broad.

The exhibition of England in leather articles was extremely good. Austria shone best in the smaller ware, in articles made of stag-horn, and in the specialty of meerschaum pipes.   Meerschaum, though popularly supposed to be made from the froth of the sea, is, in reality, a fine clay, found principally on the coasts of the North sea, and is composed of hydrate of magnesia combined with silex.   It is easily and cheaply imitated.

There were six American exhibitors in this class.   The beautiful skeleton leaves of Mrs. Hanxhurst, the meerschaum pipes of Kaldenberg & Sons, and the wax flowers of Mrs. Bloodgood, were excellent specimens of conscientious and thoughtful skill.

GROUP IV.

CLOTHING—INCLUDING FABRICS AND OTHER OBJECTS WORN ON THE PERSON.

CLASS 27. COTTON YARNS, THREADS, AND TISSUES.—CLASS 28. FLAXEN AND HEMPEN YARNS, THREADS, AND TISSUES.—CLASS 29. COMBED WOOL AND WORSTED YARNS AND FABRICS.—CLASS 30. CARDED WOOL AND WOOLLEN YARNS AND FABRICS.—CLASS 31. SILK AND SILK MANUFACTURES.—CLASS 32. SHAWLS.—CLASS 33. LACE, NET, EMBROIDERY, AND TRIMMINGS.—CLASS 34. HOSIERY, UNDER-CLOTHING, AND MINOR ARTICLES.—CLASS 35. CLOTHING FOR BOTH SEXES.—CLASS 36. JEWELRY AND ORNAMENTS.—CLASS 37. PORTABLE ARMS.—CLASS 38. TRAVELLING AND CAMP EQUIPAGE.—CLASS 39. TOYS.

The articles included in this group are of vital importance to nations, constituting, indeed, the most active source of industry and wealth. There is hardly a country in the world that is not, in our days, affected by the interests radiating from the cotton trade; yet it is hardly more than a hundred years that cotton goods were regarded as a luxury. It was known long before having been introduced into Europe as a produce of India, in the time of the Romans, but the earliest traces of the employment of the raw material do not go beyond the sixteenth century. The manufacture at that time was almost exclusively French, the cotton being obtained from the Levant. In 1770 the consumption of raw cotton in France was only 1,600 tons a year. In England it had reached 2,500 tons; though the manufacture had been introduced later, it had already made more rapid progress. In that year America sent to Europe her first venture in raw cotton. It was a ton! Before the rebellion, in 1859, that is, in 90 years, the export from America had reached the incredible quantity of 600,000 tons.

Since that time cotton has been cultivated in almost every quarter of the globe, and with more success than could have been anticipated. Owing to this circumstance the production of cotton goods was barely interrupted by the war.

The English manufacture in 1865 was of a value of more than £80,000,000 sterling, of which £52,000,000 were exported. The quantity of cotton consumed by all the other countries of Europe and by the United States, collectively, was about one-fifth more than that required for Great Britain alone, where nearly a million of persons are employed in this branch of industry.

It was natural, under these circumstances, to have anticipated a large display in the British section; but those who had this idea were doomed to disappointment. There were but 30 exhibitors, against 210 in France. Even these 30 made but an indifferent effort at display. "The exhibition of British cotton goods," says Mr. Murray, in his official report, "was chiefly remarkable, as to its contents, for the absence of many important and essential departments; and as to its arrangements, for the absence of the practical common sense one is accustomed to expect in connection with that large and active manufacture. The goods were, for the most part, in glass cases, where they could neither be seen to advantage nor tested by the touch. The display was mute and useless to the practical visitor, and quite unattractive to the general public. Among the absentees were nearly all the leading houses of the trade."

The Scotch manufacturers entirely abstained from making a display, and thus several lighter branches of the trade were entirely unrepresented. There were no plain or printed muslins, no Jacquard muslin curtains, no muslin linings, no ginghams, no handkerchiefs. Even from Manchester, whence the principal exhibitors came, most of the leading branches failed to appear. Yarns, with a single exception, were conspicuously absent. Calico, of which England exports £23,000,000 worth a year, was represented only in one branch. Fine shirtings, another immense branch, and that of prints, of which she spreads far over £16,000,000 over the world, declined to appear with remarkable unanimity. Excepting the articles of sewing thread, which was well represented, and of which the exports are £750,000, and the calicos just mentioned, the Manchester exhibition consisted of a few minor branches, in most cases imperfectly represented, which, as exports, do not sum up, altogether, a million a year. This remarkable absence is ascribed to the operations of a tariff, which maintains a protection of 10 to 20 per cent. against British goods.

The French display of all kinds of cotton goods contrasted with that of England, greatly to the disadvantage of the latter, not only in completeness and arrangement, but in the particular that everything was left open, to be touched and examined by all comers.

The Swiss collection was well arranged and attractive, especially in the particular of Turkey red. A conjunction of favorable circumstances, plenty of pure water, cheap labor, and steady, determined industry, have given the Swiss the lead of the world in this branch. Cheap, but well printed and effective calicos, were also exhibited by several firms, competing successfully with the best goods of the same class in the French department.

Germany has many exhibitors. The most remarkable of the cotton goods were those intended for men's clothing, imitating woollen cloths, of which Armitage showed various specimens intended for the American market. Some excellent specimens of velvet and velveteen were also sent by the Power Loom Company of Linden.

Belgium occupied a very important position, and held her own against all competition with true gallantry. The quiltings and piqués were the most successful articles in the display. The calicos were both good and cheap, and cloths for men's clothing similar to those exhibited in the German court were plentiful.

There were four American exhibitors in this class.

Mr. Murray, to whose report we have already made reference, concludes his survey with these frank words:

" Few practical and reflective observers will glance, even as hurriedly as we have done, round these competitive displays of industrial ability in cotton manufacture, without feeling, however long and largely England may retain the leadership, anything like an extensive empire or undisputed sway in the cotton trade is no longer possible. The superior education of continental workmen in certain branches, or the better position of foreign merchants in regard to certain articles, already reduce us (England) to a secondary position in some respects. If, in all countries, as excellent a system of public education and as independent a spirit prevailed as in Switzerland, our position would soon be menaced in many more directions. These exhibitions of the rapidly developing powers of so many rival centres of production must quicken our efforts, by education, by political development, by co-operative interests, by every means in our power, to bring every latent energy of our population to bear in maintaining our position. While we are hovering round the question of national education, and hesitating over the petty interests of parties in regard to it, the industrial sceptre is imperceptibly slipping away from us; and, with practical obtuseness, we shall refuse to see it till the fact is accomplished and it is too late to mend."

CLASS 27.—COTTON YARNS, THREADS, AND TISSUES; AND CLASS 28.—FLAXEN AND HEMPEN YARNS, THREADS, AND TISSUES.

The following statistics relating to classes 27 and 28 are extracted from the official catalogue:

The districts in France where these yarns and fabrics are manufactured may be divided into four groups: 1. The Haut Rhin and Vosges, whose centre is the town of Mulhouse, produces all these articles, but particularly the more common sorts, such as calico, cambric, muslin, jaconet and prints. 2. Normandy, which comprehends the departments of the Seine Inférieure, Calvados, and Orne, and in the towns of Rouen, Flers, Condé-sur-Noireau, Evreux, &c., are manufactured cotton cloths, handkerchiefs, jeans, prints, checks, and other articles, in which the price of the cotton employed bears a large proportion to the cost of the manufacturer. 3. In the group formed by the departments of the Nord, Aisne, and Somme, containing the towns of Lille, Roubaix, St. Quentin, and Amiens, are principally to be found manufactures of cotton yarn for net and lace of thin fabrics, figured muslin, curtains, and cotton velvet. 4. Tarare produces tarletan, muslin, and embroidered muslin curtains; Roanne, colored fabrics, and checks. The cotton employed by the French manufactories for the last forty years has been almost completely derived from the United States of America, which produced yearly from 700,000,000 to

1,000,000,000 kilograms. This market has been entirely closed for the last four years. India, China, Egypt, the Brazils, and the coasts of the Mediterranean have developed their production during this time, and have alone furnished cotton to the whole world. The price of middling New Orleans cotton, which, before the war, was 1 franc 80 centimes the kilogram, rose in 1864 to 7 francs, and is now worth 3 francs 40 centimes. Good Indian cotton costs, generally, about a franc less per kilogram.

Machinery has everywhere replaced manual labor in the cotton spinning trade, which employs more than 6,250,000 spindles. The weaving is also, in a great measure, done by machinery, especially that of the more usual articles of consumption. In the departments of the Haut Rhin and the Vosges, where 50,000 looms are employed, about 9,000 only are worked by hand. Hand weaving is still maintained for the manufacture of those fabrics which, subject to the changes of fashion, demand great variety of style and pattern, such as the thin tissues of St. Quentin and Tarare, piqués for waistcoats, and the other miscellaneous articles of the department of the Seine Inférieure. Machinery, by reducing the prices of the productions, and thereby enlarging the demand, employs a greater number of workmen than did the hand-looms. About 80,000 power-looms and 200,000 hand-looms are worked in France. In those departments where machinery is principally used the workmen work together in large manufactories; where, on the contrary, hand labor predominates, the weavers usually work at home. About 600,000 hands are employed in the cotton trade and are mostly paid by the piece. Out of this number about 200,000 work in their own dwellings.

The produce of the cotton trade is sold in the central towns of the different manufacturing districts. Mulhouse is the market of the eastern department, while Rouen is that of the western. There are also smaller markets: Flers for jeans, Amiens for velvets, St. Quentin for piqués and figured muslins, and Tarare for tarletans and embroidered muslins. Most of the manufacturers have a depot at Paris, sometimes dealing directly with the public and at others through the medium of a large wholesale house. This makes Paris one of the principal markets of the cotton trade.

The importation of cotton from different sources during the year 1866 amounted to 120,000 tons, of the estimated value of 420,000,000 francs. The yarns and woven fabrics produced amounted to 105,000 tons, of the value of 800,000,000 francs, the cost of manufacturing which may be set down at 320,000,000 francs. The export was 21,000 tons.

The committee of admission of this class make the following reports upon the progress of the cotton trade in France during the last 12 years :

"1. All the machinery employed in the preparation and spinning of cotton has been much improved. For the old spinning machinery have been substituted self-acting machines,which make thread of all sizes from No. 1 to No. 200, the first measuring 1,000 metres and the second 200,000 metres to the pound.

"2. The almost universal use of power-looms in the manufacture of heavy fabrics, the invention of the fast-working looms, throwing the shuttle no less than 240 times a minute, and making from the coarsest to the finest fabrics; the bringing into general use of sizing machines.

"3. Numerous improvements in the details of cotton printing; the employment of new colors; the introduction of new machines which, receiving between their rollers a white material, deliver it up printed in ten or twelve colors. During the last 12 years the French manufacturers have renewed their machinery, and well-organized mills, which were the exception in 1855, have now become the rule. The treaties of commerce which have led to a wholesome rivalry with foreign countries have accelerated this improvement. The employment of Indian cotton has necessitated a change in the machinery, and permits the use of part of the raw produce which was formally rejected, leaving but small amount of waste.

FLAXEN AND HEMPEN YARNS, THREADS AND TISSUES.

"The linen trade comprises the preparation, spinning, and weaving of various textile materials, such as flax, hemp, jute, China grass, &c. We have only to treat here of the spinning and weaving of these fibres from which are made cambric, lawn, coarse and fine linen of all kinds, damasks, diapers, and various tissues of thread mixed with cotton and silk.

"The principal seats of the French linen trade are: Lille, Dunkirk, Boulogne-sur-mer, Amiens, Abbeville, Valenciennes, Cambray, Chollet, and Lisieux. Hempen fabrics are made especially in the departments of Sarthe and Finistère. Lisieux and Noirmoutier are famous for white sheetings. Flax, hemp, jute, and China grass are grown in various countries. The flax used in France comes principally from the north of France, Belgium, Picardy, and Normandy. The flax grown in the department of the Nord and in the environs of Bernay (Eure) is of superior quality, but not equal to that produced near Courtray in Belgium. Russia also supplies us with pretty good flax, but which can only be employed for the lower numbers of yarns.

"Flax is very variable in price, but we may take 1 franc 70 centimes the kilogram as about the average price of No. 30 of good current quality. Flax is cheaper than hemp; the best kinds come from Picardy and Champagne. The average price of heckled hemp is about 90 to 120 francs. Jute comes from the East Indies in large quantities; its price for some time has been about 45 francs the 100 kilograms. China grass, the name of which indicates its origin, is a textile fibre which is likely in future to become of considerable importance in our trade. The methods of preparation and working are very nearly the same for all kinds of textile matters. The plant is first submitted to the operation of rotting, which is generally performed by allowing it to soak in water or to expose it on the ground until the gummy matter which it contains is dissolved. Next comes the operation of beating, the object of which is to separate

7 U E

the fibres from the rest of the plant. These two operations belong to agriculture. The spinners of flax and hemp purchase their materials of salesmen who travel about the country and act as middlemen between the farmers and the spinners. These materials are ready to be submitted to the operations of the spinning mills, from which manual labor may be said to have been banished entirely, except in the case of yarns of exceptional fineness used for the production of cambric.

"The number of spindles has increased from 90,000 in 1842 to 600,000 in 1865. Power-looms are being substituted more every day for hand-looms, as allowing of a more rapid and economical production. Of the whole number of persons employed in the flax and hemp mills two-thirds are women; but in power-loom weaving the proportion is only about one-half. In each case the female work-people gain 2 francs to 2 francs 50 centimes per day, and the men 2 francs 50 centimes to 4 francs. The organization of the linen trade is now very powerful in France. Some large manufacturers sell their goods directly to retail dealers or agents. The business increases daily in extent, and the importation, especially of table and toilet linen, has become insignificant. The prices of the various kinds of fabrics are extremely various; very low for certain qualities and certain widths, and very high for the finer sorts and widest kinds. Linen cloth, for instance, varies from 80 centimetres to 3 metres in width, and in price from 75 centimes to 15 francs. The manufacture of linen or hempen cloth and jute tissues has increased largely during the past few years, as the following figures show: The imports of flax and tow, which were only 19,200 tons in 1862, had risen to 48,000 tons in 1865. The importation of raw jute rose from 6,300 tons in 1862 to 10,650 tons in 1865. On the other hand, the exportation of heckled flax and tow fell off from 7,037 tons in 1862 to 6,068 tons in 1865, while the exports of yarns rose from 497 tons in 1862 to 2,374 tons in 1865. The exportation of plain linens rose from 2,054 tons in 1862 to 3,254 tons in 1865. It must be added that these results were principally due to the cotton crisis; but they owe something also to the improvements made in the machinery employed in its manufacture. Some very happy modifications have been introduced of late years into the machines employed in combing flax and preparing tow. In weaving, as we have already said, self-acting power-looms are replacing those worked by hand, and thus the quantity produced has been increased while the cost of labor has been diminished. Some very important establishments for spinning and weaving have been set on foot. It is right to add, in justice to the linen trade, that most of the great works are constructed and arranged in the most favorable conditions with regard to the welfare of the work-people employed in them."

CLASSES 29–30.—COMBED AND CARDED WOOL AND WORSTED YARNS AND FABRICS.

These two classes, embracing the most extensive and ancient form of industry known to the world, were represented competitively by all the manufacturing countries of Europe and by seven exhibitors from the United States. The range of articles being very large the display was naturally of great importance, particularly to experts. As a matter of interest to the eye it was unattractive, and there was little in either class that could engage other than a technical pen. Coats and pantaloons in the concrete have no innate charm, and wool and worsted, although comfortable to wear, are unsuggestive in a literary point of view. It will be readily understood that this manufacture does not admit of much scope for artistic design. It depends on a successful blending of colors and an ascertainable degree of perfection in texture and finish. The French excel in fine and fancy articles; the English in plain tissues; and the German and Belgian makers in imitations, having cheapness for their main end. During the past ten years shoddy has come greatly into use, and it is said that as much as 60 per cent. can be employed advantageously in cheap materials. Shoddy is the woolly part of old garments cleaned and prepared by processes that are daily being improved. By utilizing material that was formerly cast away as waste, great progress has, of late years, been made in the production of cheap cloths.

We give below the following details of the trade in France:

"Class 29 includes: 1. Combed wool; 2. Woollen yarns combed and carded; 3. Tissues of pure combed wool; 4. Flannels and fancy stuffs of carded and slightly fettled wool; 5. Tissues of wool mixed with other materials.

"The principal centres of production for these articles are: Rheims, Roubaix, St. Quentin, Amiens, Mulhouse, St. Marie-aux-Mines, Rouen, Fourmies, and Le Cateau, in the Nord; Guise, in the Aisne; and, lastly, Paris.

"In 1855, French wool held a more important place in the supply of our manufactories than it does at present. At that period, but little was known of Australian wool, of which 23,000 tons was imported in 1865. On the other hand, the imports from Belgium, Spain, Germany, Turkey, Algeria, La Plata, and other countries have not diminished in importance; they amounted, during the same year to nearly 50,000 tons. It is Australia, however, which has principally met the increased demands of our trade. These various wools are now combed and woven by admirably constructed machinery; the weaving of woollens by power-looms, which was scarcely tried in 1855, has acquired of late years, and particularly since 1862, a rapid development, and is increasing daily. Still, hand weaving has not diminished in importance; but it has remained nearly stationary; and the increase in production is due to the employment of mechanical means.

"The situation of the work-people employed in the manufacture of woollen fabrics is improved. Those engaged in combing and spinning works have not suffered from want of work, and their wages are generally high. The same has been the case with the power-loom weavers; but in spite of the importance which power-loom weaving has already assumed, the number of power-loom weavers is still very inferior to that of the weavers who work by hand at their own houses; and, in the 'Aisne,' the proportion of the former to the latter does not exceed five per cent. The proportion of women employed in combing and spinning, as well as in the weaving of woollen fabrics, whether working in factories, or at home, varies greatly according to local conditions; it may be safely estimated that it amounts to one-half in some places, and two-thirds in others. Nearly all the woollens, whether pure or mixed, manufactured for consumption in France, are adopted by other nations; the prices have been much reduced since 1855, in spite of the maintenance or the increase in price of the raw materials. The growth of the manufacture has been very favorable to the maintenance of the quotations with respect to these matters; but the same development has often, on the contrary, produced a depreciation in the value of the manufactured articles by the superabundant supply. Thus, the manufacturer has been compelled to look for his profit in the continual improvement of his methods and industrial processes. To this frequent over supply, and consequent increase of the stocks, must also be attributed the incessant efforts of the manufacturer to place himself in direct communication with the retailer and the exporter, and thus avoid the middleman.

"The home and export trade, and the means of production have grown rapidly. In 1855, the imports of raw wool only amounted to 68,000,000 francs; while, in 1865, they reached 247,000,000 francs. The exports of woollen of all kinds have followed the same rapid course, having risen from 165,000,000 francs in 1855 to 396,000,000 francs in 1865, in which amounts yarns and stuffs of combed wool represented 279,000,000 francs. Remarkable improvements have, moreover, contributed since 1855 to the development of the production and exportation. New methods of combing and spinning; ingenious means of facilitating the work of the operative or the machinery; the application of the products of aniline as coloring matters; and lastly, the introduction into France of new methods of dressing, have enabled the manufacturers of combed wool to make successive reductions in the price of their fabrics, while losing none of their superiority.—*From the translation of the Introduction to the Class by Gustave Larsonnier.*

CARDED WOOL AND FABRICS.

"The products exhibited in class 30 form four principal series:

"1. Black and colored broadcloths, livery cloths, billiard and coach cloths, black satin cloths, eider-down cloths, and castors.

"2. Fancy paletot and ladies' cloths.

"3. Fancy trouserings.

"4. Articles for jackets and fancy suits. These productions are manufactured in five great groups in France:

" 1. The group of Normandy, the centre of which is the town of Elbeuf, and which includes the departments of the Seine Inférieure, Eure, and

Calvados. The towns of Elbeuf and Louviers produce nearly all the descriptions of goods cited above. Vire, Lisieux, and Romarantin produce cheap fabrics especially, such as pilot cloths, fancy trouserings, and velvet cloths for ladies' mantles.

"2. The Ardennes group, the centre of which is Sedan, and where are manufactured principally the black tissues, such as satin cloths, cashmeres, eider-downs, fancy paletot cloths, and ladies' velvet cloths.

" 3. The Isère group, of which the centre is Vienna, and which produces mostly low-priced goods for trousers, paletots, and complete suits, as well as ladies' cloths.

"4. The Haut Rhin and Moselle group, the centre of which is the town of Bischwiller, and which produces satin cloths, paletots and black fancy cloths; the coarse stuffs for country wear are principally made at Nancy.

"5. The southern group, comprising the towns of Carcassone, Mazamet, Saint Pons, and Bedarieux, which produce generally all the kinds of cheap goods mentioned above. The town of Chateauroux, which supplies the cloth for the army, may be added to this group.

"The wool employed by the French manufacturers is indigenous or imported from Germany, Australia, Russia, Buenos Ayres, and Spain. The price of washed undried wool varies from 5 francs to 12 francs. Mechanism has been almost everywhere substituted for hand labor; handloom weaving is only now employed in the manufacture of articles in which the designs following the caprices of fashions demand great variety. Of these articles are the stuffs for trousers, paletots, jackets, and ladies' clothing. Mechanical labor, by reducing the price of the goods, induces large consumption, and, consequently, the employment of more workmen. It may be estimated that the labor and the general expenses, taking the average of winter and summer clothing, adds one-third to the cost of the raw material. Where steam power predominates, the operatives work in the factories; where, on the contrary, hand labor is still employed, the majority work at home; in both cases they are generally paid by the piece. About two-thirds of the whole of the work-people are engaged in factories; the proportion of women employed is about two-fifths. The manufactured goods are sold in the various centres of the trade. Elbeuf is the great market of the western department, and after it comes Sedan, Louviers, Vienna, Lisieux, Vire, and Bischwiller. Generally the large manufacturers sell their products directly to large houses of business in Paris and the departments; the latter send their travellers through France and other countries to dispose of the goods.

"The great mass of the wool used in the making of cloth comes from abroad. French wool is principally employed for common fabrics. The export, in 1865, amounted to 5,500 tons, of an approximative value of 71,000,000 francs. The annual production of France is about 250,000,000 francs.

"The committee of admission of class 30 notice, among the improvements which have taken place in the trade, during the last 12 years—

"1. The washing of wool by machinery.

"2. The improvements in the machines used in the preparation of the wool, such as beating, teaseling machines, &c., which allow of the use of wool from all sources; the new system of cards and of looms.

"3. The almost universal employment of power-looms for weaving broadcloths, satins, paletot and fancy cloths. Power-looms with several shuttles are yet but little used for weaving fancy cloths.

"4. In scouring and fulling, the conjunction of steam-engines with hydraulic motors, to prevent the works being brought to a stand-still during very dry seasons.

"5. In dressing, the employment of machines in place of hand-beaters and mechanical tenter frames."—*From the Introduction by Vanquelin to Class* 30.

In the United States section the principal exhibitors were the Webster Woollen Mills of Massachusetts, which sent broadcloths, doeskins, castors, and muskowa; the National Association of Wool Growers, John L. Hayes, secretary; and the Mission Woollen Mills, San Francisco, California. This establishment made a very fine exhibit of cloths, cassimeres, and flannels, and particularly of blankets. The following descriptive notice of the Mission Woollen Mills in California is extracted from the Commercial Review of the Pacific States for 1866:

THE MISSION WOOLLEN MILLS.—CALIFORNIA.

"The Mission Woollen Mills are located at the head of Mission creek, in the southwestern portion of San Francisco. With the exception of a very large two-story stone warehouse, used for the storage of wool, all the mill buildings are of wood. The grounds used comprise some 10 acres, a portion of which, probably three acres, is covered with the different buildings. The mills were first put in operation in the autumn of 1860, starting with a capacity equal to the employment of 40 men. The works have been increased until, at the present time, running night and day, they employ 400 operatives, 300 of them being Chinese. In 1865, these mills consumed 1,200,000 pounds of wool, which was manufactured into 32,000 pairs blankets, nearly 500,000 yards of flannels, and over 100,000 yards of cloths, cassimeres, tweeds, and cloakings. Since then the machinery has been increased, and the estimated consumption of wool for 1866 is set down at 1,900,000 pounds. The business sagacity of the proprietors of these mills has made them keenly alive to improvements in machinery, with which they have supplied their works as soon as known. The present working power of the mills consist of two engines, each of 150 horse-power, which drive 11 sets of cards, 4,000 spindles, and 50 broad-power looms, which will be soon largely increased. To give some idea of the extent of their manufacture, we may say that, during the month of August, there were completed from the raw wool to the finished cloths 15,270 yards cassimeres, tweeds, and cloakings, 35,475 yards flannels, and 6,270 pairs blankets. The Mission Woollen Mills were the first on the Pacific coast to manufacture varieties of woollen goods, besides blankets. Their blankets (in common with the Pioneer Woollen Mills) have made a reputation for California manufacturers the world over, wherever known, they exceeding in fineness of wool and finish the best blankets made in Europe. One feature in the working of these mills in San Francisco is the employment of Chinese operatives, who, being intelligent and industrious, at low wages, enable successful competition to be had against white labor in the manufactories of the eastern States and Europe. Without this cheap labor, mill-owners state that they would be unable to manufacture with profit. Very large quantities of goods are shipped to Montana Territory, where they are preferred to the manufactures of eastern mills, which pay less freightage by way of the Missouri river."

CLASS 31.—SILK AND SILK MANUFACTURES.

"The material exhibited in class 31 may be divided into three principal sections: silk and yarns, silk tissues, and ribbons. The first section includes silkworms' eggs, new and dried cocoons; raw silks; thrown, unbleached, and dyed silks, designated by the names of weft, organzine, grenadine, &c., for the manufactures of tissues; twisted silks for sewing, embroidering, hosiery, trimming, guipure, and lace, and waste and floss silk; and these last products carded, combed, and spun into single, double, twisted, unbleached, and dyed yarns. The silk tissues include velvets; plain and figured stuffs for dresses and furniture; bolting tissues; tissues for men's and women's hats; sarcenet and lutestring for linings; plain and printed foulards for dresses and handkerchiefs; shawls, neckerchiefs, and cravats; crapes and tulles. The ribbons comprise plain and figured ribbons; galloon, binding, and trimming for dresses and bonnets.

"The principal centres of production are: for the spinning and throwing of fine silk the departments of Ardèche, Drôme, Gard, Herault, and Vaucluse; then come those of Isère, Var, the Lower Alps, Rhône, Bouches du Rhône, and Tarnet Garonne. The strong silks that are imported raw from abroad, and especially from Persia, China, and Japan, are manufactured in the departments du Rhône, Gard, Loire, and Indre et Loire; and above all in the departments of Oiseet Ere, from whence the Paris manufacturers mostly supply themselves. The principal spinning mills for waste silk are in England, Switzerland, and France. For stuffs, Lyons and its environs; then Tours, where the furniture stuffs are principally manufactured. For ribbons, binding, and galloon St. Etienne and St. Chamond. There are also a few manufactories in Moselle and Haut Rhin. The cocoons used in the French spinning mills were almost entirely supplied by the silkworms of the fine breed of France; but, since the year 1863, an almost universal epidemic has successively attacked the silkworms in every part of the world. To remedy these disasters eggs have been imported from those countries in Europe where the disease had not penetrated, and afterwards from the east, to which is due, in a great

measure, the improved result of the last few yields.   Before the invasion
of the disease, from the year 1846 to 1852, the average yield in France
was generally valued at 24,000 tons of cocoons, producing 2,000 tons of
silk, and representing a sum of about 120,000,000 francs.   After the
appearance of the disease the amount fell to one-half, to one-third, to
one-quarter; and in 1865 they had become reduced to one-fifth of the
ordinary yields.   The average price has risen from four to six francs,
and has even surpassed the latter sum.   The effect of this is, of course,
to raise the price of the silks, which are employed according to their
qualities.   Thus the silks of France and Italy, and of the Broussa and
Syrian spinning mills, are used to make the best tissues, plain and figured.
The silks of Japan, China, Bengal, and Persia are employed, according
to their sizes and worth, in the manufacture of plain and figured tissues
of current qualities, in making foulards, and sewing and embroidering
silks.

"Machinery is everywhere rapidly replacing manual labor in the spinning and working of silks; machines are substituted for hand-work,
even for twisting and sewing silk, which was formerly done only by hand.
The silk goods, properly so called, are always woven by hands.   In the
manufacture of foulards, and of nearly all the stuffs which can be woven
with raw silk, the power loom has replaced the hand loom for weaving
as well as for the warping and other processes.   A great many attempts
have been made to use machinery in the manufacture of the stuffs which
are woven with prepared silk, and they have been successful for the light
satins, which are dressed, and to a certain extent with the black silks of
light quality.   At St. Etienne manual labor has been continued in the
ribbon trade, while the power loom has been adopted for the galloons and
bindings.   One or two manufacturers in Haut Rhin are using machinery
for plain ribbon-making with some success.

"The cocoons are spun and the silk prepared in the south of France
by women and girls, who work by the day in workshops belonging to a
principal, under the superintendence of foremen.   The system is the same
for the winding off by machinery, but when the winding is done by
hand, the people work at home and by the piece.   The twisting is generally done by men.   For the silks, the organization of the workshops
varies according as the weaving is done in the towns or in the country,
in private workshops or in manufactories.   At Lyons, for example, the
material prepared for weaving is delivered by a manufacturer to the
master of a workshop who possesses a certain number of looms.   This
latter furnishes the premises, the looms, and all the tools necessary to
the manufacture ; then, for the hire of the workshop and the looms, he
retains from the weavers the half of the price of manufacture paid by
the manufacturer.   In the country the manufacturer treats directly with
the weavers; he furnishes all the implements to the workmen who work
at home, and pays them 55 per cent. on the price given at Lyons.
The salaries for the work done in manufactories are from 45 to 50 per

cent. lower than those of Lyons, the premises being supplied by the manufacturer. The workmen are always paid by the piece. The fresh cocoons, silks, and waste are sold at the various centres of production.

"Marseilles is the great market for dyed cocoons as well as for the silks and waste imported from abroad; Lyons, for the fine silks of the south of France and Italy; Paris, for the foreign strong silks, unbleached and dyed; Paris and Lyons, for the spun waste. As regards the manufactured products, it is Lyons, St. Etienne, and Paris which supply England, America, Germany, Belgium, Russia, Turkey, Spain, all those countries, in a word, which employ silk tissues. The importation of all the productions from silkworms' cocoons, raw and thrown silks, floss silk, in hanks, and spun, &c., is estimated for the year 1865 at 297,000,000 francs, and the exportation of the same articles at 126,000,000 francs. The importation of silk tissues and floss silk, ribbons included, of all productions for the year 1865, were upwards of 10,000,000 francs. The exportation of the same articles amounted to 400,000,000 francs.

"The committee of admission for class 31 point out among the improvements realized in the silk industry since the year 1855:—1. As regards silks and waste, considerable improvements in the spinning and twisting machines: 2. As regards tissues, a more intelligent use of the very varied materials which the rarity of our beautiful silks of France and Italy have forced us to have recourse to:—3. The new process of dyeing by the application of colors with aniline and fuchshine bases; new processes for printing on warps and foulards; and, finally, the improved systems of figuring silks." *Report of the Committee of Admission.*

The beautiful fabrics exposed in this class exhibited, perhaps, to the best advantage the skill and taste employed in French manufactures, relatively, too, in the manufactures of other countries, for there were silks from all parts of the world. But France unquestionably maintained her rank as the foremost among the producers of these attractive articles. The exhibition was unusually large and interesting. It commenced with the cocoon and ended with the finest triumphs of the loom. Specimens were also shown of vegetable silk produced from a bulb which is common at the Cape of Good Hope. Wild silk is abundant in India and China. It is produced by an insect like a caterpillar, which forms the cocoon in a shrub, yet it does not die there, but escapes and becomes a butterfly. Though, in many respects, totally different from the silkworm, the silk is much prized on account of its strength.

The principal exhibitors of raw silks were France, Algiers, Italy, Austria, Spain, Portugal, Russia, Turkey, India, and Victoria. The spinning of silk is becoming more and more a manufacturing industry, and everywhere large spinning mills are being erected, worked by steam power. Some of the best specimens in the French department were from Ardèche; in the Italian, from Milan. Austria, too, had a good exhibition. The Russian silk is produced in the Caucasus.

The fertile imagination of French designers, stimulated by the means placed at their disposal in the new colors obtained from aniline and other chemical products, has given a fresh impulse to the silk trade, and led to very beautiful results. In appropriateness of design, happy contrast of color, and excellence of fabrication, the French silks were admittedly the best on exhibition.

Switzerland had a good display of light and low-priced silks. They were of the simple kinds that find a market everywhere. An excellent feature in Swiss silks is the extreme beauty of color, and the fact that the lightest and most delicate tints are altogether unaffected by the touch of the hand or the impurities of the atmosphere.

Italy had a large exhibition, including the famous Genoa velvet, and much fine silk brocade and tapestry.

Austria displayed tapestry silks of great beauty. Spain sent some fine specimens from Valencia and Barcelona. Tunis exhibited good tissues of silk and silver. India had some rich samples of silk tissues and silk with gold and silver. Russia had some excellent silks, from Moscow; and Prussia and the Zollverein made a fine display of silk velvets and ribbons. The British display was good in certain heavy specialties, such as moire antiques, &c., but in other respects it was meagre. A novelty was exhibited in this section. It consisted of a material for curtains, composed of silk and fine threads of glass, woven in the usual way, and producing a very charming effect; it is called *tissues de verre*. In general respects there was a marked inferiority in British silks, and a notable falling off since the exhibition of 1862.

There were two exhibitors in this class from the United States.

CLASS 32.—SHAWLS.

"Class 32 comprises figured shawls of all kinds—that is to say, Cashmere shawls, woollen shawls; shawls of wool, cotton, and spun silk, mixed, and silk shawls. The shawl manufacture exists in but three districts of France—Paris, which makes, or causes to be made elsewhere, rich, middling, or cheap shawls; Lyons, which produces chiefly shawls of moderate and low price; and Nimes, which manufactures cheap shawls only. The greater part of the shawls sold in Paris are produced in Picardy, chiefly at Fresnoy-le-Grand, at Bohain, and in the environs of the latter place. The Parisian makers have always maintained a *bona fide* superiority in the manufacture of rich shawls, by means of their taste and inventive spirit; and we may say that all the happy innovations which have perfected the invention of Jacquard are due to Paris. The designers of Paris enjoy a well-earned reputation. Foreign countries which manufacture shawls, such as England and Austria, obtain their patterns and even have them placed on cards in Paris, especially for shawls of a certain price.

"The materials which enter into the manufacture of shawls are: The Cashmere hair, which comes by way of Russia, and is principally obtained in Thibet from a peculiar variety of goat; wool of various countries, but particularly of Germany; raw silk, or the organzine of the south of France; spun silk, and even cotton. The price of the yarns made with these various materials, and used in the shawl manufacture, varies from 10 to 70 francs the kilogram. The weaving of shawls is performed by the Jacquard loom, which has been greatly improved since the time of its illustrious inventor. The shawl manufacturers, in the first place, have their designs produced either in their own establishments or out of doors. The pattern, once settled, is put on the cards by the designer, revised, and handed over to the reader. This latter operation, which is generally performed by special workmen, consists in translating, as it were, from the design-card to the cards of the machine, each of which represents one of the little squares of the former and each of the colors which has to be produced in the loom. For the weaving, the workman or the foreman receives the warp, dyed or prepared, and also the material for the weft. When the shawl is woven it is handed over to the dresser, who cuts it, shears it by mechanical means, and finally washes and dresses it. The rich shawl is the type of all the other classes. It is generally woven on a warp called cashmere, but composed of a thread of Cashmere twisted with a thread of organzine or of raw silk; the weft is of pure Cashmere, of excellent quality. The manufacturers of rich shawls are, and must always necessarily be, the originators of new types as regards design and colors; it is upon this condition only that they can obtain a remunerative price for their productions. Their novelties are usually copied by the producers of inferior shawls, and, finally, by the makers of low-priced articles.

"The persons employed in the shawl manufacture consist of foremen, heads of shops, designers, composers, carders, readers, warpers, and wefters; a few women are employed in the weaving shops. The greater part of the shawl weavers work at their own houses; they generally employ workmen, to whom they give two-thirds of the price they receive from the manufacturer. In Paris these master weavers are the proprietors of their looms, but in Picardy they generally possess only the framework of the loom, and not the Jacquard machinery and accessories. The wages of the weavers are not very high. Those who possess their own looms may earn, in Paris, from 5 francs to 5 francs 50 centimes per day; the under weaver earns from 3 francs 50 centimes to 4 francs per day; the boys and girls employed as assistants earn from 1 franc to 1 franc 50 centimes per day.

"The shawl manufacturer sells directly to the retail dealer, who sells the shawls again in the same state in which he receives them. The prices of French shawls are very various; they range from as low as 12 francs to the most elevated rates; certain long shawls, for example, sell for 1,000 francs, and square shawls from 400 to 500 francs. The export trade is carried on through the medium of commission agents, or directly with the representatives of foreign houses, who come over and buy in the markets of Paris, Lyons, and Nimes, and sometimes, also, by travellers representing the producers abroad. The value of the shawl trade in France may be estimated at 20,000,000 francs per annum. Paris alone furnishes about 15,000,000 francs' worth, and sells nearly a quarter of this amount to foreign countries, especially to North America, Belgium, Germany, and England. Very promising attempts have been made, since 1855, to substitute paper for cardboard in the pattern, which would make a notable reduction in the heavy expense of reading, in order to weave by steam-power a stuff which would rival, in relief and color, the shawl of India."—*Translation of the report of Mr. Herbert, jr., member of the committee of admission of class 32.*

The shawl is, perhaps, the most universal article of dress in the world, and, from its extreme beauty, is an object of admiration in all countries. It is the perfection of eastern skill, and years were often consumed in the manufacture of a single cummerbund or scarf for the waist. Shawls were intended for the male sex, but the fairer portion of creation quickly appropriated an article which possessed such manifest attractions. In the east the shawl is still a principal article of dress—on the head as a turban, and over the head as a hood; twisted round the neck, folded round the shoulder, or wrapped round the waist as a girdle; at times forming the entire dress, and at other times being but an adjunct of luxury, falling in graceful folds on the person; in every way it is suitable, becoming, and popular. The shawl used for the turban is of extreme length, often as much as 60 yards.

The finest shawls are still made, as of yore, in the beautiful valley of Cashmere, the Oriental Eden, which is shut out by precipitous mountains from all surrounding countries. The Cashmerian is industrious, intelligent, and lively. It is only in Cashmere that production is organized on principles nearly akin to the economical plan of Europe. The shawl is the glory and pride of the country, nearly the entire population being engaged in its production. The goats of Thibet, from which tremendous steeps separate it on the north, supply the silky wool which alone is used in the tissue; none other can surpass—none has yet equalled—it in softness. The downy substance found next the skin, and below the thick hair, is the part employed; it is of exquisite fineness. So jealous is the Maharajah of Cashmere to maintain his reputation, that he has recently taken steps to prevent any deterioration in the quality of the shawls manufactured. The Indian display of these articles was exceedingly fine, and of great value.

Next in interest and importance were the French imitations of the Cashmere shawls—the most beautiful tissue which mere machinery has yet produced. The machinery itself is one of the triumphs of human ingenuity, producing in a few days what in the valley of Cashmere would take years to produce. The French Cashmere has none of the softness of the Indian; it has a smooth, firm texture, hard and cloth-like to the touch, without the knottiness of the hand-worked Cashmerian, and the folds which it makes are more angular; but the designs are very beautiful and the colors exceedingly brilliant and varied. In the latter respect, it may be doubted if France has not already surpassed her eastern rival. There was a magnificent display of these shawls.

Austria is also famous for its imitations of Cashmere, and made an excellent display. There is more regularity and clearness in the patterns, but the colors are neither so harmonious nor so good. Prussia and England also exhibited extensively in this class.

There were three American exhibitors.

CLASS 33.—LACE, NET, EMBROIDERY, AND TRIMMINGS.

The products comprised in this class form four distinct groups: 1. Lace made by hand, with bobbins, and with needles, including Alençon point, white and black lace, guipure, and Chantilly, Mirecourt, and Puy lace.

2. Plain, figured, and embroidered net and its derivatives, comprising machine-made silk and woollen lace, known by the name of French tulle, and net of Valenciennes, India, Lama, &c.; white silk blond; plain and figured silk net, and cotton guipure for upholstery.

3. Hand and machine-made embroidery on various tissues; embroidery on civil and military uniforms; church embroidery, and embroidery in silk and wool on canvas.

4. Trimmings of all kinds, for upholstery, religious and military ornaments, men's and ladies' garments, carriages and liveries, and fancy miscellaneous articles.

Almost every civilized country in the world produces the article called lace—the most difficult and delicate result of skilled labor. There are, however, only two or three countries that have given any original impulse to the trade. The others have simply followed in the trains of events, taking what had been done as a model, and imitating it to the best advantage. To two nations—France and Belgium—belongs the credit of prosecuting this trade with vigor. The laces of Alençon and Brussels are of so complicated a nature that each process is assigned to a different lace maker, who works only at her special department. Formerly a piece of Alençon lace would pass through eighteen hands before completion; the number is now somewhat diminished. Valenciennes lace is also of most elaborate workmanship; the pattern and ground are made together, with the same thread on the same pillow. One exhibited with the lace in progress had no fewer than 1,200 bobbins.

There is a legend regarding the introduction of this manufacture into Flanders. A poverty-stricken but pious young girl was dying of love for a young man whose wealth precluded all hopes of marriage. One night, as she sat weeping at her sad fate, a beautiful lady entered the cottage, and, without saying a word, placed on her knee a green cloth cushion, with its bobbins filled with the fine thread which on autumn evenings float in the air, and which the people call "*fils de la Vierge.*" The lady, though of romantic bearing, was a practical manufacturer.

She sat down in silence, and with her nimble fingers taught the unhappy maiden how to make all sorts of patterns and complicated stitches. As daylight approached the maiden had learned her art, and the mysterious visitor disappeared. The price of her lace soon made the poor girl rich. She married the man of her choice, and, surrounded by a large family, lived happy and rich, for she had kept the secret for herself. One evening when the little folk were playing round her knee, by the fireside, and her husband sat fondly watching the happy group, the lady suddenly made her appearance among them. Her bearing was distant; she seemed stern and sad, and this time addressed her protégé in a trembling voice. "Here," she said, "you enjoy peace and abundance, while without are famine and trouble. I helped you; you have not helped your neighbors. The angels weep for you and turn away their faces." So the next day the woman arose, and, going forth with a green cushion and its bobbins in her hands, went from cottage to cottage, offering to all who would be taught to instruct them in the art she had herself miraculously learned. So they also became rich, and Belgium became famous for this manufacture.

The most recent improvement in the production of lace is the introduction of shaded tints in the flowers and patterns, giving them the relief of a picture. This effect is produced by varying the application of the two stitches used in making the flowers—the "toile," which forms the close tissue, and the "grille," employed in the more open part of the pattern. The system is successfully applied to the laces of France and Belgium, but it is in France that it has been adopted with the greatest success.

The species of lace which is peculiarly French is the "point d'Alençon," properly the "point de France," the manufacture of which was introduced by Colbert to avoid the annual importation of lace from Italy and Flanders, which in his day was employed in the dress of both sexes. A fine and very early specimen of this lace was exhibited. It was a piece of a flounce about two and a half feet long, displaying exquisite design and workmanship. It belonged in the day of its prime to the celebrated Madame de Pompadour.

The exhibition of modern French lace was of the highest order. Alençon maintained its reputation, and the black pillow lace of Bayeux was unrivalled in elegance and beauty. A dress of the former material, consisting of two flounces and trimmings, was shown, the cost of which was 85,000 francs, or $17,000 in gold. A "point" or half-shawl by the same maker was estimated at 10,000 francs.

Belgium, as a lace-producing country, is the most formidable rival of France. It was difficult to say which nation shone to the greatest advantage, but perhaps in delicate manipulation of design and ready and graceful taste the French makers were a little ahead of their energetic and intelligent neighbors. Belgium had a magnificent display of her manufactures, Brussels and Grammont, to which must still be added Mecklin, the prettiest and lightest of its fabrics, but the fashion for which has died away, and there is little made at the present time. The specimens in some cases were of great value, and all displayed the highest order of workmanship.

After the productions of France and Belgium, there was little to admire in the lace exhibition of other countries. England had some good samples of Honiton lace, but the patterns were heavy and inartistic, and detracted materially from the excellence of the workmanship.

Nottingham and Saint Pierre-les-Paris are the principal seats of the bobbin, net, and machine-made lace manufactures. Since the application of the Jacquard cards to the making of lace, many imitations of great beauty and very low price have been produced. The Calais manufacturers exhibited imitations of every kind of lace, cotton, silk, and mohair; Valenciennes, Cluny, colored laces, blondes, white and black, silver and gold. The manufacturers of Nottingham exhibited many admirable articles of the same character, and Belgium maintained her pre-eminence, closely pressed by Lyons, in the well-known article of tulle.

No particular nationalities are concerned in the production of embroideries and trimmings, of which the infinite variety almost exceeded enumeration. There were specimens from almost every quarter of the globe. Germany, perhaps, pays more attention to embroidery. There are government schools for teaching the art, and the frugal peasants of the mountainous regions practice it as a means of livelihood. Colored embroidery comes mainly from the east. In the matter of trimmings France is the centre of the trade, and sent a large display to the Exposition.

PRODUCTION IN FRANCE.

The following complete *resumé* of the productions in this Class, in France, is from the translation of the Introduction to the Class by Felix Aubry, President of the Committee of Admission.

LACE.

"Lace is generally made in the country; it takes the name of the town which is the central market, and the principal seats of its productions are: 1. Alençon, where the magnificent *'point à L'aiguille,'* (made with needles,) known by the name of *'points de France,'* are made. These laces are sometimes veritable works of art.

"2. Chantilly, Bayeux and Caen, whose products are similar, and include black silk laces of large dimension for dresses, flounces, shawls, and veils. This manufacture, particularly at Bayeux, has been brought to the highest perfection.

"3. Lille and Arras, where pillow lace is made of excellent quality, but is little used in the present day.

"4. Bayeux, which furnishes that very durable sort of lace known as Valenciennes.

"5. Mirecourt, (Vosges,) celebrated for its new creations; the productions from this locality, specially manufactured according to the exigencies of fashion, are much sought after by the general public and imitated largely in foreign countries.

"6. Auvergne, where the Puy laces and guipures are made at very low prices; the manufacture of this description of lace employs a very large number of hands, and the amount produced is enormous.

"All the raw materials used in textile fabrics are employed in this manufacture, and are spun specially for it. The flax yarn comes from Lille; the cotton from the north of France and from England. Lyons furnishes white and black silk yarn, as also the gold and silver thread; the woollen yarn (goat's hair and mohair) is spun at Bradford. The price of the raw material amounts to from six to twenty per cent. of the value of the production. Pillow lace is made on a sort of frame, very light and simple in construction, and which is held on the knees of the lace-maker. The shape of this pillow has varied little for the last 300 years, and is called a 'coussin or carreau;' it is invariably the property of the workwoman.

"The total number of lace-makers is estimated at 200,000 women and girls. They gain, on an average, 1 franc 25 centimes per day; some who are particularly skilful and industrious earn as much as 3 francs 50 centimes for 10 hours' hard work. Lace-makers are for the most part peasant women, who all, without exception, work in their own homes, often quitting their pillows and babes to attend to household duties, or to work in the fields. Lace making has the advantage of being carried on at home, and therefore not depriving agriculture of too many able hands. French lace is sold at all markets—to the United States, the Brazils, Russia, Germany, Italy, Great Britain, the East, and to India. Paris is the principal centre of consumption.

"The annual production of this trade is valued at 100,000,000 francs, but it is very difficult to arrive at any precise calculation, as lace is not only sold as a simple production, but is used in so many different ways in the several departments of trade. We may note among the recent improvements the invention of many new tinted designs and stitches, as also the production of that thick kind of lace, with the pattern in high relief, which imitates, at a comparatively low price, the old Venetian point, as well as that of Flanders and Raguse.

NET.

"Saint Pierre-les-Calais, and Calais, are the principal centres of the cotton net and silk blonde manufactories; plain, embroidered, figured, and damask silk nets are made at Lyons; figured and plain net for upholstery, at Lille; Saint Quentin and Inchy produce white cotton net, plain and figured; Amiens supplies machine-made lace in silk and mohair, (goat's hair.) Cotton, silk, and wool are employed in the manufacture of machine-made lace, as also in that of all kinds of net; the cotton comes from Lille, the silk from Lyons and England, and the wool from Bradford, where they succeed in spinning a kilogram of goat's hair into 300,000 metres of yarn. Machine-made net and lace is generally woven in manufactories by the aid of steam power, acting with wonderful automatical precision. The machinery is very complicated and expensive, and represents a value of no less than 25,000,000 francs; a great many different systems are adopted, but the most general are the pusher, lever, and circular machines.

" The manufacture of net employs about 25,000 hands, both male and female. The women earn from 1 franc to 2 francs per day, the men from 3 francs to 6 francs. The latter alone work at the machinery; the women are occupied in the preparation and arrangement of the raw materials. The net trade has made great progress in the last 10 years, thanks to the numerous improvements that have taken place in its method of production, and especially to the manufacture by machinery of silk blonde and lace. The productions of Calais and Lyons are now in universal demand, and the principal markets are the United States, Italy, Germany, Spain, the East, India, and even England—cotton and silk net, blonde and lace, made by machinery being of an infinitely lower price than hand-made lace, and is of much more general use; this section of the trade doing business to the amount of about 75,000,000 francs, yearly. Among the most recent improvements, we may point out the wonderful imitations of real silk blonde, the production of very exquisite silk and woollen lace, and numerous modifications in the machinery which permit the attainment of great variety of effects and beauty of detail.

EMBROIDERY.

" Embroidery is carried on in all parts of France, and the chief centres of production may be divided into four principal groups :

" 1. White embroidery for clothing and upholstery comes from the departments of Vosges, Meurthe, Meuse, Moselle, Haute-Saône, Rhône, and Calvados, and also Paris.

" 2. Gold and silver embroidery, artistic and fancy embroidery for military uniforms, church ornaments and vestments, upholstery and other garments, are made in Paris and Lyons.

" 3. The principal seat of production for that embroidery in silk and wool called tapestry work is Paris, and the departments of Eure, Yonne, Lot, Doubs, &c.

" 4. Tarare is especially celebrated for its large articles of upholstery, such as curtains, &c., embroidered on net and muslin.

" Embroidery is, so to speak, the raising of one fabric on another; a multitude of different kinds of articles are used to embroider with, such as straw, jet, beads, and gold and silver thread; but cotton, silk, and wool, are those most commonly employed. Embroidery is prepared by hand and by machinery; the former is worked merely on the fingers, on

8 U E

canvas, or on tambour frames. For braiding the sewing machine is much used. Within the last six years embroidery has been worked by machinery, and the new embroidering machines, though still rare in France, (about 100 having been erected,) have superseded hand-work in many cases. The number of women and girls employed at embroidery in France is estimated at 100,000; they all work at their own homes. There are but few workshops, and the work done there is confined to the production of special articles. The wages of the workwomen are very variable; those who do the artistic description of embroidery with gold and silver thread earn from 3 to 5 francs a day; the others from 1 to 2 francs.

"Paris is the principal centre of this trade, and the most important manufacturers have all a depot there. French embroidery is prized for the beauty of its manufacture and the novelty of the design. It is exported to the United States, Italy, the east, Russia, and British India. The importance of this trade, on account of the great number of hands employed, is considerable. The value of the raw materials used, and even that of the fabrics on which the embroidery is done, is often inferior to the value of the workmanship, so that it is impossible to accurately separate the value of the workmanship from that of the manufactured material on which it is based, and thus arrive at a correct estimate of the worth of the entire product in a commercial point of view. However, the wages of the workwomen amount yearly to a sum of more than 30,000,000 francs. Among the latest improvements we must distinguish the new embroidery machines, which will more than double the production; the invention of new stitches for gold and silver embroidery, and that of tinted and shaded needle-work embroidery.

TRIMMINGS.

"Lyons is celebrated for its gold and silver military gimps, cords, and trimmings; St. Etienne for its fashionable trimmings for dresses and outer garments, and for all those fancy articles which are created and changed with the fashion. At Nimes, St. Chaumond, and Rouen may be found excellent manufactures of cord, braid, and elastic fabrics; but Paris is the active and important centre of the trade. For trimmings all the textile fabrics are employed, principally wool, silk, and cotton, and sometimes straw, gold, silver, aluminium, &c. Each different description of trimmings demands a special kind of manufacture; some are made with the needle, and some on looms of high and low warp, with and without Jacquards; the more ordinary productions are made by means of steam machinery. The manufacture of trimmings occupies more than 30,000 hands, and, after that of lace-making and embroidery, is the trade which employs the largest number of women and children. The wages are variable, as they depend not only on the skill of the workman, but upon the nature of the work. Men earn from three to eight francs per day, and the women and children from one to three francs.

This branch of industry, very considerable from the great number of hands employed and the vast capital it represents, is very prosperous at the present time. All the foreign manufacturers buy the new designs from Paris, for the purpose of copying them. The exportation is very large, principally to North and South America, India, the East, England, Russia, Spain, and Italy. The entire production is supposed to exceed 100,000,000 francs yearly. The chief improvements we have to point out are: great improvements in the different kinds of looms, which has much increased the production; considerable diminution in price, and an enormous development of the whole trade."

CLASS 34.—HOSIERY, UNDER-CLOTHING, AND MINOR ARTICLES.

The productions exhibited in this class formed twelve distinct groups: 1. Hosiery; 2. Buttons; 3. Braces, garters, and buckles; 4. Gloves; 5. Fans; 6. Umbrellas and parasols; 7. Canes and whips; 8. Cravats; 9. Shirts; 10. Ladies' and children's ready-made under-clothing; 11. Stays; 12. Petticoats and crinolines.

Full and entertaining particulars of the trades interested in these branches are given below. They are extracted from the official catalogue, and relate to France. Hosiery was largely represented by other nations. France was rich in fancy articles, such as ladies' silk stockings with open lacework and embroidery, mittens, scarfs, and veils. England excelled in articles of a more substantial make, but in cotton and woollen hosiery she was without a rival. In almost all the other groups France maintained the first position, and was quite undisturbed by competition. There were four American exhibitors in this class.

HOSIERY.

"Hosiery is made in almost every part of France, it being manufactured in no less than 500 communes, but principally in the departments of the Aube, Marne, Oise, Somme, Gard, Herault, Seine, Calvados, and Upper Garonne.

The textile fabrics employed are principally cotton and wool, while silk, floss silk, flax, and the down of the Thibet goat (Cashmere) are used, though less generally. The cotton employed in the manufacture of hosiery is carded, combed, and spun in France, and is derived chiefly from America, Egypt, and India. The price of the cotton varies according to quality: that of India (No. 10 to 24) is worth from 4 francs to 5 francs 50 centimes the kilogram. That of America and Egypt (No. 16 to 150) from 7 francs to 36 francs the kilogram. The wool is furnished by France, England, Russia, Italy, Germany, Australia, Spain, and Africa, and is carded or combed and spun in France. The price varies from 4 francs 50 centimes to 20 francs, according to the number, of 6 to 80,000 meters to the kilogram.

The silk is spun, but the cocoons are obtained chiefly from the Levant, and also from Italy and France. The common qualities are worth from 75 francs to 90 francs the kilogram. The finer qualities from 120 to 130 francs. Floss silk is spun in France and Switzerland. In 1866 the price per kilogram of the more ordinary sorts varied from 25 to 36 francs, and for the superior qualities from 45 to 60 francs.

Flax thread is very little used now, and that only by some few manufacturers at Pas-de-Calais. It is sold at from 3 to 18 francs the kilogram.

Cashmere goat's hair is but seldom employed. The finer qualities are worth from 26 to 60 francs the kilogram, and the more ordinary from 18 to 20 francs. The use of the hair of the rabbit has been entirely discontinued of late years.

Nearly all the machines for making hosiery are worked by hand; however, steam machinery is being gradually introduced into some of the principal French manufactories, and hand machines are, for the most part, employed by those men who work at their own homes. Knitting is no longer required, except for some few fancy articles, and then it is always performed by women, who work at home.

In France the small manufactories of hosiery are very numerous, while there are but few large ones. The men and women who work in their own houses—which form by far the greater proportion, being 90 per cent. of the entire number employed in the trade—earn 30 per cent. less than those who are occupied in the factories. Forty-five per cent. of the employés are women, their occupation consisting in sewing the seams, embroidery, getting up the various articles, and knitting and crocheting different fancy goods.

The greater part of the manufacturers have depots in Paris, which constitutes it the principal market for French hosiery. The town of Troyes is the chief manufacturing centre, and, at the same time, an important market. About half the home trade is carried on directly between the manufacturer and retail vender, the other half through the medium of wholesale houses. Exportation is mostly undertaken by commission merchants. The annual production amounts to about 100,000,000 francs, of which 15,491,722 francs are exported. As much as 549,788 francs of hosiery was imported into France this year. Great progress has been made in this branch of industry since 1865.

Firstly, as to the means of production, we must mention the automatic rectilinear looms, of different kinds, and both of English and French invention, which allow a workman to produce at one time six stockings, and even twelve, of different fineness, whereas the old-fashioned small machines only made one at a time. Also, the circular machines made according to a new system, of every size, and by which no less than thirty rows can be woven with one revolution of the machine; and again, the machine for taking up the stitches and sewing the stockings in the greatest perfection. Secondly, the productions themselves are of a much more equal quality, and made with greater care. The amount exported has much augmented this branch of the trade, being now carried on to a very large extent. The number of mills where yarn is prepared for working hosiery has greatly increased. The salaries of the workmen have risen about 30 per cent. since 1855.

BUTTONS.

"Paris is the principal centre of this industry. Buttons of all sorts are manufactured there in metal, silk, mother-of-pearl, horn, enamel, and also those fancy kinds which serve to ornament outer garments. The department of the Oise is the seat of manufacture for buttons in shell, mother-of-pearl, vegetable ivory, bone, ivory, &c, besides silk buttons, which are one of the most staple productions of France, and which are exclusively made in this department. China buttons are made in great quantities at Briare, in the department of Loiret, at Montereau, Seine-et-Marne, and at Creil, Oise. As to the other places, where only horn, mother-of-pearl, and bone buttons are manufactured, they are of so little importance, comparatively, that we shall not make any particular mention of them.

The raw materials made use of in this branch of industry may be divided into five sections: gold, silver, aluminium, German silver, copper, tin, zinc, iron, steel, &c.; silk, wool, linen, cotton, velvet, and various other fabrics; china, enamel, glass, crystal, beads, imitation stones, mosaics, &c. The amount of metals used annually in the first series may be estimated at 2,500,000 kilograms, representing a sum of 4,000,000 francs. The silk and other fabrics of series No. 2 may be valued at about the same amount. The raw materials of the third series, nearly all derived from the tropics, are employed to an extent of 3,000,000 kilograms, exceeding 5,000,000 francs in value. The annual consumption of horns, hoofs of mammalia, and the other articles of the fourth series, is not less than 1,500,000 francs' worth; while the glass and china manufactures of the last series are entirely of French manufacture, and represent a yearly value of 2,500,000 francs.

Steam and hydraulic machinery has come into more general use since 1855, and is principally employed for cutting out the raw materials, and for stamping metal buttons. The other kinds are made by hand, with the help of small machines and tools. Certain kinds of silk buttons are the only ones made entirely by hand.

The number of hands employed in France in the manufacture of buttons is 22,000; of which 8,000 are men, 10,000 women, and 4,000 children, some of whom work at home and others in manufactories. The wages may be estimated at 4 francs 25 centimes per day for the men, 1 franc 85 centimes for women, and 1 franc 10 centimes for children.

All kinds of French buttons are exported to Great Britain, Italy, Russia, and North America, while South America, Mexico, and China buy a large quantity of the common sorts. The annual production may be valued at 45,000,000 francs, of which three-fourths are sent to foreign countries.   This branch of industry, which was comparatively insignificant thirty years ago, has made extraordinary progress since 1855, and we may safely say that France furnishes to the whole civilized world by far the greater part of the button manufacture.

BRACES, GARTERS, AND BUCKLES.

"The manufacture of garters and braces has been created in France since 1834.   From Paris, where it was first established, it was removed to Rouen, where it now almost exclusively flourishes.   Buckles in copper are principally made in Paris; those in steel at Rancourt, (Ardennes.) The raw materials consist of cotton, India-rubber, and silk for ornamentation.

For these articles, buckles are chiefly manufactured in copper and stéel; the metals employed in making buckles for trousers, waistcoats, and shoes, are steel, iron, copper, zinc, lead, tin, &c.   Rouen employs relatively but a smaller number of hands in the manufacture of garters and braces; all the work is done by machinery.   In Paris, hand labor predominates.   For the manufacture of buckles by means of cutters, machinery has, since 1836, quite replaced hand labor.   The workmen and apprentices who weave the braces are employed in manufactories, while the women who mount them work at their own homes.   The men earn 5 francs 50 centimes per day, and the women 3 francs.   The apprentices are not paid, but are provided with board and lodging.   The greater part of the men who make buckles work in manufactories; but the productions are finished off by workmen in their own homes.   The men's wages vary from 2 francs 50 centimes to 5 francs; the women's from 1 franc 50 centimes to 2 francs 50 centimes, and that of the children from 75 centimes to 2 francs.

French braces and garters are exported to nearly every part of the world; principally to England, Russia, and America.   The buckles made in France compete advantageously with those of the best foreign manufacturers.   The annual production of braces and garters in France amounts to about 10,000,000 francs in value.   The manufacture of buckles in Paris alone amounts to 2,000,000 francs, of which one-third is exported. Among the improvements of the last 15 years we must mention the weaving machine of Mr. Fromage, producing 80 dozen pairs of braces per day, of which the price of manufacture does not amount to more than six centimes per dozen; also the invention of the hygienic braces, which differ from the other kinds, inasmuch as the stress of the elastic fabric is entirely removed from the shoulders, and only bears upon the lower part of the braces.   Lastly, the common kind of buckles have been replaced by those called à pont, of a much more convenient form.

KID GLOVES.

"The principal manufactures of kid gloves are to be found in Paris, Grenoble, Chaumont, and St. Junien, (haute Vienne;) the lambskin gloves are made chiefly at Lunéville and Niort, and those in deer-skin and chamois-leather (called castor) at Rennes. Many provincial towns make gloves for local use, as Lyons, Nancy, Strasburg, and Rochefort. The manufacture of gloves employs kid, lamb, and sheep-skins, as well as the skins of the lamb, the deer, and the reindeer, (dressed chamois fashion.) Almost all these skins are procured in Europe; but the finest are to be met within the centre of France. The best kids come from Switzerland, the north of Italy, Tyrol, Austria, Bavaria, Saxony, and Silesia; those of northern Europe are, in general, very inferior.

The skins are first tanned, then dyed and cut out to make gloves. Dressing by means of machinery has been tried, but does not seem to have succeeded for any but the thickest kind of skin. The "dollage," or process of equalizing the thickness of the skin, is done either by hand or by means of stone cylinders moved by steam. The fingers are divided by means of dies moved by a screw. The remainder of the processes are performed entirely by hand. The dyeing of the skins is likewise a manual occupation. Two sorts of dyes are used for skins; that which is applied with a brush, and that into which skins are dipped. After the skins have been dressed, they are sorted and appropriated to the use for which they seem most fitting; then comes the dollage, the cutting, &c.

The working tanners earn daily, in the Paris workshops, from 4 francs 50 centimes to 5 francs; and in the provinces from 3 francs to 3 francs 50 centimes. The *palissoneurs*, who work by the piece, can earn as much as 6 to 8 francs a day in Paris, and from 5 to 6 francs in the provinces. The wages of those who dye the prepared skins, and who also work by the piece, amount, in Paris, to 5 or 6 francs per day; and in the provinces to 4 or 5 francs. Skins prepared in the chamois style are chiefly manufactured in Milhau and Niort, and the hands employed in this process receive about the same amount of wages as the tanners.

The glove trade occupies in France about 50,000 or 55,000 work-people, of whom 40,000 or 45,000 are women. Those men who cut out and prepare the gloves, and who work in their own dwellings and in the workshops of their employers, can earn from 6 to 8 francs per day, and even as much as 10 francs; the wages in the provinces may be estimated at one-fifth less. Some workwomen, especially those that cut out, earn from 3 to 4 francs; others from 2 francs 50 centimes to 3 francs. The women who sew and stitch the gloves, and who work principally in the country, seldom earn more than 1 franc per day.

The productions are sold directly to retail vendors for home consumption, and to commission merchants for exportation. Some of the principal glove manufacturers have established houses in the principal foreign markets, especially in England and America. The number of gloves made up annually in France amount to 1,800,000 or 2,000,000 dozen pairs, which represents a sum of nearly 70,000,000 francs. With the gloves of first quality the materials are worth about two-thirds of the value of the production; the price of manufacture, the general expenses, and the profits

account for the other third. The materials of the second and third qualities cost a little less, and the manufacture and general expenses a little
more in proportion. At least two-thirds of the produce of the glove trade
are exported; England and America consume nearly the whole of this
amount. The French glove trade has made no marked progress since
1855; but it still remains immeasurably superior to all foreign manufactures.

FANS.

"Fans are composed of two parts: the mountings are made in certain
communes of the department of Oise, and the upper part in Paris; and
it is there also that the two are joined together, so that Paris may be
considered the principal manufacturing centre. The raw materials used
in the making of fan-mountings are:

1. White mother-of-pearl, called poulette, which comes from Madagascar, and costs nine francs the kilogram; another sort of white mother-
of-pearl, called Franche, which comes from the same place, but costs 11
or 12 francs the kilogram; black mother-of-pearl, brought from Sid-
ney, and sold at seven francs the kilogram; oriental mother-of-pearl, and
the green délotide kind, which is found in Japan, and is worth about 35
centimes the shell; the Burgot variety, found also in Japan, and worth
85 centimes the shell; brown tortoise-shell, from India and China, and
light-colored tortoise-shell, the former costing 60 francs the kilogram
and the latter 200 francs; also ivory, sold at 40 francs the kilogram;
and lastly, bone, largely used for fan mountings, and furnished by Paris,
Bordeaux, and Rouen, at 50 centimes the kilogram.

2. Oriental woods—ebony, mahogany, rose, satin, and lemon-tree
wood, and in general all the hard woods of Africa, Ceylon, and Madagascar, which cost, on an average, 60 centimes the kilogram; sandal
wood, which comes from Japan, and which is worth 2 francs 25 centimes
the kilogram.

3. Indigeneous woods, such as plane tree, acacia, beam tree, wild
cherry, plum, apple, and pear tree, cost from 15 to 20 centimes the kilogram.

The materials of which the coverings of the fans are made vary according to taste and fashion; silk, crape, lawn, paper, feathers, and kid,
are used.

Machinery has replaced hand-work in the cutting of the mountings,
except for ivory and tortoise-shell. Designers make the drawings for
the fan coverings. These drawings are lithographed or engraved on
copper, steel, or wood, and then printed, pasted, colored, or painted,
mounted, bound, edged, spangled, riveted, and examined. Some workmen work in shops by the day; all the others work by the piece, at their
own houses, with their wives and children. The fan trade occupies, in
France, 4,000 workmen of different trades; of which 1,000 are in Paris
and 3,000 in the department de l'Oise. Workers in bone, ivory, &c.,

gilders, looking-glass workers, paper makers, feather mounters, painters, embroiderers, goldsmiths, jewellers, engravers, chasers, carvers, &c., all combine in the manufacture of these articles, whether they be plain or ornamented, superior or ordinary. The wages of the workmen are, on an average, 5 francs a day; those of the workwomen vary from 2 francs 50 centimes to 4 francs 50 centimes.

Paris, [Japan] and China monopolize the fan trade. Spain, Italy, Portugal, and England are the principal foreign markets in Europe. The Brazils, Mexico, Havana, St. Thomas, Chili, Peru, Buenos Ayres, and North America may be considered merely as tributaries. Some fans are also exported to the East Indies, and as far as Manilla; but in those ports the rivalry with China for the ordinary articles is maintained with difficulty. The annual production amounts to 10,000,000 francs, of which three-fourths is exported. The progress made in the fan trade since 1855 consists in the use of mechanical processes for the production of current articles and in the more developed application of art to industry, the improvement in certain mechanical processes, such as (to cite but one example) a machine for ornamenting by heat, capable of executing, on a mounting of two francs, a work which could not be attained by hand at any price.

UMBRELLAS AND PARASOLS.

The principal centres of production are Paris, Angiers, Bordeaux, and some less important towns of France. For umbrella and parasol covers cotton tissues are employed, which are produced at Rouen, and worth from 40 centimes to 1 franc 25 centimes the metre; silk, made at Lyons, and worth from 2 to 20 francs; and alpaca, imported from England, and costing from 1 franc to 3 francs 50 centimes the metre. Parasols are ornamented with lace of various kinds, made at Alençon, Puy, and Caen, worth from 25 francs to 1,500 francs the cover, or with imitation lace, produced at Lyons, Calais, St. Pierre, &c., only costing from 3 francs to 25 francs. The embroidery and trimmings are made in Paris.

Colonial woods, of all kinds, only form about one-twentieth part of the material used in the manufacture of umbrella and parasol handles. The price of bamboos and laurel wood vary; for bamboos, from 10 to 35 francs the hundred, and for the laurel from 20 to 50 francs the hundred. Algiers supplies the myrtle, of which the prices vary from 20 to 50 francs the hundred; and Guiana nearly all the colonial wood, costing from 30 to 170 francs the hundred for sticks of 90 centimetres to 1 metre in length. The woods of French growth are beach, yoke-elm, oak, sycamore, maple, beam tree, hazelwood, wild cherry, cornelian tree, medlar, and holly; the prices vary from 5 to 45 francs the hundred handles, ready rounded. Umbrella and parasol handles are made in considerable quantities, of various materials, of which the principal, besides woods of all kinds and from all parts of the world, are bullock, buffalo, ram, and rhinoceros horn; bone, ivory, and tortoise-shell are also employed in the manufacture.   The prices vary to infinity: bullock horn from 25 centimes to 1 franc 50 centimes each; buffalo and ram horn from 50 centimes to 3 francs; and rhinoceros horn, ivory, and tortoise-shell from 2 to 100 francs per piece.   The most ordinary prices for these last are from 6 to 20 francs each.   Umbrella and parasol frames are made in steel, ratan, and whalebone.   The steel wire used is worth, according to the size, from 1 franc 80 centimes to 2 francs 10 centimes the kilogram, all prepared; that is to say, drawn, cut into lengths, and tempered.   Ratan, which is used for common mountings, comes from India.   The prices vary from 1 franc to 1 franc 50 centimes the kilogram, cut, squared, turned, pressed, and varnished.   Whalebone is becoming exceedingly scarce; its price has risen to 15 francs the kilogram; that is to say, it has tripled during the last quarter of a century.   The wood is cut up and rounded by machinery; the ornamentation, carving, and varnishing are performed by hand.   A part of the frames are made by machinery, but hand-work is employed in the great majority of instances.   Sewing machines begin to be applied, with great advantage, to the sewing of the seams.

The workmen employed in the workshops receive one-third of the total profits; those who work at home, for others, two-thirds.   All the women work at home.   The average wages for the men are 5 francs per day; those of the women, 3 francs.

The trade is principally wholesale, and confined to the dealers in the provinces, through the medium of commercial travellers, and directly to those who negotiate personally with the manufacturers every half year, the foreign trade being conducted solely through the medium of export agents.

The Paris shops treat directly with the producers, the purchases comprising a considerable moiety of the entire trade.   Cotton parasols are worth from 1 franc 25 centimes to 5 francs; those in silk from 4 to 40 francs.   Cotton umbrellas are worth from 1 to 10 francs; those in silk from 3 to 150 francs.   The principal foreign markets are Spain, Greece, Italy, Turkey, Austria, Switzerland, Prussia, Russia, Holland, Belgium, England, and her dependencies, Asia, Egypt, North and South America. The umbrella and parasol trade does business, annually, to the amount of about 35,000,000 francs.

Few changes have taken place since 1855 in the manufacturing processes, excepting the introduction of sewing machines.   The average wages of the workmen, workwomen, and others employed, has risen about 20 per cent.

WALKING STICKS, RIDING AND DRIVING WHIPS.

Paris is the principal centre of production.   Canes, ratans, and stiff and flexible bamboos, worth from 10 to 400 francs the hundred pieces, are imported from British India, China, and Japan.   Palm, myrtle, orange, and locust-tree wood, worth from 40 to 100 francs the hundred pieces, come from Algeria. Dog-wood, thorn, oak, elm, ash, and wild cherry-tree woods, costing from 10 to 50 francs the hundred pieces, are produced in France, (Alsace, Lorraine, and Nivernais.) Whalebone, ivory, tortoise-shell; rhinoceros, buffalo, and rams' horns, worth from 90 centimes to 30 francs the kilogram; gold, silver, brass, white metal, gold-plated on copper and silver, jasper, cornelian, lapis lazuli, malachite, &c.; cotton, silk, and catgut, are bought in the Paris and London markets.

All the articles are made by hand, except the plaiting of the whips, which is done by machinery. After the moulding of the horn and tortoise-shell, and the laying of tortoise-shell on ram's horn, the principal operations are the planing and varnishing of canes, stoving, cutting and carving wood and ivory, chasing and engraving metal.

The workmen working in shops form a third of the whole number of persons engaged in this industry. This does not, however, include the women. The work-people who work at their own houses include men, women, and children, and form the other two-thirds. The wages of the work-people are very good; the men earn from 3 francs 50 centimes to 7 francs, and the women 2 francs 50 centimes to 3 francs.

The manufacturers sell directly to the retailers in Paris, and to the provincial dealers through the medium of travellers. The exportation is carried on by commission agents.

Walking sticks are worth from 25 centimes to 100 francs each; riding whips from 25 centimes to 50 francs; and driving whips from 1 to 50 francs. These articles are exported to all parts of the world.

The value of the industry, divided between about 60 manufacturers, living in Paris, is between about 3,500,000 and 4,000,000 francs.

The principal improvements to be noted since 1855 are the employment of gas for coloring and dressing the wood, and the use of machines for plaiting two whips at a time. Nevertheless, the competition of Germany has become serious as regards common and low-priced articles. This state of things must be attributed to the rise in wages, which are 8 to 10 per cent. higher in Paris than a short time since.

CRAVATS AND SHIRTS.

This trade originated in Paris, and has greatly developed since 1848. It has increased to a large extent, particularly during the last few years. The makers of cravats especially employ silk stuffs, from the lowest to the highest prices. The number of work-people occupied can be estimated at about 10,000, and the average of their daily wages at 2 francs 50 centimes.

This industry employs, at the present moment, a considerable number of sewing machines; and the low prices of the articles made render them acceptable to all classes of the population.

Shirt-making for men possesses considerable importance at the present moment. It includes the making of shirts, shirt fronts and collars, waistcoats, drawers, flannel bands and shirts, and linen and cotton drawers, The cotton stuffs employed in this trade come from the manufactories of Mulhouse and Rouen; England, Ireland, and France supply the linen fabrics. This industry, which has been scarcely 30 years in existence, has greatly extended latterly.

The shirt-makers can be divided into two categories: those who deal directly with the retail houses and those who make for the wholesale trade and for exportation. The work-people employed under the first category earn, on an average, two francs per day; the second category of dealers have their articles made in the provincial workshops, in asylums, and convents, and rarely give more than one franc a day to the work-women employed. The number of women supplied with work by the shirt-making trade is estimated at about 30,000. This total includes the cutters, needlewomen, mounters, embroiderers, and laundresses.

The amount annually produced reaches 70,000,000 francs, of which 25,000,000 francs' worth are delivered for exportation. The manufacture of shirts, flannel bands, and drawers, has also increased to a great extent during the last 10 years, in consequence of the orders given for the army. The total value of this branch of business amounts to about 10,000,000 francs. Articles in flannel are made by the same persons, as are also similar articles in cotton and linen; and although there is some difference in the methods employed, the mode of manufacture may be looked upon as the same. The flannels used in this trade are produced in the town of Rheims; and it should be added that the quality and lowness of its price causes it to find great favor in the foreign market. As a last consideration, it should be mentioned that the industries above referred to have, in one respect, a very interesting aspect, as they allow the women employed in it to work at their own homes, and thereby give their attention at the same time to the duties of the household.

UNDERCLOTHING FOR WOMEN AND CHILDREN.

This is another trade, which has been greatly developed during the last few years. Among the various articles which it includes may be mentioned chemises, jackets and drawers for ladies and children, and which, with many secondary articles, comprise the childbed linen and marriage trousseau. The manufacture of these various articles demands, especially in the case of the more elegant kinds, much experience and taste, and great skilfulness and care on the part of the women employed in it. The Parisian seamstresses earn from 2 francs 25 centimes to 3 francs a day; and those who work in the provinces, in the convents, asylums, &c., from 1 franc 25 centimes to 2 francs 25 centimes per day. The number of women employed in the made-up linen trade is about 10,000. Sewing machines have had a powerful influence in developing this business; the women who work them earning from 3 francs to 3 francs 50 centimes per day. The ready-made linen of France, and above all, that of Paris, enjoys a good reputation abroad, and the export trade has increased in a notable manner since 1855. The articles known under the name of Paris hosiery are in great demand in England, Germany, Spain, Switzerland, and the two Americas. Nearly the whole of the fabrics which are used in the ready-made linen trade are of French origin, and especially from the factories of Quentin, Tarare, and St. Etienne. Alsace and several departments of the north supply linen cloth and plain cotton tissues. That kind of lace known as Valenciennes is supplied by Belgium. The value of this trade is estimated at about 30,000,000 francs per annum.

Since 1855, the stay-making trade has progressed in a very considerable manner, and the value of the business has increased in a very notable degree. Stays are divided into two very distinct categories: stays with seams, and stays without seams. The sewn stays are made by hand or with sewing machines. The seamless stays are woven on the Jacquard power-loom. It is especially in Paris, or in the principal provincial towns, that the sewing stays are made. The seamless stays are produced in large establishments at Bapaume, (Pas-de Calais,) Bar-le-Duc, (Meuse,) and at Ehézy, (Rhône.) The raw materials employed in stay-making are white and gray drills, which are supplied by the manufactures of Flers and Evreux, real or imitation whalebone, sewing-cotton spun in France and in England, and sewing-silk, which is exclusively bought of Paris houses.

Women are alone employed in the making up of stays, and work in tacking and sewing workshops; they earn, according to their capacity, from 1 franc 75 centimes to 4 francs 50 centimes a day. For the making of stay bones of all kinds men are employed, whose daily wages are from 3 to 5 francs.

At the present moment, the stay-making trade, in consequence of the new patterns introduced since 1862—the results of which have been, in most cases, to prevent the great inconvenience, or rather the serious danger proceeding from imperfectly manufactured stays—merits encouraging notice. It has entered into a rational path, and pays still greater attention to the laws of health and nature. On this point we must not omit to mention the judicious innovations of clasps in the place of laces, the absence of gussets, &c. This branch of industry is daily establishing itself in the estimation of scientific men, who are not influenced by the unhappy and injurious exigencies of fashion.

The crinoline trade, comparatively recent in its origin, and which occupies such an important place in ladies' dress, comprehends two rather distinct articles, skeleton crinolines and crinoline petticoats. These articles are made in all the towns of France, but the sale of them is especially active in Paris; and the Parisian makers owe this not only to the elegance of their productions, but also to the continual creation of new patterns.

The principal materials used in the manufacture are the bands of steel, rolled and cut, which are made in France and England, and which are encased in a cotton covering, produced by machinery. The woollen and cotton stuffs employed in crinoline-making come especially from Roubaix, Amiens, Tarare, and Saint Quentin. The diaper, cambric muslin, and other fancy tissues are supplied by the manufactories of Mulhouse, Rouen, and Saint Marie-aux-Mines; the tape and sewing cotton are made at Bernay. To these various products must be added the silk and velvet trimmings, and the buttons of all sorts, made in France and Germany. The construction of petticoats without steel forms a special branch of the trade in question. The Parisian work-people are the most skilful in the making of these various articles, and their salaries vary from 3 francs 50 centimes to 4 francs a day. The value of this trade is estimated at 20,000,000 francs, of which half is due to the export trade.

CLASS 35.—CLOTHING FOR BOTH SEXES.

The articles exhibited in class 35 may be divided under nine different heads: 1st. Clothing for men; 2d. Clothing for women; 3d. Bonnets and head-dresses for women; 4th. Artificial flowers; 5th. Ornamental feathers; 6th. Men's hats; 7th. Men's caps and other head-gear; 8th. Boots and shoes; 9th. Fancy hair work.

The familiar objects embraced in this class need no detailed description in an official report. For the most part they have been already described in the newspapers devoted to fashions, and, at all events, a walk through any fashionable thoroughfare will convey a better idea of what was shown than any labored effort of the reporter. France was again upon her own ground, and distanced all competition. We give below the latest French particulars of the curious branches of industry included in the class. America had nine exhibitors.

Men's clothes are made almost everywhere, but the principal establishments, both for fashionable as well as ready-made garments, are in Paris. Low-priced articles are, for the most part, made in the provinces. Many important houses have their principal workshops in the departments of the Nord, Pas-de-Calais, Gironde, Gard, &c. Tailors and clothiers employ a great variety of fabrics, and consequently of all prices. Tailors and clothiers also use a considerable quantity of trimmings and buttons of all kinds. A few years ago tailors' work was done altogether by hand. Now, sewing machines are used to an immense extent; in fact, it may be said that the greater part of the seams of garments are sewn by these machines.

The cost of the workmanship of men's clothes amounts to about one-fifth the value of the goods. The workmen employed by the tailors and clothiers are naturally divided into two different categories: those who prepare, cut out, and arrange the work, and those who put it together. Five-sixths of the tailors work at home, while the rest are employed in the tailors' work-rooms. As to the workwomen—who, in Paris, are only half so numerous as the workmen—five-sixths of them work at home.

The men, working either by the day or by the piece, earn from three to six francs a day, though some more industrious and skilful gain from eight to ten francs. The women earn from two to three francs fifty centimes, and a few from five to six francs. The tailor and clothier deal directly with the purchaser. The tailors generally do their own cutting out, but the vendors of ready-made goods employ cutters, who prepare the work for the sewers. The business of exportation is generally made through the mediation of agents. It is almost impossible to ascertain the extent of the production of men's garments; but it must be considerable, as the tailors and clothiers in Paris alone do business to the amount of more than 150,000,000 francs per annum. The articles exported do not amount to the tenth part of the whole.

The business has made great progress since 1855. The use of sewing machines increases every day. Many foreign governments have now recourse to French clothiers for the equipment of their troops. This new branch of the business has rendered great service to the workmen, enabling them to obtain employment in all seasons of the year; and also to the great cloth manufactories, by giving them extra work, or helping them to get rid of unsalable articles.

CLOTHING FOR WOMEN.

Paris is the great centre for the making of ladies' clothing. This branch of trade employs an immense quantity of stuffs of all prices, from common printed cotton to the most expensive velvet. Articles for summer wear are principally made of the light fabrics of Rheims, Elbeuf, Sedan, and Roubaix, Scotch cashmeres, and French merinos; while those for winter are made of the thick, strong stuffs of Sedan, Elbeuf, and the south of France. Pillow and machine-made lace, as well as Paris, St. Etienne, and Lyons guipures and gimps, are used for the trimmings of ladies' clothes. The clothiers give the stuffs, cut or uncut, to dressmakers or ladies' tailors, who employ from four to forty workwomen besides those who work at home. The articles are generally mounted and sewn by hand, the sewing machines being used for the trimmings. The sewing of ladies' outer clothing is done almost entirely by women, and females are generally employed for the sewing machine also. At this trade men earn in Paris on an average five francs a day and women two francs twenty-five centimes.

The export of ladies' ready-made clothes is very considerable, the principal markets being England, Belgium, Holland, Russia, Spain, Italy, Turkey, North and South America, and Australia. The articles principally exported are known by the names of paletots, talmas, pelisses, mantelets, embroidered shawls, scarfs, and jackets. Dresses, hoods, and children's clothing are also exported. The wholesale houses where these articles are made furnish the small provincial linen drapers and commission merchants, while the principal linen drapers in Paris and the provinces generally buy the patterns and have the articles made up for themselves.

The production of these articles in the whole of France is estimated at 100,000,000 francs, (£4,000,000.) Paris alone makes to the amount of 40,000,000. Five-sixths of the whole are used in France, and only one-sixth is exported. Ready-made articles are sold from three francs to four hundred francs each; embroidered shawls, for instance, vary in price from eight francs to three hundred. This branch of industry increases daily; new patterns are continually produced, and are remarkable for taste and originality. Here, again, the growing use of the sewing machine must be noted. The business of the dressmakers—that is to say, those who make ladies' clothes, and particularly dresses to order—is daily becoming more important. This is one of the Parisian trades that demand the greatest amount of taste and invention; it is, naturally, almost exclusively followed by women. However, there are in Paris many important houses whose business is confined to the making of dresses where men are employed. The greater number of dressmakers are paid by the day, though some work by the piece; the wages are not high; they earn, on an average, two francs twenty-five centimes per day. Dressmakers do not work exclusively for home use; a certain quantity of handsome articles are sent to foreign countries. The export trade of Paris amounts to about the twelfth part of the whole production.

LADIES' HEAD-DRESSES.

Millinery is essentially a Parisian trade; it is in Paris that all those novelties are created which, at the commencement of each season, decide the fashions. The materials used in bonnet and cap making, such as buckram, wire, whalebone, various stuffs, flowers, and lace, are obtained from special manufacturers. The milliners, so to speak, only arrange and combine these materials. There is no fixed method of manufacturing articles of millinery; it is altogether a matter of taste and ingenuity. The workmanship forms only a small item in the value of the whole. Three-quarters of the working milliners are Parisian; about an eighth part are natives of Belgium and Germany; the rest come from the provinces, especially Angoulême, Tours, Nancy, and Dieppe. Part of these workwomen board and lodge with their employers, and earn, on an average, two and a half francs a day. This trade employs very few men; a great number of young girls work as apprentices. Nearly all milliners sell direct to the purchaser. Some houses make up articles specially for exportation, and these alone employ under-milliners, who receive the requisite materials for a certain number of bonnets and head-dresses, and prepare the work by tacking the various stuffs upon the ready-made shapes which they furnish. The ribbons and flowers are always added by the milliner herself.

It is difficult to estimate the exact amount of bonnets and head-dresses annually made in France, but it must be considerable, as the Parisian milliners' returns amount to nearly 20,000,000 francs, (£800,000;) the export amounts to about a tenth of the whole. Paris millinery is sent chiefly to America, England, Spain, Belgium, Holland, Germany, Russia, and the French and English colonies.

ARTIFICIAL FLOWERS.

The fabrication of artificial flowers occupies a conspicuous position among the various and interesting Parisian industries, and may be called artistic. The materials used in the manufacture are very numerous; for the leaves and blossoms, jaconet, nansouk, cambric, muslin, velvet, crape, satin, silk, French cambric, feathers, paper and wax are made use of; for the stems, berries and fruits, wire, silk, cotton, floss silk, paper, starch, gum, gelatine, wax, paste, chenille, quills, whalebone, gauze, chopped wool and glass balls are employed. For mounting the flowers, silk, paper, gauze, iron, and brass are required. Artificial flower-makers always use the same instruments—goffering irons, stamps, &c.; the galvano-plastic process is sometimes employed.

The cost of the workmanship amounts to about the four-tenths of the value of the productions, and the materials employed to about three-tenths; the remaining three-tenths represent the profit of the producer. The manufacture of artificial flowers is divided into a great many different branches; for the preparation of the colors there are special workshops. The manufacture of artificial flowers is generally carried on at the homes of the work-people; such is the case in, at least, 1,500 of the 2,000 Paris flower-makers. This trade employs 15,000 people, of whom nine-tenths are women and girls. The men earn about four francs a day; the women two francs twenty-five centimes. The mounting and sale of artificial flowers is carried on, for the most part, in handsome shops and show-rooms, where all kinds of flowers are generally sold, as well as the different sorts of ornamental feathers. Three-quarters of the whole amount of artificial flowers are exported through the medium of commission agents. The extent of the trade is about 18,000,000 francs per annum. Artificial flowers are exported principally to America, England, Belgium, Russia, and Germany.

ORNAMENTAL FEATHERS.

Feathers are prepared and mounted in Paris, which enjoys a justly-earned reputation for the preparation, bleaching, dyeing, and arrangement of this article. The most beautiful and *recherché* feathers are those of the ostrich and marabout, which are imported through Leghorn and London. Next come the feathers of the birds of paradise, the cassowary, and those known by the name of aigrettes, and bastard ostrich feathers, called vulture's plumes. Cock's feathers, the down of the white turkey, and the feathers of the various kinds of exotic and indigenous birds, are also made use of. The different preparations to which feathers are submitted consist merely in arrangement, bleaching, and dyeing, though we must not forget to mention a mechanical process by which.goose's feathers are made to imitate different kinds of grasses..

9 U E

The ornamental feather trade employs few men, but a great number of women and girls. The dyeing of feathers is all that is done by men. The great part of these feathers are exported, through the medium of commission merchants, to America and the colonies; but Paris furnishes also the principal milliners of Europe. Ornamental feathers are prepared to the amount of 10,000,000 francs, of which about 8,000,000 are exported. The manner of dyeing and preparing feathers has undergone little modification since 1855, only a method has been discovered of turning black feathers into gray, which allows of their being dyed of various colors.

MEN'S HATS AND CAPS.

French hatters manufacture silk hats, black and white, short nap beaver hats, fancy dressed felt hats for country wear and for travelling, and soft felt hats. Paris, Lyons, Marseilles, Aix, Toulouse, Bordeaux, and some other southern towns, are the centres of the hat trade. Caps of various kinds are principally made in Paris, Rueuil, Châlons, and Condom. The principal materials used in the manufacture of hats are the skins of the beaver and muskrat, imported from Canada; that of the Gondin rat of the centre of South America; the fur of the hare, furnished by France, Germany, and Russia; that of the rabbit, so abundant in France, and wool of different kinds used for making cheap articles. France alone supplies annually rabbit and hare skins to the amount of 70,000,000 francs, and exports 35,000,000 worth. The average price of rabbit skins is 40 francs for 104 skins; hare skins are worth one-third more. The manufacture of hats may be divided into two distinct sorts, the manufacture of soft and firm felt hats and that of silk hats. Workmen, whose special business it is to cut the hair from the skins, furnish the makers with their raw materials. The manufacture of felt hats includes several operations. The fur is first beaten, either by hand or by a machine. By this process a bag of felt twice the size of the hat is produced; this is then fulled, either by hand or by a special machine used for the purpose. Arrived at this point of its manufacture, the hat is scraped with a knife, to take off the long hairs, rubbed with pumice-stone, then stiffened or not, as required. It is then dyed, blocked into forms, bound, and finally the leather and head-lining are added. The manufacture of silk hats is different. First of all, the form is made of various fabrics, stiffened with gum-shellac, and upon it is placed a kind of silk plush, and within it a fabric which serves for lining. A great many silk hats are made with the adhesive linings, in which case the interior becomes part of the solid form. The working hatters are generally well paid; some earn as much as 10 francs per day, but the average is between 40 and 50 francs per week. The men work by the piece, and are under the direction of foremen, chosen from among the best workmen. The latter earn from 2,000 to 3,000 francs per year. Women do not earn more than from 18 to 24 francs per week. Nearly all the men and women employed in this trade, and especially the men, work in the factories.

The productions of the French hatters are exported to nearly all parts of the world, sales being effected through the medium of commission merchants. The prices of hats vary greatly; they are sold from three or four francs to 25 and 30 francs each. Opera, or spring hats, in particular, are exported in considerable quantities. The manufacture of hats alone, without taking into consideration various kinds of caps, amounts to the sum of 24,000,000 francs, or nearly £1,000,000 a year, out of which, at least, 10,000,000 worth of felt and about 2,000,000 worth of silk hats are exported.

Since 1855 a great many ingenious tools and machines have been invented to facilitate the manufacture of hats. The materials employed remain the same, but the wages of the workmen have greatly increased. The hat manufacturers now make and completely finish their goods, so that the hatter who puts his name into the crown of the hat is only an agent between the producer and consumer.

The principal places where caps are made are Paris, Toulon, Lyons, Limoges, Lille, Bernay, &c. The manufacture of men's caps employs a great variety of fabrics, from silk and fine cloth to the commonest stuff. Even old materials dyed and turned are made use of. The manufacture of the better sort of caps has been greatly improved during the last few years. This is partly due to the sewing machine, which does the sewing very neatly, besides doing a great variety of embroidery at comparatively low prices. The women who make caps sometimes work at the shops and sometimes at home. One set of workwomen join together, with the aid of the sewing machine, the several pieces of the cap, which is then padded, if required, stitched, and embroidered; women press the seams, put on the peak, and complete the work. Most of the workwomen work at home, and earn from 2 francs and 25 centimes to 2 francs and 50 centimes per day.

Most of the caps made are sold at home, but a certain number are exported to America, Spain, Portugal, Holland, Germany, and Italy. This trade is carried on through commission merchants. The value of the caps made amounts to about 20,000,000 francs annually, and a small portion, as already stated, is exported. The cap called "képi," which, since 1848, has been introduced into the army, the national guard, the public schools, and administrations, forms a considerable item in the manufacture.

The workmen's wages have greatly increased since 1855. They are all now pretty well remunerated. Connected with the general cap trade is that of the Greek cap or "fez." These are either knitted or made of felted cloth. The principal fabrics of fez caps are produced at Orleans, Paris, Rueuil, Chalons, and Condom. A considerable portion of these caps are exported.

BOOTS AND SHOES.

Shoemaking may now be divided into three classes—sewed boots and shoes, which represent a large amount of business; those put together by pegs or nails; and those put together by screws. Sewed boots and shoes are mostly made in Paris, Nantes, Marseilles, Bordeaux, and Fougères; pegged boots and shoes in Paris, Liancourt, Romans, Blois, and Angers; while those made with screws are only manufactured in Paris. Shoemakers generally use ox and cow hides for the soles, while the upper leathers are made of calf, kid, goat, and sheep skins. Woollen, silk, and woollen mixed, cotton and linen, and elastic fabrics, are also brought into use.

France produces about eight-tenths of the whole amount of hides employed for making soles, five-tenths of the calf skins, five-tenths of the kid and goat skins, and nine-tenths of the sheep skins used for upper leathers. As to the different woollen, silk and woollen, cotton and thread fabrics employed in the manufacture of boots and shoes, they are nearly all produced in France. However, only five-tenths of the mixed fabrics of wool and cotton, and eight-tenths of the elastic fabrics used in shoe-making, are of French manufacture. The various kinds of lining are made in France. No machinery is employed in the manufacture of those boots and shoes which are made to order. On the other hand, ready-made boots and shoes are partly manufactured by machinery, and those soles which are made with pegs and screws are put together by machinery. The raw materials are cut out by means of paring-knives and cutting-out machines of various kinds.

The men employed in the manufacture of boots and shoes are divided into three classes—the foremen, receivers, and cutters. The proportion of workmen and workwomen working in their own homes is eighty-five to the hundred; the rest are employed in the manufactories. Half of the people employed in this trade are women; their work consists in binding, tacking, stitching, and joining the upper portions. Women earn, on an average, 2 francs per day, and men 4 francs. The ready-made boot and shoe trade is in France carried on by commercial travellers, who sell to the provincial dealers. Commission merchants buy for exportation. The average price of good boots and shoes is 16 francs for men, 8 francs for women, and 6 francs for children; the commoner sort of boots and shoes for men are sold on an average at 8 francs, for women at 5 francs, and children at 3 francs a pair. These productions of the French trade are principally exported to the Levant, North and South America, the East and West Indies, England, Italy, and Switzerland. Paris alone produces boots and shoes to an amount of 100,000,000 francs; the provinces also contribute largely to this trade, and about 40,000,000 francs' worth of boots and shoes are exported.

Since 1855 the use of sewing machines for the putting together of the upper leathers has become very general, and the various other mechanical means for saving labor are being employed on a large scale. Workmen's wages have risen 20 per cent. in the same space of time.

HAIR-WORK.

The hair trade is now one of considerable importance. It is in Paris that hair is particularly well prepared, and it is also in Paris that wigs and false hair are made in the greatest perfection. Hair is, in France, chiefly obtained in the following departments: Puy de Dôme, Cantal, Corrèze, Lozère, Deux Sèvres, Vienne, Allier, Manche, Côtes du Nord, and Ille et Villaine. Italy, Belgium, and Germany also furnish a large amount of human hair. A great deal of beautiful hair is obtained from the convents. The hair from the western departments is superior to that from the south and midland departments. The price of hair not prepared and sorted is, on an average, 50 francs the kilogram; but in 1865 it rose to nearly 65 francs, and it is even supposed to have been sold at 100 francs.

In these productions there are only two kinds of manufacture—that of wigs, fronts, &c., and that of false plaits and curls. The beauty of the article varies according to the skill of the hair-dresser. In such matters the form is everything. The condition of the women and workmen employed in this trade is becoming better every day. The wages have risen considerably during the last few years; the working hair-dressers now earn on an average 5 francs a day; those women who are employed in making false plaits, &c., receive on an average 2 francs and 25 centimes per day, and those who make wigs, &c., 3 francs. Hair-dressers deal directly with the public.

The hair trade is carried on by large wholesale buyers, who obtain the hair from the travellers and small itinerant dealers, and then, after its having undergone different preparations, sell it to the hair-dressers. The average price of prepared hair is 140 francs the kilogram; wigs cost, on an average, 40 francs the piece, and chignons 15 francs. In France are sold annually 68,000 kilograms of hair, of which 40,000 are French and 20,000 from Italy, Belgium, and Germany; 8,000 kilograms of refuse hair are gathered in the hair-dressers' rooms and other different quarters. Great Britain, America, and Russia buy from France 30,000 kilograms of hair; 25,000 kilograms are employed in France in making wigs, &c., and about 13,000 are exported into different other countries. During the last few years a great many new kinds of fabrics in which to implant the hair for the making of wigs have been invented. They are made of all kinds of materials—silk net, cotton net, and silk gauze; there is even a fabric woven of white hairs. All the wigs, plaits, &c., are made by women, as well as the watch-chains, bracelets, and other fancy articles in hair.

CLASS 36.—JEWELRY AND ORNAMENTS.

"The articles exhibited in this class form two principal divisions, and comprise: 1. Fine and imitation jewelry; 2. Trinkets, including gold ornaments, decorated or enriched with precious stones or enamels; plated jewelry; copper gilt jewelry, which may be decorated with imitation pearls, coral work, steel and black ornaments.

"According to the distinction made in France, gold and silver are mere accessories in jewelry, but precious stones are essential; while the contrary is the case with respect to trinkets or bijouterie, (which terms do not correspond with those used in England.) Paris is the chief seat of the jewelry and ornament trade of France; after the capital city comes Lyons, Marseilles, and the departments of Cantal, Puy de Dôme, and Ariège. Watch cases are made especially at Besançon. The lapidary trade, both for precious and other stones, has become a very important branch of industry in the Jura.

"The chief materials used in the manufactures of jewelry are diamonds, precious stones, pearls, or imitation gems. The prices of pearls and precious stones are very variable, on account of weight, color, and quality. The principal sources are India, the Indian archipelago, Siberia, and the central regions of the new world. The raw materials of bijouterie are: gold, of the value of 2,600 francs the kilogram; and silver, worth 200 francs the kilogram. The chief sources of supply are Australia, Siberia, and North America. The jeweller receives the cut gems from the lapidary, who, from his experience, is enabled to add greatly to their value. The cutting is performed with the aid of a mechanical process. The business of the jeweller is to mount the gems or other substances, his trade being especially one of taste. The workman models and chases the precious metals, and enriches them with enamels, or with gems or stones. The elements employed by both, such as the bezels or settings, bodies of rings, and other parts, are produced with the aid of cutting presses, rollers, and other machines. Plated jewelry work is executed, with the aid of machinery, with such perfection that sheets of copper, upon which are soldered plates of gold one-twelfth the thickness of the former, are transformed into ornaments of all kinds without exhibiting a trace of the existence of copper. In these trades, with the exception of a few special cases, the masters rarely employ more than eight or ten workmen, and on one-quarter of the whole work alone only employ one journeyman and apprentices. The operatives sometimes work in the shops of their employers and sometimes at home; the last-named representing about one-tenth of the entire body. Women are also employed to the extent of about 20 per cent., but chiefly in polishing. The principal markets for the exportation of jewelry and trinkets are Spain and her colonies, the United States, Brazil, Turkey, Egypt, Italy, Switzerland, England, and Russia.

"The quantity of gold annually employed by the jewellers and goldsmiths of France is equal to 17 tons, and of the value of 44,200,000 francs, (£1,768,000.) The silver amounts to 89 tons, of the value of 17,800,000 francs. The workmanship adds 60 per cent. to the value of the gold and 40 per cent. to that of silver. The total value of the production is therefore 95,640,000 francs, (£3,825,000.) The exports of gold jewelry are equal to two and a half tons, and of the commercial value of 10,400,000 francs; and of silver work to eleven and a half tons of the value of 3,150,000 francs, (£126,000.) The diamonds, pearls, and other gems are nõt included in the above estimate. The trade is carried on by 1,250 manufacturers, who employ 20,500 persons, of whom 12,500 are workmen, properly so called. It is estimated that 2,000 wholesale dealers and 1,000 merchants are engaged indirectly in disposal of the produce of the trade.

"The employment of machinery has become general since 1855, and has reduced the cost of production without detracting from the perfection or finish of the work. Plated gold jewelry is without a rival abroad, and enables French commerce to compete with the low-standard gold work of England and Germany. The formation of a company for sweep-washing, and the reduction of the products, the initiation of a syndical chamber, and of an association formed of masters and workmen, have greatly favored the progress which the jewelry trade commenced to make in 1855."—*Introduction to the class; official catalogue.*

ENGLISH AND FRENCH JEWELRY.

The principal exhibitors in this class were France and England. In the absolute merit of the goods exposed—speaking only of the finer sorts—it would be difficult to say which excelled the other. English jewelry, in accordance with English taste in general, is characterized by solidity and massiveness. French jewelry, on the contrary, aims at lightness of effect and beauty of design. The English try to make their precious stones secure; the French to make them fascinating and also secure. The precious stones of a piece of jewelry are let into small cells cut for each individual stone out of a solid piece of silver or gold. The stones when inserted into these have to be secured that the portion of the edge of the stone held by and therefore concealed in the setting may be the smallest portion possible consistently with firmness of grip. To avoid the vulgarity of heaviness and the insecurity of lightness requires the nicest skill. Precisely as the artist triumphs over these difficulties does he produce a work of excellence and durability.

There was a dazzling display of diamonds particularly noticeable in the French court where the exhibitors were together, and the opportunity for contrast and study were most readily commanded. In the English section the exhibitors occupied little stalls of their own like sentry boxes, and isolated themselves as much as possible from their neighbors.

The productions of Mr. Massin and Mr. Froment Meurice were remarkable for perfection of workmanship and richness of materials. One of the most beautiful examples, by the first-named jeweler, was a sprig of wild rose executed in diamonds, presenting the lightness, the suavity of curve, elasticity of bough, and other characteristics of nature itself. In perfection of symmetry and radiant simplicity it was almost faultless.

A pair of earrings shaped as rosebuds, whereof the bud was a pink pearl, formed an exquisite adjunct to a parure of similar treatment representing a rose branch, while a charmingly rendered water-lily and a cornflower of sapphires and diamonds may further be pointed out, as lovely specimens of Mr. Massin's handling of the forms of nature. These works were carefully scrutinized by the jury and found to be as firm as they were light and elegant.

The value of Mr. Meurice's jewels was mainly in the workmanship. A head ornament of colored diamonds, a shell with a sprig, was the most elegant and probably the most valuable ornament of the kind in the Exposition. It is extremely difficult to match colored diamonds. These were extremely well mated and worked up. Among them were some pink diamonds, which, with more brilliancy, had almost the depth and color of the ruby. There were also green, yellow, and brown diamonds among them.

The fashions seem once more to incline to colored gems. Lemoine made an immense display of dark pearls in all shades of green, red, pink, yellow, and brown. They are used mainly in combination with diamonds.

Boucheron, another maker of repute, had a good display of diamond jewelry. A single pair of earrings in this case was worth $120,000 in gold. They were set with simplicity, but each weighed 23 carats. The same maker exhibited many articles where the skill of workmanship exceeded the value of the material.

In a different, but not much cheaper way, were the agates and rock crystals of Duron set in the style of Benvenuto Cellini and his school; works of art that would worthily occupy a place in any collection.

In an adjoining case were two necklaces of half crystalized black diamonds, a mineral phenomenon little known, and remarkable more for its rarity than beauty.

The jewels of the Countess of Dudley, exhibited in the English section, were the finest in the Exhibition, both as regards the size and color of the stones employed. They were shown, with many other splendid specimens of work, in the cases of Messrs. Hunt and Roskell.

Mr. Harry Emmanuel and the other British makers amply sustained their world-wide reputations.

The effort of the present time in the manufacture of diamond jewelry is to give movements to the different parts by which the reflections and refractions are increased.

DIAMOND-CUTTING ESTABLISHMENT.

The illustration of the processes by which precious stones or other artificial imitations are wrought was excellently given by Mr. Coster, of Amsterdam, who had a diamond-cutting factory erected in the Park, fitted up with the customary machinery and occupied by his regular workmen. The first rough shaping of the more important facets of the brilliant is performed by operating with two diamonds at once, each firmly secured in a handle and bruising each against the other, angle against angle. The dust that falls from the stones is preserved for the subsequent processes of grinding and polishing those facets that distinguish the many-sided brilliants from the dull original crystal of diamond. It is used, mingled with oil, on a flat iron disk, which revolves rapidly by means of steam power, the stone being laid upon this disk or wheel and pressed upon it by means of a weighted tool, which the attendant watches carefully. Skill of eye and hand, only attainable by great practice, is needed for this work. But more curious still, and requiring equal or greater skill, is the cleavage, or splitting of the stone. A little notch is scratched in the diamond by means of a knife pointed with the same material. A steel blade is then inserted in this opening, and a tap cleaves the stone in the direction required. The process is rapid and based on mathematical rules which govern the splitting of the stone. The diamond, when a blow is struck on an edged tool placed parallel to one of the octahedral faces of the crystal, readily splits in that direction. It was not the less remarkable to see the process so aptly performed.

There were other objects of interest in Mr. Coster's exhibition. For the first time the diamond was exhibited side by side with the minerals that accompany it in the river-bed of Brazil; and there were very rare examples in which crystals of diamonds were included within a mass of quartz crystals, having all the appearance of having been formed simultaneously with the diamond, but believed by some of the mineralogists to be artificial combinations. But the most extraordinary curiosity possessed by Mr. Coster was a rose-pink diamond of some 29 carats, endowed with the marvellous property of becoming perfectly bleached by an exposure of some four minutes to the effect of the atmospheric light. It recovers its rose color at a gentle heat and retains it for any length of time in darkness.

It may not be uninteresting in this place to give some particulars of Mr. Coster's Amsterdam establishment, which employs 316 lapidaries, assistants and apprentices, 88 cutters, and 21 splitters, forming an aggregate of 425 workmen, and receiving from $5,000 to $6,000 per week for wages. The annual importation of the diamond in the rough state amounts to nearly 1,000 pounds troy. Of this immense quantity Mr. Coster has received nearly half. For the finer varieties of diamond, averaging in weight under half a carat, a price equivalent to $50 or $55 (always in gold) a carat is now paid; and the price has doubled since 1848, at which date $22 or $25 would have purchased diamonds for which $50 or $55 have to be paid now. Thus a diamond of 2 carats weight, worth then some $150, is now worth from $300 to $350, and sometimes more; while a perfect brilliant of 4 carats is now worth from $1,000 to $1,500. When Jeffries wrote his book on the diamond, a century and a half ago, a carat diamond now worth $85 was valued at $40.

Two of the three great existing historical diamonds were cut by Mr.

Coster. These were the Koh-i-noor, of 103 carats, and the Star of the South, a Brazilian stone, slightly brown in hue, of 125 carats. The third, known as the Pitt or Regent diamond, the well-known crown jewel of France, weighs 135 carats, and was cut in the last century.

Among the curious uses of diamonds, rubies, and other fine stones, may be mentioned that of using them for the purpose of drawing fine wire. They are drilled to the requisite diameter, and answer the purpose better than any other material. Precious stones are also used for the working points of watches, for pointing drilling machines, and many other purposes. Indeed, the increased price of the diamond may be ascribed to the fact that it is rapidly becoming a tool as well as an ornament. Thus, while it is superseding steel, steel revenges itself by stepping into the polite domains of the diamond. It has even created a sensation there. Much of the fine steel jewelry exhibited in the Exposition was second only in delicate faceting and brilliancy to that of the diamond itself.

There was but a single exhibitor in this class from the United States. It was to be regretted, inasmuch as in a medium style of cheap jewelry the United States can, owing to the great use of machinery, be compared favorably with the best.

CLASS 37.—PORTABLE ARMS.

The manufactures included in this class form three distinct series: 1st. Sporting and gallery fire-arms, comprising fowling-pieces, rifles, pistols, revolvers, duck-guns, blunderbusses, and military arms for exportation; 2d. Side-arms and other arms, such as sabres, swords, foils, poignards, bayonets, axes, maces, casques, shields, cuirasses, &c.

The principal centres of production in France:

1st. For fowling-pieces and highly-finished fire-arms, are Paris and St. Etienne. The latter place, producing the largest quantity, may therefore be considered as the chief seat of the trade. Paris is famous for its highly-finished arms. Her models are sought by all nations, and the arms produced by her manufacturers are justly renowned for the finish of the work, the perfection of the details, and the elegance of the forms.

2d. Military arms for exportation are produced almost entirely by the directors of the factories of the state, at St. Etienne, Chattellerault, (Vienne,) Tulle, (Corrèze,) and Mutzig, (Bas Rhin,) and by private makers in Paris and Maubeuge, (Nord;) certain detached portions are made at Charleville, (Ardennes.)

3d. Swords and other side-arms are made at Chattellerault, Kingenthal, (Bas Rhin,) and St. Etienne, but the whole of the mountings, scabbards, and accessories are produced in Paris, where they form a special industry, remarkable for artistic workmanship and finish.

4th. The manufacture of percussion caps, priming, and cartridges, is confined exclusively to the metropolitan departments of the Seine and Seine-et-Oise, and is in the hands of five or six manufacturers, who are enabled not only to supply the entire home demand, but also to export a considerable quantity.

The iron and steel employed in the manufacture of fire and other arms are produced on French soil. For the finer arms the iron is derived from the department of the Vosges, and costs, on an average, 66 francs the 100 kilograms, (33 francs the hundred weight.) The steel comes from Isère and Loire, and costs from 120 to 150 francs. The mountings and accessories of swords and other side-arms require, also, copper, horn, leather, ivory, mother-of-pearl, silver, gold, and other materials. Steel forms the chief material of military arms; it is derived from the basin of the Loire, at Rive-de-gier and Ferminy, and is delivered, on an average, at 95 francs the 100 kilograms for cannons, and at 160 francs for sabres, bayonets, and cuirasses. The price of the iron employed in the making of military arms is 65 francs the 100 kilograms. Steel forms about three-fifths of the material of the arms manufactured in France, and its amount is estimated at 2,500 tons per annum for all kinds of arms. The walnut wood used for gun-stocks is produced at Auvergne and Poitou; the price of the wood when cut up is about 2 francs per piece for military arms, and 8 francs for fowling-pieces. Percussion caps, priming, and cartridges are manufactured with Chilian copper, which, when refined and rolled out, is worth in France, on the average, 250 francs the 100 kilograms. Brass wire only costs 225 francs. The makers of these articles use about 500 tons of copper per annum. The fulminating powder is composed of mercury, from the mines of Spain, costing 5 francs 40 centimes per kilogram; alcohol, mineral acids, nitrate of soda, and chlorate of potash, at the price of 4 francs. These last materials are of French production. The making of cartridges absorbs annually 200 tons of paper, made in France, and costing from 60 francs to 170 francs the 100 kilograms. Lastly, the materials of felt wads cost 28 francs the 100 kilograms.

Machinery occupies daily a more conspicuous place in the production of the barrel, the stock, and certain other parts of military fire-arms. For the barrels, the principal means employed are the tilt-hammer, for faggoting iron and steel; rollers, the lathe, and slide-rest, to replace the file; drilling, boring, and rifling machines. The barrels of highly-finished arms are manufactured with the aid of the same machines and tools, to which, however, must be added the soldering furnace, for uniting double barrels. Lathes, countersinking tools, and planing machines are used in the manufacture of the stocks, the lock-plates, and other portions of military arms. Lastly, the use of machines has been adopted in the making of revolvers, which have been in great demand during the last few years. Hand-labor, on the contrary, is still employed for the adjustment and fitting of the various parts of arms. The same is the case as regards highly-finished arms; and, with the sole exception of the barrel, it is easy to comprehend that the application of machinery would be almost impossible in the case of elaborate and fancy arms, which require delicate ornamentation and great variety in the form.   In the case of swords and other side-arms, the operations of sharpening and mounting are also performed by hand.   This portion of the manufacture which partakes of the arts of the engraver, chaser, gilder, and arm-goldsmith, in their highest phases, can never be executed without the hand of man.

Each maker of percussion caps, priming, and cartridges has his own peculiar machines and materials.   These are driven by steam, and include cutting and stamping presses, rollers, filling machines, &c.

Men, women, and children are employed in the making of arms.   The manufacture of percussion caps, and particularly of cartridges, employs a large number of women.   Generally the work-people are employed in the shops of the manufacturers, and, under the direction of the latter, a certain number work at home.   Some are paid by the piece and others by the day.   The number of workmen employed in this trade is about 15,000.   The arms are sold generally where they are manufactured, but especially in Paris and at St. Etienne.   Paris is the principal market; the armorers there supply not only the arms called Parisian, but also many others made elsewhere, and which are sometimes finished by Parisian workmen.   Paris is also the market for side-arms, mounted and finished, for the officers of the army, the officials who wear uniforms, for the provincial dealers in arms and military equipments, and finally for exportation.   Paris is thus the great *depôt* for arms, as well as for cartridges. The whole production of arms, cartridges, &c., in France may be set down at about 15,000,000 francs in value.   This amount, of which St. Etienne represents about 6,000,000 francs, may be divided as follows: Fire-arms and bayonets, 10,000,000 francs; side-arms, 1,000,000 francs; caps, priming, and cartridges, 4,000,000; total, 15,000,000 francs.

Among the improvements introduced into the manufacture of arms since 1855, the committee note the following: Planing machines, which allow more perfect workmanship in the barrels, as regards their finishing and boring; the many improvements made in the methods of breech-loading; the adoption of small calibre and breech-loading for military arms; the use of cast steel in place of iron for the barrels of rifled and other arms; and, lastly, the introduction, already referred to, of mechanical processes which tend to replace manual labor more and more every day. England and America preceded France for some time in these respects; but manufacturers have followed boldly the examples of these two nations, and march in their footsteps with courage and success.   We must point out, also, the introduction of more ingenious methods in the manufacture of cartridges, with the view to obtain the most complete and effective combustion.

The manufacture of arms, considering its importance and the amount of trade, is not yet sufficiently developed; but sensible progress has already been made, and the impulse which circumstances have recently given to the production of military arms is aiding us to bridge over more rapidly the distances which separate France from the more advanced nations. Moreover, it is right to repeat that, in the matter of highly-finished arms, French manufacturers stand in the first rank; it defies all competition as regards artistic taste, elegance of form, and the ability of workmen. Similar superiority is also to be noted in the case of priming and cartridges. This class of manufacture, which is not of French origin, and not yet 40 years old, has grown with great rapidity, and its products, which have nearly doubled since 1855, are sought by all nations, on account of their perfection and low price. The rapidly-increasing use of breech-loading arms, and the improvements which are constantly introduced, are opening up new sources for this manufacture, and promise it an almost unlimited field.—*Alexander Fouquier, member of the committee of admission of class* 37.

It may be presumed that the invention of portable arms was almost coeval with the creation of man, inasmuch as the term means anything from a stone, or club cut from the neighboring thicket, to the electric pistol, which does not even put a murderer to the trouble of pulling the trigger. To make a weapon that would protect the bearer from the onslaughts of wild beasts was an early necessity. He did it rudely at first, but improved as he found a finer beast—man—to kill. No art has made such rapid strides as this, and no art has left behind it such unquestionable traces of its growth and progress. The tiller who whistles at his plough, in our western lands, turns up at each furrow some indication that a race less agricultural, but vastly more belligerent, has preceded him. In the gallery devoted to the so-called "history of labor" were specimens of all the earlier weapons, commencing with the flint, and progressing rapidly to the metal. The progress, even in the remote past, was rapid; in the present day it has exceeded that of any other art. Especially is this the case in fire-arms. Skill can never be abolished, either in warfare or in sporting; but a very small amount of skill is all that the manufacturer of to-day requires. The soldier blazes at his enemy at 5,000 yards until he "pots" him, or undergoes the process of being "potted" himself. The sportsman pursues his covey with cartridges that place themselves in a breech-loader, and require nothing at his hands except a touch of the finger. Well may Captain Majendie, R. A., bewail this state of things. "It is impossible," says this officer, who was also a reporter on the class, "not to feel that the interest which has hitherto attached to this class of arms has, in a great measure, departed since the general adoption of breech-loading rifles. The occasions henceforth must be comparatively rare in which hand-to-hand contests will be possible." This opinion, indeed, found practical expression in a bayonet exhibited among the English arms by Mr. Scott Tucker. It was hardly half the length of the present bayonet, and Mr. Scott Tucker suggests its adoption, on the distinct ground of "the chance of crossing bayonets being materially lessened by the introduction of breech-loaders." He claims for it the advantages of being comparatively light, cheap, strong, handy to draw and return, less easily parried, quick for thrust and withdrawal, free from chance of locking, and out of the way when skirmishing.

Of swords and spears and hand-lances the variety was almost without limits. Solingen, as of yore, distinguished itself in these branches. During the early months of the rebellion Solingen blades were regarded almost as articles of luxury; they commanded extravagant prices. Here are the terms upon which they are supplied in Europe: An English infantry officer's regulation sword, with steel scabbard, may be obtained wholesale for 22 francs; a French line officer's sword, for 24 francs. The process by which the Solingen makers impart to their sword blades a gilding of peculiar permanence is said to be a secret; so, also, with the precise combination of metals used in their manufacture. A good Solingen blade can be wound round the body, and when released will straighten absolutely. The blades of Toledo, much more expensive, do no more.

Swords of every shape and quality were found in almost all the sections of the building, remarkable either for excellence, cheapness, or decoration. Among the latter class may be included nearly all weapons that came from the east. Turkey, for instance, had no fewer than 102 exhibitors, and about the same number of articles.

In long-range arms of precision there has been but little progress made since the introduction of the conical gas-expanding bullet of 1855, but an infinite variety of small improvements have been introduced for the better throwing forth of the projectile. The American rifles were considered among the best exhibited. The object of late has not been to secure greater length of range, but to obtain a quicker rate of discharge; hence the breech-loaders. Weapons of this make were naturally the features of the Exhibition. Twelve years ago breech-loading had been largely applied on the continent of Europe to sporting guns, but it has not been applied to any considerable extent (except in Prussia) to rifles, either for military or sporting purposes. In 1867 it was the accepted principle of all military arms. The battle of Sadowa abolished all theories on the subject. The principal result showed that the breech-loader was the most deadly weapon on that memorable day. Every government is now supplying its troops with new pieces on this plan, or altering the old ones in conformity with it. Models adapted for either purpose were exhibited in every court. The superiority of the breech-loaders having been accepted, a new question has arisen, namely, whether the central fire, or the pin or rim fire, is the best. It is a matter which has been fully discussed in the report of the American commission appointed to inquire into it.

Nearly all the small-arms on the revolving principle were made in accordance with well-known American inventions. The American display of these weapons was very good.

A novelty, already referred to, was exhibited in the American department, consisting of a gun fired by electricity. The apparatus is concealed in the stock. It is claimed for this mode of ignition that there will be less danger in preparing the cartridge, inasmuch as it contains no fulminating powder, and that, for a like reason, there will be no risk in the transportation of ammunition; also, that the arm cannot be fired accidentally, and that, as there is no blow of the hammer, there is no inevitable deviation in the aim of the person who fires the piece. The rapidity of fire is claimed to be greater, and the escape of gas less, than in other arms. All these points have yet to be proved in practice.

Another curiosity was a muzzle-loading cap gun, having two bullets for each barrel, the piece having two barrels and four hammers. Each barrel is loaded with two charges and two balls; the front charges are fired by the front set of hammers; when these have been fired, the second pair of hammers are brought into play to fire the hinder charges. The manufacturer claims that there is no danger of the first charge exploding the second.

And still a last curiosity was a shield in the English collection, from which projects, in place of a spike, the muzzle of a breech-loading pistol. These shields formed part of the equipment of the body-guard of Henry VIII, dated 1530.

CLASS 38.—TRAVELLING AND CAMP EQUIPAGES.

This class includes four principal divisions, which again include many distinct industries.

The first division, camp equipage, comprises two branches: Articles for soldiers and articles for officers; for agents and workmen engaged in the construction of railways, of the canal of the Isthmus of Suez, &c., and scientific explorers and travellers. During the Crimean war the French government caused a great number of tents and other objects to be made, which served first for that war and afterwards in the Italian campaign. Since that time the demand has been arrested in that quarter; but foreign governments buy their camp equipage of French manufacturers, and this trade amounts on an average to about 5,000,000 francs per annum. As regards equipments for officers, agents and others, the trade amounts to 2,500,000 or 3,000,000 francs, principally for abroad.

The second series, travelling equipments, is divided into three branches: Articles in iron work, leather trunks and portmanteaus, wooden chests and ladies' travelling bags. Fifteen years since the first-named articles were all imported; at present the home manufacture supplies all demands, and, out of a production which represents about 1,200,000 francs per annum, about one-fifth is exported. The manufacture of leather trunks and wooden boxes, which took its rise in France about thirty years since, has, since the opening of railways, assumed an importance which grows every day. The value of the trade amounts to 5,000,000 francs a year, and about one-third of the production is exported. Twelve years since the manufacture of ladies' leather bags was unknown in France. It soon afterwards was established, and has since grown with considerable rapidity. The value of the manufacture is 8,000,000 francs per annum, of which three-fourths are exported.

The third series, sporting equipments, includes, besides the articles which come properly under that denomination, certain other items, among which are water-proof and waxed cloths. This class of manufactures, the centre of which is at St. Sylvain (Calvados), has followed a constantly increasing rate for several years. The manufacture amounts to 1,000,000 francs per annum, and one-third of the whole is exported. The trade in oiled and water-proof cloths is not very old in France, but at the present moment the production not only suffices for all demands of the home markets, but allows three-fifths to be exported. The raw materials employed in the construction of the above-named articles are cloths and drills, which are produced in the departments of the Nord, Sarthe, Orne, and Mayenne; leather, which is furnished chiefly from Paris; card-board, principally supplied from Lyons; poplar wood, which comes from various parts of France; and linen tissues, the produce of the Nord, Seine Inferieure, Haut Rhin, and Rhône.

Hand labor predominates in the manufacture of camp, travelling and sporting equipages, but machines have been used for a long time to perform the sewing, especially in the case of tents and bags, and their employment has greatly reduced the market price, and at the same time allowed a production of four times the quantity in the space of time. The greater part of the workmen are engaged in the shops of the manufacturers, but a certain portion work at home. Women are employed, particularly in the tent trade. In Paris those who work the machines earn from 3 to 4 francs a day; others only realize from 2 francs to 2 francs 50 centimes. The men earn from 5 to 6 francs. The iron-workers obtain 5 to 8 francs. In the other branches the wages vary from 4 to 6 francs for the men, and from 2 francs 50 centimes to 3 francs 50 centimes in the case of the women. The total number of persons engaged in the trade is between 3,600 and 4,000; during the Crimean war the number was as high as 10,000. Paris is the chief, if not the only, centre of manufacture for camp and travelling equipments. In the departments some towns, such as Lyons, Toulouse, and Tours, a few makers can be found, but they are very limited in number and work principally to order. The distribution of the trades connected with the production of sporting requisites has been already shown.

It will have been seen, by what has already been said, that the raw materials used by the manufacturers of the articles which come under Class 38 are almost exclusively of home production. The value of the trade amounted in the year 1865 to about 25,000,000 francs, and one-half of the amount produced was exported. The date of the commencement of these various branches of industry is but recent; they had scarcely arisen at the time of the Universal Exhibition in Paris in 1855. Since that period, however, they have developed rapidly; and, at the present time, in consequence of the improvements brought to bear in every part of the manufacture and the good taste and finish of the articles, France has nothing to fear from any rivals whatever, and can even cope with them in their own markets.

The fourth series comprises blankets and rugs. This trade, although confined to a small number of houses, has assumed very great importance in France, and employs a considerable number of workmen. The trade not only supplies the home market, but exports a part of its productions. The chief materials used are wool and cotton; hair of different kinds is also employed, but there are no examples of this in the Exhibition. Wool performs the most important part in the manufacture. That of France is the most esteemed, but wool from Africa, the Levant, and La Plata is also employed. Algerian wool is capable of being bleached to almost absolute whiteness, but its quality is not equal to that of France. The manufacture of white woollen blankets presents great difficulties; like that of cloth, it has to pass through all the operations of spinning, bulling, &c., without possessing the resource of dressing to remedy imperfections. The employment of cotton blankets is less common, but the simplicity of the method of manufacture and the moderate price of the raw material gives them a special interest, as coming within the means of the less wealthy classes. It is to be regretted that the manufacture of railway rugs is not represented by French producers, especially as that industry, which has become very important, is essentially remarkable for the improvements that have been introduced, and which give the productions incontestable practical advantages over those of other countries.

There was a good display of French articles in this class, but considering the importance that travelling has assumed of late years, and never more so than in 1867, the competition, if it can be dignified by the name, was strikingly poor. There was no novelty worthy of record. America had nine exhibitors, among them the Quartermaster General of the United States army, who exhibited the material in use for transportation, clothing, and equipment in camp and in garrison.

CLASS 39.—TOYS.

"Class 39 comprises: 1. Automatons, (mechanical figures and animals.) 2. Toys in general, including an immense variety of articles, of which dolls form the most important branch, and among which may be mentioned, besides kitchen utensils, dinner and tea things, card-board boxes and other articles, dressed figures, animals and arms. The greater number of toys are manufactured in Paris. The common wooden playthings form the special trade of the town of Liesse, Aisne. Limoges supplies the China services, which are ornamented in Paris; at Nevers and Sarreguemines are made these same articles in various kinds of earthenware. The manufacture of the different kinds of toys necessitates the use of the greater portion of the raw material known, and the co-opera-

10 U E

tion of nearly every trade. Nearly all the Paris toys are made by hand, by men and women; children are not employed, being unable to bestow upon the work the excessive patience and minute attention which it demands. Cutting and stamping presses and lathes are used for the metal and wooden toys. Few makers employ more than 20 work-people. In Paris, the larger number of the men work at the shops; the number of women is about equal to that of the men. The making of dolls' clothes alone occupies several hundred women, of whom half work at home. The wages are, on an average, 4 francs a day for men, and from 1 franc 75 centimes to 2 francs a day for women. The makers deliver their products to the retail dealers and to agents for exportation. Very few among them export directly to other countries. The manufacture of small wares occupies about 2,200 people, and business is done in it to the amount of 10,500,000 to 11,000,000 francs (£440,000) a year, of which two-fifths are realized by the exportation trade. The toy manufacture is making rapid progress; the models are more varied and have more taste and elegance; greater attention is paid to the work and the prices have nevertheless diminished. Automatons and mechanical playthings have been brought to great perfection, and the singing birds are made to imitate nature so far as to deceive the most practiced ear. Certain instructive toys, moved by electricity, can, without danger, be placed in the hands of children. Numbers of dolls are made whose trousseaux show so much taste and are so elegant, that they are constantly used by dress and bonnet makers as types of Parisian toilettes."—*From the translation of the official catalogue—France.*

It is not necessary to add anything to the above *résumé*, except that the few exhibitors in other countries seldom rivalled and never excelled the French makers.

GROUP V.

PRODUCTS, RAW AND MANUFACTURED, OF MINING INDUSTRY, FORESTRY, ETC.

CLASS 40. MINING AND METALLURGY.—CLASS 41. PRODUCTS OF THE CULTIVATION OF FORESTS AND OF THE TRADES APPERTAINING THERETO.—CLASS 42. PRODUCTS OF SHOOTING, FISHING, AND OF THE GATHERING OF FRUITS OBTAINED WITHOUT CULTIVATION.—CLASS 43. AGRICULTURAL PRODUCTS (NOT USED AS FOOD) EASILY PRESERVED.—CLASS 44. CHEMICAL AND PHARMACEUTICAL PRODUCTS.—CLASS 45. SPECIMENS OF THE CHEMICAL PROCESSES FOR BLEACHING, DYEING, PRINTING, AND DRESSING.—CLASS 46. LEATHER AND SKINS.

CLASS 40.—MINING AND METALLURGY.

This class included:

Collections and specimens of rocks, minerals, and ores, ornamental stones, marble, serpentine, onyx, hard rocks, refractory substances, earths and clays.

Various mineral products; sulphur, rock salt, salt from salt springs, bitumen and petroleum.

Specimens of fuel in its natural state and carbonized; compressed coal.

Metals in a crude state; pig iron, iron, steel, cast steel, copper, lead, silver, zinc, &c.; alloys.

Products of washing and refining precious metals, of gold-beating, &c.

Electro-metallurgy; objects gilt, silvered, or coated with copper or steel by galvanic process.

Products of the working of metals; rough castings, bells, wrought iron, iron for special purposes, sheet-iron and tin plates, iron plates for casing ships and constructions, copper, lead, and zinc sheets.

Manufactured metals; blacksmiths' work, wheels and tires, unwelded pipes, chains, &c.

Wire-drawing; needles, pins, wire-work and wire gauze; perforated sheet iron.

Hardware, iron-mongery, edge tools, copper, tinware, &c.; other metal manufactures.

As almost all the countries that participated in the Exposition were able to send raw materials, the number of exhibitors in this class was very great, as will be seen by the following list, which shows the number of exhibitors from each country of importance:

France	349	Belgium	104
Algeria	33	Prussia	515
Holland	7	Hesse	2

Bavaria	25	Sweden	97
Wurtemberg	9	Norway	19
Austria	182	Russia	91
Switzerland	14	Italy	262
Spain	183	Turkey	210
Portugal	39	United States	68
Greece	79	Great Britain	137

Colonies and dependencies of Great Britain:

Barbadoes	3	Newfoundland	12
Canada	68	Nova Scotia	22
Cape of Good Hope	8	Queensland	10
Malta	6	South Australia	17
Natal	15	Victoria	6

It should be observed that in the case of Prussia, and perhaps some of the other countries, each mine, or company, or individual furnishing specimens of minerals to the collections was enumerated as an exhibitor, thus repeating many times the same product. If the names of all the donors to the collections from the United States had been sent in, the list would, in like manner, amount to many hundreds.

In neatness and careful preparation the Prussian mineral collection was the finest in the Exposition. The specimens were all rather large, but were uniformly trimmed and well arranged. The whole was illustrated by numerous well drawn and colored maps and sections of mines, and by models of furnaces. The collection comprised the products of the mines and quarries of the country, and was systematized and arranged under the orders of the minister of commerce and public works by Doctor Wedding, mining engineer, who received a silver medal from the jury as a recognition of his labors. It was accompanied by a special printed catalogue. The principal minerals shown were coal, iron ores, copper ores, and argentiferous lead ores. The salt mines of Stassfurt were represented by a quantity of the salt cut from the mine in large blocks and built up in the Exposition building into the shape of a half dome. A very interesting series of salts of potash found above the salt bed were also shown. These potash salts are now largely used for the preparation of manures and for other purposes requiring potash. The supply is believed to be practically inexhaustible, and it has already greatly diminished the demand for wood ashes.

Spain made an exceedingly interesting exhibit of its ores in a building erected specially for the purpose and for the agricultural products, in the Park. The prominent objects were blocks of cinnabar from the famous mine of Almaden, which is still largely worked.

In the Russian section there was an interesting suite of models of famous meteorites, and many pepites and nuggets of native platinum from the Siberian mines of Prince Demidoff. The display of wrought and sheet iron was very good, but the chief attraction was the variety of rough and polished precious stones, and large vases and candelabras made of malachite, jasper and rhodonite, (described more particularly under class 15, Group III.)

Among other notable objects was a mass of beautiful malachite, very solid, weighing over two tons, from the mine of Prince Demidoff. This mine was discovered in 1814 and has yielded 35,000 *pounds* of malachite, about 700 tons since 1840, besides a large amount of copper. A mass of native copper sent from the Kirghiz steppes, Siberia, and weighing about 1,500 pounds, much resembles the specimens from Lake Superior, and like them contains some native silver. The most remarkable exhibit of graphite was made by J. P. Alibert, of Mount Batougol, near Irkoutsk, Siberia. A large glass case was filled with blocks of the graphite cut and fashioned into various forms and exhibiting a fine polish.

MINERALS FROM THE UNITED STATES.

The mineral exhibition of the United States was very creditable. The coal, iron, lead, copper, zinc, gold, silver, and quicksilver, and the petroleums of Pennsylvania and California, were all represented. Among the iron-ores may be noted a large mass from the Iron mountain of Missouri, blocks from Lake Superior, and masses from the iron region of Lake Champlain. There was also a large representation of our limonite ores, and of the franklinite from New Jersey. The recently discovered *black band iron ore* was there also, and was doubtless thought by most persons to be coal, it is so black and coal-like in appearance. This ore is said to form a bed five feet thick directly below the anthracite coal, or only 150 feet from it. It contains 39 per cent. of iron and 35 of combustible matter. Its enormous value can hardly be estimated.

Several large and very rich masses of silver ore from the Poorman lode, Idaho, attracted much attention, and received recognition from the jury by the award of the gold medal. These masses occupied the summit of a pyramidal mass of ores in which were found blocks of iron ore from Missouri Iron mountain and Lake Superior, copper and ingots from Lake Superior, coal of Pennsylvania, silver ore from California, and rock-salt from Louisiana.

The copper of Lake Superior was well represented by specimens of the crystallized metal and of the minerals which accompany it, sent by Mr. Bigelow of Boston.

The gold-bearing quartz of California, and the ores of copper, quicksilver, lead, iron, manganese, and other metals, together with the salt, borax, sulphur, coal, and petroleum from the same State, were exhibited by a collection classified and sent forward by W. P. Blake, the State commissioner, and which received a silver medal. The gold ores of Colorado were shown by a large and brilliant collection arranged by J. P. Whitney, of Boston, commissioner from the Territory, who received a gold medal for the display.

The Chester Iron Company, of Massachusetts made a very interesting exhibit of the ores of iron and emery, with the associated minerals from

Chester, Massachusetts. This emery was discovered a few years since by Dr. C. T. Jackson, of Boston, in the ores of the company, and it is now largely explored.' The jury signalized the value of this emery, and of the discovery, by awarding a silver medal to the company and a silver medal to Dr. Jackson as the discoverer.

The minerals of the United States section were catalogued by Mr. D'Aligny, commissioner, to whom the jury awarded a silver medal.

IRON AND STEEL.

Wrought iron, in all its forms, figured largely in this department of the exhibition. Enormous bars, plates, and girders were to be seen in the English, French, Russian, and Prussian sections. The iron ores of Sweden were represented by large blocks which formed the base of a pyramid of iron bars and rods, square, round, and twisted, together with samples of the various grades of pig iron and bars of steel. .

The manufactures of the cast-steel works of F. Krupp, in Essen, Prussia, were represented by a cast-steel ingot, locomotive tires, railway axles, junction rings for steam boilers, crank shafts for marine steam engines, and plates or girders, besides several breech-loading rifled guns, all of cast steel.

The ingot of steel is the largest ever made. It is about 10 feet high and 56 inches in diameter, and weighs 40 tons. The upper end of the block is forged into an octagonal shape, and the whole is to be forged under a 50-tons hammer into a marine crank shaft. Cast-steel tires for locomotives form a very considerable portion of the manufactures of this establishment. Nearly 40,000 are made each year. They are all forged out of ingots without welding. This is accomplished in the following manner: The ingots are first forged out into long plates, from which rectangular pieces of the weight of the intended tires are cut off. A slit is then made in the centre of these pieces and the opening is enlarged by wedges until a ring is formed, and this ring is ultimately worked into a tire in a rolling mill.

One great attraction of the Exposition was the enormous steel gun from this establishment. This was 210 inches long, 14 inches bore, and weighed 50 tons. It is intended for the arming of coast batteries to defend them from the attacks of plated ships. It required sixteen months' work day and night to complete this monster gun.

The works of Mr. Krupp cover about 450 acres, 200 of which are roofed over. He employs 8,000 men, and 2,000 more at the iron and coal mines. The value of the yearly production of these works is said to be over $7,500,000.

The exhibition made by the Bochum Company, Westphalia, is also exceedingly interesting. Four bells of cast steel weigh, respectively, 1,800, 4,500, 9,000, 14,750 kilograms. The largest is nearly 10 feet in diameter at the mouth. This company also show a string of 22 car-wheels all cast in one piece connected at the hub, the whole weighing nearly 10,000 kilograms; and also the cylinder of a locomotive engine with steam pipes and box and flanges, complete in one piece of solid cast steel as it came from the mould.

The exhibitions of steel, iron, &c., in the French department were remarkably fine and complete. The largest ingot of cast steel weighed 25,000 kilogram, or about half as much as the great Krupp ingot. Beautifully finished steel cannon and fittings were shown in connection with thick armor plates for ships, some of which had been indented by pointed shot from heavy rifled guns. These guns and steel ingots were placed in buildings in the Park, where a great variety of the products of iron and steel works were arranged together.

One of the most interesting exhibits in the whole Exposition was the collection of ornamental iron castings from the foundries of Durenne at Sommevoire, consisting of bas-reliefs, busts, statuettes, and figures of the size of life. The success which has there been attained in the reproduction of fine works of art is wonderful. The castings are beautifully smooth and sharp, and when covered with copper, by the galvano-plastic process, they have the appearance and durability of bronze.

In the English section, Messrs. Johnson, Matthey & Co., of London, exhibited an extensive collection of the precious metals, and of large stills made from platinum for the use of manufacturers of sulphuric acid. One of these stills was sufficiently capacious to concentrate eight tons of acid a day, and was valued at $12,500. These stills are made without joints, soldered with gold as formerly, the joints having been formed by autogenous soldering. There was also shown in the same case an ingot of platinum forged into one mass large enough to make a five-ton still. The collection contained many of the rarest metals, such as ruthenium, osmium, iridium, &c. The total value of the exhibit was estimated at $100,000.

CLASS 41.—FOREST PRODUCTS AND INDUSTRIES.

The collections of specimens of forest trees, of timber, boards, and of ornamental wood, were very extensive. France, Austria, Canada, Norway, and Sweden, Russia, Brazil, and Australia were the principal exhibitors. Austria, by the Administrations of the forests of the different states, sent the trunks of oak, fir, pine, and other trees, with a great variety of planks, boards and timber for building purposes. The trunks of the large trees were sawed into lengths convenient for transportation, and were afterwards united upon the ground. The great size of these objects prevented their being received into the buildings, and they were placed in a group together in the Austrian section of the Park.

The exportation of Austrian woods is increasing, and has already reached the total value of 75,000,000 francs. Full statistics are given in a brochure prepared under the orders of the minister of commerce and political economy, and entitled "Les Richesses forestières de l'Autriche et leur exportation" Vienna, 1867.

The Canadian exhibit attracted much attention by the size of the hewed timbers of fir and pine, and the beauty of the specimen slabs of the walnut, maple, oak, ash and other forest trees. This collection was prepared under the direction of the Abbé Brunet, and was accompanied by a complete descriptive catalogue, forming a pamphlet of 64 pages. A gold medal was awarded by the jury.

The Brazilian woods were tastefully arranged in a room, with the walls and ceiling painted in imitation of the forests of the country.

The saw-mills and lumbermen of Norway and Sweden united in sending samples of their sawed and planed lumber suitable for building. There are in Norway 3,300 saw-mills, and the annual production of lumber is said to be worth about $12,000,000. The exports in 1865 reached a value of 45,600,000 francs—about $9,120,000. A little over one-half of these exports consist of sawed lumber, and the remainder is in the form of logs and timbers; the latter are sent chiefly to Holland and England.

The State of California failed to send the cross-section of the great tree *Sequoia gigantea*, as proposed. It was found that a cross-section of a tree 30 feet in diameter would weigh several tons, and that it could not possibly be transported from the forest in season for the exhibition. There were several samples of a beautiful ornamental wood from that State, a species of *arbutus*, the "madrona" or "laurel," which were remarkable for their beauty of grain. An ornamental door made of this wood by J. D. Boyd, of San Francisco, was very beautiful in this respect, and also in finish. There were contributions from the States of Illinois, Missouri, Kansas, and Wisconsin.

FORESTS OF FRANCE.

The French exhibition was beautifully arranged in a saloon of gallery V, at the end adjoining the main avenue. Sections of all the principal kinds of forest trees of the empire were ranged around the walls, and the interspaces were filled with moss. The tables in the centre supported models of the mills, and of the machines and tools used in cutting and preparing the timber for market. The following general exhibit of the extent of forests of France, and of the industries immediately connected with them, was prepared by the committee of admission of the class, and is extracted from the translated catalogue.

"The objects shown in this class fall under four principal divisions:

"1. Collections and models, including specimens of all kinds of timber and woods indigenous to or naturalized on the soil of France; the tools, implements and machines used in the forest, and in the various occupations carried on there; models of habitations and buildings, such as keepers' lodges and cottages, establishments for sawing, for the inspection of timber and other operations; plans, in relief, of various works executed in important localities; the replanting of the Alps, and the most remarkable methods employed for clearing forests situated in hilly countries; saw-mills, water-mills, &c.

"2. The more important products of forest industry, such as cork, fibres, tanning materials, charred wood and charcoal.

"3. Timber cut up and prepared; mouldings, shaped woods, clapboards, staves and other split wood.

"4. Coarse basket work, wooden-shoe making, &c.

"Amidst these various products is exhibited the forest chart of France, which shows in the most striking manner the importance of the woodlands of the country, and the remarkable relation which exists between them and the geological constitution of the soil. The collections of class 41 will be completed on the most interesting manner, by a series of specimens of destructive forest insects, with a selection of timber ravaged by the fructifications of a certain number of exotic coniferæ which must be regarded as naturalized; lastly by a series of publications on practical or scientific questions relating to sylvaculture. The collections of transverse sections of trees, and of other specimens of timber, will exhibit the marvellous productiveness of the soil of France in ligneous matters. The many kinds of timber and other woods will be represented by numerous samples from various parts of the country where the trees which furnish them exist under the most varied conditions. The examination of these specimens will show how the qualities of the same species of tree vary according to the fertility, the exposure and the mineralogical composition of the soil.

"The most important wooded spots of France are: in the north, the forest of Fontainebleau, 17,300 hectares;[1] Compiègne, 14,000; Rambouillet, 13,000; Villers Cotterets, 11,500; Mormal, 9,000; in the east, the forest of Chaux, 11,500 hectares; La Harth, 14,500; Hagueneau, 15,000; Dabo, 11,000; Haye, 7,000; Grande Chartreuse, 6,200; in the west, the forest of Lyons, 10,500 hectares; Bercé Perceigne, 10,500; Ecouves, 7,500; in the centre of France, the forest of Orléans, 37,600 hectares; Troncais, 10,500; Viezron, 5,200; Chateauroux, 5,100; Bertranges Guerigny, 5,300; in the south, the forest of La Maitrise de Quillan, 11,000 hectares; Soule, 7,000; Lannet, 5,000.

"The woodlands of the empire amount to 8,900,000 hectares, divided as follows:

"1. 1,100,000 hectares belonging to the state, of which 49 per cent. is in timber—539,000 hectares, and 51 per cent. in coppice, with or without timber—561,000 hectares.

"2. 2,200,000 hectares, the property of communes or of public establishments, of which 36 per cent. is covered with timber—720,000 hectares, and 64 per cent. coppice, with timber or coppice alone—1,280,000 hectares.

"3. 5,800,000 hectares, the property of private owners, in timber, 17 per cent., 980,000; in coppice, with or without timber, 4,814,000.

---

[1] A hectare is equal to 2 acres, 1 rood, 35 perches.

"The annual products of these forests are in the following proportions: 3 for the state land, 2.75 for those of the communes, and 2 for those belonging to private owners, giving a gross total of about 20,000,000 cubic metres, viz: timber and working woods, 2,000,000 cubic metres; wood for fuel, 18,000,000 cubic metres.

"These resources are on the increase in consequence of the numerous improvements in management and of the construction and amelioration of the means of transport undertaken upon a large scale during the past ten years in the forest lands of the state and the communes. But the production is still far from sufficient to supply the demands of consumption.

"The annual consumption in France is as follows:

"1. In timber for constructive purposes and wood used in manufactures. The naval and mercantile marine, 118,000 cubic metres; artillery and engineering, 30,000; railways, 600,000; building, 1,600,000; lath wood and espaliers, &c., 3,700,000; river navigation, carriage buildings, furniture, utensils, &c., 4,300,000. Total, 10,348,000 cubic metres.

"2. Fire-wood, 30,000,000 of steres,[1] and charcoal, 15,000,000 of steres. The consumption thus exceeds the production by the following quantities: In timber and wood for manufacture, about 8,000,000 cubic metres; in fire-wood, 15,000,000 steres. The balance is principally drawn from Norway, Russia, Germany, and Italy. The importation of common woods of all kinds, which in 1855 did not amount in value to 70,000,000 francs, was 154,000,000 in 1865. During the same period of ten years, the importation rose from less than 9,000,000 to 31,000,000. The necessarily restricted limits of this introduction renders it impossible to mention all the manufactures in which wood is employed. The number is very large and the entire catalogue of the exhibition furnishes the most complete inventory of the usages to which wood is applied. We shall pass in review, successively, the articles directly connected with forest products, and which are specially represented in class 41.

CORK.

"Cork is the substance lying beneath the true bark of a particular kind of oak, called the cork-oak, and which grows principally in Italy, Corsica, Algeria, Spain, and the south of France. The tree begins to furnish cork at the age of from twelve to fifteen years; but the first cork is of poor quality and only fit to make floats and other coarse objects, and Spanish black, which is nothing more than cork burned in closed vessels. After the first layer has been removed the cork bark is deposited with more regularity, and then yields materials fit for the finer purposes, such as the making of wine and other corks, sheets, and other well known objects used for many purposes. From the period already mentioned, the cork may be removed from the tree regularly once in 8 or 10 years, and the same tree will yield cork 12 or 15 times. Raw cork, or that which has

---

[1] A stere consists of 35.3174 feet.

merely been rasped, comes principally from Italy, Spain, Portugal, and Algeria. Spain supplies nearly the whole of the manufactured cork of commerce. Seville is the most important entrepôt of this product. Its principal application is in the making of bottle corks; but floats and a thousand small articles, in which lightness is a necessity, are also made of the same materials. The importations into France were, in 1855, 532½ tons, of the total value of about 257,224 francs. In 1865, they had risen to 3,855½ tons, of the value of 2,502,696 francs. The export amounted to 169½ tons in 1855, and in 1865 had risen to 1,319½ tons, of the value of 1,236,900 francs.

CHARCOAL.

"Charcoal is the result of the slow and imperfect combustion of wood. It is manufactured in two different methods. The first and most general is that which is practiced in the forest itself in mounds or stacks containing from 40 to 50 steres. This process yields hard, sonorous charcoal, which lights with difficulty, but which, once in a state of ignition, gives great heat and burns for a considerable time. The second process consists in distilling the wood in closed retorts; but the charcoal thus obtained has not the qualities of the former. It is friable, very light, very porous, and highly inflammable. The quality of the charcoal (valued according to its density) corresponds pretty closely with that of the wood from which it is made. The method of burning, whether fast or slow, the age of the wood, the nature of the soil in which the tree grew, all affect the quality and weight of the charcoal. Charcoal-making is the object of an extensive industry in many European countries. In France it employs a great number of workmen, who sometimes pass whole years in the forest. Still our production is not equal to the demand, and from 150,000 to 200,000 cubic metres of charcoal are imported annually from Belgium, Germany, and Italy. In 1856 the imports were 204 tons, of the total value of 3,670,128 francs. In 1865 they had fallen to 151 tons and 2,876,000 francs. In the same period of 10 years the exports grew from 1,209 tons, of the value of 108,800 francs, to 6,698 tons and 602,800 francs.

TANNING BARKS.

"This expression is applied generally to the bark of indigenous trees used in the tanning of hides and skins. Such bark is furnished by the oak, beech, chestnut, willow, white birch, and fir trees. The departments of Ardennes, Moselle, Meuse, Meurthe, Bas-Rhin, Nièvre, Yonne, Saone-et-Loire, Cote-d'Or, Ille-et-Vilaine, Deux Sèvres, Gironde, Vaucluse, Hérault, Bouches-du-Rhône, Var, and Corsica, supply nearly all the bark for the tanner's purposes. The last named departments, especially, supply oak bark, which is almost entirely consumed in the neighborhood. Algeria now furnishes considerable quantities of tanning bark, which is exported, and of which France alone consumes annually about 2,500 tons. In 1865 the importation of tanning bark into France amounted to 7,678 tons, of the total value of 930,000 francs.   In the same years the exportation was 15,900 tons, valued at 1,900,000 francs.   In 1855 the import was only 2,216 tons, and the export 558 tons.

RESINS.

"The maritime pine tree is the only tree in France from which resin is extracted.   The cultivation of this tree constitutes the principal, if not the only, wealth of the district lying between Bordeaux and Bayonne. According to the nature of the soil, the pine is tapped for resin between the ages of 20 and 40 years. The operation consists in making long incisions in the trunk, whence the resin exudes and is collected in various ways. The natural results of bleeding the pine trees, are : the soft gum or resin, which by distillation yields turpentine ; the galipots, an almost solid substance, which, by means of evaporation, forms in stalactites all down the tree ; the crottas, a mixture of the two former products ; the barras, which are the galipots entirely dry and adhering to the tree.   A pine tree 60 to 70 years old furnishes, on the average, about six or eight kilograms of raw material, of which about one-third is galipots and barras.   The American war gave a great impulse to the resin trade.   The following statistics will give an idea of the results in 1855.   The exports of French resins did not amount to more than 4,133 tons, of the total value of 2,250,000 francs.   In 1865 they had risen to the enormous total of 5,250 tons, worth 27,200,000 francs.   The importations amounted in 1865 to 2,960 tons, of the value of 2,400,000 francs.

BASKET-MAKING.

"Coarse basket-work, which alone is included in class 41, includes bakers' baskets, hampers, hottes or creels, &c.   The osier is the chief material used in this trade, which is principally exercised in the valleys, of Ver, Aubeaton and Hirson, in the Aisne, where osiers grow in large quantities. In the arrondissement of Vervins alone there are 3,000 families engaged in basket-making, who produce more than 2,500,000 francs' worth per annum, and of which two-thirds are exported to England and America. The importations amounted in 1855 to 105 tons, of a total value of 321,000 francs, and in 1865 only to 59 tons, of the value of 53,000 francs.   The importation of osiers in bundles, which in 1855 was 105 tons, had risen in 1865 to 180 tons, of the value of 22,000 francs.   The exports grew in the same period from 59 tons to 1,700 tons, the value of the last-named total being estimated at 370,000 francs.

COOPERING.

"Class 41 includes the works of the cooper, but the dimensions of the articles exhibited precluded their admission within the building, and they were placed beneath a shed in the Park.   This annèxe contained a vat and various specimens of cooperage from different localities.   The wood employed in this manufacture is called merrain; that is to say, oak or other wood split, according to the natural grain of the tree, into planks of various sizes, by means of a special tool, called a coulter. Merrains are produced of all dimensions, from 8 to 117 inches in length, from 3 to 10 inches wide, and from one-eighth to three-eighths of an inch in thickness.

"The chief places of production are Germany, Russia, Turkey, and the United States. The exports of the last-named country supply the greater part of the European cooperage. The best woods for making merrains are oak and chestnut. In Languedoc they also employ the white mulberry.

"The importation of merrains amounted in 1855 to 15,600,000 pieces, estimated at the value of 10,900,000 francs. In 1865 it had grown to 37,000,000 pieces, and 26,300,000 francs. Nearly the whole of the split wood imported into France is consumed in the country. The total exports in 1865 only amounted to 630,000 pieces, of the value of 390,000 francs."

CLASS 42.—PRODUCTS OF THE CHASE AND FISHERIES—UNCULTIVATED PRODUCTS.

(REPORT OF THE COMMITTEE OF ADMISSION FOR FRANCE.)

"This class includes a large number of natural products, having undergone but slight preparation. It also comprehends skins and furs, which represent a very complicated trade, and demands, particularly for the made-up furs, much special knowledge, a great deal of taste, and, to a certain extent, that creative power peculiar to the Parisian manufacturers in matters of dress and furniture.

"The natural products belonging to class 42, and which demand but slight preparation, are elephant and hippopotamus teeth, sponged tortoise-shell, mother-of-pearl, horse hair, the various kinds of hair employed in hat-making, fish oils, and collections of dried plants. We must also mention the collections of stuffed animals, for the study of natural history, the preparation of which belongs both to the domains of art and science. As to the trade in and preparation of skins, and the making up of furs, we think it necessary to make it the object of a special notice; because this kind of a product represents not only a considerable trade, but also a very difficult branch of industry on account of the dyeing and making up of the skins.

FURS.

"Paris first, and Lyons next, are the principal centres of commerce for skins and made-up furs. These articles are employed both for dress and domestic use. Of furs are made muffs, cloaks, tippets, coat collars, boas, cuffs, pelisses, carpets, cushions, trimmings for dresses, foot muffs for carriages, &c. The Parisian trade employs the most beautiful, as well as the most ordinary skins—from the rarest kinds of sable down to

glossy rabbit skins, of lowest price. Articles of fur are also made with
swan, grebe, and goose skins. The trade of clipping hair for hat-making also brings into use rabbit and hair skins. It is in Paris that articles in fur for dress and furnishing are most exquisitely made up. The
various operations that skins are submitted to are—firstly, dressing,
glossing, and dyeing. These preparations are done by special workmen,
who work by the piece. When the skins arrive at the shops they have
undergone no preparation whatever. They are dyed with the brush or
in the vat when the skin is to be dyed also. The number of women
employed in the trade is about equal to that of the men. The workmen
are divided into dressers, glossers, fullers, cleaners, and cutters; the
women are seamstresses and mounters. The salaries of the men vary
from 6 to 7 francs; those of the women from 3 francs to 3 francs 50 centimes daily.

"The fur trade in France includes three classes of dealers : 1. The col-
orers of skins; 2. Wholesale skin and fur merchants; 3. Furriers or
makers-up of furs.

"The price of furs varies to infinity—from the rabbit skin, worth about
50 centimes, to the Siberian sable skin, the price of which rises to 500
francs. The greater part of the best furs are sold in Paris; the rest are
bought in the departments and abroad.

"Paris possesses about 30 hair-clipping establishments, which produce
annually material amounting in value to 20,000,000 francs. The horse-
hair trade is also very important, and gives rise to a considerable movement.

"A large amount of business is also done in sponges. The dealers who
are engaged in this branch of trade are at once fishers, importers, and
cleaners of sponges. The preparations to which sponges are submitted
increase their value from six to eight per cent.

"It is difficult to state precisely the value of the furs made up in
France; but it is very considerable. It is believed to attain, in Paris
alone, the sum of 20,000,000 francs, including the sale of glossy rabbit
skins. About 5,000,000 worth of the total is delivered for exportation.

"Since 1855 the fur trade has developed to a considerable extent. This
flourishing condition is due to the fur dealers of Paris, who are incessantly creating new patterns in all kinds of new made-up articles in fur,
and who have thus maintained that supremacy in taste and design which
have so long been accorded to France in all matters pertaining to articles of dress and fashionable requisites.

"There were about 750 exhibitors in this class, displaying naturally a
vast and heterogeneous mass of objects. Although many of these articles were of the highest interest in a philosophical, geographical, and
social point of view, they were not of a nature to require much description. Furs of all sorts were, from their value and beauty, the principal
attraction of the class. There was a fine collection from the French
colonial possessions in New Caledonia, Guadaloupe, Caboon, &c. The

raw industrial products of these little-known regions were shown with great taste and skill. But of the finest sorts of furs there were hardly any important specimens. The material is so easily damaged by dust that manufacturers hesitate to expose their better classes of goods. The French exhibitors, who had less to risk in the way of transportation than any other nation, made the best display. The taste of their work and the labor bestowed upon it left nothing to be desired. A grebe mantle, of extraordinary workmanship, was exhibited by Mr. Delmar, of Paris. Each single feather was sewed in separately on a basis of silk, the darker feathers being formed into patterns of wreaths round the skirt. One of the most interesting novelties, or rather revivals of a method which was in use many years ago, was the *galonnement* of furs. Some of the finest and most expensive furs, especially those from northern regions, such as the sable and the silver fox, are almost too close and heavy in the original state of the skin. To obviate this, the pelt is cut up into strips about a third of an inch in width, and between two of these is inserted a strip of equal width of silk. The strips are carefully matched in color and united with the greatest dexterity, so that the fur completely closes over the seam. A fabric is thus produced more open, light, and better toned than the original skin, while the latter is economized by the substitution of silk for a portion of the fabric. It is claimed that the articles thus made are more wholesome and much cheaper.

"Russia had a fine assortment of furs; but, for the reasons already given, it was, except perhaps in individual specimens, inferior to that exhibited in the French court.

"Messrs. Gunther, of New York, exhibited some fine specimens of North American furs and a collection of fur-bearing animals, very well prepared.

"The fleeces of rabbits and hares, used by hatters, made a large display. They are shorn and prepared with great skill. Up to the beginning of the present century hatters prepared the materials of their manufactures, from the crude skins, on their own premises. About the year 1826, owing to the steadily-increasing demand, the process of cutting and preparing the fleeces for making the felt was separated into a distinct trade. The material is used principally in the manufacture of soft hats.

"The increasing scarcity of whalebone has led to many interesting experiments in the way of providing a substitute. Buffalo's horn seems to answer the purpose satisfactorily. From the close similarity in the structure of the two substances there seems to be no reasons why this substitute should not answer most of the purposes to which whalebone is now applied. The horns, after undergoing a special process, are cut into strips, which are compressed and straightened and rendered suitable to every purpose of the dress-makers' art. There is another imitation of whalebone in compressed cane, but it does not seem to answer the purpose so successfully."

CLASS 43.—AGRICULTURAL PRODUCTS, NOT USED AS FOOD, EASILY PRESERVED.

Many, indeed most of the articles referred to in the following and suc-ceeding classes of this group, have been or will be referred to under other heads. We continue to quote from the catalogue simply because the particulars are interesting and late:

OLIVE OIL.

"The most important article in class 43 was the oil derived from all sources. Oil, in some way or other, plays a most important part in the domestic economy of Europe. It is not only the source of light, but, to a great extent, of life itself. A large portion of southern Europe would perish were the olive crop to fail.

"Oils are obtained from an immense variety of nuts, grains, fishes, and minerals. All these are more or less edible, but real olive oil is the one which most readily agrees with sensitive stomachs, and which, for cen-turies, has partly taken the place of meat and butter with large and intelligent populations.

"The range within which the olive grows corresponds with the zone within which maize and rice can be cultivated. It is much more con-fined than that of the cereals. Its northern limit may be roughly placed in the most southern provinces of France, and it does not extend far into the interior of Africa. Spain, Italy, the islands of the Mediterranean sea, and the Greek archipelago, are the most civilized places where it flourishes. Asia Minor and Syria, and the whole northern coast of Africa from Morocco to the borders of Egypt, are covered with it.

"The olive requires but little attention, and is content with a poor, stony soil. This hardiness is rewarded by long life. Olive trees live far beyond the memory of man, and some indeed pass the ordinary limits of tradi-tion. At Pescio, in Italy, there is a tree which can be proved, histori-cally, to be more than 700 years old; and the trees on the Mount of Olives, if not those that witnessed the Passion, are at least the sprouts from their roots. Olive trees grow hollow as they grow old, the trunk splitting into fantastically-shaped masses, which unite higher up. To support them it is often found necessary to fill up the interstices and build up the trunk with stones.

"The best olive oil is that of Lucca and Tuscany. The province oil, known as the oil of Aix of commerce, is the most esteemed for the table. It owes all its merits to the admirable manipulation of the manufactur-ers, for the olives which yield it are the poorest of any country.

SEEDS, FLAX, HEMP, AND WOOL.

"Class 43 comprehends an immense variety of products, for, with the exception of cereals, fruits, cattle, and forest produce, it represents the whole of the productions of the soil. All these can, nevertheless, be arranged in 11 principal series:

"1. Seeds, which include the collections of the various seeds employed in agriculture and horticulture. 2. Textile materials of vegetable origin, and principally flax and hemp. 3. Wool in the fleece, washed and unwashed. 4. Cocoons of the various kinds, of silk-worms and raw silk. 5. Tobacco in the leaf or prepared, for the various uses of consumption. 6. Hops. 7. Plants for forage. 8. Oils of all kinds. 9. Honey and wax. 10. The various agricultural products employed in trade, such as the dyestuffs and the different plants which supply materials applicable to the arts and manufactures. 11. The mass of products which exhibit the progress and condition of the rural and agricultural industry of a district.

"The trade of seeds for sowing is becoming more and more important in France. On one hand, the agriculturists pay greater attention to procuring varieties remarkable for certain qualities; and on the other, the taste for horticulture is increasing every day. Some houses have acquired a European reputation for the care given to the selection of seeds, and many establishments have cultivated largely, exclusively with a view to produce seeds of first-rate quality.

"Flax and hemp are the two plants most cultivated in France for the manufacture of stuffs. The cultivation of flax especially, favored by the dearness of cotton, has increased, during the last few years, to a large extent. The principal centres of production are Flanders, Picardy, Normandy, and Brittany. Hemp continues to be cultivated not only wherever flax is grown, but also in several other provinces of the centre of France, and particularly in Touraine. The process of retting in running water is almost abandoned for retting on the spot. Inventors continue to occupy themselves in discovering new processes of retting, and trials on a large scale have shown that the problem is, to say the least, about to be solved. Attempts have also been made to cultivate cotton. Various interesting experiments have given rather remarkable results.

"Wool, in spite of foreign competition, which, during the last 10 years, has lowered the average prices, continues to be one of the great products of agriculture. Strenuous endeavors have been made to increase the weight of the fleeces; and to accomplish this object without deteriorating the quality of the meat, or the abundance of the wool, and at the same time to maintain an average strength and length of the staple. The improved merino race is in the highest repute in France. Chatillonais, Brie, Beauce, and Soissonais have even supplied breeders for all parts of the world. The Rambouillet type is in demand everywhere. The wools of Naz and Mauchamp, also, still occupy the attention of the breeders. The exhibition of French wool deserves the notice of visitors, by reason of the numberless efforts which have been made to improve this branch of industry.

11 U E

"The silk producers have suffered terribly in France during the last 12 years.   In the principal departments, where the culture of the mulberry was a source of considerable wealth, general desolation reigns; however, many efforts, some of which have been successful, have been made to produce eggs which will yield worms capable of resisting the disease. The small establishments, and particularly those situated in districts where the culture of the mulberry tree is not very extensive, have fortunately not suffered from the ravages of the epidemic.   Lastly, efforts to secure the acclimation of other silk-worms than those indigenous to the ordinary mulberry districts have been, in most cases, successful."

TOBACCO AND HOPS.

"The cultivation and manufacture of tobacco has followed, step by step, the constantly increasing consumption in France.   The cultivation is now pursued in 18 departments, as well as in Algeria.   The directors of the state manufactories exhibit specimens of the material grown in these localities; and they show also many varieties of tobacco in the various stages of vegetation.   Out of 36,000 tons used in the state manufactories, 23,000 to 24,000 tons are of indigenous growth, and the rest is imported from abroad.   The 17 tobacco manufactories actually in work employ about 17,000 work-people, of whom from 14,000 to 15,000 are women. Few manufacturing industries supply women relatively with so much work.   There are, besides, upwards of 2,000 persons, of whom more than half are women, employed in the tobacco-growing establishments.   There are also manufactured in France snuffs of every description; tobacco for chewing; smoking tobacco, called *tiscaferlaté;* tobacco in rolls; cigars made of Havana tobacco; cigars sold at 10 centimes and called *etrangers;* five-centime cigars, called *ordinaires;* and cigarettes of all kinds.   These establishments also sell cigars imported from Havana, Manilla cigars, cigarettes, and other articles in great demand by the trade.   The refuse tobacco, which is applicable to agriculture, consists of the waste and ashes.   The directors of the state manufactories also exhibit all the various products which can be extracted from tobacco.

"The cultivation of hops increases in France, particularly in the north and in Alsace; it had successively spread into several other regions, and has acquired a certain importance in Burgundy.   The qualities of the French hop begin to be appreciated in the most important centres of consumption.

"The evident necessity for a continual increase in the supply of animal food has led to a large extension in the cultivation of plants for fodder. Several attempts have been made to introduce new plants, or at any rate better varieties of plants under cultivation.   Instead of leaving the meadows to themselves, they are now cultivated, dressed with manure, and sown with selected seeds.   Important improvements have also been introduced in the gathering and in the mode of preserving fodder.

"The cultivation of oleaginous plants has also considerably increased during the last 12 years, the farmers appreciating more and more the importance of cultivating some industrial crops side by side with cereal and garden crops. The progress of industry, as well as the general advance of civilization, tends moreover to augment the demand for oils of various qualities. A special exhibition of oils offers a particular interest; analogous therewith will be found colza oils, linseed oils, red poppy oils, nut oils, &c., from the superfine oils used in horology to the coarsest oils employed for the lubrication of machines and the manufacture of common soap."

AGRICULTURE IN FRANCE.

"The taste for agriculture is gradually diffusing itself throughout the country. Great care is bestowed on beehives, so as to multiply the swarms, and insure a much larger quantity and better quality of wax than formerly. France produces plants of the most various perfumes, and others which give most brilliant and durable colors. Besides the aromatic and dyeing plants, are to be found those used for pharmaceutical and tanning purposes. The researches for plants capable of yielding fibres suitable for paper-making occupy the attention of scientific men; and from this point of view these various agricultural products deserve attention. Rural cultivation produces almost every description of crop, so that it is impossible to properly appreciate a system of cultivation by a single product; the whole must be examined. For this reason many eminent agriculturists have exhibited collections of the plants which their lands yield, as well as some products of their cattle-sheds and poultry-yards. Moreover, in many cases, rural trades are so essentially and peculiarly associated with the culture of the land, that it is absolutely necessary to impart a knowledge of the special method of cultivation in its various phases; and it is only in collecting all the products of a district that its riches, its fertility, and the results arising from the labor expended, can be fully appreciated. With this object several agricultural societies, committees, and other associations, were desirous of sending collective exhibitions. In studying these exhibitions it will be seen what differences are presented by the various localities of the three great agricultural circles. During the last 12 years evident progress has been made in every district of France. The agricultural produce of France has certainly increased, on an average, fully one-tenth, in spite of the difficulties which have resulted from the advance in wages, and from capitalists quitting rural enterprises and devoting themselves to industrial and commercial speculations. Agriculture demands, moreover, a supply of manure equivalent to the amount of its cultivated produce; in proportion as its products increase it needs larger quantities of fertilizing materials. The multiplication of rapid and cheap means of transport has at last favored these objects, and gives more activity to rural occupations."

CLASS 44.—CHEMICAL AND PHARMACEUTICAL PRODUCTS.

"Under the general appellation of chemical products, class 44 comprises almost every mineral and vegetable matter which chemistry has been instrumental in transforming and adapting to the use of the various branches of industry. Generally these products are manufactured entirely in the laboratory; but sometimes they are simply extracted from natural substances, in which they exist ready formed. Chemical products furnish to a great number of other industries the material necessary for their existence and working; consequently a new discovery or a remarkable improvement is a fact of importance of which the tributary trades should take special note.

"Chemical works are distributed in various departments of France, according to the convenience or locality of the proximity of the materials for manufacture. They form an important branch of commerce in Paris, Lille, Marseilles, Lyons, and Rouen; but St. Gobain, Bouxvillier, Dieuze, Thann, the island of Carmargue, and the coast of Brittany, possess establishments which are not surpassed in importance by those of any of the great towns. The aggregate trade in these productions represents an annual value of 1,200,000,000 francs, (£48,000,000.) The manufacturers of sulphuric acid, soda, soap, and stearine candles alone give circulation to 600,000,000 francs; and if to these be added dyestuffs, products applied to the bleaching of tissues, paper-making, painting, glass manufacture, calico and other printing, to manuring purposes, electro-metallurgy, photography, the gilding and silvering of metals, &c., the estimate of 1,200,000,000 francs will be a moderate one. The exports amounted, in the year 1863, to 53,000,000 francs.

"The workmen employed in these manufactures work under the direction of foremen or superintendents. In order to carry on such operations with economy, large premises are required, as well as special buildings and costly apparatus, and consequently an expenditure of capital which excludes small undertakings. Nevertheless, some workmen, having special manipulative powers, have succeeded in manufacturing certain products more advantageously than the large factories; this, however, is the exception to the rule. The improvements which have taken place in chemical manufactures since 1862 are: The invention of new coloring matters, obtained from toluidine and methylic aniline, and, consequently, an improvement in the quality and the reduction in the price of colors then exhibited in London; the conversion of naphthaline into benzoic acid, a substance derived from the vegetable kingdom; and the commercial production of magnesium, now so usefully employed as a means of illumination in photography. But the fact which merits the greatest attention is the large increase in the production of chemical matters—a certain sign of progression in all other branches of industry. If we take into consideration the extent and value of the service of chemical science, it would appear advisable, in order to maintain France in the industrial rank which she occupies, to give greater development to practical instruction in this science, either by increasing the resources of the laboratory established by the initiative of his excellency Mr. Duruy, and conducted at the Museum of Natural Sciences ('Jardin des Plantes') by Mr. Fremy, member of the Institute of France, or by the creation of similar establishments in all the great industrial centres of France. In Prussia the fact that theoretical teaching is insufficient to make good chemists has been so thoroughly recognized that immense laboratories have been established at Bonn and Berlin, where pupils are instructed in those practical experiments without which theory remains fruitless. The interest which attaches to pharmaceutical products is somewhat dwarfed by that which is created by commercial chemical products. It is true that in pharmacy progress is slow, especially under a system of excessive restrictions, which practically sets aside individual action. When the apothecary has once given a guarantee of his practical knowledge by submitting to the examinations for his diploma, he has a full claim to liberty of action in the commercial exercise of his profession. Such a new state of things would certainly give a great impulse to the importation of French medical preparations, which are highly esteemed in the commercial world."

[Signed by Menier and Forçade, members of the admission committee.]

CLASS 45.—SPECIMENS OF THE CHEMICAL PROCESSES FOR BLEACHING, DYEING, PRINTING, AND DRESSING.

I. The products exhibited in this class and in the five classes belonging to Group IV—class 27, cottons; 28, flax and hemp; 29, woollens; 30, cloths; and 31, silks—are: 1. Wool in the fleece, washed and dyed, for the manufacture of cloths; 2. Combed and carded woollen yarn, bleached and dyed, for the manufacture of shawls and garments, and furniture stuffs; 3. Cotton, linen, hempen, and other yarns, bleached, dyed, and dressed; 4. Silk yarn, bleached and dyed; 5. Cotton, linen, and hempen tissues, plain and figured, bleached for printing, or bleached and dressed; 6. The same tissues, dyed and dressed; 7. Mixed and unmixed woollen tissues, dyed and dressed; 8. Clothes dyed in the piece; 9. Cotton, linen, woollen, and silk tissues, plain and figured, mixed and unmixed, printed and dressed, dresses for the general trade, superior fancy tissues, furniture stuffs, printed shawls and carpets, and tissues printed on the weft; 10. Cotton, hempen, and linen furniture stuffs, glazed, gummed, and waxed, plain and printed; 11. Cotton cloths, waxed and grained, in imitation of Morocco leather; 12. Textile fibres of various kinds, reduced to pulp, bleached and dyed, of all colors, for the manufacture of paper-hangings.

II. The principal centres of production are Paris, Lyons, Rheims, Rouen, Mulhouse, Amiens, St. Quentin, Roubaix, Cambray, Elbœuf, St. Etienne, Sedan, Lisieux, Mazamet, Lodève, Laval, Bischwiller, Ste. Marie-aux-Mines, &c.

III. The raw materials the most in use are the following: Chemical products, starchy materials, neutral animal matters, essences, oils, greases, insoluble mineral colors, dyeing stuffs, (indigo, cochineal, madder, orchilla, dyewoods, extracts, lakes, &c.,) and artificial colors produced from coal tar, (red, violet, blue, green, yellow, brown, black.)

Among the improvements introduced since 1855 may be mentioned:

1. Improved methods of engraving the designs.

2. The application on an extensive scale of the pantograph and electric pile to the engraving of the rollers.

3. The considerable economy resulting from the application of the galvano-plastic process, or the covering with copper of cast iron, steel and bronze rollers, which now replace the solid copper rollers previously employed.

4. The restoration of old engraved rollers.

5. The improved method of, and economy in bleaching; the improvements of the dressing and washing machines, (economy of power and of water;) the more advantageous use of the power employed in working the printing machines.

6. The improvements made in the manufacture of Turkey red; the application of pyrogallic acid in the production of black for grounds; the discovery of new colors, principally those derived from coal tar, and their application on a vast scale to dyeing and printing processes; the great extension given by these discoveries to the manufacture of chemical products; the novel process, by means of animalizing vegetable fibres, to render them more susceptible of taking colored matter; the use of zinc in dyeing with aniline colors. The most important fact is the decrease of the net cost, coincident with the improvement of the products and the increase of the wages of the operatives.

CLASS 46.—LEATHER AND SKINS.

"The products exhibited in this class comprise eight divisions:

"1. Tanned leathers, including strong sole leather and leather intended to be curried. 2. Curried leathers for boots, shoes, saddlery, and machinery. 3. Black and colored varnished leathers, for boots, shoes, and saddlery. 4. Goatskin and imitation morocco, for boots, shoes, bookbinding, furniture, and small articles. 5. Tawed leathers for boots, shoes, and gloves. 6. Chamois leather. 7. Tanned Hungary leather. 8. Parchment.

"Paris is the most important centre of the trade for all kinds of leather. Givet, Chateau-Renault, and Strasburg especially manufacture strong leather; Nantes and Millhau, leather for vamps; Grenoble and Annonay, tawed leather for glove-making; Niort, chamois leather. The principal seat of the morocco manufacture is in Paris, which also furnishes strong leathers. The leathers called '*à la Garouille*' are produced by a special method of tanning, and come from the southeast of France. The leather manufactured in the French tanneries has two distinct sources of supply: the slaughtering of cattle at home, and the importation of raw hides from England, Ireland, Holland, Germany, La Plata, Peru, Brazil, Mexico, the West Indian islands, Madagascar, India, and Australia. The price varies with the locality from whence it is derived, the nature, the quality, and the state of the merchandise on its arrival. The leathers exported into France come in the salted, dry, and dry-salted forms. The imported hides, as well as those of home produce, include the skins of various kinds of animals. Each year the import of raw hides amounts to about 220,000,000 francs. The tanning materials are generally of French derivation. The export of tanning bark increases every year. Algeria supplies a great quantity of kermes oak-root bark, necessary for the tanning of leather '*à la Garouille*.'

"Mechanical aid is being introduced daily in the leather factories; still manual labor continues at the present moment the base of the industry. Machinery has been brought into use in a great many large tanneries, but it is applied especially to the preliminary processes of tanning and currying. The greater part of these operations have not been well performed by mechanical means, and recourse has, consequently, hitherto been had to manual labor in spite of the promise held out by some machines, and, among others, by those intended for the fleshing of hides, whether calf or morocco. As to the processes for rapid tanning they have not yet yielded satisfactory results. The workmen are generally employed in the tanneries. There exist, however, a few small manufacturers, of limited means, who employ sometimes one or two workmen. They either undertake one special department of manufacture, such as currying or morocco work, or only certain operations, such as flushing. There are, especially in Paris, many large establishments which work for the retail skinners and curriers.

"The leather trade has a central market, France in each centre; still, Paris, where so many beasts of the first quality are slaughtered, is at once the greatest market both for fresh raw hides and for manufactured leathers. Havre, Marseilles, Nantes, and Bordeaux are the great seats of the import trade in raw hides. The provincial manufacturers, who do not dispose of their products in the district, send them to Paris, either to the leather market or to the houses of commission merchants. A certain number of manufacturers have depots at Paris, but sale by commission is the mode most generally adopted by second-rate manufacturers.

"The consumption of meat in France is constantly on the increase, and the supply of the raw material of the leather trade augments in like manner. Between 1850 and 1863 the imports have increased one-half. The total importation of raw hides salted, dry, or salted and dried, was, in 1865, 48,646 tons, and the total value nearly 100,000,000 francs. The exportation amounted, at the same time, to 6,685 tons in tanned, curried, morocco and varnished leather. To this must be added dressed skins, which figure on the returns of the Douane for 3,168 tons, of the total value of 81,223,902 francs, thus increasing the gross total to 147,198,106 francs.

The committee regrets that it cannot point out any great improvements in the trade during the last 12 years; the leather manufacturers of France generally being very chary of innovations.

"As regards tanning, we have already said that several improvements have been introduced to accelerate the process, though the results have not been satisfactory. A new tanning substance has also been introduced in place of oak bark, namely, the wood of the chestnut tree reduced to shavings; and still further efforts have been made to do away with the use of tannin entirely, and prepare hides by means of turpentine alone, but subsequently a certain portion of tannin extract has been used with the turpentine in order to produce a better result. In the currying trade there is scarcely any actually new invention to be found, unless we consider as such a system of working which the *Société d'Incouragement* judged worthy of reward. This method has for its object the rendering the leather more supple by the improvements in the details of the manufacture, and to render it impermeable by means of a thick coating of gutta-percha. The true progress made in the trade is the increased skill brought to bear on the various operations. As regards the morocco trade, we must not omit to mention the attempts made, with the view to a more general and ordinary application of the magnificent colors produced from aniline and its derivatives, now so numerous. These are new resources for the morocco worker, who is thus enabled to assimilate more nearly the color of the skin and that of the stuff intended to be incorporated. In all the different branches of the leather trade the committee has pointed out the necessity for accelerating and improving the manufacture by the constant introduction of improved plans and utensils and improvements in the workshops in which the various operations are performed."

GROUP VI.

APPARATUS AND PROCESSES USED IN THE COMMON ARTS.

CLASS 47. APPARATUS AND PROCESSES OF MINING AND METALLURGY.—CLASS 48. IMPLEMENTS AND PROCESSES USED IN THE CULTIVATION OF FIELDS AND FORESTS.—CLASS 49. IMPLEMENTS USED IN THE CHASE, FISHERIES, AND GATHERING WILD PRODUCTS.—CLASS 50. APPARATUS AND PROCESSES USED IN AGRICULTURAL WORKS AND FOR THE PREPARATION OF FOOD.—CLASS 51. APPARATUS USED IN CHEMISTRY, PHARMACY, AND TANNING.—CLASS 52. PRIME-MOVERS, BOILERS, AND ENGINES SPECIALLY ADAPTED TO THE REQUIREMENTS OF THE EXHIBITION.—CLASS 53. MACHINES AND APPARATUS IN GENERAL.—CLASS 54. MACHINES, TOOLS.—CLASS 55. APPARATUS AND PROCESSES USED IN SPINNING AND ROPE-MAKING.—CLASS 56. APPARATUS AND PROCESSES USED IN WEAVING.— CLASS 57. APPARATUS AND PROCESSES FOR SEWING AND FOR MAKING-UP CLOTHING.—CLASS 58. APPARATUS AND PROCESSES USED IN THE MANUFACTURE OF FURNITURE AND OTHER OBJECTS OF DWELLINGS.—CLASS 59. APPARATUS AND PROCESSES USED IN PAPER-MAKING, DYEING, AND PRINTING.—CLASS 60. MACHINES, INSTRUMENTS, AND PROCESSES USED IN VARIOUS WORKS.—CLASS 61. CARRIAGES AND WHEELWRIGHTS' WORKS.—CLASS 62. HARNESS AND SADDLERY.—CLASS 63. RAILWAY APPARATUS.—CLASS 64. TELEGRAPHIC APPARATUS AND PROCESSES.—CLASS 65. CIVIL ENGINEERING, PUBLIC WORKS AND ARCHITECTURE.—CLASS 66. NAVIGATION AND LIFE-BOATS, YACHTS AND PLEASURE-BOATS.

The twenty classes of this group embraced nearly 100 sections, any one of which was sufficient to furnish a report of the length of the present. Indeed, on some subjects the best informed talent of the world has employed itself for many years. The result in literature is a library; in practice, a million new processes whereby the increasing wants of the age are supplied. The intent of this report being general and not special, a few points of public interest only will be dwelt upon. Following the remarks thus offered will be found the usual extracts from the French official catalogue, containing the latest local data on the special sections.

CLASS 47.—APPARATUS AND PROCESS OF MINING AND METALLURGY.

COAL-MINING IN FRANCE.

Among the plans and models exhibited in class 47 was one of great interest to the French people. It represented, in a map, the newly discovered coal mines of Pas-de-Calais. Fuel of all kinds is expensive in France. The country has been denuded of its trees, and coal, until recently, had to be imported from neighboring countries, and was, in consequence, a luxury which the poor could not command. It may safely be said that the scarcity of fuel has in a great measure affected the domestic habits of ordinary life.   It has compelled the masses to seek warmth and life in the various *cafés*, where these cheering influences are always conspicuously displayed.   Any prospective cheapening of the article of fuel is therefore an object of particular concern to the French, and has occupied the best attention of the government.

Another map of the Pas-de-Calais, by Mr. Coince, was on the scale of 1 to 10,000, and gave a fair idea of the prodigious perseverance and energy which have been bestowed on the opening of pits which were at one time supposed to be chimerical.   It was only in 1846 that a boring for water at Oignies, not far from Douay, gave rise to the theory that there was a deflection of coal in that direction.   Between 1850 and 1864 concessions of land were made to various companies, mostly in the Pas-de-Calais, and extending in that department over a length of 35 miles.   Some 40 pits have already been sunk, averaging in depth from 100 to 350 yards. The amount of coal produced from this hidden and accidentally-discovered source—for it had no geographical indication—has risen from 5,000 tons in 1851 to upwards of 1,600,000 tons in 1866.

There were maps of other coal mines in France, exhibited by the French ministry of public works.   The best were those of the Loire.   These pits, less than 20 years ago, were on the point of being abandoned, the obstacles in the way of their being worked seeming to be almost insurmountable. The government instituted an inquiry into the subject, and detailed its best engineers to examine thoroughly the nature of the ground.   Their reports presented in a clear and practical light the difficulties that had to be encountered.   These were in due time conquered.   At the present time over 3,000,000 of tons of coal are obtained from the basin of the Loire.

A large model of the ravine of the Grande Combe, whence the south of France begins to obtain its supplies, was also exhibited, and demonstrated the great amount of exploration done during the past few years, and the very minute and accurate record which is kept of all the phenomena of mining.   The Grande Combe is the third district, in point of productiveness, in France.   It now averages 1,200,000 tons.

These maps and models, and others exhibited in the same department, demonstrate the fact that coal beds in France diverge from the pit with singular sinuosity—the workings in several places being far beneath the overlying strata of the Trias.   Such mines in wealthy coal districts would be almost disregarded, but skill, even more than necessity, has rendered them valuable and remunerative.   When this is not the case the very wealth of the seams presents unusual difficulties.   The coal-fields of central and southern France, although individually of small extent as compared with those of England and Belgium, are remarkable in this respect.   The tolerably regular beds of coal at Blanzy and Montcean run to 50 feet, and even to 60 feet, in thickness; at Creusot, where the bed stands in a vertical position, it varies from a few feet to 50 feet, 80 feet, and ascends to as much as 130 feet; and the great seam of Decazeville (Aveyron) often extends to 100 feet in thickness.   The vast vacuities which must necessarily be produced in working these mines lead necessarily to very serious engineering obstacles, which have only been surmounted by an extraordinary display of skill, and by the adoption of a plan which, while it involves labor, almost amounting to a double operation of mining, seems at all events to insure safety. This consists of packing all excavated places, except the passage ways, with rubbish carried down from the surface. A change of hands is required for this purpose, the colliers being absent from the mine. In some districts a particular shaft and line of roads and special wagons or tubs are set apart for the work, and in certain mines fully one-third of the hands employed are engaged in the business of filling up.

The maps and models from Belgium were also singularly exact and instructive. France imports, mainly from Belgium, 7,100,000 tons of coal. Her own production had reached 12,000,000 tons in 1865, and is undoubtedly greater.

PRESSED COAL.

The progress of manufactures requires a constantly increasing supply, and the scarcity of wood, as before remarked, renders fuel in any shape a luxury. The navy, too, requires inexorably its rations. To provide these economically has been the study of many practical men, and a result has been obtained which is worthy of record. The dust of coal is used. It is pressed into cakes by a variety of processes, nearly all of which seem to be in favor. For naval use this kind of fuel possesses advantages. It is asserted that in the carriage of the little bricks there is a loss of only one per cent., instead of six to ten per cent., as in lump coal; and when stored abroad they are found after two years' exposure to be scarcely at all injured, while ordinary coal would have suffered to the extent of 50 per cent. It is claimed for them also that they are free, or comparatively free, from ash, and can be made from the refuse of almost every kind of coal, and in such a ratio as to produce the best effect in getting up steam, and maintaining it. The bricks are exceedingly compact. They are produced by hydraulic pressure and require but a small percentage of extraneous, gummy, or resinous matter to make them stone-like and thoroughly durable. The best approved process, or rather the one which seems to give the best results, is that adopted by the company of La Chazotte. The machine used has 16 cylinders disposed as the radii of a circle, in which the coal slack, after being heated by a current of steam and mingled, by the means of very ingenious apparatus, with pitch, is pressed by pistons, and formed either into cylindrical or hexagonal blocks of convenient length. The prices of compressed fuel are as follows:

First quality, containing only 2.10 per cent. of ash, 28 francs per ton; second, containing 5 per cent., 26 francs; the other sorts range from 23½ to 9½ francs. A single manufacturer produces no fewer than 175,000 tons of this agglomerated coal per year. There are several others of

almost equal extent. The slack or waste of the coal mines is thus economized, and an article produced which, apart from the question of cheapness, possesses special considerations which seem to adapt it for general
use in stoves and furnaces of every kind.

BORING SHAFTS AND DRILLING ROCKS.

An interesting display of maps and models illustrating the process of
boring was made in the French department. Two important public
works are now in progress in the city of Paris, and the contractors were
the principal exhibitors. The French capital obtains its best and purest
water from artificial sources, namely, the artesian wells of Grenelle and
Passy. Two additional wells are now in process of being sunk; one in
the suburb of the Chapelle in the extreme north of Paris, by Messrs.
Degousée and Laurent, and the other at the Butte aux Cailles in the
extreme south of the city, by Messrs. Dru. These celebrated firms
exhibited the apparatus by which they make all kinds of borings, ranging
from four inches to five feet.

Pure water being an object of great concern to every community, it
may be well here to give a few particulars of the two artesian wells now
in successful operation. At Grenelle the surface is 121.3 feet above the
level of the sea; at Passy 305.2 feet; the depth of bore-hole at Grenelle is
1,800.7 feet; at Passy 1,923.7 feet; internal diameter of tube or lining of hole
at Grenelle approximately 9 inches to 6 inches at bottom; at Passy 2.4
feet. The full diameter of the Passy bore-hole was 1 metre, or 3.28 feet.
The new ones are to be in one case above five feet, and in the other about
four feet; whilst it is proposed to sink much deeper than heretofore, in
order to open new sources of supply and avoid drawing too extensively
on the old ones.

Examples were shown of the application of boring to the ordinary process of mining, such as the excavation of shafts. It is often difficult and
sometimes impossible, owing to the watery character of the soil, for workmen in the usual way to penetrate to the requisite depth and perfect the
casing of the pit. When this is done it is at a great cost of labor and money.
By the machinery used for ordinary boring it is done with comparative
ease, and the casing is always perfect, because it is a tubing which, if
necessary, can be filled in against the side of the pit with concrete and
other preparations more or less impermeable. The expense by this process is not more than one-quarter of what it would be under the usual
way. Sections of two pits at St. Avold, France, were exhibited. The
first of these was sunk through 426 feet of permeable red sandstone, and
coal was found at a depth of 1,036 feet on the 4th of April, 1867. The
second had progressed to the depth of 521 feet on February 3d. The
diameter of the cutters used in boring these pits was 13 feet.

Several machines for working under ground and superseding hand
labor in drilling rocks were displayed, many of them of ascertained value.

One of these, by General Haupt, was from the United States, and was characterized by simplicity and directness of action.

The diamond-pointed drill of Mr. Leschot, exhibited in the French section, whatever may be the original cost, is claimed to be the best and cheapest in the end. It works with great rapidity and is utterly indifferent to the stubbornness of the material against which it is placed. A drawing showed the way in which it is proposed to arrange several of these implements moved by steam power for boring tunnels. The boring tool is tubular and admits a jet of water through the middle into the hole; its face of soft iron is studded with eight pieces of black diamond carefully set in the iron, and the incomparable hardness of the adamant is so little affected by contact with the hardest granite, that the engineer stated the cost of the abrasion of diamond for a hole of half a yard deep to be less than four cents.

A machine for channelling and quarrying marble and other stone for building or ornamental purposes was exhibited by the Steam Stone Cutter Company of New York, and is in use at the marble quarries of Rutland, Vermont, and is the invention of Mr. Wardwell. It is asserted that this machine reduces the labor cost of the production of marble, and cuts it from the quarry with much greater cleanness.

The safety lamp for working in coal mines was exhibited in many forms, but the principle was always that of Davy. Intended for the preservation of the workmen's lives by the prevention of explosions, it seems curious that the only impediments in the way of its fulfilling this duty are the workmen themselves. Nearly all explosions of fire-damp are caused by incautiously opening the Davy lamp. There seems to be a fascination about doing so, for locks are in vain, and are picked or broken when the workman wishes to get at the flame. An ingenious invention possessing strength as well as other merits was exhibited by Mr. Arnould of Mons, who so inserts an iron pin that the lamp can only be unlocked by placing it in a proper position over the poles of a powerful magnet.

The objects exhibited in this class form five principal sections: 1. Plans in relief and drawings of mineral deposits; 2. Boring tools and machines; 3. Mechanical apparatus employed in mines for extraction, ventilation, &c.; 4. Apparatus serving for the after treatment of the materials extracted, such as apparatus for the mechanical preparation of ores and the agglomeration of combustibles, machines for foundries and forges, &c. Lastly, numerous drawings of metallurgical establishments and special apparatus.

FRANCE.

"It is principally in the departments of the Nord, Saone, Loire, Seine, and Seine Inferieure that the objects contained in this class are produced. The supply has generally sufficed for all the wants of home consumption; we may even say that small exports are made to England, Italy, Spain,

Africa, and the two Americas. The French coal mines recently opened, especially those of Pas-de-Calais, can be compared, as regards their method of working, exhaustion, and ventilation, to the great establishments of Newcastle, Belgium, and the basin of the Ruhr, and can vie advantageously with all foreign countries.

"As regards the elaborations of mineral combustibles, no country is so advanced as France, and there exist none where such a large proportion of small coal extracted is submitted to purification by washing, by means most varied, and by more improved apparatus; none where the processes of agglomeration—a branch of industry which is, moreover, of French origin—have been more studied; none, lastly, where the making of coke is accomplished with less loss of combustible materials. The progress which the committee of admission of class 47 can point out, since 1855, is: 1. For the working of mines, the improvements in the processes of sinking shafts in loose and aqueous soil; the general improvement of apparatus, with a view to increase the productive power of the mines. 2. For the mechanical preparation of ores and combustibles, the employment of a great number of new apparatus, with a view to render work still more mechanical, and thereby economize hand labor; the application of improved methods of construction to those apparatus which have hitherto been executed in a rough manner. 3. In the development of metallurgy generally, the increase of the individual production of the blast furnaces; the more general and judicious use of selected fuels and ores, a use which is facilitated by the increased means of transport; the substitution—each day more marked—of the coal iron for the charcoal iron, in consequence of the new applications that coal iron has found in the production of improved pig iron, and in the invention of improved or entirely new methods of refining, (Bessemer process;) and, finally, the increased power of the machinery and tools used for hammering and rolling, augmenting every day in dimensions, such as armor plates, large iron for buildings, iron plate, &c."

CLASS 48.—IMPLEMENTS AND PROCESSES USED IN THE CULTIVATION OF FIELDS AND FORESTS.

The objects included in class 48 were exceedingly important, comprising: 1. Implements and machines for forest cultivation; 2. Agricultural machines and implements; 3. Plans of agricultural works, and reports relating to farms which have obtained the prize of honor or other prizes, and which offer incidents worth studying and good examples for imitation, either as regards rural construction or other matters, such as irrigation, drainage, plantation, &c.; 4. Commercial manures, which supply agriculture with matters of great utility in preserving or increasing the fertility of the soil.

There were exhibitors in one or all of these sections from almost every country on the face of the earth. It was curious to observe in the glittering courts of eastern nations the rude appliances for tilling the soil, appliances which have barely changed their form since the commencement of the Christian era; and thence to go to the annèxe, or better still to the island of Billaucourt, and see the huge progress that has been made since the application of steam, and the general knowledge of mechanics which was its natural result.

Traction engines were conspicuous in the English department. They are intended mainly for drawing hay or wood over the ordinary surface of the country, but by the application of belting they can be used for any other purpose connected with agriculture. A good traction engine can draw 30 tons at a mere trifling cost per mile.

Reapers and mowers were the specialties of America. They came out triumphant at the two trials which were made at the Emperor's farms at Vincennes and Fouilleuse. Several of these admirable machines were ordered by the Emperor.

The inventions and contrivances in other branches of agriculture were innumerable. They indicate clearly that the day is not far distant when the historical plough-boy will disappear from the field—whistle and all—and be replaced by an intelligent engineer.

FRANCE.

"Among the practical improvements which have been made in the articles included in this class, during the last 12 years, may be cited: First. The more general employment of machines and implements for turning over the soil, and especially the invention of the Valleraud plough, which serves to bring the sub-soil to the surface; the increased use of the threshing machine; the employment of steam power as a motor in the more advanced agricultural undertakings; the application of the drills to the sowing of cereals in line; and, finally, some attempts at steam cultivation, and the introduction of a multitude of reaping machines, which have but rarely fulfilled the expectation of those using them. Secondly. As regards the several methods of cultivation, and the progress made of late years in rural architecture, there exist a large number of farms which, by the general arrangement and details of their buildings, possess commodious and ingenious arrangements, having the effect of economizing hand labor and facilitating the connection of various operations. In reference to commercial manures, we may mention principally the fossil sulphates, which, being extracted from French soil and submitted to simple and inexpensive processes, supply the agricultural community, at a low cost, with valuable means of increasing their crops. Thirdly. Like other industries, those connected with the forest obey the law of progress. Instruments for cutting wood, such as axes, billhooks, saws, &c., have latterly shown considerable improvements. The use of the plough, in aiding natural reproduction in coppices, produces marvellous results, and its employment cannot be too much recommended. The pruning of the trees, practiced for a long time by most faulty methods, is now carried on in a superior and efficacious manner, the value and importance of which have been placed beyond all question by numerous and conclusive experiments. Interesting experiments are being carried on relative to the barking of oak, a matter of the greatest interest to the tanning trade. Lastly, strenuous efforts have been made, during the last few years, principally by the forest administrations, to effect the entire rewooding of the denuded mountains of France."

CLASS 49.—IMPLEMENTS USED IN THE CHASE, FISHERIES, AND GATHERING WILD PRODUCTS.

The objects exhibited in this class form five principal series:

1. The implements and engines of the chase include, except fire-arms, all the other apparatus used for the capture of game, such as nets, snares, decoys, &c., equipments for sportsmen, such as game-bags, powder-horns, shot-pouches, and cartouche-boxes. 2. Fishing implements and tackle, including lines, hooks, fishing-rods, harpoons, nets, bait, and the materials used in the manufacture of these articles. 3. Implements used in collecting natural and uncultivated material. 4. Apparatus of pisciculture: arrangements for hatching spawn, for raising the fry, and transporting fish; aquariums, apparatus intended to stock rivers with fish, such as salmon ladders; lastly, plans of piscicultural establishments, and scientific works treating on such subjects. 5. Apparatus for diving or for submarine industry, such as the collection of sponges, coral, and pearls, for submarine construction, the closing of water sources, the raising of sunken vessels, &c.

The whole machinery of fishing was exhibited in class 49, even to human fishing in the shape of divers, and their complicated accoutrements. France possesses several establishments for the artificial cultivation of fish, and the subject of pisciculture has attracted the attention of the government. Large quantities of fish are now bred artificially, and with the best pecuniary result. Streams that have been emptied of salmon have been repopulated with that delicacy, by means of the pisciculturists, and oysters which were in a fair way of dying out on the coast are now submitting to the same quiet mode of increase. Experiments leave no doubt that fish can be cultivated as profitably as any other article of food. With the smallest amount of state protection, the salmon fisheries could again be established on the principal rivers of America.

"Sporting implements are mostly manufactured in Paris and exported to all the world. The leather of French origin is worked by mechanical processes. Stamping presses and sewing machines are used for making shot-bags, cartouche-boxes, game-bags, &c. The nets are made by hand. The trade in sporting necessaries in France is estimated at from 3,000,000 to 4,000,000 francs. No very striking innovation has been noticed in this branch of industry since 1855; but the methods of manufacture have been so much improved that French productions now leave nothing to be desired as regards the excellence and finish of the work. Fishing-tackle and implements are made in the immediate vicinity of the principal fisheries—Angers, Bordeaux, Boulogne, Dieppe, Dunkirk, and Nantes; the hooks were formerly obtained entirely from abroad, but are now produced partly in France, and particularly in the departments of the Bouches du Rhône and the Côtes du Nord. Fresh water fishing-tackle is made in Paris and its environs. The raw materials employed in the trade are very various and are derived from almost all countries on the face of the globe. Hempen yarns are obtained from Angers, and those of flax from Lille. Rushes and reeds are obtained from Fréjus. China, Japan, and India send us bamboos and silk. The so-called Florentine horse-hair comes from Spain and Italy, and we borrow from innumerable birds the feathers with which to form artificial flies. All the delicate articles are made by hand; nets alone are in part produced by looms. The products of this industry amount in number to about 1,000,000 a year, and are exported to all countries. The trade now obtains at home that supply which, before 1855, it used to obtain from foreign markets. The apparatus employed in gathering wild products has no special characteristic that demands notice.

"The trade in piscicultural apparatus has extended, since 1855, in a very marked manner. From 500,000 to 600,000 francs' worth of such apparatus is annually sold for home consumption and export. The slate of Angers and the plate-glass of St. Gobain are laid under contribution for the construction of aquariums. The apparatus for restocking rivers with fish, such as salmon ladders, constructed at the instance and under the superintendence of government, have produced great improvements in the productiveness of our streams.

"Diving apparatus is also manufactured in Paris. Copper, lead, leather, India-rubber, with Laval thread and Rouen cottons, are the chief materials used in the manufacture. These apparatus, which are in increasing demand in every part of the civilized world for fisheries and hydraulic works, amount in value to about 400,000 francs or 500,000 francs per annum. Since the Exhibition of 1855 the apparatus have undergone great improvements which fit them for submarine exploration at great depths."

CLASS 50 TO CLASS 54.—MACHINES AND APPARATUS IN GENERAL.

The classes 50 and 51 contained: 1. Apparatus and processes used in agricultural works, and for the preparation of food, such as making pipes for drainage, making manure, making sugar, brewing, &c.; and 2. Apparatus used in chemistry, farming, and tanning, such as apparatus and utensils for laboratories; instruments for making tests, &c. Both classes were interesting to experts.

Class 52 included the machinery, &c., used for the purposes of the Exhibition. It is fully described elsewhere in an extract from the French report. There were seven American exhibitors in this class.

Class 53, machines and apparatus in general, contained detached pieces

12 U E

of machinery, supports, rollers, slides, eccentrics, cog-wheels, &c. There were six American exhibitors.

CLASS 54.—MACHINE TOOLS.

Class 54 embraced all the articles comprised under the head of machine tools, such as lathes, planing machines, and other instruments used in the working of wood and metals. No more important class was to be found in the Exposition; indeed without this, many other classes could not have existed. The principal nations exhibiting were France, England, Prussia, and America. At the former exhibitions of 1851, 1855, and 1862, the English were almost without rivals. On the present occasion they made but a small display, and were vastly outnumbered by France and Prussia, while in point of novelty of form and excellence of workmanship America was admitted to be on a par with any nation. In the French section were tools of every possible description, many possessing a high degree of excellence, and some of extraordinary size. Prussia was represented in the fullest manner, and her progress in this branch of manufacture excited general remark.

In the American department the display made by Sellers, of Philadelphia, was highly commended. Among other articles was a machine for cutting the teeth of wheels, which when once set in motion, is completely automatic until it has gone completely round, when it stops of itself and calls for the attendant. The Sellers planing machines were equal to the best in the Exposition, and were remarkable also for many novelties. So, also, the steam hammer with its new mode of manipulating the steam valve. The same firm, says Mr. J. Anderson, civil engineer, in his report to the board of council, has a fine display of screwing apparatus entirely of a new character, and all constructed on a sound principle. By this system screws of all sizes are the same in the form of the thread, namely, an angle of 60 degrees; six cutting tools for any size of screws, if placed together, will form a complete circle. The depth of the thread, the amount to be taken off the sharp point of the cutting-tool, are all derived from the diameter, or the pitch, or from each other, on a well defined principle. These screws, when complete, are what is technically termed "flat-top and bottom," and although this system may be objected to by those who are accustomed to and prefer "the round-top and bottom," yet it is very evident the flat gives greater facility for measuring the diameter with extreme accuracy. Altogether, adds the same writer, the collection (of tools generally) exhibited by Sellers probably contained more originality than that of any other exhibitor in class 54.

The lathes of Harris, and of the American Tool Company, possessed several novelties which were interesting to the experienced eye. Brown & Sharp exhibited a machine for making any description of screws out of the rough bar. . When the screw to be made is once determined upon, every instrument necessary to its production is placed in suitable holders therein provided; the wire or bar passes through the centre of the revolving spindle, when tool after tool is successively brought into operation, and screws of perfect identity are thereby produced with facility. Bement & Dougherty's machine tools displayed many points of excellence. The American exhibit in all respects, although not large, was extremely praiseworthy, and was a matter of surprise to many toolmakers who heretofore have had a sort of monopoly in this business.

FRANCE.

The machines exhibited in class 54 may be divided into four principal sections:

"1. Machine tools for working metal, such as simple lathes, mechanical lathes, parallel lathes, spherical lathes, facing lathes, and lathes with four points, axle turning lathes, lathes of precision, counter-sinking lathes and rose engines, lathes for cutting screws and forming heads of bolts, &c., for turning the wheels of carriages and the driving-wheels of locomotives, planing machines of all kinds, filing machines, mortising and drilling machines, whether horizontal or vertical, machines for shaping the heads of bolts and nuts for boring cylinders, forging, rivet making, punching, shearing, chamfering, centering, riveting and pipe-drawing, and lastly, machines for pounding and for polishing.

"2. The machine tools employed in working wood, such as reciprocating, continuous and circular sawing machines, planing, moulding, turning, and mortising machines.

"3. The various tools used in machine construction shops, such as rules, squares, trusses, bevels, chisels, glass and sand paper and cloths, &c.; blocks and tackle, and other apparatus used in mounting machines.

"4. Machines for pressing, crushing, mixing, sawing, and polishing, are comprised under the general denomination of machine tools, although they are, in fact, manufacturing machines. Such are also rolls for flattening the precious metals, cutting and stamping presses, nailing machines, brick and tile making machines, stone-breaking machines, machines for grinding plaster and colors, for bending and welding the tires of wheels, for cutting paper, for piercing hard and precious stones, and for diamond cutting.

"Machine tools used for working iron and wood, and the greater part of the machines comprised in this class, are manufactured principally in the departments of the Seine, Seine Inférieure, Nord, Haut Rhin, Bas-Rhin, Bouches du Rhône, and Somme. Paris, Rouen, Mulhouse, Graffenstaden and Hâvre, are the chief places of production of tools and machine tools. Fécamp manufactures wood-working machinery. The small machine tools used for metal-ware manufacture form the object of an important trade at Albert and Manbeuge. Machine tools, properly so-called, are constructed of metal; cast-iron is generally employed for the purpose. The preference is given to Scotch iron, at the price of about 15 francs the 100 kilograms, as presenting a uniform quality and not being too hard. Castings of moderate size cost about 35 francs, and large castings only about 25 francs.   The other metals used are mostly of French origin. Rough iron costs 24 to 25 francs the 100 kilograms, and forgings from 70 to 80 francs; pieces of small dimensions submitted to great strain, and therefore requiring superior power of resistance, are made of special iron, costing from 50 to 60 francs; but its production diminishes daily. The Bessemer steel does not yet offer sufficient guarantee to allow of its being used for these parts.   Until this is the case, the manufacturers are compelled to employ steel which costs them from 90 to 150 francs the 100 kilograms.   Case-hardened iron is substituted for the former in the case of small pieces of machinery, and for very small pieces malleable cast-iron is preferred.   The parts of machine tools are nearly all produced by machinery, in large workshops abundantly supplied with all the means of giving with rapidity the form required before the parts are put together, and with the view of increasing production without adding to the extent of their establishments.   Many constructors now produce the castings and large forged pieces on the very spot where the iron is produced. This principle of the division of labor and of the setting apart of certain workshops for special purposes is being adopted more and more every day in our large towns.

"The cost of hand work varies greatly according to the locality and the ability of the workmen.   In Paris the average wages of the operatives who work by hand or direct machine tools is five francs a day.   In other central towns more favorably situated, such as Mulhouse, for example, the average is not more than three francs.   First-class hands, however, earn much higher wages, sometimes as high as nine francs per day.   The constant increase in the machinery of construction shops tends incessantly to improve the condition of the workman by diminishing his bodily labor and giving him time and opportunity for making numerous arrangements which have the effect of increasing his earnings, especially when he is engaged on piece-work.   The machinists generally construct the machines which they produce after their own models, but they are often obliged to modify them according to the demands of the purchaser. For several years the great houses have established depots at Paris for the machines in most general use, such as lathes, drilling, planing, punching, and shearing machines.   These depots render great services to the manufacturers, who are often obliged to increase their machinery at a moment's notice.   Even public establishments often take advantage of this arrangement.   The greater part of the products of this trade are for home use, but of late important business has been done with Italy, Spain, South America, Russia, Turkey, and even Japan.

"The productions of France, which a few years since were very limited in extent, may be valued at about 12,000,000 francs.   Although the prices of the raw materials have submitted to considerable diminution since the treaty of commerce came into effect, the selling price of the machinery has remained almost stationary, in consequence of the increasing dearness of labor and the constant augmentation of the weight of the machines, in order to diminish vibration and to simplify the arrangement of the foundations."

RECENT IMPROVEMENTS.

"The committee deem it their duty to point out the following improvements as having taken place within the last 12 years:

"1. More solidity of construction, simplicity and perfection, and more frequent adoption of automatic motions.

"2. Forms better adapted to the materials employed.

"3. Constantly increasing tendency towards mechanical production, and the completion of parts by the use of machinery alone.

"4. As regards metal working machines, the introduction of machines which allow of several operations being performed on the same piece without dismounting it; as, for example, universal drilling and planing machines, working horizontally and vertically; lathes upon which parts having a different axis to the principal piece are worked by means of cutters having a compound rotating and traversing movement; bolt cutting machines; mortising machines with revolving tools, and counter-sinking machines.

"5. As regards wood-working machines, the construction of portable and locomotive machinery for sawing wood in the forest, the application of the endless handsaw to the cutting up of round timber; the employment of helicoidal blades in planing; the modification of the tools used in boring and planing, and the increase in the rapidity of rotation given to these tools.

"6. As regards the tools themselves, a general improvement in the execution of small tools used in connection with the machines, and the differential pully, which causes the load to remain in the same place when left to itself.

"7. Generally, as regards machines of all kinds, we may point to many simplifications in the means of transmitting motion, and specially the mechanical imitation of proceeding by hand; the employment of mechanical means in the working of fly-presses; an increase in the production of brickmaking machines, and, lastly, a tendency towards the suppression or diminution of previous working of the clay by the augmentation of the pressure employed and by the greater dryness of the clay employed."

CLASS 55.—APPARATUS AND PROCESSES USED IN SPINNING AND ROPE-MAKING.

In class 55 were comprised all the machinery and apparatus used in the preparation and specimens of textile materials, of which cotton, wool, flax, hemp, and silk are the most important. The materials and machinery used for rope-making were also included in this class, together with ropes and cordages of all kinds. It may be here mentioned that some excellent specimens of cordage were exhibited, made from the fibre of the aloe.

America had four exhibitors of machinery for preparing cotton and wool.   There was a considerable display from other countries, but without a complete knowledge of the technology of the trade it would be impossible to describe the peculiarities of the various machines.   The subject belongs to the specialist.   The following, from the introduction to this class in the official catalogue, will prove interesting:

FRANCE.

"This class includes the machines and apparatus destined to manufacture textile fabrics, of which cotton, flax, hemp and silk are the most important.

The machines for spinning, twisting, and weaving are constructed in different industrial centres of France.   The machines employed in the silk trade are principally made at Lyons; Alsace manufactures for the cotton, woollen, worsted and spun silk trades; Lille is principally engaged with flax and hemp machinery; Rouen furnishes the cotton trade especially; Louviers, Elbeuf, and Sedan the machines used in the cloth trade; Troyes and its environs produce hosiery looms.   Paris combines all these branches, but particularly those appertaining to the class now under notice.

"In consequence of the multiplicity of their forms and of their masses, textile materials require several series of spinning machinery or arrangements.   Cotton is, at the present time, worked upon two systems very distinct from each other, and according to whether the yarns are to be carded or combed.   For wool, four series of machines, corresponding with the denominations, carded yarns, combed merino yarns, long combed yarn, and mixed or combed carded yarns; lastly, a new apparatus, recently introduced, gives woollen yarns by felting instead of spinning.   Hemp, flax, and jute are prepared by two principal descriptions of machines—one for long fibres upon the combing principle, the other for short fibres or tow prepared by carding.   Each of the two branches of the flax and hemp manufacture has other modifications in its machinery, according to the special character of the raw material employed and the strength of its fibres.   Nor are the means the same for producing fine and coarse yarns.   The former requires not only a special machine, but the application of water at various temperatures. Of all spinning machinery employed, that which is used for the most costly material is the most simple in its construction; but the winding of silk from the cocoon, although apparently so extremely simple an operation, is in reality so delicate that with cocoons of the same quality the value of the silk may be doubled by the ability of the "durder" or winder.   The manufacture of spun silk, which increases daily with the cost of the raw material, also requires a variety of combs as well as cards.   The spinning machinery for cashmere, alpaca, and goat's hair is identical, with a trifling exception, with that employed for wool.

"This class includes, also, the machinery and products of the rope and twine manufacture. The machines of class 56 act upon yarns as upon a raw material of common origin to transform them into fabrics. The loom changes with the nature of the tissue to be produced. The same parts apply to the weaving of cotton, wool, flax, silk, &c., but are modified according as the tissue to be produced is plain or figured. Knitted articles, tulle, bobinet, net, and lace have each a machine specially constructed for it, and which changes according to the form of the mesh. The looms of class 56 are divided, then, into:

" 1. Looms to make plain fabrics with close threads. 2. Looms to weave fancy stuffs, plain, napped, or with velvet pile. 3. Frames for knitting tulle, nets, lace, &c.

" The apparatus employed in the preparation of yarns for weaving, and also for the dressing of tissues, are included in this class. Some, such as calenders, presses, clipping machines, &c., indispensable in all textile manufactures, are applied, with certain modifications, to each of the branches; others, such as fulling mills, are used only in one branch of this vast group. The purification of wools has been greatly improved as regards not only the economy but the perfection of the process since the general adoption of machinery for washing, scraping, and other operations.

" Special modifications introduced in the apparatus for the preparation of cotton have produced unhoped-for results with the common productions of India and China. The fine cottons of Georgia, Egypt, and Algeria are largely indebted for their present position in the market to the application of combing machines. The same principle has produced greater results still in the woollen manufacture. Spinning machinery, mill-jennies, and their contingents receive constant improvements, which allow increased speed to be given to the spindles, and consequently an acceleration of their productive power. The spinning of silk itself, in spite of its simplicity, is the object of many experiments, with a view to the preservation and preparation of cocoons, as well as the improvements of the mills for twisting and organzine. The machines for milling, dressing, and preparing the yarns for warps, and the wefting machines, have been modified with great success, and have thus contributed to the extension of power-loom weaving in those special articles where a substitution for hand-weaving seemed very difficult. The several parts of power looms have been the object of careful study, which has brought about many improvements, such as governors, instantaneous stopping on the breaking of a thread, either warp or weft; cages with a number of shuttles, &c. Increased care in construction has produced improvements in the movements, and a proportional increase in the production. In the weaving of figured stuffs attempts have been made to substitute paper for card in the Jacquard machine. This idea, which is far from new, has no chance of success until the various organs of this ingenious mechanism can be made to work with absolute precision. Another important fact consists in the happy combination of a Jacquard loom in which the same part produces various effects in succession, so as to simplify the machine and produce an important economy in the results. Straight looms for hosiery which work by hand produce no more than 5,000 meshes a minute, while machinery produces nearly 50,000 in the same period. In circular knitting machines the number of meshes is raised from 50,000 to about 500,000. Dressing machines do not seem to call for similar remarks, and present but few special modifications. Improvement in this direction depends, in fact, more upon the ability of the workmen than upon the principle of the machine.

"France employs annually from 80,000 to 85,000 tons of cotton, the necessary machinery and material for spinning, weaving, and dressing amounting in value to about 400,000,000 francs. The woollen manufactures involve about the same aggregate expenditure; the mechanical spinning of hemp and flax and the weaving of linen about 100,000,000 francs; lastly, the silk trade furnishes an amount very nearly approaching to 200,000,000 francs, giving the total value of the material employed in these several industries as 1,100,000,000 francs.

" It may be calculated that the amount spent annually for construction represents one-twentieth of the above total—that is to say, a sum of 58,000,000 francs, without taking into account exportation, which greatly exceeds the importation.

"The character of the improvements now in course of realization may be thus summed up:

" 1. A more precise acquaintance with the special constitution of the raw materials, and, consequently, a better arrangement of the means by which they are transformed.

" 2. A more rigorous application of the mechanical laws in the execution of all the parts of the machines."

CLASS 57.—APPARATUS AND PROCESSES FOR SEWING AND FOR MAKING UP CLOTHING.

"The machines and tools exhibited in this class form three distinct series:

" 1. Sewing machines applied to the different works of sewing and embroidery. 2. Machines employed in shoe-making. 3. Machines and apparatus used in felt-hat making.

"The articles exhibited in this class show the advance made in the trade. The first sewing machine which was worked for trade purposes was invented by a Frenchman named Thimonier, a tailor at Amplepuis, (Rhone,) the invention being patented 17th April, 1830, and improvements therein registered 21st July, 1845. Until the year 1855 the use of these machines was very restricted. They were only applied in a few special ways, and it is since that time, and particularly since 1862, that they have come into general use in France."

MACHINES USED IN SHOE-MAKING.

"Machines have long been used in shoe-making, and the principal aim of the experiments latterly has been to replace sewing by screw pegs. The exhibition in this class shows the mechanical apparatus for this kind of work in movement, such as cutting presses, mounting and screwing machines, shears, grindstones, piercers, &c. The mounting and screwing machines, worked by steam and guided by women, admit of a rapidity of execution and an economy of hand work which enables the makers to deliver the products for consumption at a much lower price and of an equally good quality. We must also mention the special apparatus called dressing machines, intended for the mountings of the upper leathers, which was done hitherto by hand."

APPARATUS FOR FELT-HAT MAKING.

"The machines serving for the manufacture of felt hats have accomplished a complete transformation in the trade during the last few years. Previously the workman shaped, fulled, pounced, and pressed by hand. This system produced much inequality in the work, and, above all, great slowness in the production. At the present time machinery replaces hand work in general. The several machines working in class 57 serve for forming, fulling, and pouncing felt hats, and for shaping straw hats. They show a considerable progress as regards the regularity of the work, and, by rendering the manufacture more easy, admit of the productions being sold at a lower price, and thus meeting in a much better degree the demands of the home consumption and the extended sales for exportation."

CLASS 58.—APPARATUS AND PROCESSES USED IN THE MANUFACTURE OF FURNITURE AND OTHER OBJECTS FOR DWELLINGS.

"The productions exhibited in class 58 form four principal sections:

"1. Tools for wood-work, including ribbon saws, reciprocating treadle saws, vertical moulding machines, planing machines with helicoidal blades and with disks, mortising, engine-turning, and carving machines, and collections of tools for hand work. 2. The worked produce of these machines. 3. Engraving machines and portrait lathes. 4. Saw blades and collection of wood-working tools."

It embraced the apparatus and processes used in the manufacture of furniture, and other objects for dwellings—familiar machinery for the most part that had very little interest for the general public. America exhibited several ingenious contrivances, among which may be mentioned a gauge lathe for turning the legs of chairs, &c. It is a lathe with a slide-rest traversed by a screw. This rest carries two tools; one, a chisel, is fixed and roughs off the work; the second, a V-shaped cutter, cuts out the pattern and is guided by a template fixed to the bed of the lathe. A knife whose edge is molded to the form to be produced, moves vertically in a frame behind the lathe; as the slide-rest passes along, this knife descends and smooths off the pattern produced by the first two cutters. In this manner chair legs are produced from a rough square log with an accuracy equal to that attained by hand, and with immense rapidity and cheapness.

FRANCE.

"In the French section a lathe for copying medals was shown by Messrs Barrere & Caussande. The work and the original revolve slowly at the same speed. A tracing-point moves from the circumference of the model, so as to describe a spiral track over the surface to be copied, and rises and falls as it meets with elevations and depressions. The vertical motion of this point is communicated to a drill which moves in a similar manner over the work. Reduction or enlargement is produced by causing the horizontal movement of the drill to be slower or faster than that of the tracing point by means of change wheels.

"The principal centres of manufacture are in the department of the Seine, but a certain number of machines are furnished by other departments. The raw materials, such as cast and wrought iron and steel, are derived almost entirely from French sources. The price of cast iron varies from 24 francs to 26 francs the 100 kilograms, and that of iron castings from 30 to 35 francs the 100 kilograms; charcoal iron, plate, and cast iron of the second fusion are preferred.

"Mechanical labor has taken the place of manual for sawing up and shaping the wood; and in the forest, when it is impossible to bring out the rough timber, saw-mills are used, which cut up the timber on the spot, and convert it into pieces for parquetry, staves, &c. It took some years to break through the old routine, but the perseverance of constructors has triumphed at last, and now almost all the works are provided with machines of all kinds. Even carving by machinery has come into practice. The engraving machines and portrait engines have made sensible progress. Lathes are used, which reproduce, with the utmost fidelity, and on steel, all kinds of models, without the slightest alteration of form. Saw blades and cutting tools have undergone considerable modifications, and complete the machinery in a satisfactory manner. The blades of ribbon saws have now arrived at great perfection. Hand-tools for woodwork leave nothing to be desired in any respect. The number of woodworking machines in actual operation may be estimated at 10,000.

"The employment of machine tools has not had the effect, as might have been expected, of superseding manual labor, the production having considerably increased. Simple laborers have become directors of machines, and workmen of the first class, following the same profession, have given to this kind of work the impulse that it required. Wages have increased in large proportion. The workmen who formerly earned three francs a day now obtain five francs. Good workmen have become masters, and established saw-mills, which now form a very important branch of commerce. Machine tools are sold principally, in France, for cabinet-making, inlaid work, furniture-making, cutting-out stuffs, bones and ivory. They are also sent abroad for forest works. The number of machines working since 1855 is estimated at 10,000, and the average cost of each, 2,000 francs, making a total of 20,000,000 francs. Each machine represents the power of four workmen, from which has to be deducted the conductor of the machine. The saving effected is therefore equal to three-fourths of the whole.

"The committee of admission has to point out the following instances of progress made during the last twelve years: Ribbon saws, with moulded or cast-iron frames and columns, which may be placed on a simple slab of stone, and worked without the slightest trepidation, the diameter of the pulley so much enlarged that wood of a metre in diameter may be cut up; moulding and mortising machines, worked with greatly increased rapidity; machines with helicoidal blades, for working wood across the grain and for planing knotty wood in all directions; planing machines, with disks working vertically, by which wood may be worked square or obliquely, according to circumstances; the improvement of hand tools, reciprocating saws, worked by treadle at the rate of 250 cuts per minute, and which move so easily that the workman is in no way occupied with the action of his foot; the arris handsaw, especially useful for cutting tenons, for square cuts, and mitreing. Saw-blades and cutting tools are manufactured in Paris; the largest articles are circular saws, and the smallest ribbon saws. The products of these, exhibited in specimens of cutting, which are models of precision and patience, show the perfection of the ribbon saw; the specimens of carving and ornamental cutting and of carton work also exhibit a high degree of perfection."

CLASS 59.—APPARATUS AND PROCESSES USED IN PAPER-MAKING, DYEING, AND PRINTING.

The exhibits included in this class are manufacturing machines employed especially in the making of paper, in dyeing and printing of all descriptions. They form six principal series: 1. Printing machines and presses, apparatus for stereotype and type-founding, and for composing by machinery; 2. Lithographic printing presses; 3. Machines for various kinds of printing and decoration on paper, roller and scraper machines for copperplate and other incised engraving, and for the cheap printing of children's copy books; machines for the rapid printing of railway tickets; self-cutting, stamping and registering machines. Among the many tools used for paper work, folding machines and powerful paper-cutting machines ought to be mentioned: 4. Machines for paper-making; 5. Apparatus for printing paper-hangings; 6. Accessories of calico and other printings; pricking machines; singeing mchaines; stretcher for dyeing dyed fabrics, &c.; accessories of printing on paper; processes of engraving with the aid of galvanic deposits; seal engraving, &c., &c.

"Paris and Mulhouse are the two principal centres of production for machinery and apparatus belonging to this class. Some of them, and particularly those used in stereotype work and type-founding, are modifications of American models. The precision of the machines for fine printing, for printing from wood-blocks, and the rapidity of production for ordinary works, and especially for newspapers, are the principal objects of the labors of the constructors. The problem of lithographic printing by machinery, at prices similar to those of typographical printing, has, within the last few years, been practically solved. The Exhibition contains many specimens of machines which have been adopted by the trade. The annual value of machines belonging to the first series amounts to 1,500,000 or 2,000,000 francs, and their success is attested by the exportation of nearly one-half the amount.

"We have to draw attention to the improvements made in the machines for reducing pulp, with regard to form, dimensions, and mode of construction, which are shown in the Exhibition. The machines exhibited consist of pulp-engines of large dimensions, of one of new construction, and lastly, of the accessories of the paper machine, dryers, wire-cloth, felt, &c. The machines for engraving, rollers for printing by means of circular cutters, engine-turning, electricity, and the employment of the pantograph, &c., are valuable auxiliaries placed in the hands of the roller-engravers. The printing machine figures in the Exhibition with its last improvements, and a special motor, adopted on account of its simple action."

Class 60 was devoted to machines, instruments, and processes used in various works. It included among the objects from the United States a machine for dressing printing types, and machines for cutting files. In the French section there were watch-makers' and jewellers' tools, and machines for making envelopes. Many of the machines in this class are described elsewhere.

CLASS 61.—CARRIAGES AND WHEELWRIGHTS' WORK.

Class 61 was devoted to carriages and wheelwrights' work, comprising carriages entire and in parts. The display of the former was exceedingly good, especially in the English and French departments. There were a few light wagons from America, but neither in style nor variety of style was the exhibit worthy of this important industry. Russia displayed several specimens of her carriage work, which, in the lighter sort of road vehicles, is obviously borrowed from American models. There were fine specimens also in the Austrian, Prussian, and Spanish sections. The English exhibit was characterized by elegance of form, brilliance of varnish, and graceful poise. There were but few novelties. The most important had in view the better and quicker opening and shutting of barouches, so as to afford immediate protection in case of rain. This is done very rapidly from the driver's seat by means of a crank, which winds it up without requiring to stop, or any derangement to the occupants. A similar contrivance, but worked by springs, and balanced to the greatest nicety, is operated from the interior. A touch of the strap raises the cover.

FRANCE.

"The productions exhibited in this class comprise: 1. Carriages of various kinds and forms, such as landaus, calashes, broughams, victorias, phætons, omnibuses, American trotting carriages, fancy vehicles of all sorts, and children's carriages; 2. Detached parts employed in the manufacture of carriages of all kinds, such as wheels, axles, springs, boxes, shafts, specimens of forging, &c. The principal manufactories for the production of dress carriages are in Paris, but there are some also at Lille, Lyons, Bordeaux, Toulouse, Caen, Abbeville, Colmar, Boulogne-sur-mer, &c. Each district, as a rule, builds carriages in ordinary use in its own part of the country. The dimensions, form, mounting, and the accessories of these vehicles are necessarily modified according to the nature of the ground, the state of the roads, and the quality of the horses of the country. As to the detached parts of carriages, carts, and other vehicles, their production is spread over the whole extent of the country; they exhibit, however, a tendency to concentrate themselves round certain centres, where, with the aid of machinery, they are produced in large workshops amply provided with means, with great rapidity and economy. The materials used by the coachmaker and wheelwright are principally wood, iron, steel, leather, cloth, galloons or coach lace, silk and woollen fabrics, horse-hair, morocco, colors, varnish, &c. For a long period French industry depended on foreigners for many of these items, especially springs and varnish, which came from England; but for some time the French makers have found nearly all they required at home. In consequence of the great variety of forms, coach-building cannot be effected by mechanical means. Such processes are only used in the case of certain detached parts, such as springs, axles, and wheels. The work is divided amongst a large number of workmen; one class of workmen make the wheels and the carriages; a second, the bodies; smiths and fitters make the springs, the axles, and all the iron-work; saddlers and stuffers provide the furniture of the interior, the seats, and also the exterior parts in which leather is employed; and to those must be added the platers, the painters, the lamp-makers, the lace-makers, the carvers, &c. Besides the great establishments in which the carriages are produced complete, and in which all the classes of workmen are employed, there are shops which confine themselves to the fitting and mounting of coach bodies purchased in the rough state; others are specially organized for painting only; and, lastly, certain persons devote themselves entirely to the production of designs and models. The products of the French carriage trade are not only sold all over France, but exported to other countries. The number built in France may be estimated approximately at about 5,000 annually, and of the value of about 15,000,000 francs. But this does not include the work of repairing and keeping in order, which surpasses considerably in amount the cost of the new work. The inquiry instituted in 1860, by the Chamber of Commerce of Paris, proved that, including the whole of the coach, carriage, and wheelwright's works, lamp-making, iron-work, painting, &c., the trade of Paris alone amounted to 36,000,000 francs. Although the treaty of commerce has offered great facilities, this is at present almost *nil*. The exportation, on the other hand, is on the increase; this scarcely exceeded 1,000,000 francs from 1847 to 1856, but it amounts, at the present time, to four times that sum. French carriages are exported mostly to Spain, Russia, Egypt, Portugal, America, Turkey, and the colonies; a certain number are even sent to England.

"The principal improvements to be noted in the trade may be summed up: 1. The carriages, being manufactured of the very best materials, are more solid than formerly; 2. The models are more varied, both as regards elegance, and also to suit the many different employments for which they are destined; 3. The manufacture is more rapid, in consequence of improvements introduced in the tools employed, and of a better distribution of the work in factories."

CLASS 62.—HARNESS AND SADDLERY.

The productions exhibited in class 62 comprise: 1. Harness of all kinds, coarse and fine; 2. Collars on different systems; 3. Saddlery work; 4. Driving and riding whips and sticks; 5. Detached parts which enter into the structure of the preceding articles, and which furnish employment to special workmen.

The display was by no means remarkable, and the contest was mainly between France and England. The latter country, some few years since, had almost a monopoly of this business, but French ingenuity and skill have made such rapid progress that she can no longer boast of occupying the same position. Both nations, however, manufacture superb articles in this branch. In the Spanish department was exhibited a magnificent set of state harness, in which material workmanship, taste, plating, &c., seemed perfect. It was for eight horses and took many years to make. Spanish leather is famous.

FRANCE.

"Paris is the centre of production for dress harness, saddles, whips, riding whips and sticks. Common harness, such as that used in trade and agriculture, is made in all parts of France; its forms, which are very various, adapt themselves to the wants of the several localities where it is employed. Formerly the makers of harness in France obtained their burnished leather and polished steel spurs from England; a few houses have still retained this habit, as regards certain articles, but the tanneries of Paris and Pont Audemer now produce leathers which will bear comparison in all respects with those of the best English houses. As to spur and harness-making, the makers of Paris and those of the departments of the Aisne, Eure, and Ardennes, are now able to produce all the fine articles. The materials employed are tanned leather, (bullock, cow, calf, pig, and horse.) Varnished leather is only used for the finer kinds of harness; the white, or Hungarian leather, is now only employed in agricultural harness; the ox and cow hides are employed for common and ordinary harness; pig skins are used for the making of saddlery.

"The articles composing this class are made in workshops under the eye of the manufacturer. Hand labor is still most in demand. The use of sewing machines has introduced great regularity in the manufacture of many parts of the saddles and collars, but hand work is preferable for the pieces which require great solidity. Some houses which employ themselves upon military equipments, and others who work for exportation and the omnibus and other great companies, possess large workshops, directed by foremen, and including cutters and preparers, as well as special workmen for each branch of the trade. French saddlery is exported all over the world, the principal markets being Egypt, Spain, Turkey, Belgium, and especially South America, which sends us the raw hides and receives back the finished manufactured articles.

"It is difficult to estimate the value of the trade. It appears, from a report made to the Chamber of Commerce of Paris, in 1860, that the amount of the harness manufacture in Paris alone was 12,276,000 francs; and that of saddlery, spur-making, and saddlebow-making, to 2,992,000 francs, giving a total of 15,000,000 francs and upwards, which must at least be doubled to represent all France, and that without including military harness. The exportation of French saddlery exceeds 5,000,000 francs per year. The improvements which have been introduced into the trade during the past 12 years are of two kinds; on the one hand, the forms have been modified so as to render them more simple than those formerly in use, and to get rid of heavy and ungraceful pieces; and, on the other hand, hand labor has been replaced with advantage in several branches of the trade, by mechanical means."

CLASS 63.—RAILWAY APPARATUS.

The objects ranged under this class comprise the material of railways: Locomotives, designs, and models of locomotives, railway carriages, goods wagons, signals, turn-tables, specimens of permanent way, weighbridges, models of various systems of brakes and modes of communication between passengers and guards; specimens of wheels and axles and other iron-work employed in the manufacture of railroad rolling stock. Nearly all the continental countries, Great Britain, and America, contributed to this highly important and interesting division.

LOCOMOTIVES.

The locomotives exhibited were 32 in number. Of these, France contributed 11 passenger and goods engines, and two small tank-engines for tramways; Belgium sent five; Prussia, two; Baden, one; Wurtemberg, one; Bavaria, one; Austria, three; the United States, one, and Great Britain, three passenger engines and two contractors' tank engines. Some of these were of enormous proportions. The Paris and Orleans railway exhibited a ten-wheeled tank-engine, weighing 60 tons, on a wheel base of 14 feet 10½ inches. English engines seldom have more than six wheels, and in England the inside cylinder is largely adopted. On the continent the cylinders are generally outside. The workmanship of the French engine-makers is fully equal to the best. In this industry, indeed, France has made immense strides. Only a few years have elapsed since the time when she used to import her locomotives from England. The shoe is now on the other foot. The Creusot Iron Works exhibited a remarkably well-finished express engine, made from English drawings, indeed, but intended for an English railroad, the Great Eastern. It was the 16th out of an order for 40; the first 15 having been already delivered over to the railway company, and accepted by them, the period of warranty for them having expired. Another singular instance of the way in which this manufacture is passing into new hands was furnished in the case of Mr. Kessler, of Esslingen, who exhibited a locomotive built by him, also on English designs, for an English colony, it being part of an order from the East India Railway Company for 20 engines. The workmanship was thoroughly good. These two engines, says Sir D. Campbell, afford incontrovertible proof of the possibility of getting English designs carried out in France, or on the continent, quite as well as in England, and at a cheaper rate.

The Grant locomotive of Paterson, New Jersey, attracted much attention, and was universally regarded as the handsomest piece of work in the Exhibition. The handles of the various cocks were made of ivory, and the covering of the boiler, cylinder, and chimney, were of polished brass and German silver. The engine-driver's house was of inlaid wood, and every particular of fine workmanship was carefully and beautifully wrought out.

In the Russian department of the Park was a model illustrating the working of the Mahovos system of locomotion on steep inclines from mines. The apparatus consisted of a truck fitted with a pair of 15-ton fly-wheels on an axle carried on friction rollers, which themselves rest on the wheels of the truck. Each train of loaded wagons has one of these trucks attached, and is impelled down the incline by its own weight, and the truck wheels, in revolving, transmit a rotary motion to the fly-wheels by means of the friction rollers. On reaching the bottom of the descent the rollers are lifted by means of levers, clear of the truck wheels, and then revolve freely, opposing hardly any resistance to the action of the fly-wheels. The truck is then detached, turned round on a turn-table, and attached to the head of a train of empty wagons. The friction rollers are then let down upon the truck wheels, transmitting to them the rotatory power stored up in the flywheels, which, it is claimed, suffices to draw the empty wagons up to the top of the incline. It certainly accomplished this object in the working model.

SIGNALS TO GUARDS.

Many devices were exhibited in this section for enabling passengers to communicate with the guards. They were curious to Americans, inasmuch as they show how much thought has been bestowed on a subject which has already in the United States found a very ready solution. With us, however, the conditions are somewhat different. The cars are open from one end to the other, and the guard is constantly passing through them. In Europe there are three and sometimes four different classes to each train, and the subdivisions extend to carriages of the same denomination, so that the guard is compelled to pass from coach to coach by means of the steps outside. This, however, would not prevent a simple rope passing along the entire length of the train, as with us. It would be sure to be within the reach of every one, while it seems that it is only desired to afford succor to the occupants of first-class carriages. This circumstance is the occasion of all the difficulties which European engineers have had to meet. Their effort has been to provide in first-class carriages a means by which the train could be arrested, and then to surround it with such difficulties and complications that no one, unless in extremity, would think of using it. And it may be added that a person in extremity, attacked by a maniac or a murderer, would be utterly unable to command the resources, within his reach technically, but practically out of his power. We will briefly refer to one or two of these methods. They consist of signals communicated by acoustic, pneumatic, or electrical agency. The latter form the large majority, especially in France, but it has been found so difficult to obtain an undisturbed connection that one-third of the signals fail. The way in which this scientific security is proffered to the traveller is curious. A small triangular piece is taken out from the partition which divides two compartments of a carriage, and which otherwise are strictly separate and private. This triangle is glazed with two panes of glass, one in each carriage. Dangling between the two is a ring attached to a wire, and beneath it an intimation that, in case of accident or dire necessity, the passenger may break the glass with his elbow, pull the wire, then open a window and wave his arms in the air, by which means the guard or engineer will be duly warned. Heavy penalties are demanded from those who should wantonly indulge in this luxury, but the difficulties in the case are sufficient to deter people from risking their elbows and fingers in such an exploit. In cases of real danger a powerful ruffian could accomplish his purpose long before his victim had mounted to the seat, crooked his elbow, broken the window, pulled the bell, opened the window, and called for help in the open air. A better arrangement than the one we have described is that where the suppliant pulls a peg like an organ stop. The lever thus pulled from its place cannot be put back. The guard knows who has summoned him, and

13 U E

can either succor or prosecute, according to the merits of the case. The good Samaritan always comes with a club in his hand.

The pneumatic method of Chevalier, Cheilus & Co. is by far too complicated for description in these pages, and its merits, we believe, have yet to be ascertained. Practically it is a bell rung by means of weights, which are kept in their place until otherwise disposed of by pneumatic means.

RAILWAY POST OFFICES.

In the English department the post office authorities exhibited an excellent working model of the carriages and system adopted in England for depositing and taking up the mail-bags at stations where the main train does not stop. The bags are suspended on poles, secured with a suitable catch. A net sweeps past them, and from its velocity opens the catch and sweeps off the bag, which is then put in the travelling post office, opened and arranged *en route*. This is the process on the cars. The same naturally holds good at the stations—the train holds out the bag, and the station pole seizes it. Thus, whilst travelling at a high rate of speed, letters are both delivered and received without a moment's detention.

The travelling post office consists of three carriages with a continuous communication throughout. Two of them are used for sorting the London and the country correspondence respectively; the third being devoted to the delivery and reception of the mail-bags. All projections in the interior, which are as few as possible, are covered with stuffed cushions in order to lessen the effect of collision on the officials. In these carriages the post office clerks perform their duties. There is a post box in the car, so that when a stoppage takes place letters may be forwarded up to the last moment. The latter convenience is well known and appreciated in various central parts of the United States.

On account of the special nature of this exhibition, the committee of admission to this class thought itself bound to study the statistics of the subject, and reviewed successively the phases of this important branch of French industry as follows:

RAILWAY CONSTRUCTION IN FRANCE.

"On the 1st day of January, 1866, the whole system of railways conceded to companies amounted to 21,000 kilometres, of which the part in working was 13,570 kilometres; remaining unfinished 7,430 kilometres. The total cost of the whole of the lines in work amounted to 6,824,000,000 francs, of which 5,840,000,000 francs was paid by the companies and 984,000,000 francs by the state; the expenditure remaining to be made by the companies amounting to about 1,900,000,000 francs. The cost per kilometre[1] of the completed portion was 500,000 francs, (£20,000,) and that of the remainder is estimated at 255,000 francs for the com-

---

[1] The kilometre is equal to about five-eighths of a mile.

panies' share. With the exception of some special railways and some lines of secondary importance, the whole system of French railways is divided between six great companies. The following statement will show their importance:

"The Northern Railway Company, 1,613 kilometres conceded, 1,197 kilometres completed, 549 locomotives, 1,032 carriages, and 13,123 vans and trucks, at a total expense of 92,172,022 francs for rolling stock and repairing sheds.

"The Eastern Railway Company, 3,088 kilometres conceded, 2,512 kilometres completed, 762 locomotives, 1,962 carriages, and 16,316 vans and trucks, at an expense of 115,832,561 francs for rolling stock and repairing sheds.

"The Western Railway Company, 2,520 kilometres conceded, 1,857 kilometres completed, 514 locomotives, 1,770 carriages, and 10,160 vans and trucks, at an expense of 85,734,342 francs for rolling stock and repairing sheds.

"The Orleans Railway Company, 4,199 kilometres conceded, 3,067 kilometres completed, 690 locomotives, 1,945 carriages, and 12,299 vans and trucks, at an expense of 223,770,000 francs for rolling stock and repairing sheds.

"The Paris and Mediterranean Railway Company, 5,817 kilometres conceded, 3,198 kilometres completed, 1,262 locomotives, 2,108 carriages, and 35, 659 vans and trucks, at an expense of 223,770,000 francs for rolling stock and repairing sheds.

"The Midi Railway Company, 2,252 kilometres conceded, 1,496 kilometres completed, 287 locomotives, 878 carriages, and 9,092 vans and trucks, at an expense of 70,827,885 francs for rolling stock and repairing sheds.

"Various smaller undertakings, 1,511 kilometres conceded, 243 kilometres completed; giving a grand total of 21,000 kilometres conceded, 13,570 kilometres completed, 4,064 locomotives, 9,695 carriages, and 96,649 vans and trucks, at an expense of 690,476,810 francs for rolling stock and repairing sheds, and 655,649,400 francs for permanent way.

"The cost of maintenance during the year 1865 was about 36,650,000 francs for the rolling stock, or 2,800 francs per kilometre; and about 15,000,000 francs or 1,150 francs per kilometre for the permanent way, &c., together 51,650,000 francs or 3,950 francs per kilometre.

"The work done during the year 1865 gave for the whole of the lines the following results: Number of kilometres in work, 13,239; number of persons carried, 84,025,546; average number of railway travellers, 40; total number of travellers to one railway, 3,330,639,807; total number of tons of merchandise carried, 34,049,435; average distance carried per ton, 152 kilometres; total number of tons to one railway, 5,172,847,825; receipts from passengers, 184,245,213 francs; receipts from merchandise, 314,609,184 francs; receipts from parcels, &c., 80,032,447 francs; total gross receipts, 578,856,874 francs; average cost to passengers per railway

0f.0553; average cost per ton, 0f.0608; total cost of working, 266,202,095 francs ; ratio of expenses to gross receipts (general average) 45.98 per cent.

"The employés on the French lines are divided (like those elsewhere) into the permanent staff and workmen and laborers.   On the 1st of January, 1866, the former numbered 60,160, and the latter 51,300, or, in all, 111,460 persons."

REPAIRING SHOPS OF THE RAILWAY COMPANIES.

"The companies, in general, do all that is required for the maintenance of the rolling stock in their own factories.   The number of workmen and others employed for this service amounts to about 20,000 for the whole of the lines, and the salaries and wages paid amount to about 23,350,000 francs, or an average of 1,167 francs per head.   Some companies also construct their own carriages and locomotives.   Such construction amounted, in 1865, to 32 locomotives, 37 tenders, 32 carriages, and 2,570 trucks, and cost 9,180,000 francs.   The railway companies have introduced the system of job work to a great extent in their machine shops, with division of profits amongst the members of each association of workmen, or *pro rata* wages.   This organization has produced the best possible effects, and may be regarded as a starting point of co-operative associations.

"The number of private construction shops, inclusive of the companies, is, for locomotives, six in number; two in Paris, two in Alsace, one at Creuzot, and one at Fives-Lille.   These six establishments can turn out annually at least 450 locomotives and tenders.   The factories for carriages and trucks are nine in number, namely, six in Paris, two in Alsace, and one at Lyons, and they are able to build at least 1,500 carriages and 12,000 trucks.   The total amount of the business of these establishments was, in 1865, in round numbers, 54,500,000 francs, made up as follows: 436 locomotives and 374 tenders, 26,700,000 francs; 1,439 carriages, 8,000,050 francs; and 31,056 trucks, 19,800,000 francs.   These figures include locomotives and carriages exported.   The total number of workmen employed in these factories amounts to about 10,000."

WORKSHOPS AND FORGES.

"The works are engaged in the manufacture of material for the permanent way, not including rails; they are scattered over the whole country and their number is considerable.   Some of these are established on a large scale, but they are not special, and therefore no statistics of any utility can be presented as applying to railways in particular.   As to rails, their production is nearly confined to the thirteen great furnaces situated on coal basins of France.   Of these two are in the department of the Nord, two in the Eastern, three on the basin of the Loire, two in that of Alais, two in Aubin, one in Commentry, and, finally, one at Creuzot.   The whole of these works produced together, in 1862, the period of the largest production, 205,000 tons of rails, of the total value of about 40,000,000 francs; in 1865 the produce was 184,131 tons.

"The iron works and construction shops exported in the year 1865, 193 locomotives and 174 tenders, for the sum of 11,900,000 francs, 420 carriages at 2,700,000 francs, 1,868 trucks at 3,200,000 francs. Total 19,800,000 francs. These figures, compared with those given as the result of the total manufacture in France, show that the reports equalled one-third of the whole amount produced. As regards rails, the statistics of 1865 show an export of 32,860 tons, or, in value, about 6,200,000 francs."

PROGRESS MADE IN THE MATERIAL.

"The progress made during the last ten years in the construction of railway material consists in the constantly augmenting power given to the locomotives, either with the view of overcoming the inclines of 25 to 30 in the thousand, or of running trains of 600 to 700 tons over inclines of four to five in the thousand feet. Thus the power of traction has been carried to 7,000 kilogrammes. The use of coal has almost entirely superseded that of coke by the employment of smoke consuming furnaces, or of well selected coal for the locomotives. The passenger carriages have been made more spacious and comfortable, the trucks have been increased in strength, and their tonnage has remained fixed at from eight to ten tons, with a few exceptions, in which it has been carried to fifteen tons. The construction of safety apparatus has been studied and its application persevered in. We may cite:

"1. The methods of communication by means of electricity between the guards and drivers of the train, and also between them and the passengers, the practicability of which are now being tested on all the trunk lines.

"2. The improvements introduced in the signal disks, their connection with the points of the branch lines, in order to connect the movements of the whole. The breaks have been improved, but they still act as gradual moderators of the speed, the instantaneous arrest of the train being in all cases carefully avoided. Besides possessing very powerful locomotives, engineers are giving great attention to the construction of small engines, employed on railways connected with mines, and which are intended in future to be employed in working agricultural and other local lines. As regards materials, we may mention the use of cast steel instead of iron plate in the construction of boilers. Attempts are being made also to substitute iron for wood in the frame-work of carriages and trucks, as well as for sleepers. Lastly, as regards the cost of manufacture, the following facts deserve special notice. In 1855, locomotives were paid for at the rate of 2 francs 10 centimes the kilogram; in 1866, the price was 1 franc 75 centimes; for tenders, the price was 1 franc 20 centimes and is now 90 centimes. The price of rails at the works was 320 francs per ton, to-day it is about 185 francs. These reductions will give an idea of the economy exercised in the provision of the material and general expenditure in the maintenance of railways."

PROVIDENT AND BENEVOLENT INSTITUTIONS.

"All the great railway companies have organized for their numerous employés funds for assistance in time of sickness and superannuation, and nearly all the companies vote to those funds an amount equal to that subcribed by their servants. Besides this, depots for the sale of articles of food and clothing have been established on several lines, which enable the employés to supply themselves with the necessaries of life at prices varying from 10 to 50 per centum lower than the ordinary rate. At the principal centres of railway traffic, places of refreshment, perfectly organized, have been established, where the employés, laborers, and their families may obtain food ready prepared for them at extremely low prices. The people are charged for what they consume, the amount of credit allowed being in proportion to that which is due to each person from the company. During the whole period of the high price of bread, the companies added to the wages of the workmen and laborers, and to the salaries of others whose income was below 60 pounds a year, a sum equal to the increase in the price of bread, not only for the officer or workman himself, but also for such members of his family supported by him. Lastly, the inauguration of courses of instruction for the workmen, and of schools for the children, and for all who need instruction, completes the organization of the institutions destined to improve the moral, intellectual, and material condition of those who are employed on the several railway establishments."

CLASS 64.—TELEGRAPHIC APPARATUS AND PROCESSES.

In class 64, American ingenuity and invention were conspicuously displayed. Every telegraphic instrument exhibited was more or less on the American principle, as indeed every telegraphic instrument *must* be. The practical value of telegraphy, at this day, is known in America, where it is not merely a political instrument of intercommunication, but a medium for the commonest expressions of domestic wish or want. In whatever country or whatever way a message be sent or received, instruments and methods of American origin are most in use. The fact was recognized by the imperial commission, who awarded the highest honor in their gift to Mr. Morse and to Mr. Hughes.

In the general application of electricity to mechanical purposes the French have advanced far beyond any other nation. The bell which you pull at the doctor's door, tingles so long as you keep your hand on the pull. It is a part of an electrical system which costs a trifle and acts positively, inasmuch as the bell will continue to ring so long as you keep your hand on the pull. This is the simple form. At the hotels they have an improvement on it. At the side of your bed there is a small dial, rather larger than an old fashioned-watch. Except that it is perpendicular, you might suppose that it was a compass. It is indeed supplied with a needle precisely like a compass. This needle has a limited service to perform, but it does it thoroughly. You press a button on the rim, and the needle moving on the surface of the dial tells you that the bell is ringing in the room of service. It continues to ring there until one of the domestics disconnects the wire. At that moment the finger of the dial returns to its place in the room whence the first communication was made, and the visitor knows that the servant *ought* to come. He has the basis, at all events, of a complaint against the management, if the servant does not come.

For railroad purposes, also, electricity is rapidly taking the place of human watchfulness. On many lines there are contrivances where the passing of a train is automatically announced to neighboring stations. The carriages pass over connecting wires and the train records itself before and behind, so that its progress and appearance are alike indicated.

It has been proposed, but not successfully carried into effect, to supply individuals and towns with the correct time by electricity; in other words, to lay it on like water. It stands to reason that if a perfect connection can be obtained, it is as easy to lay on or supply electricity as either water or gas. But so far practice has not come up to theory. The clocks regulated by electricity are the most unreliable in the world, and indeed the clocks of the Grand Hotel, Paris, regulated in this way have been the subject of common ridicule. There can be no doubt, however, that one of these days, companies *will* supply the time just as exactly and correctly as companies now supply the wants of lighting.

The American Commission was fortunate in having Professor Morse to report on the many interesting topics connected directly or indirectly with telegraphy.

FRANCE.

" The several processes applied to telegraphic purposes, and forming class 64, have occupied but a small space in preceding exhibitions; their importance, in fact, in spite of the services they have rendered, only dates from the time when the telegraph called the resources of electricity to its aid. Scientific men then entered upon a numerous series of experiments; and enlisting in their service a number of skilful constructors, they have arrived at results as important as they were unlooked for. The aerial telegraph of brothers Chappe has been made the subject of many improvements by the French telegraphic administration, but it could not attract the attention of scientific men, which was fixed on the discoveries and labors of Galvani, Volta, Oersted, Ampere, and Arago. The first electric telegraph apparatus was based on the action of current upon the magnetized needle, and the magnetization of soft iron under the influence of the same current. Wheatstone in England, and Morse in America, were the first to make (about the year 1839) experiments on lines of any length. The French administration adopted, in 1864, an apparatus founded on the property which soft iron possesses of becoming magnetized under the influence of an electric current; and this French apparatus, as it is called, reproduced the signals of the Chappe tele-graph. It was a useful connecting link between the old system and those which were at once more simple and more complete. At the same time, the railway companies felt the necessity of connecting together by means of telegraphs the principal stations on their lines, and placed simple apparatus for that purpose in the hands of their agent. Since 1855, the Morse system has been adopted in France, where, with the aid of able manufacturers, it advantageously replaced the old apparatus. From this period the labors of men of science and engineers have become more and more numerous, and a great number of new systems have been attempted within a short time. Subsequently a telegraphic printing apparatus was introduced, more rapid in its action than those with arbi-trary signals, and electro-chemical apparatus reproducing with great facility the exact image of the despatch or drawing confided to it for transmission. The telegraphic stations are connected with each other by metallic conductors insulated from the ground and fixed to supports, the elevation, form and dimensions of which vary according to the nature of the weight they have to support. Experiments with underground lines have been made from the very commencement of electric telegraphy, and have since greatly increased; and the results already obtained hold out a legitimate hope that the engineers who are persevering in these interesting labors will attain the object for which they are employing their time and talents. The submarine lines have been brought into successful action since the year 1850; their number has increased con-currently with the improvements which have been introduced into tele-graphic industry, and have resulted in the recent successful laying of the transatlantic cable.

"This special exhibition shows the immense resources which may be looked for in the very varied applications of electricity to telegraphy; they include not only apparatus for writing or transmitting thoughts, but also the piles or sources of electricity; and the conductors, aerial, underground, and submarine, which are their indispensable auxiliaries."

CLASS 65.—CIVIL ENGINEERING, PUBLIC WORKS, AND ARCHITECTURE.

The objects exhibited in class 65, under the general head of civil engi-neering, public works and architecture, comprised four series of groups, which with much interesting matter relating to France will be found described at length beyond. No nation is more occupied with public works involving the highest engineering skill, or possesses a better method of tabulating all that has been accomplished or is yet in progress. The French display was superb. It consisted of models, admirably got up, of bridges, viaducts, reservoirs, docks, tunnels, &c., with plans and particulars of unquestionable accuracy and minuteness. Among these were two models of the swing bridge of Brest, which has a larger span than any bridge of similar construction in the world, being 571 feet, spanned by two wrought-iron lattice frames, revolving upon turn-tables. The foundations of the piers are on the solid rock. There were models and drawings, of several other important engineering works which have recently been completed in the vicinity of Paris.

LIGHT-HOUSES.

One of the most conspicuous objects in the Park was the iron light-house constructed by Mr. Rigolet, and intended for practical use on the rocks called *Les Douvres*, situated midway between the islands of Guernsey and Brehat, off the coast of Brittany. The rock on which this light-house is to be built is in the middle of the south edge of the shoal; its summit is washed at high tide. The masonry foundations are 6 feet 10 inches high; the height of the iron column is from base to floored gallery, 158 feet 6 inches; to top of lantern 184 feet 2 inches. In plan it is a sixteen-sided polygon, 36 feet 6 inches at the base and 13 feet 2 inches at the top; the light being 174 feet above high water. Round the base of the column are the store-rooms and living rooms of the light-house keepers; above these are rooms for the accommodation of persons rescued from shipwreck. The staircase is in the centre. The chief peculiarity of this fine piece of work was that the structure depended for its strength wholly upon its skeleton; the external iron plates being merely a shell upon which no reliance is placed for strength. In wrought-iron light-houses of ordinary construction, strength is obtained by riveting together the plates by which it is composed. The light is dioptric, revolving upon 10 steel friction rollers; the supply of oil is regulated by clockwork.

In the English section was an important exhibit of the dioptric system of August Fresnel, the one now most generally in use. It consists of a structure of segments of glass enveloping a central flame, whose focal rays are parallelized in a horizontal direction and deflected, in the case of fixed lights, in meridian planes only, while in revolving lights the rays are gathered into a number of cylindrical beams, which are made to pass successively before the observer by the rotation of the apparatus.

The Trinity House corporation exhibited the application of the magneto-electric light. The machine is complicated, but it answers its purpose, and is being generally adopted.

MOUNT CENIS TUNNEL.

In the Italian quarter were plans and sections of the famous Mount Cenis tunnel, which, when finished, will connect France with Italy by an unbroken line of railroad communication going through the Alps. The works, which were commenced in 1857, were first carried on by manual labor, a slow and difficult process. They are now carried on by machinery driven by compressed air, and the progress is much more rapid. The present rate is about one yard a day on the French side. The excavation proceeds from both ends, and it is now stated that the probable time when the workmen will meet and shake hands in the middle of the Alps will be some time in 1873.

A series of plans illustrated the principal public buildings and restorations executed in Paris during the last 12 years. The importance of these works may be estimated by their cost, which exceeded 150,000,000 francs.

SUEZ CANAL.

One of the fullest, and at the same time most interesting, exhibits in the way of civil engineering was that made by the company now engaged in constructing a canal through the isthmus of Suez, by which the Mediterranean will be connected with the Red sea. The distance is 72 miles, as the crow flies, and the levels of the two seas only differ to the extent of 6½ inches. The canal will be about 100 miles in length, of which 37 miles are in cutting, while 63 miles are at or beneath sea-level. In order to obtain a sufficient supply of fresh water, an additional canal had to be constructed, bringing its supply from the Nile, a distance of 44 miles. The general dimensions of the maritime canal are: Width of water-level in embankment, 328 feet; ditto in cutting, 190 feet; width at bottom, 72 feet; depth, 26 feet 3 inches.

In the American department was exhibited a plan of the engineering scheme recently adopted for supplying the city of Chicago with water, by which the lake is tapped at a sufficient distance from the shore to insure purity of supply. It attracted much attention, as a bold and successful scheme of engineering.

FRANCE.

"The products exhibited in class 65 form four principal series:

"1. Materials, including natural and artificial stone, bricks, tiles, pottery, lime, cement, plaster, asphalt, and slate. 2. Productions of various trades, occupying a position of greater or less importance in the art of building, such as works in zinc, lead, and copper, sanitary apparatus, joiners' work, and parquetry. 3. Blacksmiths' and whitesmiths' work for building and furniture. 4. Apparatus, machines, and processes used in the execution of architectural and civil engineering works, as well as the models and samples of those works.

"Amongst the trades of this class some are of the very highest necessity, and are represented in every department of France; for instance, the contractor for public and private works, the mason, the carpenter, and the smith; the others, and especially those whose productions are executed in metal, are situated in those localities which are most favorable to their system of manufacture and to the nature of their particular occupation. Around the principal industrial and metallurgical centres are congregated the construction factories for extensive works in metal, such as that of Creuzot, Fourchambeault, and the great establishments in Paris and its environs. Iron-work, such as bolts, latches, window fastenings, screw, and other ironmongery and metal-work used in buildings, is principally manufactured in large works at Charleville, in the Ardennes; Aigle, in the Orne; Rugle, in the Eure; St. Etienne, in the Loire; Beaucourt, in the Haut Rhin; and in the department of the Somme. Black and whitesmiths' work, including locks, railings, gates, &c., is concentrated in the department of the Somme, Fouquierés, Bourg-Dault, Escarbotin, Bettancourt, in the department of Orne, Jura, Loire, St. Etienne, St. Bonnet-le-Chateau, of the Haut Rhin, and of the Haut Saone. The manufacturers of the Faubourg St. Antoine, of Paris, are famous for locks for furniture. The manufacture of objects in copper, lead, and zinc, cast or stamped, is also practiced on a large scale in Paris.

"The principal centres for the trade in cutting and other tools are Molsheim, Zornhoff, in the Haut Rhin; Pont-de-Roide and Valensigney, in the Doubs; St. Etienne, in the Loire; and Paris. Until a short time ago, the difficulty of transport obliged the contractors to supply themselves with stone within a relatively small radius. The exhaustion of the good quarries, especially at Paris, the impetus given everywhere to contractors, and, above all, the development of the means of communication, have greatly modified the old habits. Thus the circle of supply of the capital extends, at the present moment, to the mountains of the Vosges, the Jura, and the Alps. For the purpose of trying the materials, which are offered daily, and are often little known, the administration has opened special laboratories, where the materials of all kinds presented by the public are analyzed and tried gratuitously.

"The methods of manufacture peculiar to the numerous trades which contribute to the execution of architectural and civil engineering works cannot be set forth in a description at once general and abridged. All that can be affirmed is, that in no specialty has the simultaneous concurrence of science and practice, and the intelligent use of machinery, produced results more favorable to the welfare of all and the progress of public prosperity.

"The condition of the work people is as diversified as the nature of the occupations to which they apply themselves, whether sedentary or nomad. The workmen present the greatest variety of character, habits, and natural disposition. The inhabitants of certain districts seem more particularly suited to certain lines of business. The skilled operatives of the centre of France, in the department of Creuze and Correze, possess a special attribute for the masonry works called *limousinage;* the workmen from Piedmont and the neighboring mountains for mining and quarrying. The habits of periodical emigration, and the traditional and, it may be said, inherent skill with which the men of St. Etienne and of some other localities handle the file and the chisel, are well-known facts, and numerous similar examples might be quoted.

"The immense enterprises carried out lately have occasioned great changes in the old usages. The modifications which have resulted therefrom in the habits of the inhabitants of the country give rise to the gravest social and political questions. The trade in building materials is generally local; nevertheless, there are those exceptions, already noticed, occasioned by the necessity of supplying Paris and some other great towns. On the other hand, certain materials, on account of their special qualities or particular circumstances, are sought after far from the places of production. Of these are the granites of Brittany and Normandy, the calcareous stones of Caen, the marbles of the Pyrenees, the serpentines of the Vosges and the Alps, and similar stones, more or less precious, which the soil of France yields in such abundance; the slates of Angers and Ardennes; the various products in terra-cotta; the plasters of Paris, used for light objects and in-door work; the limes of Teil, in the Ardeche, particularly adapted to sea works; the cements of Passy, Boulogne, and Grenoble; the asphalts of Seyssel, &c.

"In the large workshops are constructed edifices, metallic bridges, cranes, dredging machines, lighters, &c., which are exported to Russia, Spain, Egypt, America, &c. The trade of black and white smiths' work has its principal entrepôt in Paris, and the amount of its exports is very considerable.

"To give an idea of the activity, during the last 12 years, of the branch of national work represented in class 65, it is sufficient to state, that in that time 9,000 kilometres of railway have been made in France; that the works in ports for the lighting and erection of beacons on the coasts, for the salubrity of towns, the sewers and the distribution of water, have received a proportionate impulse, and that the greater part of the large towns of France have been completely transformed by their application.

"The committee of admission of class 65 point out, among the principal technical improvements realized since 1855—

"1. The progress made in the trades of hydraulic limes, cements, artificial stones, potteries, slates, and asphalts; and in that of hammered metal, applied to the preservation and decoration of roofs. 2. The increase of the use of metal structures, which are more and more appreciated every day. 3. The increase in the number of machines employed in working wood for joiners' and other work. 4. The constantly increasing application of compressed air in places deep and difficult of access. 5. The ingenious methods of lifting heavy bridges, viaducts, and other metallic works. 6. The new system of movable dams. 7. The recently-invented and powerful dredging apparatus. 8. The application of electricity to light-houses and the new combinations made with a view to assist navigation, among which may be reckoned the creation of a system of coast semaphores."

CLASS 66.—NAVIGATION AND LIFE-BOATS—YACHTS AND PLEASURE BOATS.

The governments of France and England were the principal contributors to class 66. The English admiralty contributed a complete series of models of all the types of ships introduced into the royal navy since the adoption of the screw propeller, and a French firm exhibited a very valuable historical series of models of merchant ships, indicating the many and varied changes which have taken place since 1735.

In the English collection of models was the armor-plated steam gunboat Waterwitch, remarkable for having a hydraulic or jet propeller. By this plan she draws in the water from the sea through a sort of sieve in her bottom. The water is then taken up by a turbine wheel, or centrifugal pump, driven by steam, and thrown out aft with considerable force, the action of the water thrusting the boat forward. The Waterwitch is a double-ender, with a rudder at each end, and has attained the not very remarkable speed of 8.8 knots.

Fishing boats were largely exhibited by Norway and Sweden, and lifeboats by England and France.

Benoit-Champy, the president of the admission committee for class 66, makes the following observations upon boats for river navigation:

"The number of boats registered at the office of the superintendence of the Seine navigation is about 2,000. The continued extension of boat racing, by directing the efforts of the maker toward one special object, has almost suppressed the pleasure-boat of former days. The river sail navition makes use solely of boats of American construction, which are called centre-boards. The Margot, the first American clipper known in France, was imported in 1847, and brought about a complete change in the construction of vessels for river navigation. The plans and models have been improved from time to time since that period, and can now artistically compete with the American builder. During the last few years a great taste for yachting has sprung up among French amateurs, and the Parisians have endeavored to make their clippers of such a size as to reproduce the real yacht models. These large clippers are remarkably swift, and take a most successful part every year in their ocean regattas. Paris and Rouen are the two principal manufacturing centres for the construction of clippers. The manufacturers of Marseilles, Toulon, and Bordeaux produce more especially sea yachts. France has in its several ports 4,696 pleasure boats of all sizes, mounted by 5,776 amateurs, or registered, marines. The boats which took part in the races of the Society of the Regates Parisiennes during the years 1865 and 1866 represent alone a capital of 500,000 francs.

"The steam yachts, used for races properly speaking, are gradually disappearing, and are replaced by boats more especially designed for travelling. Their number increases daily, and the Parisian yachting possesses already three steam yachts. The use of new engines, the elegance and comfort of internal arrangements, the application of well-studied forms combined to swiftness and safety, the realization of great speed, with a reduction in the expenditure of strength, are the improvements exhibited in the recent constructions. One of the most difficult problems would be to create mixed models of steam yachts for the sea and river, enabling amateurs to undertake all kinds of excursions. Steamboats seem to be best adapted for both travelling and pleasure excursions. The tour through France by means of rivers and canals is the aim and ambition of the leading yachtmen. Paris, Rouen, and Angers have produced interesting specimens, but the most important have come from Havre."

In the United States section the model of the American yacht Fleetwing received the recognition of a bronze medel, and the same award was made to the model of the tackle for disengaging ship's boats, exhibited by Messrs. Brown & Level. There were several other exhibits including models of life-boats, life-saving rafts, fishing smacks, rudders and oars.

GROUP VII.

FOOD, FRESH OR PRESERVED, IN VARIOUS STATES OF PRESERVATION,

CLASS 67. CEREALS AND OTHER FARINACEOUS PRODUCTS, WITH THEIR DERIVATION.—CLASS 68. BREAD AND PASTRY.—CLASS 69. FATTY SUBSTANCES USED AS FOOD; MILK AND EGGS.—CLASS 70. MEAT AND FISH.—CLASS 71. VEGETABLES AND FRUIT.—CLASS 72. CONDIMENTS AND STIMULANTS; SUGAR AND CONFECTIONERY.—CLASS 73. FERMENTED DRINKS.

The objects embraced in these classes, especially the first six, although of the highest importance and even interesting when on the spot, cannot be sufficiently preserved or kept fresh for the purposes of a report, save by a professional pen, wielded for professional criticism. The display was a large one, but the specimens were rarely well arranged. People constantly imagine that the common products of their country are not worth taking pains with, when in reality it is precisely these common products that are of vital and national worth. It may be added here that, in almost every important instance, there was a restaurant connected with each country, where the various foods, &c., could be practically tested.

In the Algerian section were several good specimens of the fruit of the *Carica papaya*, or papaw. This, when young, is used for sauce, and water impregnated with the juice acquires the property of rendering all sorts of meat steeped in it tender. Chickens of excessive maturity can be mollified by feeding them on the leaves and fruit, and joints of exceeding toughness are prepared by hanging them for a sufficient time in the branches of the tree.

There was an excellent collection from the United States, consisting of all kinds of fruits preserved in spirits.

CLASS 67.—CEREALS AND OTHER EATABLE FARINACEOUS PRODUCTS, WITH THEIR DERIVATIVES.

The products which are included in this class comprise—

1. Cereals, including different kinds of wheat, rye, rice, maize, millet, buckwheat, and the productions these grains yield for making flour. 2. Vegetable flour. 3. Potato feculæ, tapioca, sago, arrow-root, salep, and other English productions. 4. Grain, ground and packed. 5. Semolinas and groats. 6. Macaroni, vermicelli, nouilles, and pâtes of all kinds of wheaten flour, pure and mixed. 7. Gluten and starch. 8. Alimentary preparations, produced either from meals, feculas, or vegetables.

CEREALS OF FRANCE.

The various kinds of corn and wheat, with their productions, form, in the French exhibition, 11 divisions, corresponding to the 11 territorial divisions:

"1. Paris and its radii, comprehending Isle of France, Brie, Bauce, Gatinais, Champagne, Hurepoix, and French Vexin. 2. The Normandy region, embracing Bessin, Avranchin, Caux, and Normandy Vexin. 3. The Brittany region, which includes upper and lower Brittany, Vendée, Poitou, and Anjou. 4. The Bordeaux region, containing Saintonge, Angoumais, Perigord, Bordeaux, Bazadais, and les Landes. 5. The Languedoc region, comprising the Basque provinces, the Small Landes, Chalosse, Condomais, Bearn, Armagnac, Foix, Roussillon, Lauragnais, Albigeois, and Narbonne. 6. The Provençal region, comprehending lower and upper Provence, Nice, Avignon, and Corsica. 7. The Lyons region, including Dauphiné, Beaujolais, Savoie, Lyons, Bresse, Franche-Comté, Bourgogne, and Niverne. 8. The Auvergne region, comprising upper and lower Auvergne, Limousin, Boulonnais, Forey, and Vivarais. 9. The Maine region, including Maine, Blaisois, Touraine, Berry, and Orleans. 10. The Lorraine region, comprehending Lorraine, Vosges, Alsace, Barrois, Messin, and Berthelois. 11. The Flanders region, in which is included Picardy, Hainault, Flanders, Boulonnais, and Artois.

"In 1820 the number of hectares covered with corn in France was 4,683,788, which have produced 54,347,720 hectolitres. In 1857, 6,543,530 hectares produced 110,462,000 hectolitres. So, from 1820 to 1857, the number of hectares sown with corn has augmented 50 per cent., and the production has nearly doubled. At the present time the number of hectares cultivated is 7,000,000; but the production has not increased since 1857. France exports much more flour than wheat. In 1864 the exportation of wheat amounted to 1,308,480 hectolitres unground, and to more than 2,000,000 hectolitres of flour. Rye is divided into two classes—March rye and winter rye. France yields yearly 20,000,000 to 22,000,000 hectolitres of rye, of which 1,000,000 is employed in the distilleries of northern France, Belgium, and Holland. The growth of rye is diminishing, and is being replaced with advantage by wheat, wherever the nature of the soil admits of it. Barley is divided into two classes: 1. Bearded barley, common barley, &c.; 2. Bare-eared barley, Celeste barley, &c. The barley harvests yield 16,000,000 hectolitres per year, of which 2,000,000 are used in distilleries and breweries. Of these 2,000,000 hectolitres more than one-fourth is sent to England. Oats are divided into two classes: 1. Winter oats; 2. Spring oats. Nearly as much oats as corn are grown in France. The harvest is valued at 90,000,000 hectolitres. Oats are rarely exported; on the contrary, they are often imported from Odessa, Sweden, and Ireland.

"Buckwheat is divided into two classes—common buckwheat and Tartary buckwheat. Buckwheat is grown to an amount of from 6,500,000 to 7,000,000 hectolitres yearly, which is consumed entirely in France.

"The production of maize is confined to three regions: the southwest region, comprehending Guyenne, Poitou, &c.; the southern region, comprising Languedoc, Provence, &c.; and the eastern region, including Bresse, Dijonnais, Alsace, &c.

"Millet is divided into two classes—millet in ears and millet in panicles. The production of maize and millet amounts to at least 6,000,000 hectolitres.

"The sorghos form only one class. The feculas are divided into two classes—that which comes from seed and that which is made from roots.

"The French production of potatoes amounted to 100,000,000 hectolitres per year before the outbreak of the potato disease, 15 years since. It is difficult now to estimate the exact product of this plant. About 10,000,000 hectolitres are planted; 13,000,000 or 14,000,000 are made into fecula; the rest is employed, one moiety for human food and the other for animals. A great part of the fecula is used for making sugar and certain kinds of syrups. It is estimated that for this manufacture alone the produce of more than 7,000,000 hectolitres of potatoes was employed during the year 1865. Since the disease the average yield per hectare has been 77 hectolitres; before that time it amounted to 110 hectolitres.

"The manufacture of *pâtes* may be divided into classes: 1. The northern regions—Paris, Versailles, Meaux, &c.; 2. The midland regions—Clermont, Auvergne, Lyons, &c.; 3. Southern regions—Marseilles, Nice, &c. The amount of pâtes consumed in France has much increased. The addition of fresh gluten is derived from the manufacturers of starch by the washing process, which allows of the richness of the pâtes being augmented at will, and has therefore tended to diminish in great part the difference of quality that existed between the French and Italian pâtes. The latter owe their superiority merely to the nature of the grain, which is richer and more glutinous than the French grain.

"France exported, in 1855, 1,100,000 kilograms of pâtes, of which a quarter was for Switzerland, and the rest for America, the Antilitos, Guyane, the United States, England, and Belgium. The price of the pâtes varies according to the price of wheat.

"The committee of class 67 point out as an evidence of the progress realized since the exhibition of 1865, in addition to the general improvement of cultivation :

"1. The extended cultivation of the best white and red corn, [wheat,] which have less bran and possess more elasticity and extensibility of gluten, and therefore produce flour whiter and of better flavor.

"2. The almost total change in the mode of obtaining starch, which, instead of being procured by fermentation, which causes the decomposition of the gluten, is obtained by the means of washing, a process which produces starch in greater quantities, and much whiter, without deteriorating the gluten; the preservation of grain by means of vacuums; the drying of the flour by mechanical apparatus, working in the open air, which produces flour well dried that can be kept a long time."

14 U E

CLASS 69.—FATTY SUBSTANCES USED AS FOOD; MILK AND EGGS.

Class 69 includes: 1. Conserved milk and the different varieties of cheese; 2. Alimentary fatty substances, such as butter, olive oil, and animal grease; 3. Hens' and other birds' eggs.

The following review of the production in this class is from the report of the committee of admission:

MILK AND CHEESE.

"The production of cow's milk is by far the most considerable, the number of cows in France amounting to more than 5,000,000. The departments of Calvados, Orne, Manche, Seine Inférieure, Loirèt, Nord, and the Vosges, are those which supply the largest quantity of milk. For Paris alone the consumption amounts to about 500,000 litres a day. Milk is sold at from 10 to 40 centimes a litre, according to the localities and the quality; from 25 to 30 per cent. of water is often added. The frauds practiced in the trade are easily discovered by means of the cremometer and butyrometer, and by the amount of sugar in the milk. The preservation of milk is obtained by the original process of M. Appert, and by new improved systems.

"The production of cheese in France is considerable, particularly in the departments of Aveyron, Seine Inférieure, Calvados, Loirèt, Marne, Seine d'Oise, Creuse, Cantal, Vosges, &c. Cheese is generally made by coagulating the caseine of the milk by means of pressure, in a temperature of 68 to 77 degrees Fahrenheit, and straining it on a cloth or in tin molds. The caseine holds the globules of butter, and constitutes the commercial products known under the name of fresh cheeses, such as those of Neufchâtel. When, on the other hand, strong cheeses are required of more decided flavors, and intended for preservation, they are packed in sea salt, and exposed to currents of air in a cool place, and care is taken to turn them often. Under the influence of cryptogamious growths the caseine becomes separated, and gives rise to various products, which communicate new properties to the cheese. The conditions of this manufacture differ according to the varieties of the cheese, thus: Roquefort cheese is made with sheep and goat's milk, in specially constructed cellars, at a constant temperature of about 53 degrees Fahrenheit; Neufchâtel cheese is prepared with milk and cream; that of Camembert with milk skimmed slightly, and with particular care; that of Brie is obtained in the form of a soft paste; and in the manufacture of double-cream cheese cream alone is employed. The importations of foreign cheeses rose in 1862 as high as 5,262 tons, and the exports to 5,027 tons, 1,660 tons of which were of our own production. The annual consumption of cheese is very considerable, Paris alone consuming 5,422 tons, and it would not be far from the truth to state that, for the whole of France, this consumption surpasses 100,000 tons. The Roquefort cellars deliver annually to the trade 2,750 tons; the sale of Camembert cheese amounts to 500,000 francs; and the quantity of Brie cheeses that is sold annually in Paris represents a sum of 1,400,000 francs."

ALIMENTARY FATTY SUBSTANCES.

"The departments of Calvados, Orne, Manche, Seine Inférieure, Indre and Loire, Loiret, Nord, Pas-de-Calais, and Brittany, are the principal places of production for butter. These fatty substances are much used in France, and are extracted from the cream of the milk by means of violent agitation at a moderate temperature in a cylindrical vessel of wood or tin. Steam engines and horse mills are rarely used. The quality of the milk has a great influence on that of the butter. The pleasant odor pervading certain butters, such as those of Isigny, is produced by the plants of the natural meadow lands. The fine butters are generally of an orange color, and possess a delicate flavor; the butter of inferior qualities being of a lighter color. To insure the preservation of butter, it is placed in stone jars, after having been washed several times, and mixed with five or six per cent. of sea salt. This is salt butter. In some localities the butter is warmed, either simply over the fire or in a vessel placed in boiling water on the fire. The scum is then taken off, and when the liquid butter is clear it is poured into stone jars. No fatty substance used for food is so much in demand throughout France as butter. The quantity of this product exported in 1862 represented a sum of 28,962,142 francs, while the quantity consumed in Paris alone amounted to 24,595,850 francs in value. If it be calculated that, for the 89 departments of France, the consumption is six or seven times as large, it may be estimated that the total production exceeds 200,000,000 francs. The market price of the various descriptions of butter varies from 2 francs 20 centimes to 8 francs the kilogram. The exportation of butter and cheese has reached, during the first nine months of the year 1866, the sum of 58,100,000 francs, and the importation to 16,600,000 francs.

"Olive oil, the best of the alimentary oils, is extracted from the fruit of the olive tree, which is grown in some of the departments of the south of France, in Corsica, and in Algeria. The oil obtained by the first expression, or cold drawn, is distinguished by the name of virgin oil. The second extraction, which is effected by heat, produces a condiment much less agreeable to the taste. Œillete oil, and oil from some animal greases, are also used as articles of consumption."

HENS' AND OTHER BIRDS' EGGS.

"Hens' eggs, of which the consumption is so immense, are principally supplied by the departments of Calvados, Orne, Somme, Seine Inférieure, Oise, Aisne, Eure et Loir, Indre et Loir, Seine et Marne, and Pas-de-Calais. In 1853 the quantity of eggs received in Paris amounted to 174,000,000; but the consumption of this article is much more considerable at the present time.   Ducks, Guinea fowls, geese, and turkey eggs are occasionally used."

CLASS 70.—MEAT, FISH, AND VEGETABLES.

Taking into consideration the close connection existing between the products ranged under the classes 70 and 71, which, in many cases, are shown by the same exhibitors, it was decided that these two classes should be united and submitted to the consideration of one jury, which accordingly undertook the control of the united classes and drew up the subjoined report:

"The products included in class 70 are meat, fish, and fresh fruit and vegetables. In the study of the organs which accomplish the digestive process in the human system, such as the active principles (diastasis, pepsin or gasterasis, pancreatic juice, &c.) which divide and dissolve the food, it is evident that man must depend for his nutriment upon animal and vegetable products. Besides the soil and the water, which promote the digestive and assimilative processes, it is certain that for complete nutrition the concurrence is needed of substances taken from the three natural kingdoms, and which comprehend four distinct classes of food, viz: azotic, fat, feculent or sweet, and saline. The chief characteristic of meat and fish in this respect is the abundance of azotized matter assimilated to our own tissues, and which supply the fortifying quality in our food. It is of the greatest importance at the present moment that we should encourage the reproduction of these elements, and it is the insufficient supply of them which most materially affects the strength and health of populations, particularly of those whose daily labor renders a reparative nutrition absolutely necessary to life. On investigating the average consumption of the alimentary products taken from the bovine, ovine, and porcine species, and those supplied by poultry, game, fish, eggs, and cheese, we find that each individual in the population of the eighty-nine departments of France consumes only 57 grams of these azotized alimentary products, while the average ration of an inhabitant of Paris amounts to 273 grams daily.[1]

"The flesh of slaughtered horses is being brought into use in France, when, after having been submitted to the inspection of the proper offi-

---

[1] On the basis of the statistical data furnished to the president of these classes by the minister of agriculture, commerce, and public works, it is estimated that the consumption of butcher's and pork butcher's meat during the year 1862, in the chief towns of departments and arrondissements, and in those towns where the population reaches 10,000, averages for each individual 53 kilograms 600c. per annum, or 146 grams a day, a fact which is still further corroborated by the following quinquennial return: Aggregate consumption: beef, 131,140,910 kilograms ; cow meat, 57,994,541 kilograms ; veal, 61,304,468 kilograms ; mutton, 62,147,482 kilograms ; lamb or kid, 5,268,614 kilograms; pork, 6,110,744 kilograms ; imported meats, 43,324,711 kilograms ; at an average price of 1 franc 18 centimes per kilogram for beef, 1 franc 4 centimes for cow, 1 franc 25 centimes for veal, 1 franc 27 centimes for mutton, and 1 franc 8 centimes for lamb and kid. The total weight of animal food consumed amounted to 422,288,187 kilograms among a population of 7,878,329, giving an average for each individual of 53 kilograms 60c.

cers, it is declared wholesome. This meat makes good soup, and when boiled has an agreeable flavor, although rather hard. Certain parts, above all the fillet, furnish excellent roasts. For some time past horse-flesh has been advantageously employed as food by the inhabitants of the north of Germany.

"By consulting the report of the committee of class 82, and of the committee charged with organizing the arrangement of live fish, crustacea, molluscs, &c., it will be seen what means are used for the maintenance and multiplication of marine and fresh water species, which furnish such abundant supplies of animal food for our subsistence.

"Ripe fruit exercises a favorable influence on the nutrition and health of mankind by introducing sweet, aromatic, azotized, acidulated, and saline principles into their alimentary rations, but it does much real harm when it is wrongly used in too large a proportion, or forms, as it does in some cases, nearly the whole of the habitual food.

"These alimentary substances, which help to vary and render our food more varied and wholesome, have increased to a very large extent in France since the cultivation of kitchen gardens on a large scale has so much developed on the coasts of Brittany, favored by the gentle and temperate climate of those maritime districts.

"A large quantity of the produce of this special culture has been lately exported to England; above all, since, thanks to the increased and rapid means of communication, the early vegetables of the southern districts of France, as well as the oranges, lemons, and various other productions of the Algerian orchards, have appeared in the markets of the metropolis.[1]

"It is well known that the cultivation of mushrooms in the vast quarries of Paris affords an abundant source of alimentary production, and a means of varying the appetizing flavor of our best culinary preparations. In this respect France is the most favored country for the growth of the delicioms and nutritious mushroom, which grows naturally in propitious ground under the shadow of oaks and beeches, but which, up to the present time, has bid defiance to every system of artificial culture.

"The extremely favorable influence of fresh vegetables has particularly manifested itself in the alimentary regime on board ship. It has been shown that sailors could maintain themselves in good health by making use of these productions to vary their diet, which is thus made more agreeable; and that, on the other hand, ships' crews deprived during long voyages of these precious sanitary resources suffered from special affections and particularly from scurvy."

PRESERVED MEATS AND VEGETABLES.

"Class 71 includes meat, fish, fruit, and vegetables preserved by various industrial processes. Preserved meat, fish, fruit, and vegetables are pre-

---

[1] In 1865 there were imported into France from Algeria: table fruits, 2,485,288 kilograms; vegetables, dried, 1,866,958 kilograms; green, 743,386 kilograms; total, 5,095,602 kilograms.

pared in four principal sections of the country. The first group has its centre in the town of Nantes, and furnishes pickled and preserved meat and fish and preserved vegetables. The second group has for its centre Bordeaux, and provides preserved fruit, vegetables, meat, and some fish. Excellent conserves of whole green olives are made in the department of Herault and Bouches-du-Rhône. The collection and preservation of truffles have extended over ten departments of France. The third group has its centre in Mans, and specially treats vegetables and some few meats. Paris is the centre of the fourth group, and prepares preserved vegetables, mushrooms, and some meats.

" The preparation of certain special products, composed of truffles and fatted goose liver, has its principal seat in Strasbourg, though the trade extends all over the southern part of the empire. All the materials employed in these preparations are produced on the soil of France or are supplied by the coast fisheries. Their nomenclature is very extensive, and their prices vary considerably from year to year. Mechanical labor is very seldom required in the preparation, which simply consists in a series of processes, nearly all of which are accomplished by manipulation. The methods of preservation are numerous. The only one which has been applied recently by the trade, besides pickling and concentration, is founded on the remarkable invention of Mr. Appert. It consists of: 1. Washing in boiling water the substances to be preserved; 2. Putting the ingredients into vessels soldered or hermetically fastened; 3. Expelling the air remaining in the closed vessel by boiling for a longer or shorter period, and at a degree varying according to the substance to be preserved. This unique system is diversified according to the nature of the products.

"The theory of preserving substances by the French method, which has been propagated in all the countries of Europe and America, appears to be founded, according to the observations of M. Pasteur, on the destruction, by exposure to a temperature of about 212° Fahrenheit, of the vitality of microphytic and microcosmic germs, which in a living state engender alcoholic, acid, putrid, and other fermentations. A new description of preserved food has latterly been introduced into France from South America; it is a concentrated extract of the meat of slaughtered animals, of which the grease and skins alone were previously used in these countries. This extract can be preserved in boxes which are not hermetically sealed, on the condition that they contain no fat, which would cause rancidity, nor gelatine, which would occasion the development of mould. It represents an amount of solid substance equal to 30 times its weight in fresh meat. It is already largely consumed in Germany, and is largely supplied to armies on service, and completes the quota of cereals and vegetables.

" The manual labor, the general expenses, and the price of the vessel are equal, on an average, to 50 per cent. of the value of the preparation when ready for use. The workmen employed in these various works are all employed in workshops belonging to manufacturers. Some of them—for instance, those engaged in cooking and preserving—are paid monthly, and are occupied permanently during the whole year; the others, such as the tinmen, work by the piece. The last category includes the peelers and others, who are paid by the day and hired by the week or the month, according to the wants or the seasons of production. The preparations produced in France are sold in the great centres of population, to the navy, and, above all, in foreign countries. The makers in the provinces sell directly to the retail dealers, and in Paris either directly or by the medium of small wholesale dealers, but to foreign countries directly or through agents.

"The production of preserved food has greatly developed since 1855, and this development is due to a more perfect knowledge of the best processes, on which depend the preservation of the alimentary substances, to their better application, and, consequently, to greater confidence on the part of the consumer, leading to the increased sale of those articles which presented before but a doubtful chance of success."

CLASS 72.—CONDIMENTS AND STIMULANTS; SUGAR AND CONFECTIONERY.

The exhibits in class 72 include sugar, confectionery, chocolate, liqueurs, condiments, and stimulants.

SUGAR.

France, Prussia, Belgium, Brazil, Austria, and the United States were the chief exhibitors of sugar. Beet-root sugar was conspicuous from the central European countries. The production of this sugar is increasing. In Belgium it is now equal to three-quarters of the whole consumption of sugar. In 1850 and 1851 there were only 28 establishments for the manufacture. In 1855–'56 there were 45, and in 1865–'66 the number had reached 100, and the production was 41,551,834 kilograms.

The following extracts from the official catalogue show in detail the condition of the manufacture of sugar and other articles in this class in France. :

"Sugars include raw and refined sugar and molasses. Raw cane sugar comes from the French and other colonies. The beet-root sugar is principally made in the departments of the north of France. The price of raw sugar is about 61 francs to 70 francs the 100 kilograms, (2 hundred weight,) to which must be added the customs duty, namely, 42 francs per 100 kilograms for beet-root and foreign sugar, and 37 francs 50 centimes for French colonial sugar. After the juice is extracted from the cane or from the beet root it is defecated, clarified, filtered, and bleached; it is afterwards evaporated in various apparatus, to cause it to crystallize, and after that it is purified more or less, according to the quality that is desired, and raw sugar and molasses are obtained. The raw sugar passes afterwards to the refinery, where it is converted into loaf or powdered white sugar.

It is first dissolved in water, so as to form a rather thin syrup, which is afterwards clarified, filtered, bleached, evaporated, crystallized, placed in moulds, and dried in stoves, to be delivered for consumption. It then sells for about 125 francs the 100 kilograms, duty included.

"Sugar-making is conducted in works directed by superintendents and foremen. The refiners buy the raw sugar either of the shippers, of the beet root sugar makers, or of commission agents. The loaf sugar is sold to wholesale and retail dealers, and they export it to England, Switzerland, America, Algeria, Italy, and Turkey. France produces 200,000,000 kilograms of beet-root sugar, and imports about the same quantity from the French and foreign colonies. The consumption is about 250,000,000 kilograms. The difference is exported.

"Since 1857 the manufacture and the refining have made great progress, and this has had the effect of producing sugar at a lower price. The principal improvements to be pointed out are, in sugar-making, the process of double carbonization, triple-action vacuum pans, and the employment of centrifugal machines; and in refining, the improvements in the system of bleaching, the employment of centrifugal machines, and the diminution of the general expenses, by the concentration of work in large establishments."

CHOCOLATE.

"Chocolate-making has become an important trade; it gives rise to the circulation of 30,000,000 francs annually, and is continually on the increase. In the year 1832 the quantity of cocoa consumed in France was little more than 528 tons; in 1863 it had gradually increased to 5,513 tons, which represented a production of 11,000 tons of chocolate, of an average value of 3 francs per kilogram. This increase was due to the employment of machinery, with the aid of which chocolate is manufactured both more cheaply and of better quality.

"The chocolate manufactories are situated in and near Paris and also in the departments of the Nord, Somme, Gironde, Loir et Rhône, and Pyrénées, and use both hydraulic and steam power. As to the establishments where the work is carried on by manual labor, they are now few in number, and are gradually approaching entire extinction. A great number of women are employed in cleaning the cocoa and wrapping up the cakes of chocolate. They are seldom out of work. Both men and women are engaged in the factories of their employers, and the amount of their wages is estimated at about 5 per cent. of the whole value of the production. The home consumption absorbs nearly the whole quantity made. Only 188 tons were exported in 1863; but this exportation would increase rapidly if the duties on the cocoa and sugar were returned on the export of the chocolate. French chocolate is in great repute in adjoining countries.

"The committee of admission have to observe that considerable progress has been made since 1855 in the manufacture, due principally to the improvement of the machinery and plans employed, and to the special pains taken in the manipulation and the materials."

CONFECTIONERY.

"The productions of this section comprise: 1. Sweetmeats, containing almonds and liqueurs; 2. Acidulated and other drops, barley sugar, apple sugar, &c.; 3. Pastiles and lozenges of gum, burned almonds, *fondants* or cream sweetmeats, nougats, drops, bonbons, figures and fancy articles, comfits, and fruits preserved in sugar.

"The principal places of production of confectionery are Paris, Marseilles, Bordeaux, Verdun, Clermont, Ferrand, Lyons, Rouen, and Orleans. The raw materials are sugar, almonds, gums, perfumes, and fruit. The sugar, principally employed in the refined state, undergoes this preparation in France, and the average value of that employed is about 127 francs the 100 kilograms, (2 hundred weight.) Within the last five or seven years sugar obtained by the improvements in the manufacture of beet root juice of the first quality, by means of the apparatus of Cail & Co., which has been employed in confectionery. This sugar, inferior to refined, is now worth about 117 francs the 100 kilograms. The almonds are in a large proportion also of French production, and grown in the departments of the Bouches du Rhône, Herault, Vaucluse, Lower Alps, and Aveyron. The average price on the spot varies, according to quality, from 140 francs to 250 francs the 100 kilograms. Italy and Spain also have latterly contributed a considerable quantity. The price of these varies from 120 francs to 180 francs.

"The gums come exclusively from Senegal and Alexandria. The prices of these at Marseilles or Bordeaux vary, according to the abundance of the crop, from 100 francs to 280 francs the 100 kilograms. The French confectioners generally make use of the most delicate perfumes, such as vanilla, the price of which varies from 40 francs to 100 francs, rosewater, orange-flowers, raspberries, maraschino, &c., to the exclusion of strong-flavored concentrated essences. All these perfumed waters are produced in the south of France, Var, and the Maritime Alps. The price varies from 1 franc to 1 franc 50 centimes for good qualities. The most esteemed fruits are those of the centre of France and Auvergne, and the price in the fresh, unprepared state varies, according to the season, from 20 francs to 100 francs the 100 kilograms.

"The manufacture of the various products of confectionery was carried on entirely by hand until 1845. Since that time apparatus of various kinds, propelled and heated by steam, have successively replaced that primitive method, which is rapidly disappearing. Men alone are or can be employed in this work, but many preparatory operations—long, but not fatiguing—such as the shelling and blanching of almonds, the preparation of fruit and gum, and packing, are reserved for women, who in number equal, if they do not surpass, that of the men. In Paris the wages vary with the importance of the work and the skill of the workmen, from 35 centimes to 65 centimes per hour, and in the case of women from 15 centimes to 25 centimes. The labor is estimated to cost about one-eighth of the value of the whole production. The wholesale trade in confectionery amounts to about 40,000,000 francs per annum, of which three-fourths are represented by sweetmeats and one-fourth by preserved fruits, jellies and jams. Although held in high esteem abroad, these productions are only exported to a very small extent as compared with the home consumption; but the export trade would rapidly assume important proportions if, as in the case of refined sugar, the consumer's tax on sugar were refunded on the export of the goods."

LIQUEURS.

"Paris, Bordeaux, Marseilles, Isère, and, to a less extent, all the great centres of population, possess distilleries. The principal materials employed in this industry are wine spirit, refined sugar, plants, and aromatic substances. The spirit is principally obtained from Languedoc, and the price varies greatly with the season. In December, 1866, it was worth 75 francs the hectolitre, (22½ gallons,) exclusive of duty. The refined sugar is obtained from Paris, Marseilles, and Nantes, at rates varying from 127 francs to 130 francs the 100 kilograms. The aromatic plants are grown in the environs of Lyons and Grenoble, and their prices varies from 50 francs to 300 francs the 100 kilograms. The aromatic substances are vanilla, cinnamon, cloves, and nutmeg, and the prices range between 50 francs and 80 francs the kilogram. The manufacture is accomplished by distillation, with the aid of steam and a special apparatus, more or less perfect, the alembic being the model on which all are based.

"The workmen are always engaged in the establishments of their employer, and their wages range from 4 francs to 6 francs per day. The greater portion of the liqueurs made are for home consumption; still this trade gives rise to an important export, in spite of the large augmentation of price, caused principally by the duty on the consumption of spirits and sugar. The trade is so divided and disseminated that it is difficult to give the precise amount of the annual production; but taking the statistics respecting the transformation of spirits into liqueurs, as given by authority, we arrive at a proximate estimate of 45,000,000 francs."

CONDIMENTS AND STIMULANTS.

"The white wines of the Loire and of the Charente are those which give the best vinegar. The price varies with the season, from 5 francs to 20 francs per hectolitre. The preparation consists essentially in setting in action the principles of fermentation in the wine, which, to that end, is exposed to a given heat in reservoirs prepared for the purpose. For some time the use of steam for the heating, and of machinery for transferring the liquor from one vessel to another, has reduced the cost of the manual labors by two-thirds. Lastly, the theories of Mr. Pasteur on fermentation have thrown light upon many questions which were heretofore obscure. The amount of the annual production of vinegar in France is about 1,500,000 hectolitres, which at the average rate of 20 francs gives a money value of 30,000,000 francs.

"Mustard seed is cultivated in many departments, and specially in the Nord, Pas-de-Calais, the Bas Rhin, and the Charente. The annual produce is 650 tons, worth 150,000 francs. Triturated in special mills, mixed with vinegar, and flavored with various condiments, it is delivered to the trade ready for the table. The quantity produced is about 3,000 tons, of the total value of 2,000,000 francs.

"Fruits and vegetables preserved in vinegar, English sauces, capers, &c., make up a total of about 3,000 tons, and a value of 4,000,000 francs."

"Spices are all imported from America, India, and China, and make up a total of 4,250,000 francs.

"The various countries which supply France with coffee, the use of which has so largely extended, are Brazil, the West Indies, India, and Egypt. The qualities vary extremely, but of all kinds known that which is cultivated in Arabia, and known by the name of Mocha, is decidedly the finest. The prices of coffee range from 2 francs to 3 francs 50 centimes, according to the country of production. The value of the imports in 1864 reached 80,000,000 francs.

"The continental blockade rendered it necessary to find some substitute for coffee, and hence resulted the preparation of chiccory, which, although possessing none of the qualities of coffee, has held its place to the present time, and even progresses in demand, on account of its low price and the similarity in color between it and coffee. The roots of the chiccory plant, cultivated specially in the north of France, and in the Haut and Bas Rhin, are first roasted, and then, after having been properly dried in a stove, are again roasted and reduced to powder. These operations are carried on in well organized establishments on a large scale. The green roots are worth from 4 francs 50 centimes to 5 francs the 100 kilograms. Sliced and dried, they fetch 18 francs to 24 francs. The powder, when prepared, is worth 40 francs to 50 francs the 100 kilograms, and in grain from 50 francs to 60 francs. The annual produce may be estimated at 7,000 tons, of the value of 3,500,000 francs to 4,000,000 francs.

"Finally, the productions which form the subject of the preceding enumeration contribute to the annual industry of France to the following extent:

"1. Sugar, 400,000,000 francs; 2. Confectionery, 40,000,000 francs; 3. Chocolate, 30,000,000 francs; 4. Liqueurs, 45,000,000 francs; 5. Condiments and stimulants, 127,000,000 francs. Total, 642,000,000 francs."

CLASS 73.—FERMENTED DRINKS.

In class 73—fermented drinks, wines, spirits, &c.—there was a very extensive collection from every quarter of the world. The importance of the department may be inferred from the fact that there were 7,700 exhibitors and 22,000 samples shown. France, with her splendid and delicate wines, maintained her known supremacy in this manufacture. The principal particulars of the trade in France, furnished from official sources, are given below.

The German wines, manufactured according to the highest principles of the art, and the produce of wines that are raised with a care which is not bestowed on any other article of human consumption, ranked very high. The best Rhine wines are white; but two celebrated brands, Assmannshausen and Steinwein, are red, and were liberally represented. Johannisberger maintained its position as the king of German wines. It is not, however, sold in the market, except in bad years, when the princely proprietor does not care to retain the wine. There is a large district called Johannisberg, but the vines are cultivated in the usual way, while at Schloss Johannisberg the most unremitting attention, utterly regardless of cost, is paid to them. The district, however, has a good exposure, and very often produces a superior wine.

Of the wines of Spain, Portugal, Austria, and Hungary it is impossible to speak. They were displayed in infinite variety, and of qualities, it may be presumed, that represented the highest kind of production. It was stated, however, by competent judges, that no appreciable advance has of late years been made in the manufacture of wine. A practical method has, nevertheless, been discovered by which undue fermentation is avoided in the case of wines intended for exportation. The wine is subjected to 60° Centigrade of heat. The exposure only continues for a few moments, but the heat effectually destroys all germs of further fermentation, without, it is claimed, injuring the wine.

The wines and beers exhibited from the United States are noticed in the Report on the United States section.

The products shown in this class are divided into four series:

1. Wine of all kinds; 2. Alcohol, eau-de-vie, and their derivatives, kirsch, bitters, &c.; 3. Cider; 4. Beer.

WINE, ALCOHOL, AND BRANDY IN FRANCE.

"Viticultural production is one of the most important in French agriculture. It extends to over 2,287,821 hectares,[1] situated in 81 departments, the yield being, on an average, 50,000,000 hectolitres,[2] of a total value to the producers of 750,000,000 francs. In 1865 the quantity reached 68,942,931 hectolitres, and considering the development that has taken place during the last few years, it is certain, that, unless checked by the grape disease, the oïdium, the amount of 50,000,000 hectolitres will generally be exceeded.

"Vineyard property is excessively subdivided. It is held by no less than 2,200,000 proprietors, so that each property, on an average, scarcely

---

[1] A hectare is nearly equivalent to two and a half acres English.

[2] A hectolitre is equal to 22½ gallons English.

exceeds one hectare. The cost of cultivation varies considerably, according to the season and the rate of wages in the various districts of France. They range from 150 francs to 570 francs per hectare, which give for the rate of wages from 1 franc 90 centimes to 4 francs, and even 5 francs per day. The trade in wine is, of course, a very considerable one. The city of Paris alone consumes annually about 3,600,000 hectolitres; that is to say, an average of 183 litres (a litre is rather more than 1¾ pint) per head for each inhabitant, and this consumption would certainly go on increasing largely if it were not impeded by the present system of taxes, and by their heavy rates. The city or octroi duties, for instance, exceed in amount the value of the greater part of the wine on which they are placed. Exportation increases every year under the influence of the new treaty of commerce. In the year 1866 the exports amounted to 3,194,104 hectolitres, of the value of 308,502,000 francs, while in 1851 the total value did not exceed 195,923,000 francs. Thus, in five years, there has been an increase to the extent of 60 per cent. The value of the exports of spirits and liqueurs amounted in 1866 to 93,970,000 francs, while in 1861 it had not reached over 52,966,000 francs. It had therefore increased to the extent of 80 per cent. in the same period. The total amount of the exports of wine and spirits in 1866 was then 402,472,000 francs. In 1866 the prices were far below those of 1865. This reduction of price, combined with the changes introduced in the English tariff, which make the duty on wine introduced in bottle the same as that imported in the wood, has increased the exports of wine from France to England from 94,385 hectolitres to 205,992 hectolitres; that is to say, an augmentation of 120 per cent. between 1865 and 1866, and it is hoped that this consumption will overcome the obstacles which arise out of the organization of trade in England and the great number of local taxes.

"Sixty-five departments have taken part in the Exhibition of 1867; they are represented by 600 exhibitors. Unfortunately, the Exhibition of the great growths of the Bordelais is far from being complete. As to Burgundy, the chamber of commerce and the agricultural societies and committees have zealously competed in the organization of a most remarkable exhibition. Various processes have been proposed and experimented on recently with the view to the improvement and management of the fermentation of wine, and particularly to make it capable of bearing changes of temperature, and more especially long sea voyages; but the most important improvement to be noticed is certainly that of an illustrious chemist, Mr. Pasteur, who has shown that the greater part of the maladies in wines arise from the development of fermentation from invisible vegetable growths, the germs of which are annihilated when the wine is exposed in closed vessels to a temperature of 60 degrees Centigrade for only a few minutes. Numberless experiments have confirmed the truth of this discovery, and have proved at the same time that this operation does not injure the flavor of the wine, but, on the contrary, very often improves it.

"The production of alcohol has averaged, during the last ten years, 1,124,872 hectolitres, but the increase has latterly been very considerable. Thus, the season of 1863–4 produced 1,278,192 hectolitres; in 1864–5, 1,305,905; in 1865–6, 1,789,474, which is divided as follows: Distillation of wine, 1,200,000 hectolitres, giving in alcohol 1,010,166 hectolitres; distillation of beet-root, 283,022 hectolitres; distillation of molasses, 307,409 hectolitres; distillation of farinaceous substances, 79,648 hectolitres; distillation of lees and fruits, 53,232 hectolitres; and various substances, 55,997 hectolitres. Total, 1,789,474 hectolitres."

CIDER AND BEER.

"The average annual production of cider during the past ten years has been 9,057,570 hectolitres; in 1866 it was 11,323,745 hectolitres, and it increases every year. The railways contribute largely to this result by transporting rapidly the cider apples from the place of production to the centres of consumption. The consequence is that the price of apples has been augmented, and that the farmers find it worth their while to extend their plantations. The consumption of cider is also larger than it was, because in many districts where nothing but water was drank they now make use of cider or beer. The best cider in France is made in the neighborhood of Calvados and La Manche, but it is desirable that the proprietors should bestow the same amount of care upon the cultivation and manufacture as the wine growers.

"We have previously said that the consumption of beer increases considerably in several parts of France where its use was very restricted a few years since. In other localities its use extends even where wine or cider is the common drink of the country. The manufacture has made great progress, and we no longer go to Germany or to England for light, agreeable, and wholesome beer. This development of the brewing trade has produced a similar progress in the cultivation of hops in the northern and eastern departments, and the Vosges and Alsace. At the present time French hops are in as great demand as the best Bavarian hops, and they might pass for them in common.

"These particulars, although very incomplete, show the importance of the trade in the industry of fermented drinks in France, not only on account of the number of persons engaged or interested in the culture of the wine, but also as regards the capital engaged in the production, home consumption, and export."

GROUP VIII.

LIVE STOCK AND SPECIMENS OF AGRICULTURAL BUILDINGS.

CLASS 74. FARM BUILDINGS AND AGRICULTURAL WORKS.—CLASS 75. HORSES, ASSES, MULES.—CLASS 76. BULLS, BUFFALOES, &c.—CLASS 77. SHEEP, GOATS.—CLASS 78. PIGS, RABBITS.—CLASS 79. POULTRY.—CLASS 80. SPORTING DOGS AND WATCH DOGS.—CLASS 81. USEFUL INSECTS.—CLASS 82. FISH, CRUSTACEA, AND MOLLUSCA.

All the classes of Group VIII were represented at Billancourt by a certain number of productions which were renewed every fortnight and divided into fourteen competitive exhibitions. The exhibition was divided as follows:

EXHIBITION OF AGRICULTURAL INSTRUMENTS.

APRIL.

First fortnight.—Ploughs of all kinds, hydraulic machines, steam engines. *Second fortnight.*—Steam ploughs, harrows, extirpating rollers, scarificators, pugmills, and apparatus for making drain-pipes.

MAY.

First fortnight.—Drills for seed and manures, hemp and flax strippers, vehicles, harness, weighing machines, churns, and dairy utensils. *Second fortnight.*—Mowing machines, winnowing machines, rakes, hay-making apparatus, and apparatus for tying and the preservation of hay.

JUNE.

First fortnight.—Competition in farriery and examination of specimens of rural establishments. *Second fortnight.*—Chaff and root cutters, horse hoes, &c., mills.

JULY.

First fortnight.—Apparatus for clipping various domestic animals. *Second fortnight.*—Reaping machine and other harvesting apparatus.

AUGUST.

First fortnight.—Threshing machines and other apparatus for the cleaning and preservation of grain. *Second fortnight.*—Portable ovens, apparatus for cooking vegetables, washing linen, and manufacturing manures.

SEPTEMBER AND OCTOBER.

Examination of specimens of various agricultural industries.

ANIMALS.

APRIL.

First fortnight.—Breeding sheep.   *Second fortnight.*—Fat animals.

MAY.

First fortnight.—Dairy cattle; breeders.   *Second fortnight.*—Sheep for wool; breeders.

JUNE.

First fortnight.—Horses and other animals for draught.   *Second fortnight.*—Poultry and small animals.

JULY.

First fortnight.—Cattle for labor; breeders.   *Second fortnight.*—Saddle horses, hunters, carriage horses, ponies, &c.

AUGUST.

First fortnight.—Dogs.   *Second fortnight.*—Draught oxen.

SEPTEMBER.

First fortnight.—Pigs, breeders.   *Second fortnight.*—Asses, mules, &c.

OCTOBER.

First fortnight.—Fat animals.   *Second fortnight.*—Animals acclimatized or capable of being so.

GROUP IX.

LIVE PRODUCE AND SPECIMENS OF HORTICULTURAL WORKS.

CLASS 83. GLASS HOUSES AND APPARATUS.—CLASS 84. FLOWERS AND ORNAMENTAL PLANTS.—CLASS 85. VEGETABLES.—CLASS 86. FRUIT TREES.—CLASS 87. SEEDS AND SAPLINGS OF FOREST TREES.—CLASS 88. HOT-HOUSE PLANTS.

CLASS 84 TO 88.—FLOWERS AND ORNAMENTAL PLANTS.

These classes were represented at the Exhibition in the French section by products renewed every fortnight, and gave rise to fourteen series of prize competitions. A special catalogue was published, and only a summary of the proceedings can be given here.

RESUMÉ OF THE FOURTEEN COMPETITIVE SERIES.

(First Series, from April 1 to 14, 1867.)

Principal exhibition.—Camelias in flower.
Minor exhibitions.—New plants reared from the seed, hot-house plants, (orchids, bromelia, ferns.) Greenhouse and conservatory plants, (erica, acacias and mimosa, herbaceous ferns, amaryllis, stocks, cinerarias, Chinese primrose, daphnes, cyclamens, mignonette.) Ligneous plants for the open air, (holly, magnolia grandiflora, yucca, ivy.) Bulbous plants, (hyacinths, tulips, saffron.) Forced shrubs, (lilacs, rose trees, and others.) Fruit and vegetables, (pine-apples, early fruits, fruits of 1866, &c.) Fruit trees pruned and trained, (pear, peach, cherry, plum, and apricot trees, vines,) standard fruit trees.

Second series, (from April 14 to 30, 1867.)

Principal exhibition.—Conifers.
Minor exhibition.—Hothouse plants, (orchids, cacti, lycopodium, selaginella, &c.) Plants grown in heated beds, (agave, aloe, Bonaparteæ, dasylirion, litzæa, yucca, rhododendrons, epacris, erica, cinerarias.) Herbaceous plants, grown in the open ground, (hyacinths, pansies, primroses, stocks, &c.) Ligneous plants, grown in the open ground, (magnolias, rose trees, &c.) Early vegetables.

Third series, (from May 1 to 14, 1867.)

Principal exhibition.—Azalea indica, rhododendron arboreum.
Minor exhibitions.—New plants of all kinds. Hothouse plants and plants grown in heated beds, (orchids, &c.) Plants cultivated for the decoration of apartments. Greenhouse bulbous plants, (ixia, sparaxis.)

15 U E

Plants of all kinds grown in the open ground, (peonies, rose trees, clematis, Gessner tulips, pansies, auriculæ, mignonette, gladiolus, &c.) Vegetables and fruit, (vegetables in season, early vegetables, pine apples, &c.)

Fourth series, (from May 15 to 30, 1867.)

Principal exhibition.—Palms and cycadeæ.

Minor exhibitions.—Hothouse plants, (orchids, iscara.) Plants grown in heated beds and conservatories, (azaleas, calceolarias, Himalaya rhododendrons, &c.) Ligneous plants, grown in the open ground, (clematis, rose trees, &c.) Herbaceous plants, grown in the open ground, (peonies, ranunculuses, anemones, daisies, and others.) Vegetables and forced fruit, (grapes, &c.)

Fifth series, (from June 1 to 14, 1867.)

Principal exhibition.—Orchids and pelargonium in flower.

Minor exhibitions.—Hothouse plants, (caladium bulbosum, &c.) Greenhouse plants, (calceolarias, verbenas, &c.) Herbaceous plants, grown in the open ground, (rhododendrons, azaleas, kalmia, rose trees, &c.) Vegetables and forced fruit, (melons, &c.)

Sixth series, (from June 15 to 30, 1867.)

Principal exhibition.—Roses and pandanæ.

Minor exhibitions.—Pelargonium in flower. Hothouse plants, (orchids, Theophrasta, clavija, maranta, calathea, phrynium, bananas, begonias.) Plants grown in heated beds and conservatories, (orange trees, lemon trees, verbenas, calceolarias.) Herbaceous plants, grown in the open ground, (larkspurs, irises, 10-week stocks, indigenous orchids, Alpine plants, peonies, &c.) Vegetables in season. Exotic and indigenous fruits, (bananas, cherries, strawberries.)

Seventh series, (from July 1 to 14, 1867.)

Principal exhibition.—Pelargonium zonale and tree ferns.

Minor exhibition.—Hothouse plants, (exotic, useful, and officinal plants, orchids, pitcher plants, gloxinia, caladium bulbosum.) Plants grown in heated beds, (petunias, rochea, crassula, saracenia amaryllis, lilium auratum.) Plants grown in the open ground, (larkspurs, mignonette, climbing roses, roses, &c.) Vegetables in season, (mushrooms and others.) Fruit, (cherries, strawberries, &c.)

Eighth series, (from July 15 to 31, 1867.)

Principal exhibition.—Pinks and hothouse plants.

Minor exhibitions.—Hothouse plants, (exotic fruit trees, gloxinia.) Greenhouse plants, (lantana, petunia.) Herbaceous plants, grown in the open ground, (officinal plants, phlox, penstemon, shot, climbing roses, gladiolus, larkspurs, phlox Drummondii, &c.) Ligneous plants, for decoration, (hortensias, &c.) Fruit bushes, (stone fruits, berries, melons. Vegetables in season.

Ninth series, (from August 1 to 14, 1867.)

Principal exhibition.—Fuchsias and gladiolus.

Minor exhibitions.—Exotic climbing plants, (passion-flowers and others.) Greenhouse plants, (heliotropes, cape heaths.) Plants grown in the open ground, (dahlias, pinks, climbing roses, phlox decussata, lilies, zinnia, lobelia, nasturtiums, hortensias, &c.) Stone and other fruit, (berries, grapes, peaches.) Vegetables in season.

Tenth series, (from August 15 to 31, 1867.)

Principal exhibition.—Aröides.

Minor exhibition.—Hothouse plants, (orchids, Gesnera, achimenes, nagelia, sinningia.) Greenhouse and conservatory plants, (fuchsias, erythrina, pelargonium zonale and pelargonium inguinans, plants for hanging baskets.) Perennial plants grown in the open ground, (dahlias, climbing roses, penstemon, phlox, pinks, &c.) Annuals, (china asters, balsams, zinnia, and others.) Bulbous plants, (lilies, gladiolus.) Ligneous plants, grown in the open ground. Aquatic plants. Vegetables in season, (melons and others.) Fruit bushes and trees, (fruits with pips and with stones, peaches, grapes, figs.)

Eleventh series, (from September 1 to 14, 1867.)

Principal exhibition.—Dahlias.

Minor exhibitions.—Hothouse plants, (dragon trees, croton, allamanda.) Greenhouse plants, (fuchsias, veronicas, pelargonium zonale and pelargonium inguinans.) Plants grown in the open ground, (dianthus sinensis and dianthus Hedewigii, china asters, balsams, and others.) Ligneous plants grown in the open ground, (rose trees.) Bulbous plants, (gladiolus and others.) Vegetables in season. Fruits with pips and stones, (peaches, grapes, figs, pine-apples.) Trees with caducous leaves.

Twelfth series, (from September 15 to 30, 1867.)

Principal exhibition.—Araliaceæ.

Minor exhibitions.—Hothouse plants, (canna, solanum, ficus, hibiscus, musa, and others.) Greenhouse plants, (fuchsias, pelargonium zonale and pelargonium inguinans.) Plants grown in the open ground, (gramineous plants, dahlias, chrysanthemums, asters, gladiolus. Ligneous plants, (roses, bamboos.) Annuals of various kinds. Vegetables in season.

Fruit, (grapes, fruits with pips, with stones, cucurbitaceæ, strawberries.)

Thirteenth series, (from October 1 to 14, 1867.)

Principal exhibition.—Fruits of all kinds, and Indian chrysanthemums.

Minor exhibitions.—Hothouse plants, (orchids and others.) Vegetables in season, (potatoes, cabbages, mushrooms, Indian potatoes, watermelons.)

Fourteenth series, (from October 15 to 30, 1867.)

Principal exhibition.—Vegetables of all kinds.

Minor exhibitions.—Ligneous plants grown in the open ground.   Vari
ous systems of multiplication for fruit trees, forest plantations, chrysan
themums, and other plants.

Special shows of bouquets and natural flowers.

GROUP X.

ARTICLES EXHIBITED WITH THE SPECIAL OBJECT OF IMPROVING THE PHYSICAL AND MORAL CONDITION OF THE PEOPLE.

CLASS 89. APPARATUS AND METHODS USED IN THE INSTRUCTION OF CHILDREN.—CLASS 90. LIBRARIES AND APPARATUS USED IN THE INSTRUCTION OF ADULTS AT HOME, IN THE WORK-SHOPS, OR IN SCHOOLS AND COLLEGES.—CLASS 91. FURNITURE, CLOTHING, AND FOOD FROM ALL SOURCES, REMARKABLE FOR USEFUL QUALITIES, COMBINED WITH CHEAPNESS.—CLASS 92. SPECIMENS OF THE CLOTHING WORN BY THE PEOPLE OF DIFFERENT COUNTRIES.—CLASS 93. EXAMPLES OF DWELLINGS CHARACTERIZED BY CHEAPNESS COMBINED WITH THE CONDITIONS NECESSARY FOR HEALTH AND COMFORT.—CLASS 94. ARTICLES OF ALL KINDS MANUFACTURED BY WORKING MASTERS.—CLASS 95. INSTRUMENTS AND PROCESSES PECULIAR TO WORKING MASTERS.

The articles contained in Group X were of a very miscellaneous character, and in fact were borrowed from twenty-one of the preceding classes, to be massed here.

The most interesting subjects for study were the school appliances, and the cheap houses for workmen. Germany, Switzerland, and the United States excelled in the former Exhibition not only the machinery of education, but the school-houses themselves. In the matter of economical cottages for laborers there were many competitors.

The Emperor was among the number, and obtained the principal prize, which was handed to him by the Prince Imperial, the president of the commission, on the day of the distribution of rewards.

CLASSES 89 AND 90.—APPARATUS AND METHODS USED IN INSTRUCTION.

"Among the institutions which concur for the physical and moral improvements of the working classes, the Imperial Commission has placed in the first rank the educational establishments which, from the *crêches* (asylums where the infants of female operatives are taken care of during the day) to the special schools, develop in the child and the youth, the apprentice, and the workman, the qualities of intelligence and character, and initiate them in that theoretical and practical knowledge which will guide them in all the phases of their existence, and render them fitted for any position in life."

The following observations on education in France were drawn up by M. Charles Robert, one of the vice-presidents of the united juries of Group X; M. Marguerin, member of the committee of admission of class 89; M. Ph. Pompée, vice-president of class 90; M. Barbier, member and delegate of class 89, was added to the commission. The report was

translated for the English official catalogue, from which it is here reproduced.

"The duty of collecting and classifying all the articles which could properly be shown as illustrating our public system of education, has been
confided to the committees of classes 89 and 90. The first was designed
for the admission of all the works adapted for the education of children
from their birth to the time when, their intelligences being developed,
they could either continue their special studies or enter immediately into
apprenticeship for the callings for which they were ultimately intended.

"The committee of class 90 is charged with the investigation of all those
institutions which tend either to recover lost time, to perfect the education already received in the primary schools, or to afford new acquire-
ments to youth or adults, which would permit them at a future period
to bring their works to the greatest perfection of which man's creations
are capable. But, if the institutions for teaching may be theoretically
divided into sections, as we have just done, they cannot be practically
so separated. The education of man is a thing complete in itself, which,
though it has its degrees, cannot, without great inconvenience, be subjected to change of direction, proceeding, or method. Thus, no sooner
had the united committee of the two classes commenced this work, than
it was found how difficult it was to determine to which class appertained
certain Exhibitors who had productions interesting both to the adult
classes and children's schools, and sometimes to every description of
scholastic institutions. An understanding between them being indispensable, a methodical and reasonable distribution of their respective
duties was arranged by a mixed commission; and while at the same
time they each separately preserved their own individuality, the two
committees of admission combined their efforts so as to give to this part
of the Exhibition the necessary unity and completeness. It is also for
this express purpose that this preface has been compiled in common by
the members of the two classes.

"Before entering into details concerning the articles exhibited, we cannot help stating that the space allotted for the two classes 89 and 90 has
been quite insufficient to present its whole development, or to give an
adequate idea of the details and ensemble of our vast system of public
instruction. However, we feel convinced that incomplete as this Exhi-
tion is, it will prove to our own countrymen and to foreigners that public instruction has made in the last few years immense progress in France,
thanks to a liberal and prolific impulse; and that our public and private
establishments are worthy of a nation so enlightened and advanced as
ours proves herself in all the branches of human activity.

" For the first time, at the Universal Exhibition of London in 1862, a
particular class was created to receive the school requisites, works, and
materials, but this was limited to infant schools and special schools for
drawing.

"The French Exhibition of 1867, however, embraces, on a much more comprehensive scale, all kinds of education—that of adults as well as of children, their professional education as well as technical education; and, acting up to its universal character, presents for examination the various evidences of the intellectual activity of the country. Therefore, whereas the Exhibition of London only numbered 180 exhibitors in this class, that of Paris possesses as many as 500 exhibitors, which, however, is less than half the number who applied to the Imperial Commission for admission.

"A rapid progress has been realized during the last five years, and a still more marked advance is in process of realization, to bear fruit in no very distant future. These are the results proved by facts in the exhibition of classes 89 and 90.

"I. The hygienic condition of school buildings, the judicious disposition of the interior, the arrangement and installation of the whole, are subjects of vital importance in educational matters. These requirements are now better understood and more ably carried into effect. The new schools are better distributed, besides affording the scholars a more ample supply of air, light, and space. A large number of old buildings have been greatly improved in this respect, and arranged in accordance with this principle. The impetus has been given, and this transformation will gradually be extended to the smallest and most insignificant villages.

"II. The institution of *crèches*, or infant asylums, which has been tried for some time, is now regularly organized, and is showing a progressive development. It is the same with the *salles d'asile*, (infant schools,) which are under an august and charitable patronage. France numbers 3,572 public infant schools; 264 were founded between 1863 and 1865, and during the same period the inmates of these schools were augmented by 34,912 children. In the rapidly increasing training schools for the education of teachers, school-mistresses are taught the use of those gymnastic exercises and games which make these dwelling places of youth more gay and wholesome to the little inmates.

"III. The progress in the management of the elementary schools is still more marked in every way. From 1863 to 1865 the number of communes possessing no school was reduced from 818 to 694; 938 new schools have been founded, and the scholars, which now number 4,436,470, have been augmented by 100,102 children. The communal or free schools for girls and boys amounted in 1865 to 69,699; and they are also attended more regularly, and are less frequently abandoned by the pupils, after their first communion. The institution of cantonal examinations, and prizes and primary instruction certificates, have had a most happy and surprising influence. On the other hand, the instruction has not remained stationary, having considerably extended, inasmuch as different branches of study, which were previously optional, have now become obligatory; it is also much improved by a more intelligent direction; agriculture and horticulture are being taught with eminently satisfactory results; in a word, the general improving tendency is to make primary instruction a broad, solid basis, on which may rest the education of adults—special, secondary, and technical—in accordance with the mental capability and requirements and the future career of the students.   Besides these improvements, the position of the instructors is much improved; they are better remunerated for their services; their interests are protected; elevated in the eyes of the population by public recognition of their services, and allowed to participate in all honorary distinctions.   They are thus more than ever encouraged to devote themselves to the diffusion of public instruction, which is the special requirement of the times and the sincere wish of the entire French population.   The improvements have been attained principally by boys' schools; with the girls' schools the results have been less marked; but, happily, the new law on female education, the project for which is now before the legislative body for approval, will soon give an impulse in the right direction.   An evident and progressive improvement in the science of training teachers, and in the methods and ways of teaching, is an unmistakeable sign of the vital interest that this question of education excites in the country. Proofs of this are everywhere shown in the French Exhibition of 1867.

"IV. The science of teaching and scholastic training, which only can be an efficient auxiliary to national education when it is thoroughly imbued with the immutable principles of religion and morality, and this truth is represented by works worthy of French literature.   It is not only taught in all the primary normal schools, but the taste for the study of this science is kept up among the schoolmasters by annual conferences—a system which is rapidly becoming a part of our scholastic education. The improvement in the methods and plans of instruction are manifest ever since the English Exhibition of 1862.   The elementary books and treatises for the teaching of reading, writing, grammar, arithmetic, history, geography, and drawing, have gained much in simplicity and clearness; they are more practical; they are more impressed with the end they have in view, while they spare the child both time and trouble. Their progress may be appreciated by the study of the productions of the pupils.   The needlework done by little girls shows that works of a fanciful and frivolous kind have given place to those of utility and family necessity.

" V. The province of education is to study the physical growth of children.   To the ordinary gymnastic games must be added rational gymnastics, which may rule and complete the first.   The apparatus exhibited show that there are plenty of means for physical education, but it is often impossible to organize them practically, while French habits make it more difficult here than elsewhere.   If gymnastic exercises have not yet managed to bring together the youthful population with the view of public recreation, singing has had the advantage of so doing.

"The new choral societies which are daily organized, the Orphéonic gatherings which take place periodically, the cheap musical publications that have spread widely, prove that music is fully established in the tastes and habits of the people. The most distinguished French composers are now engaged for the Orphéons, and classical music begins to arrest the public attention. The Central Patronage Committee of the French Orphéons, recently inaugurated, will forward this great movement, and give it encouragement and impulse. The Exhibition gives evidence of the considerable development given to singing during the last few years.

"VI. The French educational system would not be in accordance with the charitable habits of our country if it did not endeavor to ameliorate the condition of those unfortunate beings whose infirmities have long condemned them to loneliness. The Exhibition shows us many recent improvements in the contrivances for educating the blind and the deaf and dumb. By rendering study less irksome, these contrivances facilitate their communication with the world, and the possibility of employing talents which would otherwise be rendered useless. The endeavor to educate the deaf and dumb in ordinary schools is too recent yet for the result of the trial to be appreciated; the future will show what is to be expected from it.

"Lastly, even the idiots partake of the universal progress of a civilization which becomes more humane as it becomes more liberal. These unfortunates are received into special establishments, and attended to with the most ingenious care. It is no longer a hope, but a certainty, that these poor children can often recover, with a part of their moral consciousness, somewhat of the faculty of participating in the feelings, objects, and ordinary occupations of the more favored portion of the human family.

"VII. The improvement in the ordinary means of education provided for children, and the restoration of those who may be called the disinherited of nature, were deemed scarcely sufficient by the friends of progress in France. They felt it to be necessary that a great educational system, extensive, varied, open to all those who wished to teach, as well as those desiring to learn, should be made available to adults, offering the means of repairing the errors of their parents and guardians, or the negligence of youth, the means of extending the elementary knowledge received in preparatory schools, and finding in superior instruction suited to their peculiar avocation the legitimate reward of their labors.

"The ministerial orders suggesting lectures and evening schools for apprentices and grown-up people, responded to this double want. Private efforts had, it is true, in this instance preceded official decrees. Several societies had organized in various places, especially on the behalf of town workmen, means for scientific instruction. The Polytechnic Association, which dates from 1830, numbers now 22 different sections in Paris and its environs, while it has founded and endowed a much larger number in various departments, showing that individual enterprise has been in no wise idle. However, it was only an energetic will appealing from high quarters, such as that of our minister of public instruction, to the general intelligence of the country, that could, in the brief period of two years, determine this, and inaugurate such a vast educational movement which, from the 1st of January, 1864, to the 15th of December, 1866, augmented the number of adult educational institutions from 5,623 to 28,546, and thereby created a spontaneous accession of 600,000 voluntary pupils. These institutions have adopted two different methods of instruction, each useful in its way, that of lectures to open the minds of the public and enlighten them on various important subjects, and that of lessons for the purpose of imparting precise instruction. The future can alone determine how far the system of lectures will enter into the habits of our country; but it is of paramount importance to her dearest interests, to her prosperity and her dignity, that the regular education of adults, which heretofore has only been sustained by precarious resources and by the devotion of the teachers, should be systematized and established as a great public institution. This is the object of the law on public instruction now under consideration by the legislative body, whose business it will be to place adult education on the same permanent footing as the legislation of 1833 did for the institutions for the instruction of the youthful population. The education of apprentices and adults when it passes beyond the limits of elementary instruction changes its character, and enters into the arena of applied science and art. The programmes of the various societies which have for their object the education of the working man, clearly show the spirit and the limits of the enterprise. However, the recent introduction of the teaching of living languages, commercial geography, and political and industrial economy, cannot fail to tend to generalize, and to constitute for the working classes a superior order of education, nearly analogous to that adopted in special and high-class educational institutions.

"VIII. If we except some few departmental centres where public instruction is favorably endowed, the teaching of the applied arts is much better organized and more sought after than that of sciences. The practical and successful results achieved by the system of teaching adopted in the drawing and modelling schools, secured for France an honorable position at the Exhibition of London in 1862, and it has since shown still more marked improvement. Paris, which is the chief city of the world for the manufacture of the productions of industrial art, has naturally put itself at the head of the movement and set the example. The institution of a certificate of master or mistress of arts as a reward for skilled teachers, the introduction of drawing into the primary schools for girls and boys, the reorganization of evening classes for male adults, the opening of numerous lay schools for female adults, annual competitive examinations between classes of the same degree, a more enlightened and elevated object given to instruction, the renewal of models, and the formation of collections according to the rules of the most severe taste—these are the great educational advances in which the municipality and the state both participate, and which may be fully appreciated by the contemplation of the productions of the pupils, to be seen at the Exhibition.

"The objects sent by the provincial towns also exhibit most favorable results. The workmen, as well as the manufacturers themselves, are beginning to understand that the superiority of our productions in an artistic point of view must be maintained by the increased cultivation of artistic and scientific taste.

"IX. Besides the primary schools and the educational courses for adults, which meet the wants of popular instruction concurrently with the classical colleges and collegiate institutions—the studies in which are adapted only to the demands of certain social positions and limited careers—the middle classes require a system of education more accessible in its conditions, more economical in its cost, and better suited to the wants of a community in which the sciences are constantly improving all branches of industry, and which brings up its children in a liberal manner, and prepares them at once, without any limitation, for agriculture, trade, and commerce, as well as for the arts and public employment.

"The law of the 21st of June, 1865, completing former enactments, the superior primary instruction of 1833, the special education of 1847, and the professional education of 1850, inaugurated the system referred to. At the same time that it was completely established in new schools, secondary special education rallied round it under a common denomination, but on a broader basis, the establishments which preceded it, namely, the superior primary schools, which, as a rule, are not successful, the professional schools, which remain isolated, and the commercial colleges, where primary special education has not yet attained its development. A series of supplemental arrangements have established the new system on a solid basis.

"The action of the Council of Surveillance renders it easily adaptable to the several localities without injuring its original character; the diplomas conferred at the conclusion of the course of study set before the pupils what is always necessary to sustain and stimulate them in their work; the normal school of Cluny, the certificate of capacity, and the junction of literature with science, insure the services of a body of professors who will bring into their classes the habits of method and the spirit of study under which they themselves have been formed. It is, then, not too much to hope that the system of practical secondary education—that is to say, properly speaking, the education of the middle classes—is founded in our country on a definite principle. The Exhibition presents, as it were, an inventory of all this work of formation to which, during the last 30 years, the state, the municipal authorities, the chambers of commerce, the industrial societies, and, in a marked degree, private individuals, have contributed. This multiplicity of efforts has produced a great variety of combinations in the programme of studies; still the leading ideas show themselves clearly, and indicate the current of thought and national requirements.   This new system either sets aside the study of languages altogether or renders them entirely subordinate, and calls in the working element only as a means of, or preparation for, truly professional schools.   In spite of its name, it is a system of general education, that is to say, theoretical, with a marked practical character.   It leads to applications, but it does not insist upon them.   It stops where apprenticeship begins.

"X. Primary instruction, developing itself in adult classes, gives to the apprentice and artisan notions of science which they may apply in their own occupations; secondary special instruction initiates its pupils in scientific theories, of which they will find the applications in their workshops, when they become foremen and manufacturers; neither the one nor the other does away with the necessity of apprenticeship in any case.   Apprenticeship, however, has a diminishing tendency, in consequence of the conditions of modern industry.   In respect of free industry, that it should produce, according to the ancient system, good apprentices, and, consequently, good artisans, would be, in respect to many professions, a complete illusion.   This was clearly shown by the result of the inquiry opened by the minister of public works in 1863.   The natural force of circumstances has left technical education to supply the deficiency of instruction during apprenticeship.   Technical education existed, in fact, before the name was known.   The government, in order to meet the varied national wants, long since organized various establishments, where real professional apprenticeship was practically carried out.   The schools of agriculture and the farm schools, the schools of art and manufactures, the naval school, &c., are establishments for technical or professional education, which are here synonymous terms.   Private enterprise did still more, because, being unable to incur the same expenditure as the state, it acted in a more practical manner.   The inquiry has made known the useful creations of industrial societies, of large companies, of chiefs of works, of heads of free institutions, of congregational establishments, who have in opposite parts of the empire realized the apprenticeship of determined professions with more or less success. But, in face of the ever-increasing mass of wants, it was evident that it was necessary to encourage and to regulate technical education by making it general.   This is the object of a bill now before the legislature. The object of technical education differs then clearly from secondary special education.   The latter remains always general, leads to all the industrial professions, but only lends itself to practical work exceptionally.   It is essentially an education; the former, on the contrary, is particular; it prepares pupils for a fixed profession; it has recourse to education only as an assistant; it is an apprenticeship.

"XI. The diffusion of education cannot be carried out without the diffusion of books; they are the auxiliaries of education, and are, moreover, themselves teachers.   The colportage—that is to say, the sale of books by hawking or otherwise than in shops—can neither diffuse them in sufficient numbers, give adequate extension to circulation, or place them in all hands. Its business is trade, not education; and, even regulated as it is, it cannot furnish sufficient guarantees. The establishment of libraries in all the communes of France, lending or hiring out books, placing them within the reach of all, was the necessary object of the propagation of education. Set on foot by the minister of public instruction, established in the communal schools, kept by the schoolmaster, the scholars' libraries were the first established. There are at present 8,000 libraries, which lend 500,000 books per annum. But ministerial action was not enough to endow 40,000 communes with libraries, and public spirit came in aid with remarkable alacrity. A great number of free societies have been formed for this special object; some including the whole of an old province, such as Alsace, in their action; others a department, and the rest purely local in their action. Many in Paris attempt to organize for themselves centres of action from which to operate on the country around, either in giving their assistance in the formation of libraries, or in making known and encouraging good books, or by influencing the colportage. Whatever may be the extent of their operations, or the mode of their action, they all concur in maintaining a healthy agitation, which has already borne good fruit. Not only have thousands been induced to read who never before touched a book except by accident, but publishers having thus a large market open to them, and authors finding a public always ready for their works, have eliminated new features in their literary productions. The former, by more economic arrangements, have endeavored to reach the perfection of cheapness, while the latter comprehend that, in order to reach the soul of a whole nation, literature must separate itself from refined notions and elaborations of style, and that it cannot be too pure either as regards the form or the matter.

"XII. The exhibition of the progress of education in France would still be incomplete were it limited to the groups above indicated; the work would be uncrowned. Happily the Minister of the Interior has here intervened. By virtue of the Imperial decision, dated November 8, 1856, the Exhibition includes not only the acts emanating from his administration and the works of the pupils of the public schools, but also important collections from scientific missions, and a series of reports presenting a picture of the progress accomplished in France in science, as well as in letters, during the last twenty years. From the minister of the interior to the village schoolmaster, all the representatives of national education find themselves thus associated at the Exhibition of 1867 in a common responsibility, in the face of France and the whole world."

GENERAL INDEX

OF THE

GROUPS AND CLASSES

ACCORDING TO THE

CLASSIFICATION ADOPTED BY THE IMPERIAL COMMISSION.

GROUP IV.

GROUP V.

GROUP VI.

GROUP VII.

FOOD, FRESH OR PRESERVED, IN VARIOUS STATES OF PRESERVATION.

GROUP VIII.

LIVE STOCK AND SPECIMENS OF AGRICULTURAL BUILDINGS.

GROUP IX.

LIVE PRODUCE AND SPECIMENS OF HORTICULTURAL WORKS.

GROUP X.

ARTICLES EXHIBITED WITH THE SPECIAL OBJECT OF IMPROVING THE PHYSICAL AND MORAL CONDITION OF THE PEOPLE.

16 U E

REFERENCES

TO THE

PLAN OF THE BUILDING AND PARK.

REFERENCES TO THE UPPER LEFT HAND CORNER OF THE PLAN.

A.—SPAIN.—1, Moorish Farm-house; 2, Valentian Cottage.

B.—PORTUGAL.—3, Silk-worm Nursery.

C.—SWITZERLAND.—4, Fine Arts Annexe.

D.—AUSTRIA.—5, Bakery; 6, Restaurant; 7, Hungarian House; 8, Styrian House 9, Lower Austrian House; 10, Tyrolean House; 11, Stables; 12, Riding School.

E.—SCHLESWIG-HOLSTEIN.—13, General Exhibition.

F.—WURTEMBERG.—14, Annexe.

G.—PRUSSIA.—15, Annexe; 16, School-house; 17, Lake; 18, Equestrian Statue.

H.—BAVARIA.—19, Principal Annexe; 20, Annexe.

I.—NORWAY.

J.—DENMARK.

K.—SWEDEN.—21, House of Gustavus Vasa.

L.—RUSSIA.—22, Caucasian House; 23, Boiler.

M.—FRANCE.—24, Agricultural Exhibition; 25, Offices and Warehouses; 26, Restaurant.

ADDENDA.—a, Agricultural Machinery; b, Swiss Annexe; c, Russian Stables; d, Concert Hall; e, Russian Annexe.

REFERENCES TO THE LOWER LEFT-HAND CORNER OF THE PLAN.

FRANCE—Continued.—27, Conservatory; 28, Lake; 29, Marine Aquarium; 30, Freshwater Aquarium; 31, Greenhouse; 32, Temperate Greenhouse; 33, Cold Greenhouse; 34, Greenhouse; 35, Botanical Diorama; 36, 37, 38, Greenhouses; 39, Restaurant; 40, 41, Greenhouses; 42, Orchestra; 43, Tent of Her Imperial Majesty the Empress; 44, 45, Greenhouses; 46, Botanical Diorama; 47, Greenhouse; 48, Post Office and Telegraph.

N.—BELGIUM.—49, Fine Arts Annexe; 50, Exhibition of Railway Plant, &c.

O.—HOLLAND.—51, Farm-house; 52, Fine Arts Annexe; 53, Exhibition of Carriages.

ADDENDA.—f, Police and Firemen; g, Turnstile; h, Cloak-rooms.

REFERENCES TO THE UPPER RIGHT-HAND CORNER OF THE PLAN.

P.—ITALY.—54, Museum; 55, Boiler.

Q.—SIAM, JAPAN, and CHINA.—56, Chinese Tea House; 57, Theatre.

R.—TUNIS.—58, Bey's Palace.

S.—EGYPT.—59, Café; 60, Temple of Edfou; 61, Summer Palace of the Viceroy; 62, Exhibition of the Plan of the Suez Canal.

T.—DANUBIAN PRINCIPALITIES.

U.—MOROCCO.—63, Imperial Tent.

V.—TURKEY.—64, School-house; 65, Mosque.

W.—UNITED STATES.—66, Boiler-house; 67, American Farmer's House, or Illinois Cottage; 68, United States School-house; 69, Louisiana Cottage.

X.—MEXICO and BRAZIL.—70, Temple.

Y.—GREAT BRITAIN AND IRELAND.—71, Testing House, (heating apparatus, &c.;) 72, Light-house; 73, Barrack Huts; 74, Public Munitions of War; 75, Private Munitions of War; 76, Exhibition of Protestant Mission.

ADDENDA.—hh, Jurors' Meeting Room; i, United States Annexe; j, English Annexe; k, Café and Concert-room; l, Concert Hall; m, International Club; n, Restaurant; o, Gas Works; p, Exhibition of English Marine Machinery.

REFERENCES TO THE LOWER RIGHT-HAND CORNER OF THE PLAN.

Z.—FRANCE—Continued.—77, Imperial Tent; 78, Engine; 79, Exhibition of Pottery; 80, Exhibition of Cashmere Shawls; 81, Waterfall; 82, Swiss Cottage; 83, Exhibition of Glass; 84, Exhibition of Photosculpture; 85, Windmill; 86, Church; 87, Fire Engine; 88, Police and Firemen; 89, Lake; 90, Light-house; 91, Materials for Cleansing Woollen Fabrics; 92, Leather Working Machines; 93, Theatre; 94, Refrigerating Apparatus; 95, Mills and Presses; 96, French War Office Exhibition; 97, Porcelain; 98, Photography.

ADDENDA.—qq, Exhibition of Fire Engines; r, Mining Exhibition; s, Machinery; t, Railway Plant, &c.; uu, Money Exchange Offices; v, Equestrian Statues; w, French Marine Engines; x, Exhibition of Pleasure Boats; y, Restaurant.

REPORT

UPON THE

CHARACTER AND CONDITION

OF THE

UNITED STATES SECTION.

17 U E

CONTENTS.

UNITED STATES SECTION.

GENERAL OBSERVATIONS AND CATALOGUE.

SPACE OCCUPIED AND AWARDS.

The space occupied by the United States was a sector of the building in the end towards the Seine. It was separated on one side from the space allotted to China and Japan by the Rue d'Afrique, (one of the transverse avenues,) and on the other side it was separated from the portion of the building occupied by Great Britain and its colonies, by a narrow sector devoted to the products of South America.

The superficial area of this sector was 38,488 square feet. In addition to this there were the constructions in the Park, the space at Billancourt, and a long rectangular building at one side of the Park, called the Annex, provided for many objects that could not conveniently be received in the building. The total space occupied was as follows:

	Square feet.
In the Palace	38, 488
In the Park	55, 769
At Billancourt	3, 880
Total	98, 137

The total number of entries in the official catalogue was 717; but this did not show the exact number of exhibitors, for some of the objects entered being broken or damaged were not set up. The same exhibits were, in some cases, entered under different classes. The total number of exhibitors whose products were present in time and competing for prizes was 536, as shown by the following résumé:

Whole number of entries in the catalogue		717
Deduct the products broken and not set up, including the *hors concours*	17	
Repetitions of the same names, and admissions at different dates through the season, after the jury work was closed	164	
		181
Present in time and competing for prizes		536

The nature of the objects exhibited is fully shown in the descriptive catalogue which forms a part of this report.

The total number of awards to the exhibitors from the United States was as follows:

Grand prizes	5
Artists' medal	1
Gold medals	18
Silver medals	76
Bronze medals	98
Honorable mentions	93
Total awards	291

Full details of the distribution of these awards are given in the list appended to this report.

From the tabular statement prepared by Mr. Beckwith and presented in the preface to the General Report, it appears that the percentage of awards to exhibitors from the United States was 52.79; the percentage to exhibitors from France was 55.57, and to those from Great Britain and colonies, 26.10. The general average percentage of awards to all exhibitors was 34.53. Next after France the United States stands highest upon the list. Mr. Beckwith observes in the preface before cited:

" The high position conceded by the verdict of the juries to American industrial products is not due in general to graceful design, fertile combinations of pleasing colors, elegant forms, elaborate finish, or any of the artistic qualities which cultivate the taste and refine the feelings by awakening in the mind a higher sense of beauty, but it is owing to their skilful, direct, and admirable adaptation to the great wants they are intended to supply, and to the originality and fertility of invention which converts the elements and natural forces to the commonest uses, multiplying results and diminishing toil.

" The peculiar and valuable qualities of our products will be adopted and reproduced in all parts of Europe, improving the mechanical and industrial arts, and it is reasonable to expect and gratifying to believe that the benefits will be reciprocal, that our products will in time acquire those tasteful and pleasing qualities which command more admiration and find a quicker and better market than the barely useful."

GENERAL OBSERVATIONS.

As a participator in this great international display the United States labored under many disadvantages. The nation had not recovered from the paralyzing effects of the disastrous war of the rebellion, and the people were not aroused in season to an appreciation of the importance of the projected Exhibition. The manufacturing industry of the country was in a transition state. Labor was scarce and dear, and many manufacturers found it difficult to fill the orders which had been received, and thus could not undertake the preparation of goods for exhibition. The remoteness of the Exhibition greatly discouraged effort, inasmuch as comparatively few of the exhibitors could be present and attend to placing and explaining their contributions. The broad Atlantic separated our artizans and producers from the Champ de Mars, while most of the great competing nations were connected by rail directly with the Exposition building. The cost of transportation within the limits of the United States to the agency in New York was considerable, and although the contributions were forwarded across the ocean by the government, no provision was made for the repacking and return of the articles, which, at the close of the Exhibition, were to be at the risk and expense of the exhibitor, and thus many persons who would have joined in the Exhibition were deterred from taking any part in it.

Yet, notwithstanding all these difficulties, the country may be congratulated upon the success of its exhibition; and that the skill, industry, and energy of the people did not suffer by comparison in the great international contest.

Our raw materials were not excelled by any in the Exposition, and by their variety, abundance, and quality, gave convincing evidence of the extraordinary natural wealth of our States and Territories. In the display of mineral products the coal of Pennsylvania, the gold and silver ores of California, Nevada, Idaho, and Colorado, the copper and iron of Minnesota, the zinc ores of New Jersey, and the emery of Massachusetts, were especially prominent. The collection was rich; but some regions and products were disproportionately represented, and it lacked that unity and completeness which can only be attained through intelligent organized effort. Almost all other prominent displays in this class were prepared with the strong aid and authority of the governments, through regularly organized corps of engineers.

In forestry and productions of the forest the display made by the United States was meagre. Much attention was given to this class by other countries; the display made by France, Brazil, Australia, and Canada, were notable features of the Exposition. Several of these collections, as also some of the collections of mineral products, had been prepared for previous great exhibitions, and may be regarded as standard displays, which are added to and improved at each new exhibition.

The exhibitions of the cereal productions and of the cotton, tobacco, wool, and other staple products, though in some instances prominent and thoroughly satisfactory, were in general fragmentary and not on a scale commensurate with the enormous capacity of the country for their production.

The most notable deficiency in the exhibition made by the United States was in Group III, including the application of the fine arts to the useful arts. This deficiency was shown by the absence of rich furnishings, upholstery, and decorative work, and manufactures depending for their excellence upon a high degree of taste and skill in design. There was no fine display of richly decorated porcelain and faïence, encaustic tiles, and marquetry; and, with the notable exception of the bronzed iron work of Messrs. Tucker & Company, no collection of artistic bronzes, bas-reliefs, and ornamental castings, nor of highly ornamented and artistic furniture.

The bronze work of the Messrs. Tucker may justly be excepted for its novelty, intrinsic excellence, and artistic value. It attracted much attention, and the articles were in demand. Some were ordered in person by the King of Prussia.

OBSERVATIONS UPON THE GROUPS.

The following observations upon the display made by the United States in several of the groups are from a report submitted by Commissioner Freese and others.

"In Group II, 'Materials and their applications in the liberal arts,' we find among the contributions made by the 86 American exhibitors much to admire and commend.[1]   The specimens of typography were such as could not fail to be commended by any one conversant with the art, and we are pleased to observe that three of the exhibitors of book printing received prizes. Of specimens of stationery, book-binding, &c., the display is very limited, and out of all proportion with our immense trade in these articles, and yet of the 13 exhibitors no less than six received prizes, proving that what is lacking in quantity and variety of these articles in the Exhibition is more than made up in quality.   Of plastic moulding there is but one exhibitor and three specimens, one representing what is called 'Uncle Ned's School,' another called 'Taking the Oath and Drawing Rations,' and a third called 'The Charity Patient,' representing a benevolent faced old doctor compounding a prescription for a poor woman in waiting.   All these are peculiarly American, and are admirably executed. Of proofs and apparatus of photography most of the specimens are commendable, and of the ten exhibitors four received prizes.

"Of instruments of music the display, though small, adds decidedly to the character of the American exhibition. Of the nine exhibitors in this class two received gold medals, one a silver medal, and two bronze medals.

"In Group IV the contributors to the American exhibition are few in number, (only 54.)   Of yarns and tissues of cotton the contributors are six, and of these five have received prizes.   Of other yarns and tissues of linen, hemp, wool, and silk, the contributors are nine, of whom five have received awards.   Of shawls, hosiery, and clothing, the contributors are sixteen, of whom five have received recognition.   But the great feature of this group was the display of breech-loading fire-arms, metallic cartridges, and rifled cannon, of which there are fourteen contributors, of whom seven have received gold or silver medals.

---

[1] In this and the following enumerations of the number of exhibitors no allowance has been made for the repetitions of entries in the catalogue.

" To Group V American contributors have made most noble and appropriate contributions, embracing products (raw and manufactured) of mining industry, forestry, etc.

" In Class 40 of this group the following States and Territories have contributed from their mines and quarries: Illinois, Minnesota, Massachusetts, New Jersey, Michigan, Ohio, California, Nevada, Arkansas, Missouri, Louisiana, Alabama, Kansas, Iowa, Wisconsin, Pennsylvania, New York, Tennessee, Vermont, Georgia, West Virginia, Utah, and Idaho.

" In products of the forest, embraced in the next class, (41,) we find specimens of woods from Missouri, Kansas, Wisconsin, Illinois, New York, Massachusetts, California, Louisiana, and Utah.

" Passing to Group VI, comprising instruments and processes of common arts, we find in the American exhibition a larger number of exhibitors (227) than in any other of the groups, though this is accounted for in the fact that this group embraces a larger number of classes, (20,) and consequently a larger range of articles, than any other.

A large proportion of the contributions to this group add to the excellence of the American exhibition, and that some should fail to do so is scarcely to be wondered at, among so large a number of contributions.

" Of apparatus and methods of mining and metallurgy there are four contributors, two of whom have received prizes; of implements and processes of rural and forest work there are 25 contributors, six of whom received medals, of which two, for the best specimens of mowing and reaping machines, are of gold; of apparatus for hunting, fishing, &c., there is but one contributor, and the contribution is of no special value; of materials and method, of agricultural work, and of alimentary industry, there are 20 contributors, eight of whom have received recognition, and nearly all the contributions do credit to the genius and industrial activity of our country; of chemical, pharmaceutic, and tanning apparatus there are seven contributors, four of whom have received medals and honorable mention, and the other three are scarcely less worthy of a like recognition; of machines and mechanical apparatus in general, we have 38 contributions; that these should have been awarded no less than 21 prizes, (over 50 per centum of the number of exhibitors,) cannot be otherwise than gratifying to every American.

" Of machine tools we have 14 contributors, 10 of whom have received recognition, and the contributions of the other four are commendable; of cotton-ginning, cord-twisting, and burr-picking machines we have five contributions, all good, and two of which have received recognition; of weaving and knitting machines we have five specimens, three of which have received silver medals; of apparatus and process of sewing and making clothes, (which class includes our inimitable sewing machines, in which, as agreed upon by all impartial judges, we are far in advance of all other nations,) there are 18 contributors, of whom two have received gold medals, and one of them an imperial decoration; three have received silver medals; seven, bronze medals; and one an honorable mention, making a sum total of 13 prizes among 18 contributors. Of apparatus and methods of making furniture and household objects, there are 10 contributors, three of whom received prizes, and others of this class would, doubtless, have received high prizes could they have been present themselves to explain to the jury the peculiar working and intrinsic value of their inventions. Of machinery for paper making, printing, &c., there are but five contributors, and none of the great steam power printing presses, for which American inventors and manufacturers have become so justly celebrated, were included in this display; and of five exhibitors in this class, two have received medals, while another machine for dressing type, classified under the next head, received the award of a gold medal, and is every way worthy of it.

" Passing into the annex, we find two American buggies, and a street railway carriage, all three of which are fine specimens of skill and taste in carriage-building, and all of which received prizes. Near these are ladies and gentlemen's saddles, of good workmanship, which also received recognition from the jury. Here, too, we find the great American locomotive, which in workmanship and beauty of finish far excels all others in the Exposition, and to which the jury awarded a gold medal. There are eight other contributions to this class, though only one other—a railroad scale—received a prize.

" Of models relating to navigation and salvage there are 14 contributors to the American exhibition, four of whom received recognition from the jury, and nearly all the specimens do honor to the inventors.

" Next in review we reach Group VII, and find from the Department of Agriculture, Washington, D. C., 33 samples of wheat and other cereals, from as many different States and special localities, together with a large number of contributions of like character direct from the States. The specimens are invariably good, and have attracted a large share of attention from European agriculturists, and would have attracted more had their installation been better. Of the 17 private contributors to this class, nine have received awards. Of baking and pastry cooking we have but one contribution, and judging from the quality of bread, cake, and crackers produced we should call it first-rate, but for some cause it has received no recognition from the jury. Of prepared specimens of meat and fish, (including salt-cured and smoked hams, packed beef, pork, and lard, preserved lobster, canned oysters, &c.,) we have seven contributions, and every one has received an award.

" Of preserved fruits and vegetables, sugars, chocolates, &c., the contributors number 21, of whom 10 have received awards, and all the samples are commendable. Of fermented drinks, such as wines, brandies, ales, porters, and brown-stouts, the contributors number 25, of whom seven received awards.

" This brings us to Group X, the last, though not the least important. Here we find a specimen of a western primary school-house, school furniture, and school apparatus. It is safe to say that nothing in the

American exhibition has excited more general attention and commendation from European visitors, and no other of our exhibits tended to excite more general inquiry into the peculiar character of our political institutions, and especially as to the relations which those institutions bear toward our common school system. The school-room, in size, finish, ventilation, and furnishing, is superior to any other in the Exposition, and the apparatus within, though not in quantity, certainly in quality, equals those exhibited by any other nation. But as a report is in course of preparation, covering the whole subject of school-houses, school apparatus, &c., additional remarks are here unnecessary.

"In this same class we find books and apparatus for the use of the blind, contributed from Massachusetts, which are in every way quite equal to any others in the Exposition, and we are pleased to know that both these and the school building received awards.

"Passing to the next class, (20,) we find the articles to consist of surgical instruments, artificial limbs, hospital wagon, ambulance and relief material, medicine wagon, and camp equipage, such as were used by the United States Sanitary Commission, and all collected by one of the United States commissioners. To the United States Sanitary Commission the jury awarded a grand prize, and each of the other six exhibitors in this class received recognition.

"In the next class (93) we find three specimens of houses from the United States, namely, a western farmer's house, a Boston bakery, and a cottage made of Louisiana cypress. The first of these has deservedly attracted a large share of attention, and added much to the character of the American exhibition. It is decidedly American in its construction— plain, substantial, and convenient—representing thrift and comfort without display."

CATALOGUES AND OTHER PUBLICATIONS.

Three editions were published and circulated of an official catalogue of the products of the United States that were exhibited. This catalogue was printed in English, French, and German, and was accompanied by geographical and statistical notices in French upon the population, trade, and resources of the United States, prepared from data furnished by the Secretary of the Interior.[1]

A special catalogue in 8vo of the minerals of the United States exhibited in Group V, class 40, was also printed. This catalogue was compiled by Commissioner D'Aligny.

Numerous copies in English, French, German, and Swedish, of the report of the Commissioner of the United States General Land Office for 1866, accompanied by a map, were gratuitously distributed.

---

[1] The following is the title of this catalogue in full. It was printed in 12mo, pp. 160: "Official Catalogue of the Products of the United States of America exhibited at Paris, 1867, with Statistical Notices. Catalogue in English, Catalogue Français, Deutscher Catalog. Third edition. Paris: Imprin erie Centrale des Chemins de Fer. A. Chaix et Cie, Rue Bergère, 20, près du Boulevard Montmartre, 1867."

The territorial commissioner from Colorado published a beautifully printed pamphlet descriptive of the Territory and its resources, and of the large collection of the ores of gold, silver, and copper. These books were printed in French and in English, and were gratuitously distributed to those who took an interest in the display from that portion of the United States.

The State commissioner from Nevada published a small edition of a similar pamphlet, accompanied by a map of eastern Nevada.

The Agricultural Society of the State of California sent a few sets of its transactions for distribution. Illinois also sent reports of its Agricultural Society and complete sets of the reports on the geology of the State. A small volume on the mineral, agricultural, and manufacturing resources of the State of Alabama was printed in Paris, and gratuitously distributed. The colony of Vineland, New Jersey, also circulated a descriptive pamphlet.

The descriptive catalogue of the products of the United States which follows will show the character of the exhibition made in the various groups and classes. The notices of the various objects have been prepared, in part, from data furnished by Dr. Thomas W. Evans, of Paris.

DESCRIPTIVE CATALOGUE

OF THE

PRODUCTS OF THE UNITED STATES,

EXHIBITED AT PARIS, 1867.

GROUP I.

WORKS OF ART.

CLASS 1.—PAINTINGS IN OIL.

BAKER, G. A., New York.—1. Portrait of a Child, the property of A. M. Cozzens, esq. 2. Portrait of a Lady, the property of F. Prentice, esq.

BEARD, W. H., New York.—3. The Bears' Dance, the property of Josiah Caldwell, esq.

BIERSTADT, A., New York.—4. The Rocky Mountains, the property of James McHenry, esq.

BOUGHTON, G. H., Albany, New York.—5. Winter Twilight, the property of R. L. Stuart, esq. 6. The Penitent, the property of J. F. Kensett, esq.

CASILEAR, J. W., New York.—7. Genesee Flats, the property of Shepard Gandy, esq. 8. A Swiss Lake, the property of R. M. Olyphant, esq.

CHURCH, F. E., New York.—9. Niagara, the property of J. Taylor Johnston, esq. 10. The Rainy Season in the Tropics, the property of M. O. Roberts, esq. Mr. Church received the Artists' Medal, with 500 francs in gold.

COLE, J. F., Boston, Massachusetts.—11. Pastoral Landscape.

COLMAN, S., New York.—12. View of the Alhambra.

CROPSEY, J. F., New York.—13. Mount Jefferson, New Hampshire, the property of R. M. Olyphant, esq.

DIX, C. T., New York.—14. Marine.

DURAND, A. B., New York.—15. In the Woods, the property of J. Sturgess, esq. 16. A Symbol, the property of R. M. Olyphant, esq.

ELLIOTT, C. L., New York.—17. A Portrait, the property of Fletcher Harper, esq.

GIFFORD, S. R., New York.—18. Twilight on Mount Hunter, the property of J. W. Pinchot, esq. 19. Home in the Wilderness, the property of M. Knoedler, esq.

GIGNOUX, R., New York.—20. Mount Washington, New Hampshire, the property of A. T. Stewart, esq.

GRAY, H. P., New York.—21. The Apple of Discord, the property of R. M. Olyphant, esq. 22. The Pride of the Village, the property of W. H. Osborn, esq.

HART, JAMES M., New York.—23. Landscape: Tunxis River, Connecticut, the property of S. P. Avery, esq.

HEALY, G. P. A., Chicago, Illinois.—24. Portrait of Lieutenant General Sherman. 25. Portrait of a Lady, the property of W. B. Duncan, esq.

HOMER, WINSLOW, New York.—26. Confederate Prisoners at the Front, the property of J. Taylor Johnston, esq. 27. The Bright Side, the property of W. H. Hamilton, esq.

HUBBARD, R. W., New York.—28. View of the Adirondacks, taken near Mount Mansfield, the property of Mrs. H. B. Cromwell. 29. Early Autumn, the property of H. G. Marquand, esq.

HUNT, W. M., Boston, Massachusetts.—30, 31, 32, 33, 34, 35, 36. Portraits. 37. Portrait of Abraham Lincoln. 38, 39. Italian Boy. 40. Dinan, in Brittany. 41. The Quarry.

HUNTINGTON, D., New York.—42. Portrait of Gulian C. Verplanck, esq. 43. The Republican Court, time of Washington, the property of A. T. Stewart, esq.

INNESS, GEORGE, Perth Amboy, New Jersey.—44. American Sunset, the propety of Marcus Spring, esq. 45. Landscape, with cattle.

JOHNSON, E., New York.—46. Old Kentucky Home, the property of H. W. Derby, esq. 47. Mating, the property of Major General John A. Dix. 48. Fiddling his Way, the property of R. L. Stuart, esq. 49. Sabbath Morning, the property of Robert Hoe, esq.

JOHNSON, F., Brooklyn, New York.—50. The Omelet.

KENSETT, J. F., New York.—51. Lake George in Autumn, the property of R. M. Olyphant, esq. 52. Coast, Newport Harbor, the property of G. T. Olyphant, esq. 53. Glimpse of the White Mountains, the property of R. L. Stuart, esq. 54. Morning off the Coast of Massachusetts, the property of S. Gandy, esq.

LAMBDIN, G. C., Philadelphia, Pennsylvania.—55. The Consecration, 1861, the property of George Whitney, esq. 56. The Last Sleep.

LANGDON, WOODBURY, New York.—57. The Storm. 58. Out at Sea.

LAFARGE, JOHN, Newport, Rhode Island.—59. Flowers, the property of S. F. Van Chote, esq.

LEUTZE, E., New York.—60. Mary Stuart hearing the first mass at Holyrood, after her return from France, the property of John A. Riston. esq.

LEWIS, J. S., Burlington, New Jersey.—61. The Fisher Boy.

MAY, E. C., New York.—62. Lady Jane Grey giving her Tablets to the Governor of the Tower on her way to Execution. 63. Lear and Cordelia, (King Lear, Act IV, scene 7.) 64. Portrait.

MAC ENTEE, J., New York.—65. Virginia in 1863, the property of Cyrus Butler, esq. 66. Last of October, the property of S. C. Evans, esq. 67. Autumn, Ashokan Woods, the property of Robert Hoe, esq.

MIGNOT, L. R., New York.—68. Sources of the Susquehanna, the property of H. W. Derby, esq.

MORAN, T., Philadelphia, Pennsylvania.—69. Autumn on the Conemaugh, in Pennsylvania, the property of C. L. Sharpless, esq. 70. The Children of the Mountain.

OWEN, GEORGE, New York.—71. Study from Nature, New England scenery.

RICHARDS, W. T., Philadelphia, Pennsylvania.—72. Woods in June, the property of R. L. Stuart, esq. 73. Foggy Day at Nantucket, the property of George Whitney, esq.

WEIR, J. F., New York.—74. The Gun Foundry, the property of R. P. Parrott, esq.

WHISTLER, J. McNEIL, Baltimore, Maryland.—75. The White Girl. 76. "Wapping," or "On the Thames." 77. Old Battersea Bridge. 78. Twilight on the Ocean.

WHITE, E., New York.—79. Thoughts of Liberia, the property of R. L. Stuart, esq.

WHITTRIDGE, W., New York.—80. The Old Hunting Ground, the property of J. W. Pinchot, esq. 81. Rhode Island Coast, the property of A. M. Cozzens, esq.

WEBER, PAUL, Philadelphia, Pennsylvania.—82. Bolton Park, England.

CLASS 2.—VARIOUS PAINTINGS AND DESIGNS.

DARLEY, F. O. C., New York.—1. Cavalry Charge at Fredericksburg, Virginia, the property of W. T. Blodgett, esq.

JOHNSON, E., New York.—2. Wounded Drummer Boy, the property of the Century Club.

ROWSE, S. W., Boston, Massachusetts.—3. Crayon Portrait of Ralph Waldo Emerson, esq. 4. Crayon Portrait of J. Russell Lowell, esq.

CLASS 3.—SCULPTURE, DIE-SINKING, STONE AND CAMEO ENGRAVING.

HOSMER, Miss H. G., Boston, Massachusetts.—1. The Sleeping Faun.

THOMPSON, L., New York.—2. Statue of Napoleon, the property of C. C. D. Pinchot, esq.; cast by Mr. L. A. Amouroux. 3. Bust of a Rocky Mountain Trapper.

VOLK, L. W., Chicago, Illinois.—4. Bust of A. Lincoln.

WARD, J. Q. A., New York.—5. The Indian Hunter and his Dog, the property of the Central Park, New York; cast by Mr. L. A. Amouroux. 6. The Freedman, the property of John Baker, esq.; cast by Mr. L. A. Amouroux.

CLASS 4.—ARCHITECTURAL DESIGNS AND MODELS.

(For American Farmer's House, and School-house, see Group X, class 93.)

CLASS 5.—ENGRAVING AND LITHOGRAPHY.

MARSHALL, W. E.—Lincoln; engraving on steel. Washington; engraving on steel.

HALPIN, F., New York.—President Lincoln; engraving on steel.

WHISTLER, JAMES MCNEIL, Baltimore, Maryland.—Twelve etchings.

GROUP II.

APPARATUS AND APPLICATION OF THE LIBERAL ARTS.

CLASS 6.—PRINTING AND BOOKS.

AMERICAN BANK NOTE COMPANY, New York.—Specimens of bank-note engraving and printing.

AMERICAN BIBLE SOCIETY, New York.—Specimen copies of the publications of the society.

Since the formation of this society, in 1816, it has received from sales and donations $10,847,854, and has issued, in every known language, an aggregate of 22,118,475 copies of the Holy Scriptures. It has 17 power presses, and about 400 persons employed in the Bible House. During the late war 6,555,231 volumes were issued.

APPLETON, D., & Co., New York.—Books, including a copy of the American Encyclopædia. Bronze medal.

BAKER & GODWIN, Printing-house square, New York.—Typography, plain and in colors.

BOND, Professor G. P., Cambridge, Massachusetts.—Description of the Great Comet of 1858.

BRADSTREET, J. M., & SON, 18 Beekman street, New York.—Specimen of book printing by Hoe's press.

BREWER & TILESTON, 131 Washington street, Boston, Massachusetts.—Worcester's Dictionary.

BROUGHTON, NICHOLAS, Jr., 28 Cornhill, Boston, Massachusetts.—Specimens of typography from the American Tract Society.

BUFFORD, J. H., & SONS, 313 Washington street, Boston, Massachusetts.—Lithographic view of Mr. Bacon's bakery, in that establishment.

DEMOREST, W. J., 437 Broadway, New York.—Specimens of a monthly magazine, illustrated.

GALLAUDET, E. M., President of the Columbia Institute for the Deaf and Dumb, Washington, D. C.—Reports of that institution.

HOUGHTON, H. O., & Co., Riverside, Cambridge, Massachusetts.—Specimen books illustrated. Bronze medal.

ILLINOIS, STATE OF.—Reports of the State Geologist, Superintendent of Public Instruction, Adjutant General, State Agricultural Society, Chicago Board of Trade, &c.

KNEASS, N. B., Jr., Philadelphia, Pennsylvania.—Books for the use of the blind.

The exhibitor of these books is blind, and a graduate of the Institution of Teachers in Pennsylvania.

MERRIAM, G. & C., Springfield, Massachusetts.—Specimens of book printing.   Bronze medal.

MISSOURI, STATE OF.—Books; in the Farmer's Cottage, Park.

NATIONAL BANK NOTE COMPANY.—Samples of bank note engraving and printing.

PRANG & CO., Boston, Massachusetts.—Chromo-lithographs; in the Restaurant.

STATE AGRICULTURAL SOCIETY OF CALIFORNIA.—Reports and Transactions; several sets distributed through the Commissioner, but not on exhibition except in the American Farm-house.

Many copies of these publications were distributed by exchange with those most interested in these subjects.

CLASS 7.—PAPER, STATIONERY, BOOK-BINDING, PAINTING. AND DRAWING MATERIALS.

AMERICAN LEAD PENCIL COMPANY, New York.—Samples of lead pencils.   Bronze medal.

BACON, S. T., Boston, Massachusetts.—Office card rack; in the Bakery, Park.

DAY, AUSTIN G., Seymour, Connecticut.—Ordinary and indelible pencils in hard rubber cases.   Bronze medal.

FAIRCHILD, L. W., & CO., New York.—Gold pens; pen and pencil cases. Bronze medal.

FORMAN, J. C., Cleveland, Ohio.—Specimens of work executed on the American Circular Border Ruling machine.

JESSUP & MOORE, 27 North Sixth street, Philadelphia, Pennsylvania.— Specimens of paper made from wood, straw, and hemp.   Bronze medal.

MATTHEWS, W., New York.—Specimen of binding; the Colt "Memorial."

MURPHY'S, W. F., SONS, Philadelphia, Pennsylvania.—Samples of blank books.   Bronze medal.

NOONAN & McNAB, Milwaukee, Wisconsin.—Specimens of writing paper.

NORTHAMPTON INDELIBLE PENCIL COMPANY, Northampton, Massachusetts.—Indelible pencils for marking linens.

PIERCE, T. N., & CO., 427 North Eleventh street, Philadelphia, Pennsylvania.—Slates.

SAN LORENZO MILLS, San Lorenzo, California.—Paper.

SECOMBE MANUFACTURING COMPANY, 264 Broadway, New York.— Holt's improved ribbon hand stamps.   Bronze medal.

CLASS 8.—SPECIMENS OF DESIGN AND PLASTIC MOULDING APPLIED IN THE ORDINARY ARTS.

ROGERS, J., New York.—Three groups of statuettes.

CLASS 9.—PROOFS AND APPARATUS OF PHOTOGRAPHY.

BEER, SIGISMUND, 481 Broadway, New York.—Stereoscopic views. Bronze medal.

DRAPER & HUSTED, Ridge avenue and Wallace street, Philadelphia, Pennsylvania.—Photographs.

GARDNER, A., Washington, D. C.—Photographs.

GUTEKUNST, F., 704 Arch street, Philadelphia, Pennsylvania.—Photographs.

LAWRENCE & HOUSEWORTH, San Francisco, California.—Photographic and stereoscopic views, comprising 22 large photographs of the Yosemite valley, California; 4 of the mammoth trees; 21 stereoscopic views of the Yosemite valley; 33 of the mammoth trees; 40 of San Francisco; 17 illustrating the art of hydraulic mining; 43 of placer mining; 158 of scenery in California, and 29 in Nevada.

At the close of the Exposition, these views were donated by the exhibitors, through the commissioner from the State of California, to various public societies and institutions, including the Photographic Society of Paris, the Jardin des Plantes, the Geological Society of France, and the British Museum.

MORVAN, A. G., (Heliotype Company,) 90 Fulton street, New York.—Photographic engraving.

RUTHERFORD, L. M., New York.—Photographs of the moon and solar spectrum.

A remarkably large and fine photograph of the moon's surface, and another of the solar spectrum showing the dark lines with great distinctness. A silver medal was awarded to Mr. Rutherford.

VISCHER, EDWARD, San Francisco, California.—Six photographic albums, containing views of California and Washoe.

The contributions of this exhibitor were duly invoiced and shipped from San Francisco, but failed to reach the exhibition, having been lost or mislaid in the transit.

WATKINS, C. E., San Francisco, California.—Photographic views of California.

Being a complete set, 30 or more, of the celebrated views of the Yosemite valley and of the great trees of Mariposa county. These views are of large size, and were sent by the exhibitor framed in *passe partout* and ready to hang. The views of the Mariposa trees were framed in the wood of the trees appropriately carved. These photographs attracted much attention, and the jury awarded a bronze medal.

WILLARD & CO., 684 Broadway, New York.—Photographic camera tubes and lenses. Honorable mention.

WILLIAMSON, C. H., 245 Fulton street, Brooklyn, New York.—Photographs.

CLASS 10.—MUSICAL INSTRUMENTS.

CHICKERING & SONS, New York and Boston.—Pianos. Gold medal. (See a notice at the end of this class.)

GEMUNDER, GEORGE, 174 East Ninth street, New York.—Stringed instruments. Bronze medal.

LINDEMANN & SONS, 2 Leroy place, New York.—Cycloid piano. (See a notice at the end of this class.)

MASON & HAMLIN, 596 Broadway, New York; Washington, D. C., and Boston, Massachusetts.—Cabinet organs. Silver medal.

METZEROTT, W. G., & CO., Washington, D. C.—Wind instruments.

SCHREIBER CORNET MANUFACTURING COMPANY, 99 Houston street, New York.—Wind instruments of brass and German silver. Bronze medal.

STEINWAY & SONS, New York.—Pianos. Gold medal. (See a notice at the end of this class.)

WRIGHT, E. G., & CO., Boston, Massachusetts.—Wind instruments of brass and German silver.

ZIMMERMANN, C. F., 238 North Second street, Philadelphia, Pennsylvania.—Accordeons.

NOTICE OF THE PIANOS EXHIBITED FROM THE UNITED STATES.

The piano manufacturers of the United States may justly claim to have gained and preserved the first reputation in the world. The principal feature upon which that reputation is founded is the introduction of the iron instead of the wooden frame, an improvement which has necessitated and been followed by various others.

Until the third decade of the present century only European instruments found a ready market in America. It was soon found, however, that no wooden framed piano could long resist the extraordinary climatic changes of the country without requiring almost constant tuning and repairs.

In the Exhibition of 1867, two firms more especially dispute the palm of pre-eminence—Messrs. Steinway & Sons, of New York, and Messrs. Chickering & Sons, of New York and Boston. The jury readily acknowledged the remarkable qualities of the pianos of these two houses, and, pronouncing them both first-class products, gave equal awards to each, and the highest in its gift, viz: the gold medal.

By a decree of the Emperor Mr. C. F. Chickering was created Chevalier of the Imperial Order of the Legion of Honor of France.

Each of these firms has, from time to time, taken out patents for improvements. Mr. Chickering claims to be the sole inventor of the circular scale, and to have made many other improvements which have been rendered necessary from time to time by the development of musical science.

Messrs. Steinway & Sons claim the application of various important improvements necessary for avoiding the thin and disagreeably nasal

18 U E

character of tone at first possessed by the iron frame, and for supplying that solidity of construction which the gradual extension of the musical capabilities of the piano rendered necessary. They claim also the introduction of over-stringing as well as the adoption of agraffes. It will not be presumed in this notice to judge of the respective merits of the improvements or the claims as to priority of the inventions of either party, or to attempt a technical particularization of them, but it may be said that the pianos of Messrs. Chickering & Sons and of Messrs. Steinway & Sons, not forgetting the beautiful cycloid instrument manufactured and exhibited by Messrs. Lindemann & Sons, are unrivalled, and that while these instruments have a solidity of construction which withstands the deleterious influence of any climate, their depth, volume, power and delicacy of tone are fully equal to all that can be required.

CLASS 11.—MEDICAL AND SURGICAL INSTRUMENTS AND APPARATUS.

ABBEY, CHARLES, & SONS, 230 Pear street, Philadelphia, Pennsylvania.— Dentist's gold foil. Bronze medal.

The exhibition made by this firm sustains the well-deserved reputation of their manufacture. This gold foil has all the essential requisites for filling teeth, whether it is to be used in its ordinary state, or is to be rendered adhesive by heating. It has great tenacity, coherence, and ductility, and is uniform in its thickness.

ALLEN, JOHN, & SON, 22 Bond street, New York.—Artificial teeth. Honorable mention.

The pieces of continuous gums shown by the Messrs. Allen are very beautiful, and are striking imitations of nature, but being placed upon platinum plates they are rather heavy for comfort in using.

BARNES, J. K., Surgeon General United States army, Washington, D. C.— Plans of field hospitals, surgical instruments, and hospital apparatus of the United States army. In the Annex, ambulance, medicine wagon; in the Park, hospital tent and furniture. Silver medal.

BATES, R., 730 South Eighth street, Philadelphia, Pennsylvania.—Instruments to cure stammering. Honorable mention.

This extraordinary invention consists of a metallic tube which by a simple arrangement can be attached to the upper part of the mouth, thus preventing the adhesion of the tongue, and allowing the air to pass. This is intended to assist in enunciating the lingual letters. For the labials another tube is provided, and prevents the lips closing against each other by nervous contractions. For the gutturals a small band is supplied with a screw, by which a small plate can be forced against the glottis so as to keep it open, and give passage to the sounds produced by it. The neck-band is made of silk or satin, and has the appearance of an ordinary cravat.

BEALS, J. H., Boston, Massachusetts.—Improved corset.

CUMMINGS, WILLIAM, & SON, New York.—Model of a hospital car. Bronze medal; in the international sanitary department. (See notice at the end of this catalogue.)

CRANDALL, L., & SON, 470 Grand street, New York.—Crutches.

DAVIS, T. J., 64 East Fifteenth street, New York.—Artificial eyes.

FIRMENICH, J., 7 Arcade building, Buffalo, New York.—Dermic instruments for cauterization.

HOWARD, Dr. BENJAMIN, New York.—Ambulance and relief material. Honorable mention; in the international sanitary collection. (Noticed at the end of this catalogue.)

HUDSON, Dr. E. D., New York.—Artificial limbs. A bronze medal was awarded.

This exhibition was made in the international sanitary collection, and is noticed at the end of the catalogue.

JOHNSON & LUND, 27 North Seventh street, Philadelphia, Pennsylvania.—Artificial teeth. Bronze medal.

LINCOLN, M., 19 Green street, Boston, Massachusetts.—Artificial arms.

MARKS, A. A., 575 Broadway, New York.—Artificial legs.

MOODY, Mrs. S. A., 12 East Sixteenth street, New York.—Abdominal corsets.

MOORE, J. G., New Holland, Pennsylvania.—Illustrations of teeth filling.

PEROT, T. MORRIS, Philadelphia, Pennsylvania.—Medicine wagon.—Honorable mention; in the international sanitary department. (Noticed at the end of this catalogue.)

SCOTT, J., Ocala, Florida.—Improved trusses.

SELPHO, WILLIAM, & SON, 516 Broadway, New York.—Artificial limbs. Honorable mention.

STOCKTON, SAMUEL W., Philadelphia, Pennsylvania.—Mineral teeth, with porcelain pivots and new system of transverse holes. Honorable mention.

TAYLOR, CHARLES F., 159 Fifth avenue, New York.—Therapeutic apparatus. Honorable mention.

TIEMANN, GEORGE, & CO., New York.—Surgical instruments. A silver medal was awarded to Mr. George Tiemann as co-operator.

USTICK, S., Philadelphia, Pennsylvania.—Model of an apparatus for invalids.

WESTON, J. W., 706 Broadway, New York.—Artificial leg.

WHITE, SAMUEL S., 528 Arch street, Philadelphia, Pennsylvania.—Dentist's furniture and instruments. Artificial teeth. Gold medal.

The teeth exhibited by Mr. White are of superior quality, and are remarkable imitations of natural teeth. Their smooth surface, semi-opaque and enamelled, has not that appearance of vitrification so disagreeable in most artificial teeth. Their forms are excellent, preserving not only the distinctive characters of the different teeth of the upper and lower jaw, but also of the right and left side of the mouth. Their tint is a mixture of brown and yellow at the base, and a bright and clear enamel on the sharp part of the tooth. They are light and yet solid and strong.

The "block-teeth," with porcelain gums, also exhibited, are made of different sizes, so that any mouth may be fitted. Those intended for mounting in hardened caoutchouc have a pivot, with an expanded head, . which prevents the teeth being pulled away from the base.

Among the other objects is a case of dental instruments containing excellent forceps and a variety of other articles. All of these instruments are elaborate and ingenious, but they are injured by an excess of luxurious ornamentation which is misplaced, for surely it is unnecessary and undesirable to encumber instruments intended for constant use with fine stones and other ornaments.

The gold foil and spongy gold exhibited are excellent. The gold medal awarded to Mr. White is only a just recompense for the excellent services rendered to the dental art by his house, which employs a large number of operatives, and has more than 300 agents in the United States and Europe.

CLASS 12.—INSTRUMENTS OF PRECISION AND APPARATUS FOR INSTRUCTION IN SCIENCES.

BARLOW, MILTON, Richmond, Kentucky.—Planetarium. Bronze medal.

BOND, WILLIAM, & SON, 17 Congress street, Boston, Massachusetts.—Astronomical clock, chronograph and chronometer. Silver medal.

CLUM, H. A., Rochester, New York.—Aelloscope.

COCHRANE, JAMES, 64 West Tenth street, New York.—Apparatus for measuring water under pressure.

DAVIDSON, GEORGE, United States Coast Survey, Washington, D. C.—Improved sextant. Honorable mention.

DARLING, BROWN & SHARPE, Providence, Rhode Island.—Graduated rules, squares, gauges, scales, &c. Silver medal.

EDSON, WILLIAM, Boston, Massachusetts.—Hygrodeik for indicating the amount of moisture in the atmosphere. Honorable mention.

MORSE, S. E., & G. L. HARRISSON, New Jersey.—Bathometer; an instrument for measuring the depth of water.

TOLLES R. F., Canastota, New York.—Microscope and telescope glasses; eyepieces and telescope. Silver medal.

WALES, WILLIAM, Fort Lee, New Jersey.—Microscopic object glasses. Silver medal.

CLASS 13.—GEOGRAPHY, COSMOGRAPHY, APPARATUS, MAPS, CHARTS, ETC.

BACHE, A. D., Hydrographic Bureau, superintendent of the United States Coast Survey, Washington, D. C.—(Out of competition.)—Nautical · charts and apparatus, deep-sea thermometers, gauging instruments.

JOHNSON, A. J., 113 Fulton street, New York.—New illustrated Family Atlas. Bronze medal.

JOSLIN, G., Boston, Massachusetts.—Terrestrial globe.

KNIGHT, E. H., Washington, D. C.—War map of the United States.

TILLMANN, S. D., 12 Clinton place, New York.—Tonometer. New system of chemical nomenclature.

SCHEDLER, JOSEPH, Hudson City, New Jersey.—Terrestrial globes. Bronze medal.

SMITH, S., & CO., Boston, Massachusetts.—Counting-room desk; in Mr. Bacon's bakery in the Park.

USTICK, S., 108 Fourth street, Philadelphia, Pennsylvania.—Water cooler.

GROUP III.

FURNITURE AND OTHER OBJECTS USED IN DWELLINGS.

CLASS 14.—FURNITURE.

BOSTON CHAIR COMPANY, Boston, Massachusetts.—Rocking-chair on a new plan.

BOYD, JOHN D., San Francisco, California.—Door of California wood.

This door was a superb piece of workmanship, being most highly finished and polished so as to bring out the beautiful natural grain of the wood of the Madona or California laurel used in its construction. This wood has a yellowish color like satin wood, but is remarkable for the ease with which it may be stained so as to look like black walnut, mahogany, or rosewood.

BUTLER, J. L., St. Louis, Missouri.—Sofa-bedstead; in the Annex in the Park.

ENGLISH & MERRICK, New Haven, Connecticut.—Folding chairs.

GLASS, PETER, Barton, Wisconsin.—Mosaic tables and table top. Honorable mention.

These tables are said to contain no less than 96,321 pieces of wood.

PHELAN & COLLENDER, 63 to 69 Crosby street, New York.—Billiard table.

It is claimed that the cushions of this table combine elasticity and correctness in the highest possible degree. The lowness of the cushion also compared with the ball affords the player unusual advantages in regard to the facility and accuracy of the stroke, advantages unattainable except by the present improved method of constructing the cushions. With the ordinary construction a low cushion causes the ball to "jump."

ROBINSON, D. T., Boston, Massachusetts.—Model of an extension dining table; in M. Bacon's bakery in the Park.

CLASS 15.—UPHOLSTERY AND DECORATIVE WORK.

SHUSTER, JOHN, 133 Court street, Brooklyn, New York.—Chimney pieces of American marbles.

Three beautiful mantles, one of Tennessee marble, one of white marble from Vermont, and the other of the beautiful stalagmitic marble from Suisun, California.

BOYD, JOHN D., San Francisco, California.—Ornamental door of California wood. (See Class 14.)

CLASS 16.—FLINT AND OTHER GLASS, STAINED GLASS.

BOSTON SILVER GLASS COMPANY, Boston, Massachusetts.—Silvered glass table ware; in the Restaurant.

JONES, THOMAS, Centre and Franklin street, New York.—Window sash of cut and ground glass, colored sidelights.

LYON, JAMES B., & Co., Pittsburg, Pennsylvania.—Pressed glassware. Bronze medal.

PACIFIC GLASS WORKS, J. Taylor, president, San Francisco, California.— Glass bottles of various forms and colors designed for wines, preserves, pickles, sauces, &c., manufactured in California from sand obtained upon the bay of Monterey.

These samples, which compared favorably with any in the Exhibition, were donated, at the close, to the museum at Sevres.

SCHWITTER, ANTHON, 177 Broadway, New York.—Glassware engraved by a mechanical process.

CLASS 17.—PORCELAIN, FAÏENCE, AND OTHER POTTERIES.

BOCH, WILLIAM, Bochtown, Newtown, New York.—Porcelain ware.

RAMSAY, G. M., 23 Courtlandt street, New York.—Air-tight jars for preserving purposes, &c.

CLASS 18.—CARPETS, HANGINGS, AND OTHER FURNITURE TISSUES.

CHIPMAN, GEORGE W., & Co., 119 Milk street, Boston, Massachusetts.— Carpet lining. Honorable mention.

TOWNSEND, WISNER H., 20 Reade street, New York.—Samples of oilcloth. Bronze medal.

CLASS 19.—PAPER HANGINGS.

BIGELOW, J. R., Boston, Massachusetts.—Paper-hangings.

CHRISTY, CONSTANT & Co., New York.—Paper-hangings.

GRAVES, R., & Co., New York.—Paper-hangings.

HOWELL & BROTHER, Philadelphia, Pennsylvania. — Paper-hangings. Honorable mention.

CLASS 20.—CUTLERY.

BIGGS, C., 57 Beekman street, New York.—Pocket cutlery from the manufactory of Booth Brothers, Newark, New Jersey.

SHAVER, A. G., New Haven, Connecticut.—Erasers and pencil-sharpeners.

CLASS 21.—GOLD AND SILVER PLATE.

MERIDEN BRITANNIA COMPANY, West Meriden, Connecticut.—Plated table ware; in the Restaurant.

TIFFANY & Co., 550 and 552 Broadway, New York.—Ornamental plate and silver-ware in various styles of chasing; reduction of the "America" of Crawford, decorating the cupola of the Capitol at Washington; models of the steamers "Commonwealth" and "Vanderbilt." Bronze medal.

The hull of the model of the Vanderbilt is fashioned in frosted or dead silver, with a burnished streak or gunwale. The paddles are of burnished silver, tipped with gold; the tops and bottoms of the funnels are of gold; the deck is formed of polished silver; the quarter boats of gold. The just proportion of every part is preserved in the model, and every detail, even of the minute parts, has been carefully wrought in silver or gold.

CLASS 22.—ARTISTIC BRONZES, ARTISTIC CASTINGS OF VARIOUS KINDS, AND CHASED METAL ORNAMENTS.

TUCKER, HIRAM, & Co., 59 John street, New York.—Iron ornaments bronzed by new process. Silver medal.

These objects, consisting of clock stands, vases, lamps, chandeliers, brackets, &c., were much admired.

CLASS 23.—CLOCKS AND CLOCK WORKS.

NEW HAVEN CLOCK COMPANY, New Haven, Connecticut.—Clocks. Honorable mention.

FOURNIER, S., 60 Royal street, New Orleans, Louisiana.—Clocks and clock works. Silver medal.

This exhibition consisted of several large and accurately made clocks for churches and public buildings. They were set up and running during the Exhibition, and the works were in full view in an alcove or enclosed space reserved for them.

CLASS 24.—APPARATUS AND METHODS OF WARMING AND LIGHTING.

BEIDLER, J. H., Lincoln, Illinois.—Hydro-caloric light.

CLOGSTON, T. S., & Co., Boston, Massachusetts.—Steam radiator for heating buildings.

GOUGES VENTILATING COMPANY, 254 Broadway, New York.—Atmospheric ventilator.

HASKINS, D. G., Cambridge, Massachusetts.—Gas furnace.

IVES, J., & Co., 18 Beekman street, New York.—Kerosene and petroleum lamps and chandeliers.

MARKLAND, T. J., 835 Ellsworth street, Philadelphia, Pennsylania.—Coal scuttle.

MUELLER, J. U., Detroit, Michigan.—Improved stove handles.

O'NEIL, A., Portsmouth, Ohio.—Sheet metal stove boiler.

PEASE, F. S., Buffalo, New York.—Gas apparatus.

PRATT & WENTWORTH, 89 North street, Boston, Massachusetts.—Cooking stove and utensils. Bronze medal.

TUCKER, H., & Co., 59 John street, New York.—Lamps and chandeliers.

USTICK, S., Philadelphia, Pennsylvania.—Model of an improved street lamp.

WHITELY, EDWARD, Boston, Massachusetts.—Cooking range and apparatus, in the American restaurant.

CLASS 25.—PERFUMERY.

TALLMAN & COLLINS, Janesville, Wisconsin.—Perfumery. Honorable mention.

WRIGHT, R. & G. A., 624 Chestnut street, Philadelphia, Pennsylvania.—Toilet soap and perfumery. Bronze medal.

CLASS 26.—FANCY ARTICLES, TOYS, BASKET WORK.

BLOODGOOD, ANNIE DE ETTA, 127 Ninth avenue, New York.—Wax flowers.

HAUXHURST, CAROLINE, Rahway, New Jersey.—Ornaments of skeleton leaves.

KALDENBERG & SON, New York.—Meerschaum pipes. Honorable mention.

LACHAUME, J., 163 Prince street, New York.—Rustic work, baskets, stands, &c.

MACDANIEL, Miss F., New York.—Natural flowers with color preserved.

SMITH, Mde. E. W., West Medford, Massachusetts.—Wax flowers, fruits, &c.

GROUP IV.

CLOTHING, (INCLUDING TISSUES,) AND OTHER OBJECTS WORN ON THE PERSON.

CLASS 27.—COTTON YARN, THREADS AND TISSUES OF COTTON.

BELL FACTORY, Huntsville, Alabama.—Cotton fabrics. Honorable mention.

CLARK THREAD COMPANY, G. A. Clark, treasurer, Newark, New Jersey.—Cotton and cotton yarns. Silver medal.

GROLL & GRUBBS, Chicago, Illinois.—Cotton batting.

HADLEY COMPANY, Holyoke, Massachusetts.—Spool cotton. Bronze medal.

NEW YORK MILLS, Walcott & Campbell, 57 Worth street, New York.—Fine muslins. Silver medal.

SLATER, S., & SON, Webster Woollen Mills, Webster, Massachusetts.—Jaconets and cotton fabrics. Bronze medal.

CLASS 28.—YARN AND TISSUES OF LINEN, HEMP, ETC.

HARVEY, W., 84 Maiden Lane, New York.—Flax, hemp, cotton, linen, and paper twine and cordage.

HALL MANUFACTURING COMPANY, Boston, Massachusetts.—Cordage made on Bazin's twisting machine.

CLASS 29.—COMBED WOOL AND WORSTED YARNS AND FABRICS.

(No exhibitors.)

CLASS 30.—YARN AND TISSUES OF CARDED, WOOL.

HAYES, JOHN L., secretary National Association of Wool Manufacturers, 75 Summer street, Boston, Massachusetts.—Series of woollen fabrics, manufactured by the Washington Mills, situated in Lawrence, Massachusetts. Silver medal.

None of the pieces exhibited were made expressly for the exhibition, but were specimens of the daily products of the establishment. They were forwarded with a statement that they were intended to show the average styles and quality of the woollen goods then being made in the United States. To each sample a card was affixed showing the selling price in the United States. The goods exhibited consisted of eight varieties of shawls; carriage rugs; one piece of each of the following goods: fancy shirting, Nevada plaid, Italian cloth, American poplin, blue Esquimaux coating, black doeskin, tricot, Moscow beaver, diagonal coating, A. W. braid, Union broad beaver, Jansen silk mixture, blue, black and white silk mixture, Paris indigo blue coating, extra blue Washington coating, repellant cloaking, fancy cassimere; and three pieces of each of the following: sackings, mixed Scotch tweed.

KLAUDER, R., Philadelphia, Pennsylvania.—Dyed and printed zephyr.

MISSION WOOLLEN MILLS, San Francisco, California, D. McLennan, superintendent; Lazard Fréres, agent.—Woollen goods, comprising a large assortment of blankets, travelling shawls, cassimeres and flannels, all made from pure California wool at the company's mills at the Mission, San Francisco.

The blankets exhibited were remarkably fine and soft, of large size, and unrivalled in quality. The assortment contained blankets for family use, for miners, for the army, and for Indians. The family blankets were 86 by 94 inches in size, and weighed from 10 to $11\frac{1}{2}$ pounds each. The miners' blankets were 62 by 84 inches, and weighed from $9\frac{1}{2}$ to $10\frac{1}{2}$ pounds each. Those for the army were 66 by 89 inches, and weighed 6 pounds each.

The cassimeres were mixed, plaid, and plain; and the flannels were both plain and colored. The collection contained a sample of the peculiar shaggy blanketing used in sluices by miners to catch and hold the fine particles of gold and sulphurets of iron flowing from stamp-batteries. A bronze medal was awarded for this display.

SHIELDS, J., Davenport, Iowa.—Woollen goods.

SLATER, S., & SON, Slater Woollen Mills, Webster, Massachusetts.—Woollen fabrics, broadcloths, doeskins, castors and moskowa. Silver medal.

STURSBERG, H., 97 Reade street, New York.—Beaver cloth. Bronze medal.

CLASS 31.—SILK AND TISSUES OF SILK.

WILLIAMS SILK MANUFACTURING COMPANY, 469 Broadway, New York.—Silk twist for sewing machines. Honorable mention.

CLASS 32.—SHAWLS, ETC.

THE WASHINGTON MILLS, Lawrence, Massachusetts.—Shawls. Honorable mention.

TORRENCE, Mrs. J. S., 111 Broadway, New York.—Worsted Affghan.

CLASS 34.—HOSIERY, UNDER-CLOTHING, AND MINOR ARTICLES.

COHN, M., 147 Chambers street, New York.—Crinolines of various descriptions.

MOODY, S. N., New Orleans, Louisiana.—Two dress shirts.

MOUNT CITY PAPER COLLAR COMPANY, St. Louis, Missouri.—Paper collars.

SACHSE, F., & SONS, Pine street, Philadelphia.—Dress shirts. Bronze medal.

CLASS 35.—CLOTHING FOR MEN, WOMEN, AND CHILDREN.

BOUVET, J., New Orleans, Louisiana.—Hats.

BURT, E. C., 27 Park Row, New York.—Machine-sewed boots and shoes. Silver medal.

DEMOREST, Mrs. ELLEN, Broadway, New York.—Corsets, patterns, &c.

FELMEDEN, J. K., New Orleans, Louisiana.—Boots and shoes made from alligator leather.

LINTHICUM, W. O., 726 Broadway, New York.—Spring overcoat. Honorable mention.

NICELY, H. C., 34 West Baltimore street, Baltimore, Maryland.—Hats and caps.

PACALIN, O., 3 Amity Place, New York. Metallic sole fastening for boots.

WHITNEY BROTHERS & CO., Chicago, Illinois.—Boots.

WINDLE & CO., New York.—Boots and shoes with wooden soles and heels, and flexible shanks.

ZALLÉE, JOHN C., 110 Olive street, St. Louis, Missouri.—Frock coat, black doeskin pantaloons, and silk vest. Honorable mention.

CLASS 36.—JEWELRY AND ORNAMENTS.

(No exhibitors.)

CLASS 37.—PORTABLE ARMS.

ARM MANUFACTURING INDUSTRY OF THE UNITED STATES.

It was found so difficult to decide upon the relative merits of the portable fire-arms exhibited in the American section, and their superiority was recognized as so indisputable, that the international jury, as a compliment, and at the same time for the purpose of avoiding what might be construed as an invidious distinction, voted a gold medal to "The Arm Manufacturing Industry of the United States."

BERDAN, COLONEL H., 30 Bond street, New York.—Breech-loading rifle.

BONZANO, A., Detroit, Michigan.—Cannon-muzzle spikers.

COLT'S FIRE-ARMS MANUFACTURING COMPANY, Hartford, Connecticut.—
Colt's fire-arms; a Gatling gun. Silver medal. (See Gatling gun.)

FERRISS, G. H., Utica, New York.—Wrought-iron breech-loading rifled
cannon; target perforated by it.

GATLING, R. J., Indianapolis, Indiana.—Improved battery gun.

This is a breech-loading repeating gun, in which all the operations of
loading, firing, and getting rid of the debris of the case of the cartridge are
preformed by a simple rotary movement. It is fed with metallic cartridges,
each of the largest containing 15 musket balls and one conical ball, thus
throwing 16 projectiles at every discharge. Twenty discharges can be
made in eight seconds. Among other advantages may be mentioned the
absence of any gas escaping by the breech; no recoil tending to divert
the aim; great accuracy of aim, and rapidity of firing; and, lastly, lightness. This gun was exhibited by the Colt Fire-arms Manufacturing
Company, and a silver medal was awarded to this company for its manufactures.

JENKS, A., & SON, Philadelphia, Pennsylvania.—Fire-arms, and parts of
same manufactured by machinery.

MISSOURI, STATE OF.—Indian weapons, curiosities &c.

PROVIDENCE TOOL COMPANY, J. B. Anthony, president, Providence,
Rhode Island.—Peabody's breech-loading fire-arms. Silver medal.

REMINGTON, E., & SON, Ilion, New York.—Breech-loading fire-arms.
Silver medal.

ROBERTS, General B. F., Washington, D. C.—Breech-loading rifle.

Description: calibre, .58 inch; distance from muzzle to face of breech-
lock, when closed, 37 inches; length of chamber, 1.25 inch. The chamber
has a uniform taper for its entire length; maximum diameter, .64 inch,
minimum diameter, .58 inch; receiver, 2 inches in length; breech block,
.75 inch wide. Breech-block and all its appendages assembled from one
piece, 5 inches in length.

The musket presented is of the United States " Springfield" pattern,
made by machinery. The breech-loading parts, five in number, were
made by hand, and constitute " the Roberts breech-loading attachment."
The first piece is an iron breech frame or receiver, into which the barrel,
having been cut off at proper point, is firmly screwed. This receiver is
imbedded in the stock in the place of the old breach pin. The barrel is
cut off about one inch in front of the cone, and a male screw cut, reaching nearly to the rear sight of the barrel. The breech block is inserted
through this receiver, and supported against the rear end on a semi-
circular shoulder, forming the back of receiver, the centre around which
this semi-circle is described being in the prolongation of the axis of the
barrel. The rear of the breech block is turned to fit with exactness this
semicircle, and is played around it as a fulcrum. The cheeks of the
receiver support the breech block laterally. When the breech block is
in place in the receiver it forms a curved lever, the handle projecting

backward, and it then is moved about the solid abutment of the receiver, instead of being pivoted by any system of points or pins, thus affording great solidity and strength.

The forward end of the breech block has a semicircular groove cut transversely through it, for the purpose of receiving a corresponding tenon formed on a block of steel, termed the recoil plate. The front face of this block is flat, so that when in position it fits squarely against the vertical face of the chamber and the rear end of the cartridge case. A small space is left between the tenon on the rear of this block and the front surface of the breech block above the transverse groove, to admit of a slight rocking motion of recoil plate, so that it will descend to expose the breech of the barrel and admit the cartridge into the chamber. This small open space permits the recoil plate to descend perpendicularly when the rear of the lever is raised, until the top of the plate passes below the axis of the barrel, after which it swings with the arc of the circle on the rear end of the receiver. When the rear of the lever is raised the recoil plate ascends to its position by the exact reverse motion, up to the axis of the barrel on a circular motion, and afterward to close the chamber, ascending vertically and closing squarely against the head of the cartridge case and the vertical face of the chamber.

The firing pin is located on the right side of the breech block, and runs through both this block and the recoil plate, directed to the centre for centre-fire cartridges and grooved into the sides for rim-fire cartridges. It s so set on a shoulder that the force of the blow of the hammer cannot drive it a greater distance than is necessary to insure fire.

The retractor is a curved lever, fixed on the left side of the chamber, with one arm behind the flange of the cartridge case and the other operating in a vertical groove on the left side of the recoil plate. When the breech lever is raised and the recoil plate descends, the arm in the groove is not touched until the top of this plate reaches the bottom of the chamber, the shoulder at the upper end of the groove then strikes the lever and ejects the cartridge case.

SMITH & WESSON, Springfield, Massachusetts.—Fire-arms and metallic cartridges. Silver medal.

SPENCER REPEATING RIFLE COMPANY, Boston, Massachusetts.—Spencer rifles. Breech-loading, capable of being fired seven times in twelve seconds. Silver medal.

UNITED STATES SANITARY COMMISSION.—Camp material, in the international sanitary department. (See a notice at the end of this catalogue.) Honorable mention.

WINDSOR MANUFACTURING COMPANY, Windsor, Vermont.—Ball's patent repeating fire-arms. Silver medal.

WHIPPLE, H. B., Faribault, Minnesota.—Arms, curiosities, &c., of the Ojibwa and Dakota tribes.

CLASS 38.—ARTICLES FOR TRAVELLING AND FOR ENCAMPMENT.

BAIRD, H. S., Green Bay, Wisconsin.—Indian curiosities.

COLLINS, Mrs. L., New Orleans, Louisiana.—Embroidered flags in the Louisiana cottage.

MEIGS, M. E., Quartermaster General in the United States army Washington, D. C., (out of competition.)　In the park.—Material in use in the United States army for transportation, clothing, and equipment in camp and in garrison.

NOYES, J. H., Oneida, New York.—Traveller's lunch bag.

PADDOCK, W. S., Albany, New York.—Fastenings for trunks, arranged on a model trunk.

PIERCE, CARLOS, Boston, Massachusetts.—The Frémont army tent, in the Park.

This tent is so constructed that during rain storms, when the canvass shrinks from wetting, it can be lowered a little from the inside instead of loosening the pegs outside to provide for the shrinkage.

PULLAN, R. B., Cincinnati, Ohio.—Model tents.

SHORT, J., Salem, Massachusetts.—Army knapsack.

CLASS 39.—TOYS AND GEWGAWS.

MUELLER, T. U., Detroit, Michigan.—Toy puzzle.

GROUP V.

PRODUCTS, RAW AND MANUFACTURED, OF MINING INDUSTRY, FORESTRY, ETC.

CLASS 40.—MINING AND METALLURGY.

The display of mineral productions of all kinds from the vast metalliferous regions of the United States was one of the most important features of the Exposition.　The most distant States were represented there by samples of their ores and minerals.　California, Nevada, Idaho, Colorado, Arizona, Montana, Dakota, New Mexico, Oregon, and Washington, with a united area equal to the whole of Europe, nearly all sent specimens indicative of their marvellous resources in gold, silver, copper, lead, iron, coal, petroleum, and other minerals.　The most prominent collections were from California, Colorado, and Nevada.

ALABAMA, STATE OF.—Minerals from that State.

ARKANSAS, STATE OF.—Minerals from that State.

AVERY, R. D., Petite Anse, Louisiana.—Rock salt.

BALTIMORE AND CUBA SMELTING AND MINING COMPANY, C. Levering, president, Baltimore, Maryland.—Ingot and sheet copper.　Bronze medal.

BARR & COX, Beloit, Wisconsin.—Hammers and hatchets.

BARR, J., Licking county, Ohio.—Minerals, samples of coal.

BIGELOW, H., Boston, Massachusetts.—Rocks, ores, and minerals from Michigan. Silver medal.

This collection included a variety of specimens of native copper, from Lake Superior, and of the various interesting materials which accompany it.

BIGLEY, N. J., Pittsburg, Pennsylvania.—Samples of coal, limestone, fire clay.

BLAKE, WILLIAM P., California, Commissioner from the State to the Exposition.—A collection of the ores and minerals found in California and the adjoining States and Territories, intended to illustrate the mineral resources of the Pacific coast region of the United States. Silver medal.

This collection contained over 300 specimens of good size, taken from the principal gold-bearing veins of California, and from the copper, quicksilver, lead, and iron veins. The borax, salt, petroleum, and building materials were also shown. All the specimens were properly classified and labelled.

BURT, J., Detroit, Michigan.—Iron ores, iron, steel, samples of iron made from Lake Superior specular and magnetic ores.

CHESTER IRON COMPANY, (J. B. TAFT,) Chester, Massachusetts.—Emery and minerals from Chester, Massachusetts. Silver medal.

This was a very interesting and instructive suite of specimens of the massive emery stone and the minerals which are usually associated with it, together with the crushed and prepared emery and the emery cloths and papers. The presence of emery at this locality was discovered by Dr. Charles T. Jackson, of Boston, when giving some samples of iron ore found there a scientific examination. This important service was recognized by the class jury, and a bronze medal was awarded to Dr. Jackson as co-operator, for "Discovery of emery in the United States."

CHILDS, T., & Co., Hartford, Connecticut.—Skates.

CONNELL, S. G., & SON, Buffalo, New York.—Pure white lead.

DIXON, J., & Co., Jersey City, New Jersey.—Plumbago crucibles and stove polish.

DOUGLAS, J. L., 158 Broadway, New York.—Minerals from the Territory of Nevada.

DOUGLASS AXE MANUFACTURING COMPANY, D. D. Dana, treasurer, Boston, Massachusetts.—Edge tools. Silver medal.

DOUGLASS MANUFACTURING COMPANY, 70 Beekman street, New York.—Edge tools. Bronze medal.

ELSBERG, Dr. L., 123 West Fifteenth street, New York.—Prepared peat fuel. Honorable mention.

GAUJOT, R. C. E., Tamaqua, Pennsylvania.—Samples of coal, rocks, and iron ores.

GOODENOUGH HORSESHOE COMPANY, W. C. Colgate, president, 1 Dey street, New York.—Horseshoes. Honorable mention.

GOULD, J. D., Boston, Massachusetts.—Mica.　Honorable mention.

This was a fine assortment of mica, in large, clear sheets, suitable for stoves, lanterns, and for roofing.

GREEN, JAMES D., Cambridge, Massachusetts.—A column of Winooski marble, (Vermont.)

HALLIDIE, A. S., & Co., San Francisco, California.—Wire rope.

Samples of the various sizes of wire ropes, cables round and flat for mining purposes, sash cords of various sizes, &c., &c., all manufactured in San Francisco, and proving great skill in this art.　These samples, at the close of the Exposition, were donated to the Museum of Arts and Manufactures.

HARRIS, J., Sturgeon Bay, Wisconsin.—Samples of native copper from Lake Superior.

HERRING, FARRELL & SHERMAN, 254 Broadway, New York.—Crystallized iron—"Franklinite."

ILLINOIS, STATE OF.—Collection of minerals, building stones, fossils. Silver medal.

IOWA, STATE OF.—Specimens of the mineral productions of that State.

JACKSON, J. H., 155 Broadway, New York.—Minerals and fossils.　Honorable mention.

KANSAS, STATE OF.—Specimens of the mineral productions of that State.

KASE, S. P., Danville, Pennsylvania.—Coal from the Beaver Creek Coal Company.

KASSON, A. C., Milwaukee, Wisconsin.—Patent auger bits.

LALANCE & GROSJEAN, 273 Pearl street, New York.—House furnishing hardware.　Chairs, in the Annex.　Honorable mention.

McCORMICK, J. J., Williamsburg, New York.—Skates.　Honorable mention.

MERRITT, W. H., North Anthracite Coal-field, Luzerne county, Pennsylvania.—Anthracite coal.

MINNESOTA, TERRITORY OF.—Collection of minerals from that Territory.

MISSOURI, STATE OF.—Minerals from that State.

NEVADA, TERRITORY OF.—Silver ores.　Silver medal.

This was a splendid display of rich ores of silver from eastern Nevada, collected chiefly by a committee appointed by the citizens, and represented at the Exposition by David E. Buel, esq.　Many of the masses were over 18 inches in diameter, and were from the newly-discovered districts in the southeastern portion of the State.

NEW JERSEY ZINC COMPANY, G. A. Bell, president, 64 Maiden Lane, New York.—Specimens of ores, and products manufactured therefrom.

This series contained masses of the red zinc ore, of the Franklinite, and of the silicate of zinc, all from the company's mines at Stirling Hill and at Mine Hill, in Sussex county, New Jersey.　These ores are worked chiefly into oxide of zinc for paints and into pig iron, known as Franklinite iron.

PARK BROTHERS & Co., Black Diamond Steel Works, Pittsburg, Pennsylvania.—Cast-steel edge tools.

A very interesting display of superior tools, for which a silver medal was awarded.

PATTERSON, S., Mauch Chunk, Pennsylvania.—Anthracite coal.

This was an enormous single block of coal weighing three and a half tons, taken from the colliery of W. Johns. It occupied a prominent place in the mineral collection, and a bronze medal was awarded. (See following entry.)

PENNSYLVANIA, STATE OF.—Anthracite coal. (S. Patterson's.) From colliery of W. Johns, as noted above. Bronze medal.

PORTAGE LAKE SMELTING WORKS, E. D. Brigham, treasurer, Boston, Massachusetts.—Ingots and cakes of copper. Bronze medal.

PRENTICE, F., Nevada.—Ores from Nevada.

PIGNÉ, Dr. J. B., San Francisco, California.—Collection of minerals from California. Silver medal.

This was a very complete collection of ores of gold, silver, copper, lead, iron, quicksilver, &c., &c., from the principal mines of the Pacific States, all neatly classified, labelled and catalogued, and intended for the collection of the *Ecole Imperiale des Mines* at Paris.

PIONEER AND INSKIP MILL AND MINING COMPANY, D. H. Temple, secretary, 8 Pine street, New York.— Minerals and silver ores from Nevada.

RANDALL, SAMUEL H., New York.—Specimens of mica, feldspar, beryl, quartz, &c. Bronze medal.

ROBINSON, E., & SON, Boston, Massachusetts.—House hardware, in Mr. Bacon's bakery, Park.

SAFFRAY, C., 26 East Fourth street, New York.—Agglomerated coal.

SHAUB, G., superintendent of the Southern Porcelain Company, Augusta, Georgia.—Kaolin.

SHELTON COMPANY, Birmingham, Connecticut.—Iron, copper, and tinned tacks.

SHUSTER, J., 133 Court street, Brooklyn, New York.—Samples of California, Tennessee, New York, and Vermont marbles.

SIBLEY, F. K., Auburndale, Massachusetts. — Samples of emery and crocus cloths.

TEXAS CHROME MINING COMPANY, Texas, Pennsylvania.—Chromic iron ore in large masses as taken from the quarry.

THOMAS IRON WORKS, Hokendauqua, Pennsylvania.—Iron and iron ores.

UTAH, TERRITORY OF.—Minerals.

WALDRIDGE, W. D., 51 Exchange Place, New York.—Samples of gold, silver, tin, and copper from Idaho. Large masses of silver ore from the Poorman lode in Idaho. These blocks contained large quantities of ruby silver ore. Gold medal.

WARNER, G. F. & Co., New Haven, Connecticut.—Malleable iron castings.

A very great variety of small objects, chiefly carriage hardware, all neatly arranged upon a large square tablet. Bronze medal.

WEST VIRGINIA, STATE OF.—Minerals from that State; building stone.

WETHERBEE, SHERMAN & CO., Port Henry, New York.—Magnetic iron ore, iron.

WHARTON, JOSEPH, Philadelphia, Pennsylvania.—Ores and metals, nickel, cobalt, zinc. Honorable mention.

WHITNEY, J. P., Boston, Massachusetts.—Gold and silver ores and minerals from Colorado Territory. Gold medal.

A very large and brilliant collection of the pyritic gold-bearing ores of Colorado, accompanied by maps of the region, photographs, and statistics, published in three languages.

WILKINSON, A. S., Pawtucket, Rhode Island.—Horseshoes.

WISCONSIN, STATE OF.—Minerals, ores, building stones, and metals from Wisconsin. Bronze medal.

CLASS 41.—PRODUCTS OF THE FOREST.

ANDREWS, HARRIS & CO., St. Louis, Missouri.—Black moss from Louisiana

BOYD, JOHN D., San Francisco, California.—Samples of cabinet woods from California.

This exhibition consisted of masses of the trunk of the madrona, and of bundles of veneers cut from it, also of a series of panels veneered, stained, and polished, showing a grain of remarkable beauty.

CARTER, G. W., 98 Hudson street, New York.—Fret, scroll, and ornamental sawing.

EDWARDS, D., Little Genesee, New York.—Specimens of wood and clapboards.

HALL, E., Athens, Illinois.—Collection illustrating the botany of Illinois.

KANSAS, STATE OF.—Specimens of wood. Honorable mention.

LEAVITT & HUNNEWELL, Boston, Massachusetts.—Prepared peat fuel.

MEARS, C., & CO., Chicago, Illinois.—Shingles.

MISSOURI, STATE OF.—Specimens of wood from Missouri.

PAUL, J. F., & CO., 441 Tremont street, Boston, Massachusetts.—Wood mouldings, oval frames, specimens of wood. Honorable mention.

PERSAC, A., New Orleans, Louisiana.—Illustrations of American forests.

UTAH, TERRITORY OF.—Specimens of wood.

WISCONSIN, STATE OF.—Samples of wood.

CLASS 42.—PRODUCTS OF HUNTING AND FISHERIES, AND UNCULTIVATED PRODUCTS.

BELL, J. G., 335 Broadway, New York.—Stuffed birds.

GUNTHER, C. G., & SONS, 502 Broadway, New York.—Stuffed animals. Silver medal.

ILLINOIS, STATE OF.—Stuffed game birds from the Chicago Academy of Sciences.

19 U E

KANSAS, STATE OF.—Furs, antlers, and skins.

WISCONSIN, STATE OF.—Furs, antlers, and skins.

CLASS 43.—AGRICULTURAL PRODUCTS (NOT USED FOR FOOD) OF EASY PRESERVATION.

ALABAMA, STATE OF.—Samples of cotton. Silver medal and honorable mention.

BOURGEOIS, E., New Orleans, Louisiana.—Perrique tobacco. Honorable mention.

CAROLL, J. W., Lynchburg, Virginia.—Tobacco. Bronze medal.

COZZENS, FREDERIC S., 73 Warren street, New York.—Cigars. Honorable mention.

DELPIT, A., & CO., New Orleans, Louisiana.—Snuff and smoking tobacco. Silver medal.

DIEHL, I. S., 80 Broadway, New York.—Specimens of Angora wool from different parts of the United States and articles manufactured from the same.

HUMPHRIES, JOHN C., parish of Rapides, Louisiana.—Samples of cotton. Bronze medal.

ILLINOIS CENTRAL RAILROAD COMPANY.—Hemp, flax, cotton, and tobacco. Silver medal.

JOHNSON, C. G., New Orleans, Louisiana.—Specimen of cotton; in the Louisiana cottage.

JOHNSON, O., Galba, Illinois.—Samples of broom corn.

KANSAS, STATE OF.—Agricultural products from Kansas.

LEHMAN, NONGASS & CO.—New Orleans, Louisiana.—Wool.

LILIENTHAL, C. H., 221 Washington street, New York.—Snuff and tobacco. Bronze medal.

MAGINNIS, A. A., New Orleans, Louisiana.—Cotton seeds.

MEYER, VICTOR, parish of Concordia, Louisiana.—Sample of cotton. Gold medal.

MISSOURI, STATE OF.—Cotton, hemp, cashmere wool.

MONTAGNE & CARLOS, New Orleans, Louisiana.—Black moss for upholsterers. Honorable mention.

RICHARD RICHARDS, Racine, Wisconsin.—Specimen of wool. Bronze medal.

ST. LOUIS LEAD & OIL CO.—Seed and seed oils.

SARRAZIN, J. R., New Orleans, Louisiana.—Samples of tobacco. Bronze medal.

SCHERR, T., San Francisco, California.—Bale of hops.

These hops were grown on the grounds of Wilson Flint, esq., in the Sacramento valley, and were of superior quality. Samples of them were freely distributed during the exhibition.

TAMBOURY, A., parish of St. James, Louisiana.—Samples of tobacco. Bronze medal.

TOWNSEND, J., Edisto Island, South Carolina.—Superfine sea island cotton.

TRAGER, LOUIS, Black Hawk Point, Louisiana.—Samples of cotton. Gold medal.

WILLIAMS, THOMAS C., & CO., Danville, Virginia.—Samples of tobacco. Bronze medal.

WISCONSIN STATE AGRICULTURAL SOCIETY.—Specimens of wool and of seed oils. Bronze medal.

CLASS 44.—CHEMICAL AND PHARMACEUTICAL PRODUCTS.

BABCOCK, JAMES F., Boston, Massachusetts.—Rosin oil. Bronze medal.

BECKER, H. C., New York.—Extracts for culinary use.

BELMONT OIL COMPANY, 333 Market street, Philadelphia, Pennsylvania.—Crude and refined petroleum, benzine, gazoline. Bronze medal.

BRANDON KAOLIN AND PAINT COMPANY, J. W. Prime, president, Brandon, Vermont.—Specimens of paints. Honorable mention.

BUTLER, T. S., Cincinnati, Ohio.—Oil blacking.

CALIFORNIA, STATE OF.—Oils. Samples of petroleum, both crude and refined, from localities in various parts of the State.

The refined oils were from the establishments of Messrs. Hayward & Coleman, Stanford Brothers, and Charles Stott, in San Francisco.

CHICAGO GLUE WORKS, Chicago, Illinois.—Samples of glue.

DAY, AUSTIN G., Seymour, Connecticut.—Samples of hard, semi-hard, and soft India-rubber, and artificial rubber. Honorable mention.

DIEHL, J. S., 80 Broadway, New York.—Petroleum; silicated copper.

DUNDAS, DICK & CO., 110 Reade street, New York.—Capsulated medicines.

FRIES, ALEXANDER, Cincinnati, Ohio.—Flavoring extracts. Honorable mention.

GLEN COVE STARCH MANUFACTURING COMPANY, W. Duryea, secretary, 166 Fulton street, New York.—Maize starch.

GLIDDEN & WILLIAMS, Boston, Massachusetts.—Soluble Pacific guano.

HALE & PARSHALL, Lyons, New York.—Oil of peppermint.

HERZBERG, I., & BROTHER, Philadelphia, Pennsylvania.—Chronometer and watch oil.

HESS, BECKER & CO., St. Charles, Missouri.—Sample of ultramarine.

HIRSCH, JOSEPH, Chicago, Illinois.—Glycerine, albumen, &c. Honorable mention.

HOLLIDAY, T. & C., 194 Broadway, New York.—Dyes made from aniline, pigments and colors, chemicals. Honorable mention.

HOTCHKISS, H. G., Lyons, New York.—Samples of essential oils. Bronze medal.

HOTCHKISS, L. B., Phelps, New York.—Specimens of oils of peppermint and spearmint. Bronze medal.

KIEFFER, N., New Orleans, Louisiana.—Bitters.

LOUISIANA PETROLEUM AND MINING COMPANY, A. L. Fields, secretary, New Orleans, Louisiana.—Specimens of petroleum.

MAGINNIS, A. A., New Orleans, Louisiana.—Cotton seed oil, soap, and oil cake.

MARIETTA AND GALES FORK PETROLEUM COMPANY, R. K. Shaw, director, Marietta, Ohio.—Crude lubricating petroleum. Honorable mention.

McROBERTS & DICK, New Orleans, Louisiana.—Soap.

MORGAN'S, E., SONS, 274 Washington street, New York.—Family soap.

PEASE, F. S., Buffalo, New York.—Illuminating and lubricating oils, paraffine. Silver medal.

RHODES, B. M., & Co., Baltimore, Maryland.—Superphosphate of lime for manure.

SMITH, R. M., Baltimore, Maryland.—Refined burning and lubricating petroleum oils. Honorable mention.

STANDARD SOAP COMPANY, San Francisco, California.—Soap and washing powder.

The soap is represented to be made in San Francisco exclusively from materials produced in the State of California. The alkali is said to be made from the ashes of the ice plant, which grows in Santa Barbara county.

VANDERBURGH, G., 24 Vesey street, New York.—Specimens of alkaline silicates.

VAN DEUSEN BROTHERS, Kingston, New York.—Oil of wintergreen.

VOLCANIC OIL AND COAL COMPANY, of Western Virginia, Philadelphia, Pennsylvania; H. G. Moehring, agent.—Lubricating mineral oil. Honorable mention.

WAHL, C., Milwaukee, Wisconsin.—Specimens of glue.

WESTON, H., 706 Broadway, New York.—Concentrated aqueous solution of iodine.

WEST VIRGINIA, STATE OF, J. H. Diss Debar, agent.—Crude and refined petroleum. Bronze medal.

WHITE, G. E., New York.—Swan Island guano.

WHITE, M. J., parish of Plaquemines, Louisiana.—Extract of red Tobasco pepper.

UREN, DUNSTONE & BLIGHT, Eagle River, Michigan.—Water proof safety fuse.

CLASS 45.—SPECIMENS ILLUSTRATING THE CHEMICAL PROCESSES IN BLEACHING, DYEING, PRINTING, AND DRESSING FABRICS.

HOLLIDAY, T. & C., 194 Broadway, New York.—Woollen, cotton, and silk goods, dyed and printed with aniline dyes.

CLASS 46.—LEATHER AND SKINS.

BACON, S. T., Boston, Massachusetts.—Vulcanized rubber.

BROWNE, D. JAY, Park street, Roxbury, Massachusetts —Enamelled leather, manufactured by a new process. Honorable mention.

GUNTHER & SONS, 502 Broadway, New York.—Furs for ladies' and gentlemen's wear, sleigh robes.

KORN, CHARLES, 19 Ferry street, New York.—Calfskin leather. Honorable mention.

McDONALD & HURD, Winchester, Massachusetts.—Calfskin leather.

MEYER, C. F. W., Union Hill, New Jersey.—Piano-forte buckskins.

PAGE, M. W., Franklin, New Hampshire.—Samples of belt lacing made by a new process of tanning.

SCHORR, T., New Orleans, Louisiana.—Alligators' skins tanned for shoe leather.

SMITH, LYMAN, & SON, Boston, Massachusetts.—Samples of leather for cotton factory rollers.

WISCONSIN, STATE OF.—Leather and skins.

GROUP VI.

APPARATUS AND PROCESSES USED IN THE COMMON ARTS.

CLASS 47.—APPARATUS AND METHODS OF MINING AND METALLURGY.

ELSBERG, L., 123 West Fifteenth street, New York.—Model peat fuel machine.

GAUJOT, R. C. E., Tamaqua, Pennsylvania.—Apparatus and methods of mining and metallurgy.

HALLIDIE, A. S., San Francisco, California.—Samples of round and flat wire cables for mining and other purposes.

Donated, at the close of the Exposition, to the Museum of the *Conservatoire des Arts et Metiers*.

HARRINGTON, J. R., Brooklyn, New York.—Self rarefying tuyere.

HAUPT, HERMAN, Philadelphia, Pennsylvania.—Steam drill tunnelling machine. Bronze medal.

This machine is the result of the experience of ten years. The attempt has been made to construct a machine which is strong, light, compact, and cheap; so mounted as to be placed and secured at any desired elevation, and which does not occupy a great space in the tunnel of a mine. All these desirable qualities are claimed for this machine.

STEAM STONE CUTTER COMPANY, G. F. W. Wardwell, superintendent, 18 Wall street, New York.—Stone channelling and quarrying machine, full size and model of the same.

This machine was exhibited in the Annex, in the Park, near the Avenue Suffren, and received a silver medal.

CLASS 48.—IMPLEMENTS AND PROCESSES USED IN THE CULTIVATION OF FIELDS AND FORESTS.

The exhibits in this class were placed in the Annex, in the Park, near the Avenue Suffren.

ALDEN, M., & SON, Auburn, New York.—Horse hoe.

BIDWELL, J. C., Pittsburg, Pennsylvania.—Comstock's rotary spader; ploughs.

BRINKERHOFF, J., Auburn, New York.—Hand Indian corn sheller, separator and cleaner.

BROWN, J. S., Washington, D. C.—Harpoon fork, for lifting hay.

COLLINS & COMPANY, 212 Wall street, New York.—Steel ploughs. Silver medal.

The special good qualities claimed for these ploughs are, that the soil does not adhere to them, that they do not require as much power as other ploughs, and that they last longer. Any part of one of these ploughs that becomes broken or worn can be replaced without difficulty.

CLIPPER, MOWER, AND REAPER COMPANY, 189 Water street, New York.—Combined clipper, mower and reaper, and other agricultural machines.

DEERE & COMPANY, Moline, Illinois.—Steel ploughs. Bronze medal.

EMERY & COMPANY, Chicago, Illinois.—Hog tamer.

EMERY, H. L., & SON, Albany, New York.—Horse power.

FREE, J. W., Richmond, Indiana.—Fanning mill, clover sower.

FULLAM, A. T., Springfield, Vermont.—Machine for shearing sheep and clipping horses.

HALL & SPEER, Pittsburg, Pennsylvania.—Iron centre plough.

HALL, J. A., Columbus, Ohio.—Cotton clipper, strawberry cultivator and drill.

HERRING, S. C., 251 Broadway, New York.—Bullard's patent hay tedder.

LANGSTROTH, L. L., Oxford, Ohio.—Bee hives.

McCORMICK, C. H., Chicago, Illinois.—Reaping and mowing machines.

The reaping and mowing machines of Mr. McCormick are well known. Although invented as early as 1831, they were not brought to the notice of Europe until the Universal Exhibition at London, in 1851, when the Council medal was awarded to the exhibitor. In 1855 Mr. McCormick received the medal of honor at the Paris Exhibition, and in 1857 the gold medal of the Agricultural Society of New York. He has also received prizes at London, Lille, and Hamburg. About 10,000 of his machines have been made and sold in two years. Several machines have been purchased for use on the Emperor's farms. Gold medal, also, Grand prize, gained in the field trials of agricultural machines.[1]

Mr. McCormick, by a decree of the Emperor, was created Chevalier of the Imperial Order of the Legion of Honor of France.

MUNROE, H. H., & COMPANY, Rockland, Maine.—Rotary harrow.

PARTRIDGE FORK WORKS, Leominster, Massachusetts.—Hay forks, rakes, potato diggers. (Palace.) Bronze medal.

PERRY, JOHN G., Kingston, Rhode Island.—Mowing machine. Bronze medal.

SEYMOUR, J. B., Pittsburg, Pennsylvania.—Corn planter.

SEYMOUR, MORGAN & ALLEN, New York.—Reaper.

---

[1] See List of Awards.

WELLINGTON, A. H., & COMPANY, Woodstock, Vermont.—Root cutter.

WHEELER, MELICK & COMPANY, Albany, New York.—Palmer's excelsior horse pitchfork.

WOOD, W. A., MOWING AND REAPING MACHINE COMPANY, Hoosick Falls, New York.—Mowing and reaping machines.

The value of the mowing and reaping machines of Mr. W. A. Wood is shown by the large number of prizes obtained by him at the principal exhibitions in England, France, and America, as also by the immense number of machines sold—no less than 40,000 during five years, to 1867. He has wisely adhered to the wooden frame, believing that it renders a machine more elastic than when made exclusively of iron. By the admirable proportions and balance of his machines he has been able to secure that lightness of draught, power of close cutting, and portability, for which they are so remarkable. Several machines have been purchased for use on the Emperor's farms. Gold medal, also, a gold medal with a work of art. This last medal and prize was gained in the field trials of agricultural machines.[1]

Mr. Wood, by a decree of the Emperor, was created Chevalier of the Imperial Order of the Legion of Honor of France.

WOOLDRIGE, S. H., Venice, Illinois.—Plough.

AMERICAN PLOUGHS AT THE EXPOSITION.

The following notice of American ploughs at the Paris Exposition was translated for the monthly report of the Department of Agriculture:[2]

" American ploughs at the Paris exhibition, 1867, were few in number, but furnished a complete illustration of the excellent construction and solid execution of farming implements in the United States. With but few exceptions all the ploughs were furnished with beams and handles of wood, but this was of such excellent quality that wood in this instance, on account of its extraordinary toughness, withstanding the utmost amount of tear and toil, is to be preferred to iron most decidedly. With us, such an excellent material (white oak and hickory) is wanting entirely, otherwise it ought to be substituted for iron at once.

" The form of the American smoothing board has been applied with us long ago, and wherever the soil is too cohesive for the Ruchadlo plough, it always has proved to be the best, as it holds a middle place between the long, sharp, and screw-like English board and that of the Ruchadlo plough, composed of two straight sides uniting above in form of a triangle. As the English board excels in heavy, tough clay soil, while the latter is adapted best to loose, falling ground, the American share is the best for a medium soil to be turned entirely upside down. All these ploughs exhibited were swing ploughs, sometimes with a stilting-wheel attached to the fore part of the beam, as also frequently used with us, while fore-carts, (running on two wheels to rest the beam,) such as are

---

[1] See List of Awards.

[2] Monthly Report of the Department of Agriculture, May and June, 1868, p. 286.

used in England and on the continent, seem to be but of little use in America.

"The cutter is peculiar in most American ploughs; either a common cutter like ours, attached to the beam or to the share, in form of a vertical blade, as high as the plough is to go down into the ground, one piece with the share itself; or at last a revolving cutter, attached below the beam. The latter arrangement seems excellent to cut turf and roots in marshy ground that is to be broken up.

"The most interesting ploughs from America were exhibited by:

"1. Collins & Company, Hartford, Connecticut. Collins & Company's ploughs are of different sizes, from three inches to one and a half feet in depth, otherwise built on the very same plan; thus the connecting irons, screws, etc., of one size will do for all the others. Their steel smoothing-boards, cast, according to statements, in polished forms, are highly polished, so as to warrant easy work. Their extraordinary lightness is another advantage, those for seven inches depth weighing forty, and those ploughing fourteen inches deep no more than ninety-five pounds.

"2. Deere & Company, Moline, Illinois. The same as the former, except as to double or Ruchadlo shares with some numbers, on the Bohemian plan, of German, probably Westphalian steel, as the manufacturers assure us. Sometimes the whole lower part of the share and both smoothing-boards are formed of one single piece. Their depth is very uniform, from 12 to 14 inches, (destined for prairie soil.)

"3. Hall & Speer, Pittsburg, Pennsylvania, whose ploughs showed some essentially different qualities from those of other firms; rod-iron strongly-bent beams, shares with attached blade for cutter, and also a peculiar connection of the beam with the body of the plough, giving great firmness to the latter. The connection of all these parts is effected by means of screws, the heads of which are sunk so as to afford an even surface. These ploughs are constructed of very different sizes, ranging from 60 to 150 pounds each, and from 10½ to 17 dollars, respectively.

"4. Canadian ploughs, by Mahaffy in Brampton, Gray in Edmondville, and Duncan in Markham, all having rod-iron or cast-steel smoothing-boards, more like the English than like the American patterns, and instead of being concave they were convexed like those by Hornsby in England, and had very long handles. Those ploughs exhibited by Mahaffy and Gray had wooden handles and beams, while Duncan's were entirely composed of iron. Concerning their construction and technical execution, these Canadian ploughs were by no means inferior to those from the United States; their workmanship every way being worthy of imitation."

CLASS 49.—APPARATUS AND INSTRUMENTS FOR FISHING, HUNTING, AND FOR COLLECTING NATURAL PRODUCTS.

ONEIDA COMMUNITY, J. H. Noyes, agent, Oneida, New York.—Traps.

CLASS 50.—MATERIALS AND METHODS OF AGRICULTURAL WORKS AND OF ALIMENTARY INDUSTRY.

BACON, S. T., Boston, Massachusetts.—Cracker, bread, and cake machinery; (in the bakery, Park.)   Honorable mention.

The principal parts of this apparatus, which is capable of preparing 5,000 pounds' weight per day, is protected by European patents and comprises: 1. A mechanical revolving oven capable of receiving and holding a continuous supply of 600 pounds of bread or crackers.   It is claimed that this oven with a given amount of fuel, time, space, and labor, will bake at least twice as much as any oven in Europe.   2. A smoke and gas consuming furnace, the invention of Jonathan Amory, of Boston, which has been put into practical operation by Mr. Bacon.   The combustion is so perfect that no smoke issues from the chimney.   3. Various machines used in mixing, kneading, and cutting.   4. A sectional steam generator, exhibited by T. S. Clogston & Company, of Boston.   This generator consumes only 48 pounds of coke per day, and will bear, if required, a pressure of 900 pounds per square inch.   This generator supplies the Root trunk engine which drives the machinery in Mr. Bacon's establishment.   5. Clark's steam and fire regulator.   6. Grate bars by L. B. Tupper, New York, which, from their peculiar shape, effect a saving in cost of one-fifth compared with the ordinary grate bar.   7. Root's trunk engine, from J. B. Root, of New York.

BAKER, GEORGE R., St. Louis, Missouri.—Dough-kneading machine. Honorable mention.

BASSETT, J. B., & Co., Minneapolis, Minnesota.—Wooden buckets.

CHAMPLIN, J. R., & Co., Laconia, New Hampshire.—Ice cream freezer. (In the American restaurant.)

COLBY, D. C., Washington, D. C.—Flour sieve; coffee mill and can.

ELTING BOLT AND DUSTER COMPANY, Cincinnati, Ohio.—Bolt and duster machine.

GOODELL, D. H., Antrim, New Hampshire.—Apple parer.   Bronze medal.

HUDSON, C. H., 5 Barclay street, New York.—Washing machine.

LOW, D. W., Gloucester, Massachusetts.—Ice crusher.   (In the American restaurant.)

METROPOLITAN WASHING MACHINE COMPANY, R. C. Browning, agent, 32 Courtland street, New York. — Clothes wringers.   Honorable mention.

MORRIS, TASKER & Co., Philadelphia, Pennsylvania. — Wringing machine.   Bronze medal.

PALMER, S. W., & Co., Auburn, New York.—Clothes wringers, mangles, and ironers.

PURRINGTON, G., Jr., 5 Barclay street, New York.—Carpet sweeper. Honorable mention.

SARGENT, E. H., Boonton, New Jersey.—Alarm coffee boiler.

SEDGEBEER, J., Painesville, Ohio.—Grinding mills for corn and spices.

SOMERS, D. M., Washington, D. C.—Self-acting tumbler washer. (In the American restaurant.)

TILDEN, HOWARD, Boston, Massachusetts.—Flour and sauce sifter; R. Smith's tobacco cutter; champion egg beater. Honorable mention.

WARD, J., & Co., 457 Broadway, New York.—Clothes wringer. Honorable mention.

WINDLE & Co., 56 Maiden Lane, New York.—Carpet sweeper.

CLASS 51.—CHEMICAL, PHARMACEUTIC, AND TANNING APPARATUS.

BUTLER, J. L., St. Louis, Missouri.—Soda water fountain. (In the Annex.)

DOWS, CLARK & VAN WINKLE, Boston, Massachusetts.—Ice cream soda water apparatus and fountains, carbonic acid gas generators.

HOGLEN & GRAFLIN, Dayton, Ohio.—Tobacco-cutting machine. Bronze medal.

METROPOLITAN WASHING MACHINE COMPANY, R. C. Browning, agent, 32 Courtland street, New York.—Doty's clothes washer. Honorable mention.

PRENTICE, J., Sixth avenue, New York.—Cigar-making machine. (Shown in the Annex in the Park.) Honorable mention.

SCHULTZ & WARKER, New York.—Soda water apparatus and fountains. Silver medal.

One of the fountains was tested by a pressure of 15 atmospheres.

WARD, J., & Co., 457 Broadway, New York.—Washing machine. Honorable mention.

CLASS 52 AND 53.—MACHINES AND MECHANICAL APPARATUS IN GENERAL.

AMERICAN STEAM GAUGE COMPANY, Boston, Massachusetts.—(In M. Bacon's bakery, Park.) Pressure steam gauge; Bourdon's patent with T. W. Lane's improvement. Honorable mention.

ANDREWS, WILLIAM D., & BROTHER, 414 Water street, New York.—Centrifugal pump and oscillating engine. Honorable mention.

AUTOMATIC BOILER FEEDER COMPANY, G. A. Riedel, director, 945 Ridge Avenue, Philadelphia, Pennsylvania. — Automatic boiler feeder. Bronze medal.

BACON, S. T., Boston, Massachusetts.—(In the bakery, Park.) "Anti-incrustator," for steam boilers.

BROUGHTON & MOORE, 41 Centre street, New York. — Oilers, cocks, &c. Honorable mention.

BRYANT, F., Brooklyn, New York.—Grinding mill.

BRYANT, J., Brooklyn, New York.—Bushing for ship's blocks; anti-friction journal boxes.

CLARK'S STEAM AND FIRE REGULATOR COMPANY, New York.—(In M. Bacon's bakery, Park.) Steam and fire regulator. Honorable mention.

CLOGSTON, T. S., & COMPANY, Boston, Massachusetts.—(In M. Bacon's bakery, Park.) Cast-iron sectional steam generator, steam indicator and fire regulator combined.

COCHRANE, JAMES, 64 West Tenth street, New York.—Model balancing slide valve, showing method of lubricating.

COLUMBIAN METAL WORKS, J. P. Pirrson, President, 40 Broadway, New York.—Seamless copper and brass tubes.

CORLISS STEAM ENGINE COMPANY, G. H. Corliss, president, Providence, Rhode Island.—Steam engines. Gold medal.

The 30-horse power steam engine exhibited by this company was one of the most prominent objects in this class. It was much admired and appreciated, not only for its elegant and elaborate finish, but its perfect and noiseless automatic motion and the wonderful sensitiveness of its "cut-off." Its proportions and features were closely studied by many noted European engine builders.

CROSBY, BUTTERFIELD & HAVEN, 22 Dey street, New York.—Roper's hot air engine.

DART, HENRY C., & Co., New York.—Behren's patent rotary engine and pump. Honorable mention.

This remarkable invention may be used either as a motor or pump. It consists of three principal parts: a cylinder and cylinder head, two pistons with their shafts, and two gear wheels to connect the pistons. It is not liable to break down or get out of order, and, as the pump is without either valves or air-chamber, it is particularly well adapted for feeding, bilge, air, and wrecking purposes. This engine can be worked by compressed air or explosive gases. As it measures accurately the quantity of water passing through it at every revolution, it may be used as a water meter.

DOUGLASS, W. & B., Middletown, Connecticut.—Pumps of various descriptions. Bronze medal.

DWIGHT, GEORGE, Jr., & Co., Springfield, Massachusetts.—Steam. Honorable mention.

FAIRBANKS E. & T., & Co., St. Johnsbury, Vermont.—Weights and weighing machines. (In the Annex.) Silver medal.

The weighing machines shown by this company were of all sizes and descriptions, from letter-balances and apothecaries' scales up to those used for weighing canal boats and loaded trains.

HARRISON, C. H., San Francisco, California.—Steam pump.

This pump is used chiefly for wrecking, and is remarkable for the large quantity of water it will raise in a given time. It was kept running during the Exhibition, and was a conspicuous object at the entrance to the building by the Rue d'Afrique.

HICKS ENGINE COMPANY, C. D. Kellog, treasurer, 88 Liberty street, New York.—Steam engines. Honorable mention.

A report and description in detail will be found in the Report on the Steam Engineering of the Exposition. The following notice is extracted from the company's circular:

"This engine, invented by Mr. William C. Hicks, is patented in the United States, (February 21, 1865, and May 22, 1866,) and in nearly all European countries and their dependencies.

"It has many advantages over any engine now in use, its chief feature being its intrinsic and matchless simplicity. While retaining the entire principle and action of the best approved reciprocating-piston engines, and doing no violence to the convictions of our most intelligent engineers that this principle and action cannot be superseded as long as the present mode of applying steam continues, the details are so far simplified that the pistons connected directly to the crank form the only moving parts, and these with the cylinders compose the whole machine. This is done by making the pistons of suitable form and arrangement to enable them to perform also the offices of valves and cut-offs, dispensing not only with these contrivances, but also with the whole array of valve-rods, eccentrics, rock-shafts, packing-boxes, slides, levers, cross-heads, and external attachments of every kind which they necessitate. The action of the pistons is alike simple and uniform, each being a slide-valve for the one beside it. This invention, therefore, forms the most radical and entire change in steam engines which has occurred since the days of Watt, and enables us to offer a better machine, simple, compact, light, durable, accurate, and economical in operation beyond all comparison with the past, and at far less original cost than ever before attained.

" Four single-acting pistons working in the four cylinders marked B, B, B, B, are all connected to cranks on one shaft by suitable connecting rods, each piston taking steam before the next succeeding one has finished its stroke, thereby insuring a uniform and continuous motion, and avoiding the dead points which render ordinary engines so variable in their motions and difficult to start, if stopped or caught on the centre. This is in fact a double cut-off engine, without the friction of a double set of valves with their multiform attachments.

"The pistons are provided with proper ports and passages, which act in combination with ports and passages in the cylinders, to admit and release the steam, thus combining a slide-valve with the piston in one and the same piece, each piston admitting and exhausting the steam for its neighbor cylinder, as well as cutting off its own supply of steam from the boiler at any desired point. By this means the expansive force of the steam is used, and the exhaust allowed to remain open during the entire return stroke.

"These ports and passages are arranged opposite each other in such a manner that a perfect balance to the pressure of the steam is effected, and the ordinary wear and friction of cylinders, pistons, and valves almost entirely obviated. It will also be observed that the motions of the valve and cut-off are equal in rapidity to the speed of the piston, and that the cut-off works in the closest possible proximity to the piston.

"The pistons are effectually packed by a simple and convenient method, and can be tightened at pleasure. All the working parts are encased in one casting, and are in no way exposed to the action of the weather, or to an accumulation of dirt.

"The number of parts and the wearing surface being so vastly reduced, tends of course to the same decrease of wear and tear, and of the risks and costs of repairs. In this connection, the facility of repairs deserves especial notice, every part being accessible by the removal of a few bolts, and the whole machine being capable of dissection and reconstruction in a few minutes; and the parts also being interchangeable, any portion can be quickly and cheaply replaced.

"The reduction of friction; the diminished length of the steam-ports and clearances; the decrease of the surface exposed; the facility for casing the whole engine; the accuracy and perfection of the valve motions and cut-offs; the extent to which the expansion of the steam may be carried to advantage—all combined, necessarily give an unequalled economy in the consumption of steam."

HILL, W. E.—Furnace grate bars.

HOWE SCALE COMPANY, Brandon, Vermont.—Scales of various sizes. (Also in the Annex.) Bronze medal. A large and excellent assortment of well-finished and useful instruments.

JENKINS, N., Boston, Massachusetts.—Globe valves, cocks, faucets, &c.

JONES, T. J., chief engineer, United States navy, Brooklyn navy yard, New York.—Piston packing spring.

JUDSON, J., Rochester, New York.—Graduating governor for steam engines.

OLMSTEAD, L. H., Stamford, Connecticut.—Friction clutch pulley. Bronze medal.

PEASE, F. S., Buffalo, New York.—Pump for petroleum. Honorable mention.

PICKERING & DAVIS, New York.—Marine and stationary engine regulators. Bronze medal.

PLATT, J. L., Kewanee, Illinois.—Coal chute.

ROBINSON, J. A., 164 Duane street, New York.—Ericsson's hot air engine. Honorable mention.

ROOT, J. B., New York, (in M. Bacon's bakery, Park.)—Root's trunk engine. Bronze medal. See a notice under "Boston Cracker Bakery."

ROOTS, P. H. & F. M., Connersville, Indiana.—Rotary blower. Bronze medal.

SELLERS W., & Co., Philadelphia, Pennsylvania.—Injectors, dies, stocks, &c.

SHAW, PHILANDER, Boston, Massachusetts.—Hot air engine. (Special installation in the Park.) Bronze medal.

This engine is made with two vertical cylinders, with single acting trunk pistons, hung from the extremities of an overhead working beam. The beam centre on the side next the furnace is sufficiently prolonged to receive a fixed arm, from which the connecting rod runs to the crank of the main shaft. From the furnace, which is hermetically closed, the heated air and products of combustion pass over to the cylinders (to which they are admitted by suitable valves) with an average pressure of 14 pounds per square inch. While one piston is making the upward stroke, its annular face acts as an air pump for forcing cold air into a heater, whence the air passes under the grate to sustain combustion. The succeeding down stroke draws cold air into the annular space, and expels the gases just used through the tubes of the heater to the stack. By an ingenious arrangement the fine cinders are prevented from cutting the cylinders, and the cylinders are kept sufficiently cool.

Mr. Shaw's engine, though not constructed with that regard to handsome finish and elaborate polish which characterize many of the machines forwarded from the United States, was nevertheless much admired and esteemed for its originality.

SHELDON, J., New Haven, Connecticut.—Water-pressure regulator. Honorable mention.

STEAM SYPHON COMPANY, H. S. Lansdell, superintendent, 48 Dey street, New York.—Steam syphon pump, and model of a railroad station pump. Honorable mention.

STILLWELL, D., Fall River, Massachusetts.—Brushes for cleaning tubular boilers.

TUPPER, L. B., New York, (also in Mr. Bacon's bakery, Park.)—Furnace-grate bars. (See notice under head of Bacon's cracker bakery.)

WEBSTER & CO., 17 Dey street, New York.—Webster's patent ordinary wrench.

CLASS 54.—MACHINE TOOLS.

AMERICAN TOOL AND MACHINE COMPANY, G. H. Fox, president, Boston, Massachusetts.—Fox's screw-cutting lathe, with Nason's screw attachment.

BEMENT & DOUGHERTY, Philadelphia, Pennsylvania.—Bolt and nut-threading machine, with opening dies. Silver medal.

BERGNER, T., co-operator, engineer of Messrs. Sellers & Company, of
· Philadelphia, Pa.—Exhibitors of machine tools, who received a gold medal for their exhibition of tools; a silver medal was awarded to Mr. Bergner as co-operator.

BROWN J. R., & SHARPE, Providence, Rhode Island.—Revolving head screw machine; milling machine. Silver medal.

It was stated that five or more of these machines were sold in Europe during the Exhibition in Paris.

COOL, FERGUSON & Co., Glen's Falls, New York.—Barrel machines. Silver medal.

GREGG, ISAAC, Philadelphia, Pennsylvania.—Model of a brick machine, and specimen bricks.

A full-sized machine in operation was shown in the Annex of the Exhibition, Nos. 100 and 102 Avenue Suffren, and was said to be capable of making from 35,000 to 40,000 bricks in ten hours.　A bronze medal was awarded.

HARRIS, D. L., & Co.—Improved engine lathe, with Van Horne's patent tool elevator and screw cutter.　Bronze medal.

JUSTICE, P. S., Philadelphia, Pennsylvania.—Power hammer.　Bronze medal.

It is claimed that this hammer, with half-a-horse power, will work faster and better than those of the old style requiring the power of ten horses. It is a very compact machine; the hammer is suspended by a flexible attachment to a cast-steel spring moving between guides and receiving an alternate movement from a crank.

LYON & ISAACS, 9 Jane street, New York.—Self-feeding hand and power drill.

MORRIS, TASKER & Co., Philadelphia, Pennsylvania.—Pipe-cutting machines.　Honorable mention.

OLMSTEAD, L. H., Stamford, Connecticut.—Machine tools.　Honorable mention.

SELLERS, WILLIAM, & Co., Philadelphia, Pennsylvania.—Machine tools. Gold medal.

This house exhibits perhaps the finest collection of machine tools to be found in the Exposition.　Their large planer is 24 feet long and 8 feet broad, with a carriage 8 feet high; it cuts one way only, and the carriage goes back with double-quick motion.　The novelty in principle is that the bed is fixed, and the frame or carriage carrying the cross-head and two lateral tool-posts travels on V slides, and is moved by racks and pinions actuated by two worm wheels from above.　The forward and backward movements are given by racks and pinions along the sides at the end of the strokes; the reversal of motion takes place by a ring, at the end of the worm shaft, being driven in by a projecting stud from the wall, the lever gearing thus throwing off the drawing belt from a large wheel to a small one, and *vice versa*, as the motion is required to be quick backwards or slow forwards, for the cut of the tools, which are all three (one vertical and two lateral) self-acting.

The length of the stroke is given in a very ingenious way by a movable jam-nut on a vertical screw-shaft.

A small planing machine, with moveable plate seven feet long, planes the whole length of its table; and this, like all the rest of Mr. Seller's machines, has an automatic outlift of the tool, so that in the backward motion it travels clear, and the point is not ground by trailing along the work.　The feed motion is peculiar, the limits of motion being attained by means of a segment piece attached by a rod to the crank, and adjustable by a screw from central to any degree of eccentricity—the amount of eccentricity being the limit of the feed motion.　The change of movement from the quick to the slow is effected by two pivoted levers traversing within an irregularly shaped iron circular ring, by means of which each of the belts, working in opposite directions, is turned, as required, over one of the three divisions of the vertical driving pulley. The two outside pulleys run free; the middle one actuates the travelling plate, which works in V slides, and is moved by bevel gearing actuating a worm wheel set diagonally to the length of the plate. The rack is cut diagonally to suit the thread of the worm.

The automatic gear cutter is adapted for cutting cylindrical as well as bevel wheels of any size, from the smallest in ordinary use to wheels five feet in diameter. The cuts are made by interchangeable tools of the required dimensions for the tooth. The machine is fixed in an L-shaped frame upon the carriage, on which the wheel to be cut is pivoted. The cut is made by a milling tool, and after each operation the wheel is turned automatically to the required pitch for the cut, and so on, one cut at a time, turning out a finished tooth.

The 25-inch lathe has the peculiarity of friction disks for moving the carriage for ordinary turning, and for cutting special gear, which can be put in or out of contact at the will of the operator; also, a rest for long, thin work, which requires support to prevent vibration. The face plates are cast solid, in one piece, and further stiffened by ribbing at the back, so that there is never any spring.

Mr. Sellers also exhibits an excellent 500 pounds' weight hammer, of which he is the lessee, remarkable for its simplicity and easy management. By means of a handle a workman may instantly alter the height, rapidity, or force of the blow, or render the valve motion manual or self-acting.

The self-adjusting injector is an improvement upon that of Giffard, and is provided with a handle which regulates the steam supply, the increase or decrease of which corresponds to that of the water delivery. The water supply also corrects itself at all variations of steam pressure independent of the handle movement.

Mr. Sellers exhibits in addition a variety of shafting, hangers, and couplings, which show a direct saving of first cost, from their diminished weight, as well as perfection in construction. The double cone vice-couplings are easy of detachment, with double-traced ball and socket hangers, the bearings of which are light and easily adjustable; the journal boxes are long, with uniform pressure and length of bearing. Iron, not brass, is used in the pulley castings. The whole presents a very neat appearance.

UNION VICE COMPANY, A. H. Brainard, Boston, Massachusetts.—Cast iron vices.

WICKERSHAM NAIL COMPANY, A. L. Wood, treasurer, Boston, Massachusetts.—Nail cutting Machine. Bronze medal.

It is claimed that this machine can be worked at a less cost than other machines now in use, while at the same time it produces a nail superior in its holding property to those generally manufactured. As the nail is pointed like a chisel and tapers gradually through its whole length, it is easily driven and does not break the grain of the wood like a blunt or roughly jointed nail. In the second place this machine, instead of manufacturing one nail at a time, as is done by machinery now in use, can cut from a 20-inch iron plate eight two and a half inch nails at one blow, and can make three blows per second, thus giving 24 nails headed and jointed in a second. The same machine will make 160 half-inch brads per second, 40 at a time, or about 3,600 pounds per day, including all sizes of small finishing nails. As a comparison between the Wickersham machine and those ordinarily in use at other factories, it is said that a large factory with 50 machines will produce 50,000 kegs of nails per annum, whereas 50 Wickersham machines will make the enormous quantity of 75,000 per annum.

WINSOR, H., Philadelphia, Pennsylvania.—Shot and shell polishing machine.

CLASS 55.—APPARATUS AND METHODS OF SPINNING AND ROPE-MAKING.

BATES, HYDE & Co., Bridgewater, Massachusetts.—Power cotton gin; hand cotton gin.

EMERY, H. L., & SON, Albany, New York.—Cotton gin.

GODDARD, C. L., 3 Bowling Green, New York.—Mestizo burring picker. Bronze medal.

HALL MANUFACTURING COMPANY, Boston, Massachusetts.—Bazin's cord twisting machine.

SOUTHERN COTTON GIN COMPANY, Bridgewater, Massachusetts.—Saw and roller cotton gins. Bronze medal.

CLASS 56.—APPARATUS AND METHODS OF WEAVING.

CROMPTON, GEORGE, Worcester, Massachusetts.—Loom for weaving fancy woollen casimeres, two yards in width. Silver medal.

This machine will make 82 picks per minute while the others rarely exceed 65.

LAMB, J. W., Rochester, New York.—Knitting machine. Silver medal.

OPPER, M., Convex Weaving company, New York.—Power loom. Silver medal.

PROUTY, A. B., Worcester, Massachusetts.—Card setting machine.

SHAW, C. A., Biddeford, Maine.—Card grinding machine and model of the same.

CLASS 57.—APPARATUS AND PROCESSES OF SEWING AND MAKING CLOTHES.

AMERICAN BUTTONHOLE COMPANY, Philadelphia, Pennsylvania.—Buttonhole, cording, braiding, and embroidery sewing machines. Silver medal.

BARTLETT SEWING MACHINE COMPANY, 569 Broadway, New York.—Sewing machines.

BARTRAM AND FANTON MANUFACTURING COMPANY, Danbury, Connecticut.—Sewing machines. Bronze medal.

20 U E

BRUEN MANUFACTURING COMPANY, J. L. Lilly, secretary, 371, Broadway, New York.—Sewing machine attachments.

CONTINENTAL MANUFACTURING COMPANY, E. H. Smith, secretary, 18 Beekman street, New York.—Sewing machines.

ELLIPTIC SEWING MACHINE COMPANY, 543 Broadway, New York.—Sewing machines.

EMPIRE SEWING MACHINE COMPANY, T. J. MacArthur, secretary, 536 Broadway, New York.—Sewing machines. Honorable mention.

FLORENCE SEWING MACHINE COMPANY, 505 Broadway, New York.—Sewing machines. Silver medal.

FOLSOM, J. S., Winchenden, Mass.—Sewing machines.

GRISWOLD & SHELDON, New York.—Hat blocking machine.

HOOPER, N. B., Newark, New Jersey.—Hat finishing machine.

HOWE, A. B., 437 Broadway, New York.—Sewing machines. Bronze medal.

HOWE MACHINE COMPANY, E. G. Sterling, secretary, 629 Broadway, New York.—Sewing machines.

A gold medal was awarded to Mr. Elias Howe, jr., as promoter, and by a decree of the Emperor he was created a Chevalier of the Imperial Order of the Legion of Honor of France.

MUMFORT, FOSTER & COMPANY, Detroit, Michigan.—Boot trees and lasts. Bronze medal.

SHAW, C. A., Biddeford, Maine.—Knitting machine. Bronze medal.

UNION BUTTONHOLE AND EMBROIDERY COMPANY, Boston, Massachusetts.—Button hole and embroidery machine. Bronze medal.

WEED SEWING MACHINE COMPANY, 506 Broadway, New York.—Sewing machines. Silver medal.

WHEELER AND WILSON, 625 Broadway, New York.—Buttonhole machines; sewing machines. Gold medal.

Bronze medals were also awarded to Messrs. A. J. House and A. H. House as co-operators.

It is useless here to review the history, progress, and advantages of sewing machines. Every one understands their importance and appreciates their services. The various modes of construction exhibited by American manufacturers at the Champ de Mars have already been presented in preceding Universal Exhibitions, and have been explained and discussed either in the reports of the juries or in industrial publications. To Mr. Elias Howe redounds the credit of the original invention from which, with progressive variations, all the other systems are derived.

Mr. Howe's invention, in its relation to labor, is analogous to that of the Jacquard loom, effecting an enormous saving of hand labor, and although, like the loom in question, looked upon at first with distrust by the working classes, it has in the course of time equally proved itself one of the greatest benefits ever offered them; the increased facility of labor more than making up the loss occasioned by the diminution in the price of the article manufactured—a benefit falling in turn to the lot of the consumer—so that Mr. Howe may be considered not only in the light of a promoter of industry, but as a benefactor of humanity in general.

The original machine, for which Mr. Howe has obtained the gold medal, decreed by the international jury in honor of his long and useful researches in this line, was exhibited. The improvements made up to this time refer rather to perfection of form than to any great development of the actual principle. Mr. Howe, although possessing the exclusive patent for all sewing machines during a certain period of time, has generously allowed the right of fabrication to all parties inventing remarkable improvements in special branches.

Since 1855, the sewing, embroidering, and braiding machines have been considerably simplified and perfected. The only really new inventions since 1862 are those for making button holes. These complete the revolution operated by sewing machines. The machines for button holes are of two kinds, designated under the heads of special and mixed. The special ones are represented by two systems, both of which are automatical.

1st. The system of Wheeler & Wilson, invented by two brothers, James and Henry House, all the mechanism of which is enclosed in a case placed underneath the work table, and moved, like all sewing machines, either by the treadle or steam. The machine on exhibition operates with wonderful rapidity; the needles moving backwards and forwards along the button hole until the work is completed. Under the eyes of the jury it made three button holes, on heavy winter cloth, in the short space of 24 seconds. Its advantages over many other machines consist in avoiding the necessity of turning or moving the cloth along by hand. It makes button holes of every size and form; and by an ingenious arrangement can be adapted to sewing tents, sacks, and, in a word, all work which requires the solid and uniform stitching of two straight or curved borders.

The second system is that exhibited by the "Union Button Hole and Embroidery Company," Boston. In this machine the upper or superior needles move vertically, while the lower mechanism makes the button hole stitch. The system is the inverse of Wheeler & Wilson's; the cloth or material moving and turning, and the needle operating in a fixed place. The cloth is attached upon a turning plate which, first, by a rectilinear, then rotary, and, lastly, another rectilinear movement, brings all the parts of the button hole under the vertical needle. It is a very ingenious machine, and makes excellent button holes of all sizes. The only inconveniences which have been spoken of respecting it are, that it is heavy and complicated, and requires the cloth or garment to be turned and put in movement during the work.

The mixed machines are ordinary sewing machines which, by a change of certain pieces, or by certain transmissions, can be transformed into button hole machines. There are three systems under this class:

1st. Wheeler & Wilson's, which is also due to the invention of Messrs.

House, consisting in replacing the platform of the ordinary sewing machine by a particular plate, which has a double movement of oscillation and translation. The oscillating movement, combined with the action of the upper needle, serves to form the button hole stitch, while the translatory movement advances the work under the same needle. The button holes thus obtained are made fast at the two ends, and are similar to those made in linen drapery. With this system buttons may be secured on garments, not, however, with sufficient rapidity to guarantee much economy of hand labor.

2d. The system of Bertram & Fanton applied, and applicable exclusively, to the sewing machines of Wheeler & Wilson. The plate or button hole guide undergoes the same movements before mentioned, but by different transmissions.

3d. The system of the "American Button Hole Company," of Philadelphia. The machine exhibited by this company is so made that it can be used either for ordinary sewing or for button hole making. This machine makes very good button holes for the use of tailors, &c., but cannot be employed in linen drapery.

For various improvements and modifications of sewing machines we may notice among the exhibitors the names of the Florence Sewing Machine Company, New York; the Bruen Manufacturing Company, New York; the Weed Sewing Machine; the Continental Manufacturing Company; the Bartlett Sewing Machine Company, and the Empire Sewing Machine Company, of New York; as also J. S. Folsom, Massachusetts.

Special machines for shoemaking are contributed by two houses: The Howe Machine Company, which has obtained a silver medal for its machines, and the house of A. B. Howe, New York, to which the jury decreed a bronze medal. These machines, in the construction of which the Howe type is the most generally adopted, are used for all kinds of sewing on leather.

In comparing the execution of the sewing machines exhibited at the Champ de Mars, one is particularly impressed with the superior finish and the uniform accuracy of every part of the American machines. This is due to two causes:

1st. To the immense impetus which has been given to the manufacture of sewing machines in the United States.

2d. To the system of manufacture there observed. Every piece is separately made by machinery, so that any two complete machines of the same calibre are strictly identical in size and form in almost every particular, and the pieces of one accord perfectly with those of the other.

Statistics of the progressive march of this industry would be interesting; unfortunately, however, the committee is not in possession of anything like complete documents on the subject. The following table, showing the number of machines made by only one establishment, may serve to give an idea of the importance of this industry in America.

The house of Messrs. Wheeler & Wilson have manufactured sewing machines as follows:

Years.	Machines.	Years.	Machines.	Years.	Machines.
1853............	799	1858............	7,978	1863............	29,778
1854............	956	1859............	21,306	1864............	40,062
1855............	1,171	1860............	25,102	1865.........	39,157
1856............	2,210	1861...........	18,556	1866...........	50,132
1857............	4,591	1862...........	28,202		

All the machines sent from the United States possess indisputable merits, and establish the fact that the country is still far in advance of Europe in the construction and improvement of these great labor-saving inventions.

CLASS 5ᵃ.—APPARATUS AND METHODS USED IN MAKING FURNITURE AND HOUSEHOLD OBJECTS.

AMERICAN SAW COMPANY, S. W. Putnam, secretary, 2 Jacob street, New York.—Emerson's patent saw.

DAVENPORT, H., New York.—Armstrong's dovetailing machine.

FENN & FELBER, St. Louis, Missouri.—W. Zimmermann's mortising and slotting machine.

GRANIER, ÉMILE.—Dovetailing machine

MILLER, W. P., San Francisco, California.—Adjustable teeth for saws.

It is undoubtedly true that there is no one tool used in the mechanical arts of more practical utility than the circular saw. Notwithstanding their high cost and the daily expense incurred in keeping them in order, they are used almost without limit.

Formerly all saws were made by forming solid teeth on the periphery of the plate. Teeth thus made do good work, but are liable to be, and frequently are, broken off. There is no means of restoring them when broken, except by reducing all the other teeth to the same radius.

A circular saw, thirty inches in diameter, presents a cutting edge more than seven and a-half feet in extent. To reduce the saw one-eighth of an inch, and relieve the teeth the same as before, necessitates the filing away of a strip of steel one-eighth of an inch in width, by the thickness of the plate, and seven feet and ten inches in length, and by such operation the saw will be reduced one-fourth of an inch in diameter. Each filing of a solid tooth saw is attended with a like corresponding expenditure of labor and files.

To obviate this difficulty, several plans for attaching teeth to saw plates have been devised and put in use, but with little or no success, for the following reasons: First, all insertable teeth heretofore used require a thicker plate to support them than do the solid teeth. And, secondly, what is saved in labor and files by the use of insertable teeth is absorbed in their purchase. For the above, and other reasons incidental therewith, insertable teeth are not much used except in connection with large saws, and it is asserted by practical saw makers, and not a few mill men and sawyers, that there is no economy in the use of insertable teeth as heretofore made and applied.

Miller's saw teeth are annular disks with a portion cut out so as to make a cutting edge or point to the tooth. The teeth thus have the shape of a flattened ring with a portion removed. They are inserted in circular openings, or sockets, made in the periphery of the saw. This circular socket allows the teeth to be turned outward or backward within the outer edge of the saw plate.

Among the many advantages claimed for these teeth above all others the following may be cited: They are stronger even than solid teeth formed on the plate; there is more room for the chips; the saws require less power, and will make from eight to ten per cent. more of inch boards from a log than can be made by other insertable teeth saws; the teeth are self adjusting; being round, they may be turned in a lathe and be easily duplicated; they can be quickly inserted; they cannot be thrown out of their sockets, and, finally, they will last longer and thus accomplish more work than other forms.

ROGERS C. B., & CO., Norwich, Connecticut.—Wood working machines. Gold medal.

WHITNEY, BAXTER, D., Winchendon, Massachusetts.—Wood working machines. Silver medal.

WINSLOW, J. B., 110 East Twenty-ninth street, New York.—Serpentine wood moulding machine. Honorable mention.

WINSOR, H., Philadelphia, Pennsylvania.—Model of a machine for sawing timber for ships.

WRIGHT & SMITH, Newark, New Jersey.—Scroll sawing machine.

CLASS 59.—APPARATUS AND METHODS OF PAPER-MAKING, COLORING, AND STAMPING.

DEGENER & WEILER, 111 Fulton street, New York.—Printing presses. Bronze medal.

Forms may be corrected on this press without being removed. It can be worked by the foot or by steam power. From 1,000 to 2,500 impressions can be taken from this press in an hour, according to the capability of the workman.

GALLOUPE, NICHOLSON & WOODBURY, Boston, Massachusetts.—Paper collar machinery.

McLAUGHLIN, R., Boston, Massachusetts.—Morse's improved bed plate for paper making machinery.

SWEET, J. E., Syracuse, New York.—Composing machine. Bronze medal.

WELCH, PATRICK, 356 East Fourth street, New York.—Improved lower case for compositors.

CLASS 60.—MACHINERY, INSTRUMENTS, AND METHODS USED IN VARIOUS WORKS.

SMITH, H., Salem, Massachusetts.—Spring power machines.

VAN DENBURGH, G., 24 Vesey street, New York.—Emery wheels.

WELCH, PATRICK, 356 East Fourth street, New York.—Machine for dressing printing types. Gold medal.

CLASS 61.—CARRIAGES, WAGONS, AND WHEELWRIGHTS' WORK.

BLANCHARD, A. V., & Co., Palmer, Massachusetts.—Bent wood.

HALL, JAMES, & SON, Boston, Massachusetts.—Top buggy. Silver medal.

RUCKER, Major General, Washington, D. C.—United States army wagon, harness, &c.

SCHUTTLER, P., Chicago, Illinois.—Lumber wagon.

This wagon is capable of bearing a load of 4,000 pounds. The box may be removed and a rack placed upon the wagon that will hold a large load of hay or straw. Loads of timber or lumber can be drawn without box or rack. It is furnished with a spring seat, which is moveable, and can be placed to suit the convenience of the driver. The ends of the box may be removed when desired, and, lastly, the wood of the wagon is of the very best seasoned and most durable material.

SCOTT, J., Ocala, Florida.—Carriage wheel.

STEPHENSON, JOHN, & Co., 47 East Twenty-seventh street, New York.—Street railway carriage. Honorable mention.

This beautiful and highly finished vehicle, intended for India, has the wheels placed underneath, and is so constructed as to bear with ease and safety a very great weight. The carriage, which is fitted up in rich style with exquisitely painted panels, is capable of containing from 30 to 40 persons.

WOOD BROTHERS, 596 Broadway, New York.—Phæton and buggy. Silver medal.

CLASS 62.—HARNESS WORK AND SADDLERY.

SMITH, T. S., Boston, Massachusetts.—New system of bit.

STATTMAN, C., Natchez, Mississippi.—Ladies' saddles.

WELLMANN, C., 932 Broadway, New York.—Ladies' saddles; gentlemen's saddles. Honorable mention.

CLASS 63.—MATERIALS FOR RAILROADS AND CARS.

CREAMER, W. G., 15 Platt street, New York.—Railroad brake and ventilator.

EASTMAN, Z., United States consul at Bristol, England.—Model of street railway and carriage track.

FAIRBANKS, E. & T., & Co., St. Johnsbury, Vermont.—Railroad scale. Bronze medal. See a notice, also, in Class 51.

FOSTER, A., 50 John street, New York.—Graham's locomotive spring balance.

HALL, T. S., Stamford, Connecticut.—Electric switch alarm.

This invention is intended to remedy the mischief which too frequently occurs on railways in consequence of the misplacement of a switch. It has an alarm and a signal, and is worked by electricity. When the switch is on the main line the electric current is broken, but if displaced the circuit is complete and an alarm is given by the vibration of a hammer against a gong. In addition to this the lines are connected with a magnet, which operates a red and white signal, as the switch is right or wrong, displaying in the night time a red or white light. Both the alarm and the signal operate at the same time by the movement of the switch, thus affording a double security by an alarm for the switch tender and a signal for the engineer.

MYERS, G., Upper Sandusky, Ohio.—Railroad journal boxes of "silicated copper."

STAR METAL COMPANY, E. E. Childs, president, New York.—Star metal railroad journal boxes.

THE GRANT LOCOMOTIVE WORKS, Paterson, New Jersey.—In the Annex, Park. Locomotive and tender, the "America." Gold medal.

The weight of the engine, in running order, is 27½ tons, of the tender, when empty, 9 tons, or 18 tons when loaded. The engine frame is composed of the best American iron, and is light and strong. The truck of the engine is simply used to guide it, and at the same time carry the small amount of overhanging weight. The driving wheels bear the main weight of the engine, and, by means of equalizing levers, distribute it equally upon each wheel, giving the entire adhesive power of the engine. The side valves are what are termed roller valves; the boiler is composed of 5-16th iron, and is double riveted. The usual load for this class of engine is 200 tons at a speed of 40 to 50 miles an hour. An engine built by the Grant works and similar to the "America," has drawn 400 tons at a speed of 25 miles per hour during the last 14 months without as yet requiring repair. All the various operations required in the conduct of the engine can be carried on in the apartment of the engineer and fireman; and the engine, even if running at the rate of a mile per minute, can be reversed, the reversing brake being capable of being managed by a child. A signal bell communicates with the conductor as usual in American railways, and a powerful reflecting light is placed in front of the engine, and can be seen, it is said, at five miles distance. The wheels of the engine are of cast-iron and hollow, and its grate bars are composed of hollow iron tubes through which the water passes.

WARNER, H. W., Greenfield, Massachusetts.—Cast-iron railroad rail chair.

CLASS 64.—APPARATUS AND METHODS OF TELEGRAPHING.

CATON, J. D., Ottawa, Illinois.—Pocket field telegraph apparatus.

This instrument consists of a pair of helices, each two inches long and three-fourths inch in diameter, incased in a thin cylinder of hard rubber. They are wound with No. 36 insulated copper wire. The armature is $1\frac{5}{8}$ inch long, $\frac{1}{20}$ inch thick, and $\frac{1}{4}$ inch wide. The sounding lever, of brass, is $1\frac{1}{4}$ inch long, is placed horizontally, from the centre of which drops a perpendicular arm to which the armature is attached. The free end of the sounding lever plays between the milled heads of two set screws, the upper of which is inserted in the lower. This connects with a branched anvil, the two legs of which rest upon a brass sounding board, $1\frac{3}{8}$ inches diameter, which is concave beneath and is attached with three screws to the bottom of the case, a diminutive adjusting spring, actuated by a milled headed adjusting post with milled headed connecting screws. At the opposite end of the magnet is a key of very thin tempered brass, $\frac{1}{4}$ inch wide and $1\frac{3}{4}$ inch long, with ivory finger piece, connecting points of platinum, and a current breaker with ivory handle. This completes the mechanical contrivances, and the whole is enclosed in a hard rubber case, with a cover like a snuff box.

The external dimensions when shut are, length 5 inches, breadth $2\frac{1}{4}$ inches, height $1\frac{1}{4}$ inch. The ends of the box are semi-circular. The case stands upon four brass legs, $\frac{3}{8}$ inch diameter and $\frac{3}{8}$ inch long. Entire weight $10\frac{1}{4}$ ounces.

Here are all the instruments necessary for a complete telegraph office where the operator receives by sound, which is now almost universally the case in this country. No local circuit is required, but it is operated on the main circuit. The report is as clear, distinct, and audible as that of an ordinary sounder actuated by a local circuit. It is designed for use in the field or out of doors. A telegrapher will attach it to the main line anywhere in the country in five minutes, when he can send and receive messages with the same facility and accuracy that he can in a regular telegraph office. During the war Mr. Caton supplied the government with a large number of these instruments, but was unable to fill all of the orders of General Stager, who had charge of the government telegraph department. Nearly all telegraph superintendents are supplied with them, as well as very many operators, who never travel without them. Their invaluable services in case of railroad accidents may be readily appreciated, and at the West they are in constant use. An account of their services thus rendered each year would fill a volume, and really no train should ever move without one in the hands of a competent operator. These instruments are only made at Ottawa, Illinois, under the superintendence of that accomplished mechanic, Mr. Robert Heming.

COSTON, Mrs. M. J., Washington, D. C.—Coston's telegraphic night signals.

FARMER, M. G., Boston, Massachusetts.—Thermo-electric battery.

FIELD, CYRUS W., THE ANGLO-AMERICAN COMPANY, New York.—
Transatlantic telegraph. Grand prize.
HUGHES, DAVID E., New York.—Printing telegraph. Grand prize.
MORSE, S. E. and G. L., Harrison, New Jersey.—Model of a new mode of
laying and raising submarine cables.
WARD, A. F., Philadelphia, Pennsylvania.—Combination of colors for
signals.

CLASS 65.—CIVIL ENGINEERING, PUBLIC WORKS, AND ARCHITECTURE.

BACON, S. T., 1010 Washington street, Boston, Massachusetts.—Door
fastener.
BACON, S. T., Boston, Massachusetts.—Challenge lock.
BANKER & CARPENTER, Boston, Massachusetts.—Paints, for buildings.
BELCHER BROTHERS, St. Louis, Missouri.—Plan of an artesian well at
St. Louis.
BOARD OF PUBLIC WORKS OF CHICAGO, A. W. Tinkham, secretary,
Chicago, Illinois.—Drawing of a tunnel constructed under Lake
Michigan. Silver medal.
BRADSTREET, J. R., Boston, Massachusetts.—Rubber mouldings and
weather strips.
CHAPIN & WELLS, Chicago, Illinois.—Model of swing bridge. Silver
medal.
DANA, J., Boston, Massachusetts.—Faced or pressed brick.
DAY, H. H., 23 Courtland street, New York.—Model of a system of canals
without locks, for steamers, &c.
DERROM, A., Paterson, New Jersey.—Model trestle bridge.
DODDS, MACNEALE & URBAN, Cincinnati, Ohio.—Bank locks.
GREGG, ISAAC, Philadelphia, Pennsylvania.—Brick-making machine, in
operation, to be seen in the Annex of the Exhibition, Nos. 100 and
102, Avenue Suffren. Silver medal.
HERRING, FARREL & SHERMAN, New York.—Fire and burglar-proof
safes. Bronze medal.
HUSTIN, A., Bristol, Massachusetts.—Mitre box, with scale.
JOHNSON, J., Saco, Maine.—Dredging and excavating machine.
JOHNSON'S ROTARY LOCK COMPANY, 18 John street, New York.—F. G.
Johnson's rotary locks.
JOHNSON, W., Milwaukee, Wisconsin.—Bank lock.
LA MOTHE, J. B., 5 Wall street, New York.—Model of a house with
tube frame.
LOUISVILLE CEMENT AND WATERPOWER COMPANY, Louisville, Ken-
tucky.—Cement.
MILWAUKEE BRICK COMPANY, Milwaukee, Wisconsin.—Building bricks.
MORRIS, TASKER & Co., Philadelphia, Pennsylvania.—Steam-coils, pipes,
&c.
NEWMANN, H. J., Andover, Massachusetts.—American woods painted
in oil and distemper.

NICHOLSON, S., Boston, Massachusetts.—Model of wooden pavement.

PEASE, C. F., Boston, Massachusetts.—Spring-balance curtain fixture.

ROBINSON, E., & SON, Boston, Massachusetts.—Metallic roofing.

SMITH, H., 255 East Thirtieth street, New York.—Window blind and shutter fasteners.

VANDERBURGH, G. E., 24 Vesey street, New York.—Artificial building-blocks.

WASHBURN, B. D., Boston, Massachusetts.—Kingman's paint roofing. Exhibited in Mr. Bacon's bakery in the Park.

WEBSTER, W., Rochester, New York.—Plans of parks.

WESTON & PUTNAM, Boston, Massachusetts.—Graining, in imitation of American woods.

YALE AND WINN MANUFACTURING COMPANY, Sherbune Falls, Massachusetts.—Locks. Silver medal.

CLASS 66.—NAVIGATION, LIFE-BOATS, YACHTS, AND PLEASURE BOATS.

BECKWITH, E. P., New London, Connecticut.—Model of a fishing smack.

BROWN & LEVEL, Wall street, New York.—Tackle for disengaging ship's boats. Bronze medal.

This apparatus has been adopted upon many vessels and steamer lines in the United States. It is simple, reliable, cheap, and can be easily adapted to boats without change of rig. By its aid, one man, standing in the centre of a loaded boat, can detach it instantaneously from the ship, even while it is under full speed.

DABOLL, C. L., New London, Connecticut.—Fog whistle. In the Annex, Park. Silver medal.

DUFFY, J., Paterson, New Jersey.—Sectional model of iron-clad ship containing various improvements.

HUDSON, Captain J. M.—The ship "Red, White and Blue."

This little vessel, constructed by Mr. Ingersoll, of New York, which crossed the Atlantic with the two daring men, Captain Hudson and Captain Fitch, was, by special permission of the Emperor, installed in the Park. It was rigged as a three-master, 26 feet long, 6 feet beam, and registered 2 tons 28 cwt.

LEPELLY, N. D., Cleveland, Ohio.—New construction of rudder.

MANLEY, W. R., New York.—Model of a paddle wheel for steamers, with vertical floats.

PAGE, E. W., 69 West street, New York.—Oars. Honorable mention.

PERRY, E. F., New York.—Life-saving raft.

PRATT, H. D. J., Washington, D. C.—Model of a propelling apparatus attached to a small metallic vessel.

The propelling screw in this apparatus is placed under the keel.

REED, J., San Francisco, California.—Model of a life boat.

REIM, W. O., Springfield, Ohio.—Hydrostatic scale.

ROLLE, H., Boston, Massachusetts.—Model of a propelling apparatus for steamships.

VANDEUSEN, J. B., 274 Seventh street, New York.—Model of the American yacht "Fleetwing." Bronze medal.

GROUP VII.

FOOD, FRESH OR PRESERVED, IN VARIOUS STAGES OF PREPARATION.

CLASS 67.—CEREALS AND OTHER FARINACEOUS EDIBLES, WITH THEIR DERIVATIVES.

The cereal productions of the United States on exhibition were by no means sufficient to give one an adequate idea of the great grain-growing capabilities of the country. A resolution passed by both Houses of Congress in January, 1867, instructed the Commissioner of Agriculture "to collect and prepare, so far as practicable, and with as little delay as possible, suitable specimens of the cereal productions of the several States of the Union for exhibition at the Paris Exposition."

It was naturally expected that such a proposed exhibition of the finest samples of the best varieties of wheat, corn, and other cereals, would command the admiration of Europe, as it would assuredly arouse the pride of all Americans.

Notwithstanding the commendable activity of the Commissioner of Agriculture, the short time authorized for making the collection, and the multitude of unforseen difficulties which presented themselves, prevented the assembling of such an imposing variety of cereals as was desired and originally intended.

AGRICULTURE, DEPARTMENT OF, Washington, D. C.—Products from the following States: Wheat from Ohio, Indiana, Minnesota, Virginia, Michigan, Pennsylvania, New York, Washington, Vermont, Massachusetts, Michigan winter wheat; wheat from Boyer valley, Maine, Iowa, Wisconsin, Tennessee, Missouri, and Nebraska; barley from Maryland and Connecticut; cotton seeds from Georgia; wheat from North Carolina, Minnesota, Texas, Kansas, Massachusetts, and Georgia; oats from Baltimore county, Maryland; peas from Illinois, Michigan, and Vermont; beans from New York and Maine. Bronze medal.

BABILLON, HINCHMAN & Co., Detroit, Michigan.—Indian corn meal, white and yellow.

CALIFORNIA, STATE OF.—Cereals. Silver medal.

The exhibition of cereals of California production was made by Mr. Campbell, of San Francisco, Mr. Peters, of Stockton, and Mr. Perkins, of Oakland. The two former exhibited samples of remarkably fine wheat. A silver medal was awarded to the State, as above.

CAMPBELL, J. W. H., San Francisco, California.—Cereals.

A large sack, about two bushels, of California "high mixed white wheat," weighing about 120 pounds. This wheat attracted much attention and was greatly desired for seed by agriculturists. Agreeably to the directions of the exhibitor, it was donated, at the close of the Exhibition, to the Royal Agricultural Society of England.

CARPENTER, WILLIAM S., Harrison, New York.—Indian corn in the ear. Bronze medal.

GLEN COVE STARCH MANUFACTURING Co., W. Duryea, secretary, 106 Fulton street, New York.—"Maizena," a preparation of Indian corn for puddings, custards, &c. Silver medal.

Maizena is made from the Indian corn grown in the Atlantic States. It is remarkable as well for its nutritive qualities as for the many different and useful ways in which it may be employed. The exports of this article to Australia are said to amount to $60,000 annually, while England demands as much more, and on the continent it is rapidly coming into favor as an article for table use. Large quantities are also shipped to Japan and other portions of the world. It is estimated that the exportation of maizena now amounts to $400,000 a year, while in the United States perhaps even more is consumed. Three articles are manufactured from the corn: 1st, the fine flour called maizena; 2d, corn starch; 3d, a starch made from the refuse, and employed for laundry purposes.

ILLINOIS, STATE OF.—Cereals, grain in the ear, and flour. Bronze medal.

IOWA, STATE OF.—Cereals and flour. Honorable mention.

KANSAS, STATE OF.—Cereals and flour. Bronze medal.

MINNESOTA, STATE OF.—Cereals. Honorable mention.

MISSOURI, STATE OF.—Corn, wheat, barley, oats, corn in the ear.

OHIO, STATE OF.—Cereals. Bronze medal.

PERKINS, D. L., Oakland, California.—A collection of seeds of cereals and vegetables grown in California, 120 varieties in all, classified and labelled, and packed in glass.

Donated at the close of the Exposition to the *Imperial Societé de Acclimatation*. This collection was accompanied by a photograph showing the variety of vegetables grown in California.

PETERS, J. D., San Joaquin county, California.—Specimens of wheat grown in California.

SAMORY, H., Gentilly, Louisiana.—Pecan nuts.

URQUHART, J. M., New Orleans, Louisiana.—Samples of rice.

WARDER, J. A., Hamilton, Ohio.—Samples of various kinds of Indian corn.

WESTERN VIRGINIA, STATE OF.—Cereals.

WISCONSIN, STATE OF.—Cereals and flour. Bronze medal.

CLASS 68.—BREAD AND PASTRY.

BACON, S. T., 1010 Washington street, Boston, Mass.—Crackers; bread and cakes; aerated bread, Dauglish's system. Establishment in the Park. See a notice under Class 50.

CLASS 69.—FATTY SUBSTANCES USED AS FOOD, MILK AND EGGS.

CLASS 70.—MEAT AND FISH.

BORDEN, GAIL, 36 Elizabeth street, New York.—Extract of beef. Honorable mention.

BRAY & HAYES, Boston, Massachusetts.—Preserved lobster. Honorable mention.

CAPE, CULVER & CO., New York.—Manhattan hams. Silver medal.

CULBERTSON, BLAIR & CO., Chicago, Illinois.—Packed beef, pork, and lard. Silver medal.

DUFFIELD, CHARLES, Chicago, Illinois.—Salt cured, and smoked hams. Silver medal.

PORTLAND PACKING COMPANY, Portland, Maine.—Preserved oysters and lobsters. Honorable mention.

TOWNSEND BROTHERS, 79 Water street, New York.—Canned oysters. Honorable mention.

CLASS 71.—VEGETABLES AND FRUITS.

MOTT, R. C., New Orleans, Louisiana.—Sample of filé, powdered sassafras root; gumbo powder for soups.

ONEIDA COMMUNITY, J. A. Noyes, agent, Oneida, New York.—Preserved fruits. Honorable mention.

PORTLAND PACKING COMPANY, Portland, Maine.—Preserved vegetables.

SQUIRE, JOHN J., New London, Connecticut.—Preserved fruits and vegetables. Bronze medal.

TOWNSEND BROTHERS, 79 Water street, New York.—Canned fruits.

CLASS 72.—CONDIMENTS AND STIMULANTS, SUGAR AND SPECIMENS OF CONFECTIONERY.

AVERY, D. D., Petite Anse, Louisiana.—Crushed rock salt. Honorable mention.

DAVIDSON, JOHN, St. Bernard Parish, Louisiana.—Refined yellow sugar. Honorable mention.

GERMANIA SUGAR COMPANY, Chatsworth, Illinois.—Beet sugar.

IOWA, STATE OF.—Sorghum syrup and sugar.

JOHNSON, BRADISH, Louisiana.—Sugar. Bronze medal.

LAURENCE, E., Louisiana.—Sugar. Silver medal.

LOPEZ, D., New Orleans, Louisiana.—Chocolate.

PECK, O. E., Vermont.—Maple sugar.

SABATIER, G., Plaquemines Parish, Louisiana.—Sugar. Honorable mention.

STANFORD, W. L., Plaquemine parish, Louisiana.—Clarified sugar.

THOMPSON, A., New Orleans, Louisiana.—Samples of powdered and crushed sugar, and golden syrup.

TOWNSEND BROTHERS, 79 Water street, New York.—Canned fruits.

WALTEMEYER, JACOB, Baltimore, Maryland.—Preserved fruits. Honorable mention.

WALTER BAKER & Co., Dorchester, Massachusetts.—Cocoa and chocolate. Silver medal.

WILLIAMS, C. C., 314 Dean street, New York.—Hermetically sealed fruit in syrup. Honorable mention.

WISCONSIN, STATE OF.—Sorghum syrup and sugar.

CLASS 73.—FERMENTED DRINKS.

AMERICAN WINE COMPANY, Saint Louis, Missouri.—Wines. Honorable mention.

ANDERSON, W. F. & J. P., Cincinnati, Ohio.—Longworth's sparkling and still Catawba, Catawba brandy, red wine from Norton seedlings. Honorable mention.

BACON, S. T., & D. JAY BROWNE, Boston, Massachusetts.—Sorghum brandy, and brandy made from American wines and wild grapes.

BOTTLER, CHARLES, Cincinnati, Ohio.—Dry and sparkling wines. Honorable mention.

BREHM, F. C., Waterloo, New York.—Wines and brandies.

BUENA VISTA VINICULTURAL SOCIETY, San Francisco, California.—Sparkling Sonoma wine.

Two cases of quart bottles, sample of the wine put up by this society at its establishment in Sonoma valley. This wine was much liked by the committees and experts, and received the diploma of honorable mention. The company commenced operations in 1863, and in 1866 they put up 40,000 bottles, and in 1867, 90,000 bottles. The California grape is used. Honorable mention.

COZZENS, FREDERIC L., 73 Warren street, New York.—Wines and liquors.

DOWS, GUILD, CLARK & VAN WINKLE, · Boston, Massachusetts.— American bar and restaurant. Restaurant gallery.

GRIFFITH, W. M., North East, Pennsylvania.—American red and white wines; brandies made from wine and lees.

HELLMAN, A., 202 Broadway, New York.—Sparkling Catawba, made from grapes growing in the State of New York.

HUSMANN, G., Hermann, Missouri.—Wines.

ILLINOIS, STATE OF.—Wines.

KELLER, M., Rising Sun and Los Angeles vineyards, California.—California wines, brandy and bitters.

KOHLER & FROHLING, San Francisco, California.—Wines. White and red wines produced from the California grape at the vineyards, Los Angeles, California.

LE FRANC, C. H., New Almaden, California.—Red and white wines.

Four cases, of 12 bottles each, of wine made by Mr. Le Franc at his vineyards seven miles south of San José, upon the road to New Almaden.

LEICK, G., Cleveland, Ohio.—Wines.

METAYÉ, F., Jefferson parish, Louisiana.—Rum.

PLEASANT VALLEY WINE COMPANY, C. D. Champlin, secretary, Hammondsport, New York.—Sparkling wines and brandy. Honorable mention.

ROWLEY, J. & S., Hastings-on-the-Hudson, New York.—Hastings wine.

ST. LOUIS PARK OF FRUITS, St. Louis, Missouri.—Catawba wine.

SANSEVAIN BROTHERS, Los Angeles, California.—Wines.

Red and white, of several vintages, made from grapes grown in their vineyards in Los Angeles county.

SMITH, McPHERSON & DONALD, West Eighteenth street, New York.—Pale ale, porter, and brown stout. Bronze medal.

SYLVESTER, E. W., Lyons, New York.—Wine made from the American Oporto grape.

UNDERHILL, R. T., Clinton Hall, 7 Astor Place, New York.—Wines.

WERK, M., & SON, Cincinnati, Ohio.—Dry and sparkling wines. Honorable mention.

CLASS 74 TO 89.

(No exhibitors.)

GROUP X.

ARTICLES EXHIBITED WITH THE SPECIAL OBJECT OF IMPROVING THE PHYSICAL AND MORAL CONDITION OF THE PEOPLE.

CLASS 89.—MATERIALS FOR, AND METHODS OF, TEACHING CHILDREN.

HOWE, S. G., Director of the Perkins, Institute for the Blind, Boston, Massachusetts.— Books and apparatus for the use of the blind. Silver medal.

ILLINOIS, STATE OF.—Specimen of a western primary school and school furniture.

The United States school house was intended to be an exact reproduction of one of the numerous free primary schools which are erected in the country districts of Illinois. It was about 32 by 50 feet, with an entrance porch, and a place for hanging up hats and bonnets, and could accommodate 50 pupils. This modest structure attracted great attention from those interested in popular education, and it was specially noticed by M. H. Ferte, late chief of primary instruction in Paris, in a contribution to the Manuel General de l'Instruction Primaire, from which the following descriptive extract is translated: "Let us enter this modest structure of which we have spoken. We find a large room, which at first appears like all those built for educational purposes; but let us examine the details attentively, and we soon notice the excellent conditions under which it is established. First, the ceiling is twelve feet above a good floor—very necessary in a place where many children are to be

gathered.   In the second place, the ventilation is perfectly provided for
by means of sash windows upon each side and at the ends, which we
designate in France as 'guillotine;' but however they may be called,
these windows have the immense advantage over ours that they give
ventilation at pleasure, from the top or bottom, as may be found desirable.
Besides, they allow a free circulation, which, among us, is prevented by
our poor system of windows, opening inside, and which take off for this
reason nearly two feet of passage room.   Let us add that with the
American windows the breaking of glass is made less frequent, and that
the drafts produced with ours by the windows opening in the middle, by
their arrangement are easily avoided.   If, after the windows, we examine the desks for the teachers and pupils, we find them very much
preferable to those in use in France.   While we have long tables accompanied by long benches for accommodating ten or twelve pupils, who
crowd, elbow, and hinder each other; in this American school we find
the desks or tables neatly arranged for either one or two scholars, with
a seat having a support for the back of the pupil.   The teachers who
read this will understand at once the advantages of such an arrangement.   Does a scholar need to leave his seat, he can do so without disturbing his neighbor, or without being obliged, to the great detriment
of discipline, to pass before seven or eight of his fellow students, who
never fail to make good such an occasion for mischief.   It would be
highly desirable to have these American desks introduced in our schools.
The discipline would be benefited by it, the children could prosecute
their studies without disturbance, and be very much more comfortable.
We wish the same for the introduction of the inkstand with which each
table is provided.   The calculators, geometrical figures, globes, charts,
and other school apparatus resemble much those in our best schools.
Among the books we have examined we find many deserving of high
commendation.   We notice improved methods of teaching penmanship,
excellent and simple spelling, reading and drawing books, quite superior
in every respect, and also conveniences for cleaning blackboards, carrying books, and methods of object teaching, quite unknown with us.   The
desks, maps, globes, books, and school apparatus exhibited we find were
contributed by the Messrs. Sherwood and A. H. Andrews, two large and
enterprising dealers in these articles in Chicago, the principal city of
Illinois and the northwest."

MISSOURI, STATE OF, J. L. Butler, agent.—Collection of books, papers,
    photographs, maps, &c., illustrating the resources of the State of
    Missouri.   (In the Illinois cottage.)

CLASS 90.—LIBRARIES AND APPARATUS USED IN THE INSTRUCTION OF
    ADULTS AT HOME, IN THE WORKSHOP, OR IN SCHOOLS AND COL-
    LEGES.

(No exhibitors.)

21 U E

CLASS 91.—FURNITURE, CLOTHING, AND FOOD FROM ALL SOURCES, RE-
MARKABLE FOR USEFUL QUALITIES COMBINED WITH CHEAPNESS.

(No exhibitors.)

CLASS 92.—SPECIMENS OF THE CLOTHING WORN BY THE PEOPLE OF
DIFFERENT COUNTRIES.

(No exhibitors.)

CLASS 93.—EXAMPLES OF DWELLINGS CHARACTERIZED BY CHEAPNESS,
COMBINED WITH THE CONDITIONS NECESSARY FOR HEALTH AND
COMFORT.

FLINT & HALL, Boston, Massachusetts.—Sectional building, containing
M. Bacon's bakery establishment.

GOTTHEIL, EDWARD, New Orleans, Louisiana.—Portable cottage building, made of Louisiana woods.

ILLINOIS, STATE OF.—Specimen of a western farmer's house.   Silver
medal.

In the section of the Park assigned to the United States the State of
Illinois exhibited the western farmer's home, or "American cottage."   It
was constructed by Colonel Lyman Bridges, of Chicago, from plans furnished by O. L. Wheelock, esq., architect, of that city, and was forwarded
in sections by railway to New York, and was installed among other types
of residences and palaces in the Champ de Mars.

The object was to show the kind of dwelling much used in the agricultural regions of the United States—a dwelling which, while combining
beauty and comfort, is within the reach of all prudent and industrious
persons.   It was intended also to make known the fact that the farming
population of the United States may, and do very generally, own a similar or comfortable home, and that the laws give them liberal protection
in the ownership.

The building did not conform to any special order of architecture.
The plan was such that one of the rooms on the ground floor could be
first constructed and occupied as a temporary home by a new settler at
an expense of not over $300, and so that the other rooms and the spacious
hall could be added after at the convenience of the owner.   It was constructed of Wisconsin and Michigan pine lumber, in part generously
contributed by two companies, represented by Hon. W. B. Ogden, and
by Messrs. Wood & Lawrence.   The capacity of the house was sufficient
for a family of six or eight persons, it having three rooms on the first
floor and five chambers on the second floor.

The cottage occupied a conspicuous and favorable position in the Park,
and a low terrace around it was adorned with shrubs.   It became a
centre of attraction also by the distribution of documents and information there relating to the extent and resources of the United States.
The walls were lined with maps and photographic views of prominent
places, and many statistical works relating to the productions and agriculture and geology of the country were ranged upon the shelves of one

of the rooms.   Information of this nature was eagerly sought by, and was freely given to, thousands of European visitors.

This interesting and valuable addition to the United States section was secured through the exertions of James H. Bowen, of Chicago, United States commissioner.

INTERNATIONAL EXHIBITION OF WEIGHTS, MEASURERS AND COINS.

IN THE PAVILLION IN THE CENTRE OF THE CENTRAL GARDEN.

THE UNITED STATES TREASURY DEPARTMENT, Washington, D. C.— Weights, measures and coins, (in the central pavillion,) scales, (in the Palace.)

INTERNATIONAL SANITARY DEPARTMENT.

Collection of objects from the United States made by Dr. T. W. Evans. (See also classes 11 and 38.)

THE UNITED STATES SANITARY COMMISSION.—Material used in the late war.   Grand prize.

During the terrible civil war which desolated the United States for four long and bloody years, public feeling was forcibly aroused to the necessity of devising effective means for mitigating the sufferings and improving the sanitary condition of our armies.   Laudable and philanthropic efforts were set on foot for the realization of that noble purpose. All parts of the country were interested in the construction of apparatus, and the assembling of material which should contribute to the attainment of the desired end, and render the scenes and sufferings of the battle field less terrible.   As a natural consequence of these efforts the United States sanitary societies and commission sprung into existence and rendered incalculable services to the nation.

It is almost impossible to arrive at a just appreciation of the great good accomplished by the sanitary and relief societies of the United States.   They mark a new era in the history of the world, as organizations based upon acts and impulses of the noblest philanthropy ever conceived by humanity.

The happy influence of these institutions has been felt in the Old World, and relief societies, animated with the same noble and generous feelings, have been established in Europe upon the exact model of those existing in America, and have also rendered immense services during the late wars.

Many of the objects, apparatus, and inventions used by the United States sanitary societies were collected together after much effort, and exhibited on the Champ de Mars, in the name of the United States Sanitary Commission.

To mention in detail the many very useful objects composing this collection would fill of itself a considerable volume, only brief notices of some of the leading articles will therefore be given.

Under the head of Ambulances of Transport, may be noticed:

1. The Howard ambulance; made from plans furnished by Dr. Benjamin Howard, of New York.   It is a light, two-horse, four-wheeled carriage, designed to carry four persons besides the driver, two recumbent and two sitting, or eight persons sitting.   The body of the ambulance is mounted on elliptic springs, and the stretcher mattresses are furnished with inferior and lateral counterpoise springs, which modify or altogether prevent concussions, and contribute greatly to the safety and comfort of the patients transported.   There is also connected with it a special mechanical contrivance—a "sling"—for the suspension of wounded limbs when necessary.

2. An ambulance known as the Wheeling ambulance, improved by T. Morris Perot, of Philadelphia.   This is a light, two-horse, four-wheeled vehicle, intended to convey four persons besides the driver, two recumbent, two sitting, or eight persons sitting.   Perot's improvement consists in the employment of springs of caoutchouc.   It is claimed that this improvement secures for the carriage an easy and agreeable movement, and an almost entire absence of concussion, even over the roughest roads.   Aside from Perot's improvement, the ambulance is similar in its construction to those which, under the same name, were extensively used by the United States government during the late war.

3. An ambulance made by G. Brainard, Boston.   This ambulance is intended to carry six persons besides the driver, four recumbent, two sitting, or eight persons sitting.   The body is mounted on "platform springs;" the mattresses and seats are arranged on what is known as the "Rucker plan," the back of the seats being hinged on the top, so as, when opened inward, and locked, to form an upper tier of mattresses. The ambulance on exhibition was employed during the war in the hospital service for several months.

4. An ambulance, one of 30 of similar construction, given by the citizens of Philadelphia to as many fire companies of that city, and employed in the late war in conveying sick and wounded soldiers across the city from station to station.   Not less than 3,000 soldiers were thus transported in this ambulance.

5. A model of a railway ambulance, or hospital car, made by Messrs. Cummings & Sons, Jersey City, from plans furnished by Dr. Elisha Harris, of New York.   This model is a fac simile of the hospital cars employed during the war by the United States Sanitary Commission, on the railway between Washington and New York, as well as on several other military railways in other portions of the country.   The model, constructed on a scale of one-fourth, shows in detail every thing—couches, dispensary, wine closet, water closet, systems of ventilation and heating, &c., employed in the construction and equipment of the sanitary com-

mission cars, while at the same time externally it perfectly represents the construction of an ordinary American passenger car. To it is attached a patent safety break, as well as a set of self-acting ventilators, furnished by W. Creamer, of New York.

The Evans ambulance, constructed at Paris by Dr. Thomas W. Evans, was made with the view of uniting a possible capacity for four persons recumbent, with lightness, easiness of movement, facility of loading and unloading, and simplicity. It was not finished until the last of August, so late as to be even *hors de concours* in the competition for the special prizes offered for the best ambulance by the *Société de Secours aux Blessés*. Nevertheless, such were its merits that the jury of the society saw fit to award to it a second prize of 500 francs, accompanied with an expression of regret that they were unable, in consequence of the fixed condition of the *concours*, to award it the first prize.

This ambulance can carry ten persons seated, besides the driver and one or two attendants, or four lying down and two seated, besides the driver and attendants, as in the first-named instance. The seats can be used each as a mattress upon the floor of the wagon, the iron wheels with which they are furnished resting, when in position, upon springs beneath the floor, the object being to place these supplementary springs out of the way, and where when once fixed they would be secured against accidents. For the upper tier four rings of caoutchouc are attached in front and rear to the sides of the wagon, two feet nine inches from the floor, two rings to an upright in the centre of the wagon, immediately behind the seat of the driver, and two rings to a hook which may be dropped from the rear centre. By means of this arrangement, so very simple as scarcely to be observed, unless special attention is directed to it, two ordinary French, English, or American stretchers can be suspended whenever necessary, and two additional wounded transported in the most comfortable manner. This ambulance, weighing about 1,300 pounds, is slightly heavier than the other American ambulances. The forward wheels turn readily under the body of the wagon; the top is covered with enamelled cloth, and folding seats are placed at the rear end, outside, for one or two attendants. It is furnished with a double tank for ice and water, and a box for a few necesssary supplies. Two stretchers are carried overhead inside and a supplementary one outside.

AMBULANCES OF SUPPLY.

1. A medicine wagon, known as Autenreith's, the fixtures having been furnished by G. Autenreith, of New York. The wagon is intended to carry for field service a full complement of the medicines authorized by the "supply table" of the medical bureau; also a set of hand litters, as well as a light, compact amputating table. Wagons of this kind were favorably regarded and extensively employed by the United States government during the war.

2. A medicine wagon, known as Perot's, constructed by T. M. Perot,

Philadelphia. In this wagon the drawers and compartments are adapted to carrying medicines in bulk, in parcels, and in bottles; the system of packing being such as to secure the latter against fracture, in certain instances by the employment of springs, in others by the employment of columns of compressible air, obtained by a simple device. A set of hand litters is carried, as also a strong amputating table.

3. An ambulance kitchen, invented by Mr. Pinner, of New York. The special purpose of this kitchen is to furnish soldiers, particularly the sick and wounded, while on the march, or on the battle field, with hot coffee, soup, and cooked food of various kinds. While possessing all the necessary apparatus of a well organized kitchen, it can be used with great advantage at all temporary encampments and hospital stations, and is so made and furnished as to be used, if needed, as an ambulance of transport.

4. A coffee wagon, invented by J. Dunton, of Philadelphia. The wagon exhibited designs to furnish the soldier on the march and on the field of battle with hot coffee and tea, was one of several in the service of the United States Christian Commission during the last months of the war, furnishing hot coffee and tea to the wounded of both armies.

HOSPITAL TENTS.

Several square tents are exhibited, similar to those generally used by the United States army.

The hospital tent, called the "umbrella tent," made by William Richardson, Philadelphia, is claimed to occupy less space when packed, to be more readily unpacked and erected, and when erected to be more convenient and secure, than either the square wall or Sibley's tent, which have hitherto been regarded with most favor.

An officers' "umbrella tent," made by N. Walton, St. Louis, is also exhibited, and claims to possess the same advantages as the one already mentioned. Its height is 11 feet, diameter at base 13½ feet, form octagonal. It is supported by a telescopic centre pole, slender T-iron rafters, and eight light wooden props.

In this collection appear a number of horse and hand litters on improved principles; pack saddles, old and new pattern; models, plans and lithographic views of various hospitals; a great variety of beds, stools, tables, mess chests, mess kits, surgical instruments and apparatus, invalid beds, mess panniers, hospital and field knapsacks, splints, fracture and amputating apparatus, artificial limbs, clothing used by the commission, food of all kinds, liquors, &c., bandages, comforts, cotton batting, crutches, and, in a word, everything necessary for the comfort and convenience of the sick and wounded soldier.

Under the head of material, historical and co-ordinate, are exhibited a number of books, pamphlets, and documents relating to the sanitary work, &c., &c.

A grand prize was awarded by the international jury, which was handed over to the United States Sanitary Commission.

LIST OF AWARDS

BY THE INTERNATIONAL JURIES TO EXHIBITORS AND OTHERS, FROM THE UNITED STATES AT THE PARIS UNIVERSAL EXPOSITION OF 1867.

For convenience of reference this list of awards has been alphabetically arranged. In the French official catalogue[1] the names are not placed in alphabetical order.

Each medal issued was accompanied by a framed diploma, which certified that a medal had been awarded. When two or more awards of medals were made to one person or association the number of diplomas issued corresponded with the number of awards, but only one medal was issued, and this medal was always of the highest denomination decreed to the exhibitor. No medals were issued with the diplomas of Honorable Mention.

NEW ORDER OF RECOMPENSES.

For persons, establishments, or localities, which, by organizations or special institutions, have developed harmony among co-operators, and produced, in an eminent degree, the material, moral, and intellectual well-being of the workmen.

AGRICULTURAL SOCIETY OF VINELAND, Charles K. Landis, New Jersey.—An Honorable mention, unaccompanied by a medal.

CHAPIN, WILLIAM C., Lawrence, Mass.—Grand Prize, a Gold Medal of the value of 1000 francs, and 9,000 francs in gold.

ARTISTS' MEDAL.

CHURCH, F. E., New York city.—The Artists' Medal, with 500 francs in gold.—Landscape paintings in oil.

GRAND PRIZES.

FIELD, CYRUS W., and Anglo-American Transatlantic Telegraph Company.—Transatlantic cable.

HUGHES, DAVID E. New York.—Printing telegraph.

McCORMICK, C. H. Chicago, Illinois.—Reaping machines. See, also, Gold Medal.

By a decree of the Emperor, Mr. McCormick was created Chevalier of the Imperial Order of the Legion of Honor.

---

[1] Catalogus officiel des Exponants Récompennés par le Jury International. 8vo. Paris: E. Deutu, Libraire-Editer de la Commission Impériael.

UNITED STATES SANITARY COMMISSION.—Ambulances, materials, instruments, &c., for the relief of the wounded, used in the late war. See, also, Honorable Mention.

GOLD MEDAL, WITH WORK OF ART.

WOOD, WALTER A., Hoosick Falls, New York.—Mowing machines. See, also, Gold Medal.

By a decree of the Emperor, Mr. Wood was created Chevalier of the Imperial Order of the Legion of Honor of France.

GOLD MEDALS.

CHICKERING & SON, New York and Boston.—Pianos.

By a decree of the Emperor, Mr. C. F. Chickering was created Chevalier of the Imperial Order of the Legion of Honor of France.

CORLISS STEAM ENGINE COMPANY, Providence, Rhode Island.—The Corliss engine.

FIRE-ARM MANUFACTURING INDUSTRY OF THE UNITED STATES.—Fire-arms.

GRANT LOCOMOTIVE WORKS, Paterson, N. J.—Locomotive and tender.

HOWE, ELIAS, Jr.—"Promoter of the sewing machine."

By a decree of the Emperor, Mr. Howe was created Chevalier of the Imperial Order of the Legion of Honor of France.

McCORMICK, C. H., Chicago, Illinois.—Reaping and mowing machines.

According to the rule of the Imperial Commission this medal is absorbed in the Grand Prize.

MEYER, VICTOR, Parish of Concordia, Louisiana.—Short staple cotton.

ROGERS, C. B., & Co., Norwich, Connecticut.—Wood-working machines.

SELLERS, WILLIAM, & Co., Philadelphia.—Machine tools.

STEINWAY & SON,[1] New York city.—Pianos.

TRAGER, LOUIS, Blackhawk Point, Louisiana.—Short staple cotton.

WALBRIDGE, WELLS D., New York city.—Gold and silver ores from Idaho.

WELCH, PATRICK, New York city.—Type-dressing machine.

WHEELER & WILSON MANUFACTURING CO., New York city.—Sewing and button-hole machines.

WHITE, SAMUEL S., Philadelphia.—Artificial teeth, and dentists' instruments and furniture.

WHITNEY, J. P., Boston, Massachusetts.—Silver ore from Colorado.

WOOD, WALTER A., Hoosick Falls, New York.—Reaping and mowing machines.

---

[1] By the adoption of the alphabetical arrangement of the names in this list, already explained, the name of the firm of Steinway and Sons is here made to follow that of Chickering & Sons, but in the French official catalogue of awards the sequence is the reverse.

According to the rule of the Imperial Commission this medal is absorbed in the first accompanied with a work of art.

SILVER MEDALS.

ALABAMA, STATE OF.—Short staple cotton. See Honorable Mention.

AMERICAN BUTTON-HOLE COMPANY, Philadelphia.—Sewing and button-hole machines.

BAKER, WALTER, & CO., Dorchester, Massachusetts.—Chocolates.

BARNES, Surgeon General J. K., United States army, Washington.—Surgical instruments, hospital apparatus, &c.

BEMENT & DOUGHERTY, Philadelphia.—Machine tools.

BERGNER, THEODORE, Philadelphia.—Co-operator—engineer of Messrs. William Sellers & Co.

BIDWELL, J. C., Pittsburg, Pennsylvania.—Comstock's rotary spader.

BIGELOW, H., Boston, Massachusetts.—Copper and minerals from Lake Superior.

BLAKE, WILLIAM P., San Francisco, California.—California minerals.

BOND, WILLIAM, & SON, Boston, Massachusetts.—Astronomical clock and chronograph.

BROWN, J. R., & SHARPE, Providence, Rhode Island.—Screw-cutting and milling machines.

BURT, EDWIN C., New York city.—Machine sewed boots and shoes.

CALIFORNIA, STATE OF.—Cereals.

CAPE, CULVER & CO., New York city.—Hams.

CHAPIN & WELLS, Chicago, Illinois.—Model of a swing bridge.

CHICAGO BOARD OF PUBLIC WORKS, Chicago, Illinois,—Design of the lake tunnel.

CLARK THREAD COMPANY, Newark, New Jersey.—Cotton yarns.

COLLINS & CO., New York city.—Steel ploughs.

COLT'S PATENT FIRE-ARMS MANUFACTUING COMPANY, Hartford, Connecticut.—Fire-arms.

COOL, FURGUSON & CO., Glen's Falls, New York.—Barrel machines.

CROMPTON, GEORGE, Worcester, Massachusetts.—Loom for cloths.

CULBERTSON, BLAIR & CO., Chicago, Illinois.—Salted meats.

DABOLL, C. L., New London, Connecticut.—Fog-signal.

D'ALIGNY, H. F. Q.—Co-operator in the organization of the United States section.

DARLING, BROWN & SHARPE, Bangor, Maine, now of Providence, Rhode Island.—Steel measures.

DELPIT, A., & CO., New Orleans, Louisiana.—Snuff.

DIXON, JOSEPH, & CO., Jersey city, New Jersey.—Plumbago crucibles.

DOUGLAS AXE MANUFACTURING COMPANY, Boston, Massachusetts.—Edge tools.

DUFFIELD, CHARLES, Chicago, Illinois.—Hams.

FAIRBANKS, E. & T., & COMPANY, St. Johnsbury, Vermont.—Scales. See, also, under Bronze Medals.

FLORENCE SEWING MACHINE COMPANY, New York city.—Sewing machines.

FOURNIER, S., New Orleans, Louisiana.—Electric clocks.

GLEN COVE STARCH MANUFACTURING COMPANY.—New York city.—"Maizena" and starch.

GOTTHEIL, EDWARD, New Orleans, Louisiana.—Co-operator, services rendered to agriculture in Louisiana.

GREGG, ISAAC, Philadelphia.—Brick-making machine. See, also, Bronze Medal.

GUNTHER, C. G., & SONS, New York city.—Furs.

HALL, JAMES, & SON, Boston, Massachusetts.—Buggy.

HOWE, Dr. SAMUEL G., Boston, Massachusetts.—Works for the blind.

HOWE MACHINE COMPANY, New York city.—Sewing machines.

ILLINOIS CENTRAL RAILROAD COMPANY, Chicago, Illinois.—Agricultural products.

ILLINOIS, STATE OF.—Collection of minerals; farmer's house; schoolhouse. See, also, Bronze Medal.

JACKSON, Dr. CHARLES T., co-operator.—Discovery of emery.

LAMB, J. W., Rochester, New York; now of Ann Arbor, Michigan.—Knitting machine.

LAWRENCE, E., Louisiana.—Sugars.

MASON & HAMLIN, New York and Boston.—Cabinet organs.

NEVADA, STATE OF.—Silver and copper ores.

NEW YORK MILLS, New York.—Muslins.

OPPER, MORRIS, New York.—Loom for corsets.

PARK BROTHERS & COMPANY, Pittsburg, Pennsylvania.—Cast steel and edge tools.

PARTRIDGE FORK WORKS, Leominster, Massachusetts.—Steel hayforks, rakes, &c. See, also, Bronze Medal.

PEASE, F. S., Buffalo, New, York.—Petroleum oils. See, also, Honorable Mention.

PERRY, JOHN G., Kingston, Rhode Island.—Mowing machine.
This prize was gained in the field trials of agricultural machines. See, also, Bronze Medal.

PIGNÉ, Dr. J. B., San Francisco, California.—Minerals.

PROVIDENCE TOOL COMPANY, Providence, Rhode Island.—Peabody's patent fire-arms.

REMINGTON, E., & SONS, Ilion, New York.—Fire-arms.

RUTHERFORD, LEWIS M., New York city.—Astronomical photographs.

SCHULTZ & WARKER, New York city.—Mineral water apparatus.

SCHUTTLER, PETER, Chicago, Illinois.—Wagon.

SLATER WOOLLEN MILLS, Webster, Massachusetts.—Woollen fabrics.

SMITH & WESSON, Springfield, Massachusetts.—Fire-arms and cartridges.

SPENCER REPEATING RIFLE COMPANY, Boston, Massachusetts.—Spencer rifles.

TAFT, JOHN B., Boston, Massachusetts.—Emery from Chester, Massachusetts.

TIEMANN, GEORGE, & CO., New York.—Surgical instruments.

TOLLES, R. F., Canastota, New York.—Microscopes.

TUCKER, HIRAM, & COMPANY, Boston.—Iron ornaments, imitation of bronze.

UNITED STATES GOVERNMENT.—Specimen of frame house for settlers.

WALES, WILLIAM, Fort Lee, New Jersey.—Optical instruments.

WARDWELL, GEORGE J., Rutland, Vermont.—Stone-quarrying machine.

WASHINGTON MILLS, Boston, Massachusetts.—Woollen fabrics. See, also, Honorable Mention.

WEED SEWING MACHINE COMPANY, New York city.—Sewing machines.

WHITNEY, BAXTER D., Winchendon, Massachusetts.—Wood working machines.

WINDSOR MANUFACTURING COMPANY, Windsor, Vermont.—Ball's patent fire-arms.

WOOD BROTHERS, New York city.—Phæton.

YALE & WINN MANUFACTURING COMPANY, Shelburne Falls, Massachusetts.—Yale locks.

BRONZE MEDALS.

ABBEY, CHARLES, & SONS, Philadelphia.—Dentists' gold foil.

AMERICAN LEAD PENCIL COMPANY, New York city.—Lead pencils.

APPLETON, D., & COMPANY, New York.—Books.

BABCOCK, JAMES F., Boston, Massachusetts.—Rosin oil.

BALTIMORE AND CUBA SMELTING AND MINING COMPANY, Baltimore, Maryland.—Copper.

BARLOW, MILTON, Richmond, Kentucky.—Planetarium.

BARTRAM & FANTON MANUFACTURING COMPANY, Danbury, Connecticut.—Sewing and button-hole machines.

BEER, SIGISMUND, New York city.—Stereoscopic views.

BELMONT OIL COMPANY, Philadelphia.—Oils.

BRIGHAM, E. D., treasurer Portage Lake Smelting Works, Boston, Massachusetts.—Lake Superior copper.

BROWN & LEVEL LIFE-SAVING TACKLE COMPANY, New York city.—Disengaging tackle for boats.

CARPENTER, WILLIAM S., New York city.—Collection of corn.

CARROLL, JOHN W., Lynchburg, Virginia.—Smoking tobacco.

CUMMINGS, WILLIAM, & SON, Jersey City, New Jersey.—Model of a hospital car.

DAY, AUSTIN G., Seymour, Connecticut.—Indelible pencils and lead pencils in India-rubber cases. See, also, Honorable Mention.

DEERE & CO., Moline, Illinois.—Steel ploughs.

DEGENER & WEILER, New York city.—Printing presses.

DEPARTMENT OF AGRICULTURE, Washington.—Collection of cereals.

DISS DEBAR, J. H., Parkersburg, West Virginia.—Petroleum oils.

DOUGLASS MANUFACTURING COMPANY, New York city.—Edge tools.

DOUGLAS, W. & B., Middletown, Connecticut.—Pumps.

FAIRBANKS, E. & T., & Co., St. Johnsbury, Vermont.—Railroad scale. See, also, under Silver Medals.

FAIRCHILD, LE ROY W., & Co., New York city.—Gold pens and cases.

GEMÜNDER, GEORGE, New York city.—Stringed instruments.

GODDARD, C. L., New York city.—Mestizo burring picker.

GOODELL, D. H., Antrim, New Hampshire.—Apple parer.

GOODENOUGH HORSESHOE COMPANY, New York city.—Horseshoes. See, also, Honorable Mention.

GREGG, ISAAC, Philadelphia.—Model of a brick machine. See, also, Silver Medal.

HADLEY COMPANY, Holyoke, Massachusetts.—Sewing cotton.

HARRIS, D. L., Springfield, Massachusetts.—Engine lathe.

HAUPT, HERMAN, Philadelphia.—Tunneling machine.

HERRING, FARREL & SHERMAN, New York city.—Fire and burglar proof safes.

HOGLEN & GRAFFLIN, Dayton, Ohio.—Tobacco-cutting machine.

HOTCHKISS, H. G., Lyon, New York.—Oils of peppermint, &c.

HOTCHKISS, L. B., Phelps, New York.—Oils of peppermint, &c.

HOUGHTON, H. O., & Co., Riverside Press, Cambridge, Massachusetts.—Books.

HOUSE, HENRY A., Bridgeport, Connecticut.—Co-operator in the establishment of Wheeler & Wilson.

HOUSE, JAMES A., Bridgeport, Connecticut.—Co-operator in the establishment of Wheeler & Wilson.

HOWE, AMASA B., New York city.—Sewing machines.

HOWE SCALE COMPANY, Brandon, Vermont.—Scales.

HUDSON, E. D., New York city.—Artificial limbs.

HUMPHRES, JOHN C., parish of Rapides, Louisiana.—Short staple cotton.

ILLINOIS, STATE OF.—Cereals and flours. See silver medals.

JESSUP & MOORE, Philadelphia.—Papers.

JOHNSON, A. J., New York city.—Johnson's Family Atlas.

JOHNSON, BRADISH, Louisiana.—Sugars.

JOHNSON & LUND, Philadelphia.—Artificial teeth. .

JUSTICE, PHILIP S., Philadelphia.—Power hammer.

KANSAS, STATE OF.—Collection of cereals.

LILIENTHAL, C. H., New York city.—Snuff and tobacco.

LILIENTHAL, THEODORE, New Orleans, Louisiana.—Photographic views.

LOUISIANA, STATE OF.—Portable cottage.

LYON, JAMES B., & Co., Pittsburg, Pennsylvania.—Pressed glassware.

MERRIAM, G. & C., Springfield, Massachusetts.—Webster's Illustrated Dictionary.

MISSION WOOLLEN MILLS, San Francisco, California.—Woollen fabrics.

MOODY, S. N., New Orleans, Louisiana.—Shirts.

MORRIS, TASKER & CO., Philadelphia.—Wringing machine.

MUMFORD, FOSTER & CO., Detroit, Michigan.—Boot-trees, lasts, &c.

MURPHY'S, WILLIAM F., SONS, Philadelphia.—Blank books.

OHIO, STATE OF.—Collection of cereals.

OLMSTEAD, L. H., New York.—Friction clutch pulley. See, also, Honorable Mention.

PARTRIDGE FORK WORKS, Leominster, Massachusetts.—Agricultural hand tools. See, also, Silver Medal.

PENNSYLVANIA, STATE OF.—Anthracite coal.

PERRY, JOHN G., Kingston, Rhode Island.—Mowing machine. See, also, Silver Medal.

PICKERING & DAVIS, New York city.—Engine governors.

PRATT & WENTWORTH, Boston, Massachusetts.—Heating apparatus.

RANDALL, SAMUEL H., New York city.—Mica.

RIEDEL, G. A., Philadelphia.—Automatic boiler feeder.

RICHARDS, RICHARD, Racine, Wisconsin.—Wool.

ROOTS, F. M. & P. H., Connersville, Indiana.—Rotary blower.

ROOTS, JOHN B., New York city.—Steam engine.

SACHSE, F., & SON, Philadelphia.—Shirts.

SARRAZIN, J. P., New Orleans, Louisiana.—Tobacco.

SCHEDLER, JOSEPH, Hudson city, New Jersey.—Terrestrial globes.

SCHREIBER, LOUIS, New York city.—Brass instruments.

SECOMBE MANUFACTURING COMPANY, New York city.—Ribbon hand stamps.

SHAW, CHARLES A., Biddeford, Maine.—Knitting machine.

SHAW, PHILANDER, Boston, Massachusetts.—Hot-air engine.

SLATER, SAMUEL, & SON, Webster, Massachusetts.—Cotton fabrics.

SMITH, McPHERSON & DONALD, New York city.—Ales and porter.

SOUTHERN COTTON-GIN COMPANY, Bridgewater, Massachusetts.—Cotton-gins.

SQUIRE, JOHN J., New London, Connecticut.—Preserved fruits and vegetables.

STURSBERG, H., New York city.—Beaver cloths.

SWEET, JOHN E., Syracuse, New York.—Composing machine.

TAMBOURY, A., parish of St. James, Louisiana.—Tobacco.

TIFFANY & CO., New York city.—Silverware.

TOWNSEND, WISNER H., New York city.—Oil-cloths.

UNION BUTTON-HOLE AND EMBROIDERY COMPANY, Boston, Massachusetts.—Button-hole machine.

VAN DEUSEN, J. B., New York city.—Model of the yacht Fleetwing.

WARNER, G. F., & CO., New Haven, Connecticut.—Malleable iron castings.

WATKINS, C. E., San Francisco, California.—Photographs, landscapes.

WICKERSHAM NAIL COMPANY, Boston, Massachusetts.—Nail-cutting machine.

WILLIAMS, THOMAS C., & CO., Danville, Virginia.—Chewing and smoking tobacco.
WISCONSIN STATE AGRICULTURAL SOCIETY.—Agricultural products.
WISCONSIN, STATE OF.—Collection of minerals.
WISCONSIN, STATE OF.—Collection of cereals and flours.
WRIGHT, R. & G. A., Philadelphia.—Perfumery.

HONORABLE MENTIONS.

ALABAMA.—Short staple cotton. See No. 30.
ALLEN, JOHN, & SON, New York city.—Artificial teeth.
AMERICAN STEAM GAUGE COMPANY, Boston, Massachusetts.—Steam gauges.
AMERICAN WINE COMPANY, St. Louis, Missouri.—Sparkling wines.
ANDREWS, WILLIAM D., & BROTHER, New York city.—Oscillating steam engine.
AVERY, D. D., Petite Anse, Louisiana.—Rock salt.
BACON, S. T., Boston, Massachusetts.—Cracker machinery.
BAKER, GEORGE R., St. Louis, Missouri.—Dough-kneading machine.
BATES, R., Philadelphia.—Instruments to cure stammering.
BELL FACTORY, Huntsville, Alabama.—Cotton fabrics.
BUENA VISTA VINICULTURAL SOCIETY, San Francisco, California.—Sparkling Sonoma wine.
BORDEN, GAIL, New York city.—Extract of beef.
BOTTLER, CHARLES, Cincinnati, Ohio.—Sparkling Catawba wine.
BOURGEOIS, E., New Orleans, Louisiana.—Tobacco.
BRANDON KAOLIN AND PAINT COMPANY, Brandon, Vermont.—Specimens of paints.
BRAY & HAYES, Boston, Massachusetts.—Preserved lobster.
BROUGHTON & MOORE, New York city.—Oilers, cocks, &c.
BROWNE, D. JAY, Roxbury, Massachusetts.—Enamelled leather.
CHIPMAN, GEORGE W., & CO., Boston, Massachusetts.—Carpet lining.
CLARK STEAM AND FIRE REGULATOR COMPANY, New York city.—Steam and fire regulator.
COHN, M., New York city.—Crinoline.
COZZENS, FREDERICK S., New York city.—Cigars.
DART, HENRY C., & CO., New York city.—Rotary steam engine.
DAVIDSON, GEORGE, Washington.—Sextant.
DAVIDSON, JOHN, New Orleans, Louisiana.—Sugars.
DAY, AUSTIN G., Seymour, Connecticut.—Artificial India-rubber. See, also, No. 120.
DUFFY, JOSEPH, Paterson, New Jersey.—Designs for improvements in iron-clad vessels.
DWIGHT, GEORGE, Jr., & CO., Springfield, Massachusetts.—Steam pump.
EDSON, WILLIAM, Boston, Massachusetts.—Hygrodeik.
ELSBERG, Dr. LOUIS, New York city.—Specimens of peat fuel.
EMPIRE SEWING MACHINE COMPANY, New York city.—Sewing machines.

FRIES, ALEXANDRE, Cincinnati, Ohio.—Flavoring extracts.

GLASS, PETER, Barton, Wisconsin.—Mosaic tables.

GOODENOUGH HORSESHOE COMPANY, New York city.—Horseshoes. See, also, Bronze Medal.

GOULD, J. D., Boston, Massachusetts.—Mica.

HERRING, SILAS C., New York city.—Bullard's hay tedder.

HICKS ENGINE COMPANY, New York city.—Steam engine.

HIRSCH, JOSEPH, Chicago, Illinois.—Albumen, glycerine, &c.

HOLLIDAY, T. & C., New York city.—Aniline colors.

HOWARD, Dr. BENJAMIN, New York city.—Ambulance, &c.

HOWELL & BROTHER, Philadelphia.—Wall papers.

IOWA, STATE OF.—Collection of cereals.

JACKSON, J. H., New York city.—Minerals and fossils.

KALDENBERG &.SON, New York city.—Meerschaum pipes.

KORN, CHARLES, Wurtsboro', New York.—Calf-skin leather.

LALANCE & GROSJEAN, New York city.—House-furnishing hardware.

LINTHICUM, W. O., New York city.—Cloth clothing.

LONGWORTH'S WINE-HOUSE, Cincinnati, Ohio.—Sparkling wines.

McCORMICK, J. J., Williamsburg, New York.—Skates.

MARIETTA & GALE'S FORK PETROLEUM COMPANY, Marietta, Ohio.— Petroleum oil.

METROPOLITAN WASHING MACHINE COMPANY, New York city.— Clothes wringers.

METROPOLITAN WASHING MACHINE COMPANY, New York city.—Washing machines.

MINNESOTA, STATE OF.—Collection of cereals.

MOEHRING, H. G., agent of the Volcanic Oil and Coal Company of West Virginia, Philadelphia.—Volcanic lubricating oil.

MONTAGNE & CARLOS, New Orleans, Louisiana.—Black moss for upholsterers.

MORRIS, TASKER & Co., Philadelphia.—Pipe-cutting machine.

NEW HAVEN CLOCK COMPANY, New Haven, Connecticut.—Clocks.

OLMSTEAD, L. H., New York.—Machine tools. See, also, Bronze Medals.

ONEIDA COMMUNITY, Oneida, New York.—Preserved fruits.

PAGE, E. W., New York city.—Oars.

PAUL, J. F., & Co., Boston.—Specimens of wood.

PEASE, F. S., Buffalo, New York.—Pneumatic pump. See, also, No. 82.

PEROT, MORRIS T., Philadelphia.—Medicine wagon.

PLEASANT VALLEY WINE COMPANY, Hammondsport, New York.— Brandy.

PORTLAND PACKING COMPANY, Portland, Maine.—Preserved lobster and vegetables.

PRENTICE, J., New York city.—Cigar machine.

PURRINGTON, GEORGE, Jr., New York city.—Carpet sweeper.

ROBINSON, JAMES A., New York city.—Ericsson hot-air engine.

SABATIER, G., Plaquemine parish, Louisiana.—Sugars.

SELPHO, WILLIAM, & SON, New York city.—Artificial limbs.
SHELDEN, JOSEPH, New Haven, Connecticut.—Water-pressure regulator.
SMITH, ROBERT M., Baltimore, Maryland.—Petroleum oils.
STEAM SYPHON COMPANY, New York city.—Steam syphon pump.
STEPHENSON, JOHN, & COMPANY, New York city.—Street railway carriage.
STOCKTON, SAMUEL W., Philadelphia.—Artificial teeth.
TALLMAN & COLLINS, Janesville, Wisconsin-—Perfumery.
TAYLOR, CHARLES F., New York city.—Therapeutic apparatus.
TILDEN, HOWARD, Boston.—Sifter, tobacco-cutter, and egg-beater.
TOWNSEND BROTHERS, New York city.—Preserved fruits and oysters.
UNITED STATES SANITARY COMMISSION.—Camp material. See Gold
  Medal.
WALTEMEYER, JACOB, Baltimore, Maryland.—Preserved fruits.
WARD, J., & CO., New York city.—Clothes wringers. ·
WARD, J., & CO., New York city.—Washing machines.
WASHINGTON MILLS, Lawrence, Massachusetts.—Shawls. See, also, Sil-
  ver Medals.
WELLMAN, C., New York city.—Saddles.
WERK, M., & SON, Cincinnati, Ohio.—Sparkling wines.
WHARTON, JOSEPH, Philadelphia.—Nickel, cobalt, and zinc.
WILLARD MANUFACTURING COMPANY, New York city.—Photographic
  camera tubes and lenses.
WILLIAMS, C. C., New York city.—Fruits preserved in syrup.
WILLIAMS SILK MANUFACTURING COMPANY, Bridgeport, Connecticut.—
  Silk twist for sewing machines.
WINSLOW, J. B., New York city.—Wood-moulding machine.
YOUNG, ISAAC, Leavenworth, Kansas.—Specimens of wood.
ZALLÉE, JOHN C., St. Louis, Missouri.—Clothing.

SUMMARY.

Grand prizes	5
Artist's medal	1
Gold medals	18
Silver medals	76
Bronze medals	98
Honorable mentions	93
Total awards	291

CHEVALIER OF THE LEGION OF HONOR.

By a decree of the Emperor the following gentlemen were created
Chevaliers of the Imperial Order of the Legion of Honor of France:
  C. H. McCORMICK, Chicago, Illinois.
  WALTER A. WOOD, Hoosick Falls, New York.
  CHICKERING, C. F., New York.
  ELIAS HOWE, Jr.

PARIS UNIVERSAL EXPOSITION, 1867.
REPORTS OF THE UNITED STATES COMMISSIONERS.

REPORT

ON

THE FINE ARTS.

BY

FRANK LESLIE,

UNITED STATES COMMISSIONER.

WASHINGTON:
GOVERNMENT PRINTING OFFICE.
1868.

CONTENTS.

REPORT ON THE FINE ARTS.

EXTENT AND CHARACTER OF THE EXHIBITION.

THE EXPOSITION BUILDING.

The building in which the Universal Exposition of 1867 was held in Paris was singularly deficient in architectural display and merit. It can, perhaps, be best described, in homely phrase, as a series of vast sheds ranged concentrically around an open oval court or garden, and intersected, at regular intervals, by avenues radiating from the central area to the circumference. Or, it may be compared to a Roman amphitheatre, by which, it is possible, it was suggested, with a garden for the arena, and radiating passages answering to the vomitories.

There was consequently nothing salient about the building; no striking mass standing out against the sky to mark the spot where the industry of the nations of the earth was gathered, nor lofty façade to awe or impress the visitor. Built on curved lines, the interior was equally without those grand vistas and imposing effects which might have been obtained in a rectangular structure of equal proportions.

Yet for many, if not all the practical purposes and results of such an Exposition, the plan and arrangements of the building could hardly be surpassed. It admitted of the classification of the articles exhibited, not only in respect of their character, but their nationality. Each gallery or zone was set apart to a specific group or class of art or manufacture. The larger products, such as machinery and raw materials, bulky and requiring most room, occupied the larger outer galleries, while the products of the liberal and fine arts found the narrower areas of the inner ellipses sufficiently roomy for their exposition. Thus the visitor interested in machinery had only to make the circuit of the outer gallery to review in succession the achievements of each nation in that department. Or, if devoted more especially to the fine arts, he had only to make the circuit of the gallery dedicated to them. On the other hand, if desirous of studying the collective exhibition of any single nation, he could do so by following the radiating avenues of the edifice, which cut it up like the folds of a fan, one or more folds being assigned to each nation, according to its requirements or the extent of its display.

CLASSIFICATION.

Articles and objects exhibited under the classification of Group I occupied the interior galleries, and consisted of five classes, viz:

Class 1.—Paintings in oil.

Class 2.—Other paintings and drawings.
Class 3.—Sculpture, die-sinking, stone and cameo engraving.
Class 4.—Architectural designs and models.
Class 5.—Engraving and lithography.

The space assigned to this group, especially in respect of what are generally denominated the "fine arts," (painting and sculpture,) was well filled, nearly every country represented at all in the Exposition fully occupying the room conceded to it. A few countries, Belgium, Switzerland, Holland, and Bavaria, finding their space in the main edifice inadequate to what they considered a fair exposition of their paintings, erected "annexes" or supplementary buildings for that purpose in the Park, which were better adapted for showing the pictures to advantage than the main structure.

COUNTRIES REPRESENTED AND AWARDS.

In the department of painting, the following countries were represented and received prizes in the proportions expressed in the subjoined table:

Countries.	No. of pictures.	No. of artists.	Prizes awarded.			
			Grand.	First.	Second.	Third.
France	626	333	4	8	10	10
Algeria	1	1				
Holland	179	77			1	1
Belgium	186	72	1	2	1	
Prussia and Northern Germany	98	67	1		1	1
Hesse	2	1				
Bavaria	211	112	1	2		2
Baden	19	19				
Wurtemburg	11	8				
Austria	89	56		1	1	1
Switzerland	112	58			1	
Spain	42	35		1	1	2
Portugal	12	12				
Greece	14	14				
Denmark	29	19				
Sweden	54	28				
Norway	45	25			5	1
Russia	63	39				1
Italy	51	42	1		1	2
Rome	25	14				
United States	75	40			1	
Turkey	7	3				
Republics of South America	3	2				
Brazil	3	2				
Great Britain	156	124		1	1	2
Total	2,004	1,103	8	15	20	24

The jury on paintings and drawings consisted of 25 members—12 from France and 13 from all other countries, as follows:

France.—Bida, Cabanel, Français, Fromentin, Gérôme, Meissonier,

Pils, T. Rousseau, Marquis Maiton, F. Reisch, Paul St. Victor, and Count Welles de Lavalette.

Belgium.—De Lavelaye.
Holland.—T. Wittening.
Prussia.—E. Magnus.
Bavaria.—Herschlet.
Austria.—Engerth.
Switzerland.—Glevre.
Spain.—Benito Soriano y Murillo.
Sweden.—De Dardel.
Italy.—Morelli and Bertani.
England.—Lord Hardinge and Spencer Cowper.
United States.—W. J. Hoppin.

Of the 12 French jurors, eight were painters and competitors for prizes. Of the members of the jury not French, five were artists, and three of them competitors for prizes.

There were, in all, 67 prizes, viz: 8 grand medals; 15 first prizes; 20 second prizes; 24 third prizes.

Of the 8 grand medals, 4 were awarded to France, namely to Meissonier, Gérôme, Rousseau, and Cabanel, all of whom were members of the jury.

Of the 15 first prizes, 8 were awarded to France, (4 to the four French artists on the jury not obtaining a grand medal, viz: Pils, Fromentin, Bida, and Français.)

France had 333 exhibitors out of 1,103, and secured 32 out of the 67 awards.

In the department of sculpture, out of 36 prizes, 23 were awarded to France; 5 to Italy; 2 to Prussia; 2 to Spain, and 1 each to Greece, Switzerland, Belgium, and Great Britain.

THE FINE ART DEPARTMENT NOT COMPETITIVE.

In extent, the exhibition of paintings was one of the largest ever known, but it has very justly been remarked of it that it could hardly be considered as a competition, "which can only be fair when all parties are equally well represented, and enter the lists with the intention of competing, and with a careful selection of pictures by their ablest painters."

France had every inducement not only to be well but perfectly represented in the exhibition, and she had furthermore the facilities for being so represented. She had all the advantages of proximity, all the stimulus of glory and gain, and if these were insufficient to call out and display her treasures in art, there existed behind an authority capable of achieving things much more difficult. Besides, she had, in the department of painting, and in that alone, a committee of inspection, composed of men of recognized if not infallible taste, to determine what pictures should be exhibited. In all other departments the meanest and most sordid spirit prevailed toward native (French) exhibitors, and a narrow and offensive policy characterized the management of the whole affair. The privilege of placarding on the enclosure was sold for 650,000 francs; the privilege of placing chairs in the structure was sold for 70,000 francs; the right of taking photographs and of making drawings was also sold, and the visitor who endeavored to assist his memory by making a sketch of any object, however trifling, was liable to arrest. Every French exhibitor was obliged to hire the space, horizontal and vertical, that he occupied, at rates varying from 11 to 1,000 francs the square metre. In this space he might exhibit almost anything he chose, with little or no regard to its quality or merits, and without interference on the part of the managers.

But in the department of painting, as already said, space was free, and a careful criticism and sound judgment were exercised, with excellent results.

Some other countries besides France, Belgium and Russia, for instance, seem to have had a competent organization sufficiently early to exercise some direction in the choice of objects that were proffered to be exhibited as evidences of the art and industry of their people. Most European sovereigns are munificent patrons of art, and have under their control, outside of their own collections, vast public galleries, containing the best productions of modern art. From these, and the galleries of private collectors proud of the skill of their countrymen and ambitious of national *éclat*, it was comparatively easy to select a sufficient number of good paintings to make the national exhibit respectable, if not competitive.

THE EXHIBITION MADE BY VARIOUS COUNTRIES.

THE AMERICAN GALLERY.

These and the following remarks are not intended to deprecate public judgment as regards the art exhibition of the United States in Paris, which received so slight a recognition in the distribution of awards, but to show that circumstances did not permit of the United States entering as an art competitor in the Exhibition. Every picture sent from here should have had placed over it *" hors du concours."* And this for many reasons.

In the first place, the action of Congress, as regards the Exhibition, was so tardy that, almost up to the moment when all entries were to be closed, it was doubtful if any attempt at a national exhibit would be made. The little that was done was in an informal way, and even when the national commissioners were authorized and appointed, their instructions did not warrant an exercise of their functions until the opening of the Exhibition in Paris. As a consequence, they were unable to render that aid in the organization of the American exhibition here which they would have willingly extended.

The arrangements for securing works to be sent on as types of American art were left to the overtasked hands of the forwarding agent of the government in New York, who appointed a committee consisting of local patrons of art and dealers in pictures. There were no artists, or recognized authorities on art matters on the committee, and the selection was made chiefly from the galleries or sales-rooms of the members of the committee themselves. Some of these selections were good, but most of them, although by artists of acknowledged merit, were not their latest or best productions.

Here, it may be said broadly, there are no galleries of national art, no public collection of pictures that have stood the test of exhibition and criticism, from which a selection of either original or characteristic paintings could be made. What paintings we have are in the hands of individuals, scattered over a country as large as all Europe, or else in the hands of the artists. Now, few owners of pictures of recognized merit were ever asked to contribute towards making up a competitive exhibition of American art in Paris, and even among those who were applied to, few were willing to submit to the annoyance of having their pictures removed, or to incur the risks of having them sent so far from home, with no better guarantee than the word of a committee informally organized, and invested with no responsible authority.

Notwithstanding all disadvantages, seventy-five pictures, by thirty-eight artists, were sent forward from the United States and placed in the Exposition. Of this number at least one-third should not under any pretence or influence have been admitted to a place. It is doubtful if they could have obtained room in any local exhibition where ordinary discrimination is exercised in the choice of pictures. Now, we have upwards of four hundred painters, members of the different Academies of Design in New York, Boston, Philadelphia, and other cities, and it is idle to pretend that the place of the 25 mediocre or utterly worthless pictures could not have been supplied by at least creditable works of art. Many such works were accessible. Among them may be mentioned with credit the fine pictures by Bradford, drawn after careful study among the icebergs and on the coast and among the natives of Labrador. One of these, offered by the artist, he was obliged himself to exhibit in Paris, where it speedily found a sale, while the eye of the visitor to the Exhibition was offended by, in one instance certainly, no less than four so-called works of art, from a single unpracticed and obscure hand.

The American collection occupied one end of the British gallery, and the walls of the *Avenue d'Afrique* dividing this gallery from the Italian. This passage was constantly crowded, so that the lower ranges or tiers of pictures could seldom be seen, or if at all at a great disadvantage. Thus Gifford's "Twilight on Mount Hunter," Hubbard's "View of the Adirondacks," and McEntee's "Virginia in 1863," were hung in very bad light, while works far inferior had prominent places in the gallery itself.

The relative proportion of space occupied by us in the fine art department is shown by the shaded portion in the outer circle of the sub-joined diagram, which represents the two inner galleries of the building:

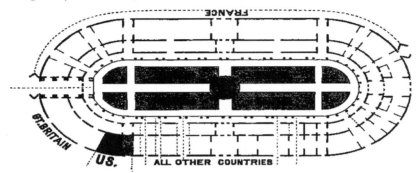

PLAN OF THE INNER GALLERIES OF THE EXPOSITION BUILDING.

Not only was the American exhibit of paintings by no means an exhibition of the various styles of American artists, but it was equally deficient as a type or representative of American art in 1867. Few of the pictures had a distinctive, still less a distinctively American character, except Johnson's well-known and justly appreciated "Kentucky Home," which compared favorably with the best European works of similar character, and attracted much attention from students. Some other small pictures by the same competent artist could hardly be called finished, and might, judiciously, have been left out of the exhibition. It is to be regretted that of character, or *genre*, pictures we had so few specimens, since, in this department, always the most popular, we might have achieved a real distinction.

The department of painting in which the United States may lay claim to highest excellence is undoubtedly landscape, and as was to be expected the largest proportion of pictures in the American gallery were landscapes—28 in all. But these were inadequate representatives of the genius of our painters in this, their favorite branch of art. They were not characteristic; for, with the exception of a single work by Bierstadt with his broad effects, one by Church, accurately studied and well manipulated, and one each by Gignoux and Gifford, they represented no important scene or combination of scenery in the United States, and might be taken as presenting views in almost any other country. Our autumn and winter scenery found no proper representation, although in depicting these we have artists of real merit. It may be said generally that our landscapes are brighter, more cheerful and pleasing than those of European artists—a not unnatural result considering our clearer and more brilliant sky and atmosphere. Our artists, as a whole, have to deal with bolder scenery, and are consequently often more effective in their results. The dull skies, long twilights and generally tame outlines of nature in European countries are reproduced in pictures also dull, leaden and monotonous, but nevertheless carefully manipulated. Yet, with all our natural advantages of subject and general success in landscape painting, it is humiliating to say that we did not rank any higher in the Exposition than Prussia, Sweden, and Switzerland.

In saying this it is not meant to be understood that the Exposition, as a whole, presented many remarkable landscapes. A number might be called good, but few could be pronounced excellent.

And here it may be observed, in parenthesis, as a matter worth remembering by aspiring artists, that landscapes have a less general or popular appeal than many other classes of paintings. Authors of the best works in this department, not in the American gallery alone, but in every other, would be astonished to see how indifferently their productions were passed over by the thousands who wandered through the galleries of the great Exhibition, while figure subjects and representations of active or historical scenes, never missed attracting a crowd of gazers, if not of critics.

The natural scenery of our country, its variety and kaleidoscopic effects cannot be surpassed. Italian sunsets and Alpine scenery have become conventional in Europe as synonyms artistically of the tropics and of grandeur in vale and mountain; and as contrasted with the dullness of English skies, and the puny altitudes of Wales and Scotland, they may be justly regarded as beautiful and grand. But Washington is in a lower latitude than Rome, and Florida is parallel with the Desert of Sahara.

Every aspect which nature exhibits, from the torrid heats of Algeria to the bitter cold of Norway, is to be found in our own country, on every scale of extent and grandeur. Our Atlantic seaboard stretches over 3,000 miles, and our Pacific line from the headlands of California to the pole. Our field of art, like our area of development, is almost illimitable, and it is no fault of ours if the wilderness in one instance be uncultivated, or in the other nature's wonderful combinations unportrayed on the canvas. It has taken almost 500 years to rear the unfinished Duomo of Milan to its present proportions. It was commenced 105 years before the discovery of America by Columbus, and yet scarcely 100 years have elapsed since the United States had a being.

Nevertheless, as already said, we have an art material that ought to inspire and develop the native artist, whatever his tastes or talents, whether as a painter of lake, river, marine, and sea-shore subjects, of mountain, prairie or forest scenery, or of the thousand striking aspects and episodes of busy and adventurous life of which the United States offers so many illustrations. The stormy Atlantic and the placid Pacific, the broad lakes of the North and the shadowy lagoons and *bayous* of the South, the turbid Mississippi, crystal Hudson, swirling and plunging Niagara, St. Lawrence and Columbia, and the Colorado flowing through the deep refts of *plateau* and mountain, are all equally subjects for the poet's pen and the painter's pencil. The severe landscapes of Maine, with steel-colored lakes framed in by dark evergreens, and reflecting the cold, stern hills, afford abundant scope for a taste gloomy and severe. The rich valleys of the middle States, green with growing crops, golden with ripening grain, or ruddy with autumnal tints, brightened with cities and villages, and streaked with railways and canals ; the smoother expanses of the South, its endless wastes of pines, broad, dreamy cotton plantations, and level horizons of rice fields, its orange and palm trees— these, too, offer their thousands of combinations to the eye of the artist, and their inspirations to his touch.   Our meteoric conditions and pheno- mena are equally varied and grand, and we have the characteristics, accessories and incidents belonging to three great and broadly defined races of men and types of human life and civilization.

We should excel in landscape painting in a degree corresponding with the variety and majesty of our subjects, and with the exceptional favor with which this branch of art is regarded in our country.   But our artists must be less timid, and catch more of the boldness and vigor which made Norway and Sweden, and even Russia, conspicuous in the Paris Exhibi- tion, and enabled them to take rank as our superiors in landscape paint- ing.

As shown in the table in the appendix, but one picture in the Ameri- can gallery was honored with an award, namely, Church's "Niagara."[1] This well-known picture has an established American, and a considera- ble English reputation, and is a faithful and effective rendering of nature. The second and perhaps more ambitious picture exhibited by the same artist, "The Rainy Season in the Tropics," received not unmerited criti- cism for the dazzling glow of its rainbow, a meretricious feature which blinds the eye to the fine effects of cliff and mountain, which constitute the chief merit of the picture.   The next largest, and perhaps in all respects the most conspicuous picture in the American gallery, was Bierstadt's "Rocky Mountains."   In arrangement of light and shadow, and in the rendering of the water, its purity and depth, this picture was probably unsurpassed by any in the entire exhibition.   And it derived signal advantage from the introduction of an element, too often neglect- ed, even when admissible in a landscape, viz: life in the foreground. The introduction of a camp of explorers, with Indians, etc., is not only effective, but appropriate, and gives a living interest to the picture with- out detracting from the silent majesty of the natural features which it was the great object of the painter to portray.   Had the solitary award made to the United States been left to the suffrages of the mass of the lovers and appreciators of art who visited our gallery, it is not improbable that that doubtful honor might have been conferred on "The Rocky mountains."

"Mount Washington," by Gignoux, is a good, effective picture, but by no means one of the best productions of that artist, and lost much of its real effect by being badly hung in the gallery.   Had this prolific painter been consulted in the matter, the American gallery might have been

---

[1] The artist's medal with 500 francs in gold.

greatly enriched, and the credit of American art much elevated. Gifford had two very excellent pictures in the Exhibition, but one of them was hung, together with Hubbard's good and well manipulated "View on the Adirondacks," in the dim passage called *Avenue d'Afrique*, where it was difficult to see it at all. "Lake George in Autumn," by Kensett, a carefully studied piece, but deficient in force in the foreground, attracted much attention and was well appreciated. "The Symbol," by Durand, was generally regarded as an effective picture by competent foreign critics, as was also "Virginia in 1863," by Mac Entee. "Autumn in the Woods," by the artist last named, is by no means one of his best works.

These were the principal landscapes exhibited; the remainder were either mediocre or absolutely poor, and if their place could not have been supplied with better works, they might, for the credit of American art, have been omitted from the gallery altogether.

The second department of painting, in which American artists are presumed to excel, and to which it is alleged that nearly all are obliged to resort as a means of support, is that of portraiture. There were 10 portraits exhibited in the American gallery, of which three only were creditable specimens of art, while the remainder ranged between bare mediocrity and absolute caricature. None except those of Elliott and Baker could be favorably compared with works of the same character in the various galleries, and even these did not rank with some in the Belgian and Russian exhibitions.

In figure and historical composition, the highest branch of painting, the American department was singularly deficient. There were but four or five pictures of this class of any pretension, and these were overshadowed by greatly superior works in almost every gallery except the Roman. Figure compositions, to be effective, require scope of canvas, and the figures themselves should be of size approximating to that of life. It is only on this scale that genius in composition can fairly exhibit itself. Single historical pictures in the French gallery covered an area almost equal to that occupied by all our pictures combined.

The "Old Kentucky Home," by Johnson, and "Marie Stuart hearing Mass," by Leutze, were probably the best pictures of the class under notice in the American gallery. The "Lear and Cordelia" of May has some of the essential elements of a figure picture, but is roughly manipulated. The largest picture of this class was "The Republican Court" of Huntington, which, however faithful in respect of portraiture and costume, illustrates no event, and tells no story; it is nevertheless carefully manipulated, but weak in color, lacks force and is more a costume picture than a historical composition.

"Lady Jane Gray giving her Tablets to the Governor of the Tower of London," by May, was accepted as very fair in composition, but crude and roughly handled.

We did not exhibit a single animal picture, which is unfortunate for our reputation, since we have very competent animal painters whose works would have taken a high place in the Exposition. Among animals we have a peculiarity in the bison, so little known in Europe; and we have also artists who, like Hays, have spent years in the Far West in the study of its habits and peculiarities, whose large pictures, truthful in drawing and color, bold and effective, would have been a feature in our collection, and compared favorably with any corresponding works in any of the competing galleries.

Our exhibit of marine pictures also was scant, and by no means representative; yet among our artists at home there are several who have made marine painting a specialty, and whose works would be an honor to any country. Their pictures, large and effective, could not have failed to arrest attention and command admiration. Of those exhibited two were by Kensett, very good little pictures, very well worked up, but not effective, and one, a very promising work, by Dix. This department might have received valuable additions from the easels of De Hass, Hamilton, and others.

We did not present a single strictly *genre* picture, although here, as in animal and marine painting, the United States is not deficient in very competent artists such as Guy, Brown, and Irving, whose unexhibited works, in this branch of art, are equal to many of the same class that were conspicuous in the Exposition. In their manipulation they may not perhaps come up to the perfection of some of the *genre* pictures in the French and Belgian galleries. It would indeed be difficult to equal and almost impossible to excel the touch and handling of Meissonier. But *genre* painters seldom grapple with much action in their composition. Thus Willems and Baugniet, whose works are numerous in the Belgian gallery, do not show any great variety in choice of subjects, nor any great amount of genius; but their representations of rich draperies and fine satins are exceedingly beautiful. Their works may be described as pleasing pictures of modern, fashionably dressed ladies, showing, however, no action and little invention.

Of the nude and classical, and in drawing and color, few pictures in the entire exhibition surpassed "The Apple of Discord," by Gray. Wier's "Cannon Foundry at West Point" was equally unique, striking, and faithfully studied, and was among the few pictures in the American gallery that commanded constant attention from visitors.

The American gallery was also deficient in pictures of still-life, fruits, and flowers, and in miniatures; although in all these branches of painting it is certain we could have made a good exhibition.

It would probably be deemed to be outside the scope of a report like this, as well as a difficult and dangerous task, to undertake to analyze and point out the defects in American painting generally, or to attempt to indicate the causes why American art has not achieved for itself a higher position. It is true that it is yet young, and it may be true that it does not find adequate and constant support and encouragement; that there are no schools of art, and that we are without public galleries in which accredited works of competent masters may be consulted and studied. But perhaps the most conspicuous cause of our acknowledged deficiency is the absence of sound and judicious criticism. We are accustomed to pay too much homage and deference to artists and the works of artists who succeeded in arresting the public attention in the infancy of art in the United States, or whose pictures exhibited singly, surrounded by green cloth and other adventitious aids, obtain undeserved and sweeping commendations from injudicious friends, who discover beauties that do not exist, and ignore or overlook defects that are real. It is only when such works are put in fair competition with other pictures, without any artificial and meretricious surroundings, that their relative merit appears, and their deficiencies become conspicuous. Nothing could exercise a more wholesome influence in American art than the necessity of our paintings appearing as a whole, year by year and side by side, with the annual productions of France, Russia, or even Switzerland. Not that such an exhibition would not be without a degree of credit, but because it would rapidly destroy the prevailing system of indiscriminate praise, by which artists are led into conceit and a consequent neglect of that study and attention by which alone their real powers can be brought out and enlarged.

All great painters must have produced works in the early periods of their careers, or at inauspicious periods, which they would gladly cancel if they had the chance, but which are nevertheless exhibited to us as works stamped, and correctly, with their great names, yet utterly beneath their genius, and which they would feel humiliated to have placed in competition with the productions of minor artists.

As regards the Paris Exhibition it is undoubtedly true, that had the American artists whose works appeared there been consulted in the premises, a number of them would have objected strongly to the figure they were compelled to make. They would have refused to appear at all, or insisted that their matured works, the results of their later taste, judgment, study and skill, should represent them.

The majority of our painters are landscape artists, and such they must probably long remain, unless they fall into the easy but wonderful style of reproducing lay figures, of which West and Alliston were masters, under the delusion that these are historical compositions. But in whatever direction their own or the public taste may lead them, there can never be serious dispute of the proposition that figure painting is the highest effort of art. In this the old masters of the Italian school excelled, and in this they established that pre-eminence they have held and seem destined to hold. Precisely in this, broadly speaking, American art is most deficient. It is possible that if our painters were able to study the human subject as easily and readily as they are able to study our undoubtedly grand and varied scenery, and if public taste were educated in this department up to the same critical standard that it possesses in landscape, figure painting might receive a stimulus and obtain a high, if not the first, position in American art.   Figure painting, outside of composition, and apart from color and effect, requires much careful and laborious study for outline alone.   To this point the old masters directed their first attention, and to this the leaders of art of our own time have given their best and most earnest efforts.   If the Dusseldorf school has gained special distinction, it is due to the care the artists who compose it have bestowed in this direction.   The tendency with us, unhappily, is to cover up and disguise bad drawing by color, and avoid grappling with the difficulties of the figure by a resort to what are called "effects."   But there is no royal road to excellence in any department, least of all in figure painting.   Genius is useless and often vicious unless directed with judgment, and unless it submits itself to a sound element- ary education it can never find true scope and expression.   The works of the earnest student alone can stand the severe tests of time.   The eagle cannot rise in his flight unless the earth from which he is to spring be firm under his feet.

Our best figure artists are unquestionably those who have gone through the very careful and conscientious training of the Dusseldorf school.

Among our younger artists there are some who show much invention and undoubtedly possess real genius, but, from lack of good art education, never rise above mediocrity.   In this consists the viciousness of our school of art, if we can claim to have a school, of which the character- istics are lack of study, haste, carelessness, and ambition for easy, mere- tricious effects.   But the greatest evil and drawback of all is to be found in want of proper tutorship in drawing, and default of patience.   Let no one suppose the orations of Cicero were spotaneous bursts of eloquence.

Turning from our own meagre and unsatisfactory gallery, a few words may not be inappropriate relative to the others, among which that of France was the only one that may be said to have been in any sense complete; that is to say, the one most judiciously selected, and suf- ficiently large to present every phase of the painter's art.

THE FRENCH GALLERY.

As already stated, the French gallery consisted of 626 pictures by 333 artists.   Many of the latter were represented by a number of pictures illustrating their various styles and capacities.   Thus Gérôme had thir- teen pictures, all highly dramatic and powerful; Bouguereau had ten; Meissonier, fourteen, etc.   Those thus honored were of course the leading artists of France, and the selections were made from their best works. In our own gallery, on the other hand, the largest exhibitor was a Balti- more artist, whose productions scarcely rose to the level of caricature in drawing, or the dignity of daub in color.

As already said, every department of pictorial art was adequately rep- resented; figure and historical compositions on a scale great enough to admit of a large treatment of groups and incidents, covering in some instances as many as sixty square yards of canvas; marine, landscape, portrait and *genre* works, and still-life, all in sufficient proportions. In fact the French gallery may be said to have been rather aggressive and monopolizing towards the galleries of nations who from proximity and artistic taste and skill were able and willing to enter the lists as competitors in art. It occupied considerably more than one-third of the space assigned to the exhibition of paintings, and to a certain extent compelled Holland, Bavaria, Belgium, and Switzerland to seek accommodation in the Park for what they deemed a proper exhibition of their works.

France has peculiar facilities for getting together, at any time, a collection of good pictures, due in great part to her extended system of art culture and art support. Besides maintaining art schools, like that in Rome, the nation is a liberal purchaser from the works exhibited annually at the Academic des Beaux Arts in Paris, and these works are freely distributed among the royal residences and in the metropolitan and provincial galleries, which are always free, and all this with the avowed purpose and real effect of stimulating artistic aspirations and forming a healthful popular artistic taste. Artists struggling to establish a reputation have the stimulus of knowing that a high position once gained, they are certain of orders from the State which will give them profit as well as fame, and lead to other honors and recognitions which probably appeal more to the French mind than to any other. If they possess good or superior capacities for historical composition, French painters feel sure of identifying themselves with the martial history of their country, through illustrations of events that are supposed to have contributed to the "gloire de la France." Some ambitious American artists have aspired to something of this sort, and most of us have recollections of very pretentious attempts at illustrating the battles and "victories" of the Mexican war, as well as some of the incidents and events of the revolutionary war and that of 1812. To know how absolutely abject these were, or are, it would only be necessary to place them side by side with the works of Pils and Yvon.

Owing to the practice of distributing first-class works among the provinces, part of that unhappy tendency to centralization of everything in the way of art, science, and literature in Paris is prevented, and local students obtain the advantages to be derived from easy access to paintings, which, if not exactly models, offer abundant and useful hints and suggestions in drawing, coloring, and effect. Thus we find the famous painting of Paul Delaroche, "Cromwell viewing the dead body of Charles the First," in the small but excellent gallery of the little provincial town of Nismes.

The French gallery in the Exposition was greatly enriched by the best pictures from the walls of the annual French exhibition in the old *Palais d'Industrie* in the Champs Elysées which closed on the 5th of June, and which, although embracing only the national competitive works of the year, numbered not less than 1,572 pictures, some of them of great merit.

2 F A

Incidentally, it may be mentioned, that in this annual exhibition there was a very large proportion of figure subjects, many of them of nude figures, and illustrating the importance that in France more and more attaches to figure drawing. These were generally of life size.

As an illustration of the extent to which the French government is the patron of art, and how far its powerful influence and resources were thrown into the scale of competition, it is only necessary to say that, out of the 625 pictures in the French gallery, 252 (almost half) were contributed by the state.

It will not be out of place to repeat here that, out of the 67 prizes of all classes awarded to painters, France secured 32, viz, 4 grand prizes, 8 first, 10 second, and 10 third-class prizes, to artists whose names appear in the appendix.

The best pictures exhibited, prior to the addition from the annual exhibition, (and which were *hors de concours*, or not competitive,) were: I. L. H. Bellange, "The Parting Salute," a scene in the trenches before Sebastopol; I. A. Breton, "Return of the Gleaners;" G. Brion, "Pilgrims of St. Odile, Alsatia;" B. Desgoff, flowers and objects of art, (Nos. 210, 213 of catalogue,) wonderful in delicacy and accuracy of manipulation; Madame E. Escallier, flowers, (Nos. 243, 244;) T. Robert-Fleury, "Warsaw, April 8, 1861;" J. L. Gérôme, "Phryne before the Tribunal," and "Duel after Masked Ball;" T. Gide, "Rehearsal of a Musical Mass;" J. F. Gigoux, "Napoleon on the day of Austerlitz;" A. A. Herbert, "Rosa Neva at the Fountain;" C. F. Jalabert, "Christ Walking on the Sea;" G. Jundt, "Returning from the Agricultural Show;" J. L. E. Meissonier, "Campaign in France, 1814," "Information," and "General Desaix at the Army of the Rhone and Moselle;" A. Prignon, two female portraits, with reflected light, admirably managed and effective; J. V. A. Rigo, "General Canrobert Visiting the Trenches;" P. L. Roux, "Rembrandt's Studio;" A. Yvon, "Taking of the Malakoff," and "Convoy of Wounded Soldiers."

The pictures that commanded most attention in the entire Exhibition were undoubtedly those of Yvon and Pils. This distinction was, in part, due to their mammoth size, but mainly to their unquestionable great excellence. There are few examples extant of equal vigor and truth of drawing, combined with breadth of effect and naturalness of tone. In contrast with these, but almost equally popular with the great public jury, were the celebrated *genre* pictures of Meissonier, and the groups of Gérôme—exquisite in every way, perfect in drawing, fine in color, and most carefully manipulated.

The nude figure pictures in the Exposition, singularly enough, were not equal to those in the annual exhibition, in which "Phryne before the Tribunal," by Boutibonne, "The Syren," by Belly, were of the very first-class of such works, and which merited the distinction they subsequently received of a place in the Exhibition. By the terms of their agreement, the French students in Rome are bound to send specimens of their progress, to be submitted to the directors of the Academy in Paris, and these specimens must possess the utmost accuracy in designing the naked human form; but as the taste for exclusively classical forms no longer exists, they are led to comply with the rules by representations of figures more congenial to modern French notions. Thus we no longer find studies of Achilles or Romulus, or other ancient heroes, but their places are occupied by paintings of Venus, nymphs, and goats, Andomedas, etc.

The paintings thus produced are decidedly more harmonious in coloring than those of "the Italian masters," and it is a relief to the eye to turn from the nude figures of the latter, with their harsh, incongruous back-grounds, to the fresh living tints, transparent shadows, and delicate back-grounds of the later works of this class, which harmonize perfectly with the flesh colors. Some of these have delicate white or gray for high lights of the drapery, while blueish grays and tender greens appear in others, bringing out, in the highest degree, the charm of harmony in coloring.

A strong tendency of French art, in sculpture as well as painting, is towards the romantic rather than the classic style. Religious paintings are now rarely produced, and only to fill special orders. As very truthfully observed by a competent English critic: "France has a school of painters in the best sense of the word, which is different from that in which we employ it in writing of Germany, a people which is all school and little more. As art is anti-scholastic to the core, and hates a common standard, so in the most varied school we find its wealthiest development. In France, better than anywhere, students learn the executive of painting; yet nowhere is art so seldom sacrificed for the sake of training, or are the results of training so obvious. The executive standard is generally high among the French, because their professional tone is high, and nothing short of peculiar power is received in place of good workmanship. With us bad workmanship need not be compensated by peculiar ability; our professional tone is so low that bad handicraft and want of purpose often appear in the best places on exhibition walls."

In landscape, French art seems to have taken a new departure, and to have made recent and rapid progress. Although this class of pictures was not numerous, yet most of the specimens were good and some of them excellent. They are distinguished less for care in manipulation than for broad effects, affording a hint which our own artists in this department might accept with advantage. This tendency is perhaps carried too far by the French, who might, on the other hand, gain much by a closer attention to finish. The happy mean applies here, as in all things else, and strong features pushed to exaggeration are not necessarily powerful, but oftener caricatures.

The high rank of French artists in animal painting is universally conceded. The works of Rosa Bonheur are too well known to need remark; and although she was awarded a second prize, it was rather as a matter of course than for any extraordinary excellence in the 10 pictures exhibited under her name, and of which the best was a "Razzia in Scotland."

There were some very fine works by Fromentin and Troyon, all carefully studied and vigorously handled.   American and English artists too often paint animals as mere accessories in quiet dreamy landscapes, and the animals themselves appear only as contemplative, dozing creatures, apparently indisposed to movement, if not incapable of it.   French artists, on the other hand, give action to their animals; one almost hears the neigh and tramp of the burly Normandy stallion; there is life in the ox, and even the sheep hurry over the heath in search of the green grass-plats, or huddle struggling to the fold before the sharp bark of the shep-herd's dog.

Among the most marvellous paintings of still-life in the Exposition are five pictures by Desgoffes, two of which belong to the Empress.   In one of them are an ewer, silver-gilt, (style of the 16th century,) a Christ in bloodstone, bust of the Virgin in rock crystal, door-knocker, statuette in box-wood by Jean de Bologne, enamelled vase, etc., grouped together with consummate skill and painted with Rembrandtish effect.   For draw-ing, management of light and shade, minute manipulation, this work is unapproachable by any other of its kind in the Exhibition or out of it. The same may be said with almost equal truth of some flower pieces by the same artist.

Although portraits were not numerous in the French gallery, yet they were almost uniformly good.   In this branch of art criticism resolves itself into few words—the French lead the world.   Portraiture is not a trade in France, it is a profession.

Reference has been once or twice made to the annual French exhibition of painting and sculpture, which was quite as largely frequented by lovers of the fine arts as the galleries of the Exposition itself, and became so associated with the latter as to be really regarded as a part of it.   No doubt the general impression of French art left upon the mind of the visitor to Paris during the summer of 1867 was quite as much due to the Annual as to the Grand Exposition.   For this reason a few words in reference to it may not be inappropriate.

As already stated, this exhibition took place in the old *Palais d'Indus-trie* in the *Champs de Elysée*.   As its name implies, this exhibition is only for works of art of the year, and no picture is a second time exhibited. The Exhibition of 1867 consisted of 2,166 paintings and 382 pieces of sculpture, besides water-color drawings, lithographs, chromo-lithographs, photographs, and engravings on wood, etc., and it may be said broadly, but with truth, that out of the 2,116 paintings there were not 20 that would be pronounced positively bad, while hundreds were of a very high order of excellence.   In fact, a majority of them were good; not merely passable and creditable, but positively good.   In looking through this vast collection, the eye failed to discover a single branch of painting that was not cleverly represented.   The area of canvas covered was quite amazing.   A single picture by Gustave Doré, "The Tapis Vert," was not less than 60 feet by 25 feet, and was by no means the only picture of that size. Two such pictures cover more canvas than all the pictures exhibited annually in New York.

Doré is one of the most active artists of the age. Not only has he made more designs on wood than probably any dozen other living artists, but has found time to design and paint, among others, the picture just referred to, a work of much study, true to life, filled with character, from the dashing belle to the sturdy English baron, and displaying many of the strongest points and features of modern French art.

It would be almost impossible to notice separately the best pictures in the annual Exposition. Among them, however, may be mentioned a beautiful Psyche, by Duval Amaury; "The Taking of the Fort of San Xavier de Puebla," a grand picture, by Beauce, excellent in design and action, perfect in drawing, and with the tone and touch of a master; "The Syren," by Belly; "Ships on the High Seas," by Bonnetter; "Death of Sappho," and "Idylle," by Bertrand; "Le Jour de la Pentecôte," by Bischoff, are all good pictures. The Phryne of Boutibonne was probably the best of all the nude figures, comparing favorably with that of Gérôme in the Grand Exhibition. Great praise is also due to "The Council of Three in Venice," by Bronnikoff; "Le Lendemain," by Broune; "Portraits," by Madame Chatillon, etc., etc.

THE BRITISH GALLERY.

Great Britain exhibited 156 pictures by 124 artists, of whom only 26 were represented by more than one specimen. All of the pictures, except three portraits, viz., of the Queen, the late Prince Albert, and the president of the Royal Academy, were contributed by art connoisseurs or artists themselves. The United States collection of pictures occupied one extremity of the gallery assigned to the English, who took commendable pains in its preparation, toning the walls, covering the floor with matting, and providing a liberal allowance of seats for the convenience of visitors. The central part of the gallery was in part occupied with screens, on which were displayed a good collection of paintings in water-colors, some of them of unexampled size. In fact, England was the only country that made this branch of the fine arts, so capable of fine effects, a distinctive and prominent feature in its collection, and well deserved the only prize awarded for a water-color drawing. It went to Mr. F. Walker. Most of the specimens were very fine studies, and some of the interiors particularly good in manipulation and effect. Without going into particulars, it may be said, generally, that in this department England was unapproached by any other country represented in the Exhibition.

The British exhibition of oil paintings, although less obviously betraying the influences of the French school, which pervade all continental art, and consequently more distinctive, was not particularly excellent. It consisted chiefly of large cabinet pictures of domestic and rural scenes, with very little incident, lacking, also, historical and figure compositions, and had no marines of importance. The portraits were very excellent, being in some instances something more than the mere outline of the features or the figure in repose. The accessories of pets or companions, by Landseer, were occasionally introduced, with good effect, giving something of life and reality to the portraits, which were generally carefully and conscientiously manipulated.

"La Gloria, a Spanish Wake," by the late J. Phillips, was, without doubt, the best work in the British gallery in invention, drawing, coloring, and effect. Next in rank may be mentioned "Baith Fayther and Mither" and "Music hath Charms," by T. R. A. Faed, excellent compositions, full of sentiment, carefully drawn, elaborate and effective in light and shadow. After these may be mentioned: R. Ansdell, "Treading out Corn," (Alhambra,) forcible and effective. J. B. Burgess, "Bravo Toro," an Andalusian bull fight, with fine types of Spanish beauty, and much spirit in the arena, good in drawing, but defective in chiaroscuro and color. F. Goodall, "The Palm Offering." R. B. Martineau, "Last Day in the Old Home," drawn with vigor, full of feeling, but deficient in color. E. Nichol, "Both Puzzled," an excellent picture, quite deserving of the prize (second) which it received. H. O'Neil, "Eastward Ho!" and "A Volunteer." Q. Orchardson, "Christopher Sly," a good design, which received a third prize. H. Willis, "The Death of Chatterton." In Landseer's "Shrew Tamed" that painter's conceded genius and established capacity were well illustrated in the figure of the horse, which is the essential feature in the picture.

Among the best landscapes was Graham's "Spate in the Highlands," a vigorous and effective composition, in which the artist has grappled with the aspects of nature in her wildest mood. It represents a turbulent highland torrent during a storm, the water—stained to the color of amber—rushing down between bold, rugged rocks from mountains scarcely discernible through the rifts of a stormy sky. The picture combines almost all the qualities of a good painter—quick grasp, free, firm drawing, and careful but not excessive manipulation—and may be taken as a model of its class. This was one of the largest pictures in the gallery, and its size gave the artist scope for the free display of his powers, as well as of the large features of a mountain landscape.

There were many other pictures in the English gallery that might be pronounced good. Some were excellent, but none that could be termed great, while there were a few the admission of which in the gallery can only be accounted for on the principle of forcing the most striking contrasts possible. On the whole, the collection was a pleasing one, and had a degree of freshness and independence to be found in none of the others; but it could scarcely claim a high place in an artistic sense. A visit to the annual British exhibition in London was sufficient to satisfy the visitor to both that the display made in Paris was a very fair exposition of the various departments of British painting, and justified the impartiality and judgment of the committee of the Society of Arts, to which the selection of the gallery was confided by the government.

BAVARIA.

Next to France, the largest contributor to the collection of paintings in the Exposition was Bavaria. She sent 211 pictures, by 112 artists, which were exhibited in a separate building, or "annexe," built by the Bavarian commissioners. More than half the number were exhibited by the artists themselves, and, next to France, Bavaria secured the greatest number of prizes, viz., one grand prize, two first prizes, and two third-class prizes—five in all.

The artistic taste and fostering care of King Louis were manifested in various ways in the Exhibition, and the Bavarian paintings were strongly marked with the classic style to which he gave such prominence in Munich. The Bavarian artists, as a rule, evince much invention, and are well grounded in true outline and drawing. At present they appear to be ranging themselves into two camps—the old professors adhering rigidly to the classical; the younger artists conceding much to that school, but refusing to be bound by all of its canons. The first are correct, and almost severe, in drawing, but lack breadth, their light and shade being too diffused, while the latter inclines to the French style, with a constantly increasing tendency, and succeeds in effect and management of light and shade. The end of this divergence is obvious and not distant. Bavarian art will speedily become a reflection of French art, more exact perhaps, and, it may be, more formal.

Exterior fresco painting, which was formerly much encouraged in Munich, where many public buildings are disfigured by elaborate and gigantic works of the older artists, is falling into disuse and fading with the colors which glowed from the walls they were intended to ornament. It is now limited to interior decoration, and commands much attention from the younger or new school of painters.

Among the pictures exhibited were several large ones, 15 by 20 feet. The best of these was "The Benediction of the Flags," by Piloty, professor of the Academy of Fine Arts of Munich, who was awarded a first-class prize. But all these large pictures, although correct in drawing, are severe in style and unsympathetic in character, as they are inharmonious and cold in color.

The best of the modern school of Munich are: F. Adam, "Groups Marching between Solferino and Vollegio, June 24, 1859;" a large excellent figure composition, full of action and marked by breadth and harmony of color. P. Baumgartner, "A Procession Surprised by a Shower." Knude Baade, "Moonlight on the Coast of Norway." F. Bamberger, "View of Gibraltar." J. Brandt, "Chodkiewicz, the great Hetman of Lithuania, fighting against the Turks." G. Closs, "Campagna of Rome." A. Eberle, "Military School during the Thirty Years' War." C. Haefner, "A Coming Storm in the Upper Alps;" the sheep well drawn and well colored. T. Horschelt, "The Russians Storming a Tscherkessian Intrenchment on Mount Gounib;" one of the largest and best of the modern Bavarian school, evincing invention, fine in drawing, coloring, breadth, and expression in the figures.   H. Hobach, "Tasting Wine in Secret."   J. Koekert, "Nuptial Cortege on a Lake in the Bavarian Mountains;" rich in color and effect.   A. Liezenmayer, "Maria Theresa Feeding a Poor Sick Child."   G. Max, "The Martyr."   A. Vischer, "National Dance of the Peasants of Upper Bavaria."   A. Wagner, "A Soldier Saving a Child in a Manœuvre."   R. S. Zimmerman, "A Nuptial Cortège."

Altogether the Bavarian pictures exhibited much careful study and a sound education on the part of the artists, who show also much inventive talent and good powers of design.

Bavaria, exhibited a large number of crayon drawings, chiefly by the pupils of the various art academies of Munich, showing careful study of outline and indicating the severe course of training through which the art pupils of that capital are obliged to pass.   Kaulbach, director of the Academy of Fine Arts of Munich, secured one of the eight grand prizes for his crayon work, "The Epoch of the Reformation," a large picture, 16 by 25 feet.

BELGIUM.

Belgium sent to the Exhibition 186 pictures by 72 artists, which were displayed in a separate building or "annexe," and which obtained four prizes, viz: one grand prize to Leys, one first-class to Willems, another to Stevens, and a second-class to Clays.

The most ambitious pictures were by the artist first named, who exhibited not less than 12 works, some of large size, and for the most part subjects from the stirring period of the struggle with Spain for civil and religious freedom in the 16th century.   They are painted in a mediæval style peculiar to this artist, but hardly consonant with modern notions of art.

The principal features of this gallery were the *genre* pictures of Stevens and Willems.   The first named sent 18 works, many of them hardly more than studies of single female figures dressed in latest fashionable styles, with little expression or sentiment.   The titles of some of these sufficiently indicate their character: "The Lady in Pink," "The Return," "A Duchess," "Miss Fauvette," "Pensive," "The Autumn Flower," etc.   It is not to be denied that a few of the 18 exhibit some invention and a capacity for better if not more profitable things.   They are pleasing and effective in light and shade, and especially in color, but are not so finely manipulated as those of Willems, of which 13 were exhibited, mostly small pictures of simple scenes of domestic life in the 17th century.   They are very broad in effect, and in texture wonderfully true to nature.   It seems unfortunate that artists of such exquisite touch and so good masters of effect should not dedicate their talents to more ambitious subjects.

Verlat is a painter of higher scope and power.   His "Dead Jesus at the Foot of the Cross," is most carefully drawn and well handled. His "Danger" is a bold effective picture, representing a party of peasants driving away wolves from a dead lamb. The animals are exceptionally well drawn, and the whole picture carefully elaborated.

"March of Animals in the Pass of Beni Aïcha," by Chev. C. Tschaggeny, is a good picture. "Sacking a Convent at Antwerp," by Roberts, is a large picture, effective in coloring, but deficient in action, which, from its subject, should be its characteristic. "Lake Lomond" by Rofflaen, a fine, broad picture, carefully manipulated.

Clays sent five pictures of Dutch coast and canal scenery, excellent studies from nature, broad in effect and well handled.

Bauguiet's "Dream after the Ball" is a gem, exquisite in color, breadth and manipulation.

Jacob Jacob's large picture, "The Falls of Sarp on the River Glommen, Norway," was the best of its class in the Belgian department, which lacked any striking figure subjects, on a large scale. Nor did the various works indicate much inventive power. In style and color it may be said of them, as of almost all modern continental pictures, they approximate to the French, but at a respectful distance.

HOLLAND.

Holland exhibited 170 pictures by 77 artists in a separate building, erected by the Dutch government. Of these 57 were figure compositions, chiefly of a domestic character or illustrative of every-day life, with little action but some sentiment. Nearly all were small or of medium size, unambitious, and with little or no dramatic interest. A large number were contributed by the artists themselves, and the rest, with few exceptions, by private individuals.

Alma Tadema exhibited twelve pictures illustrating domestic life in ancient Egypt and Rome, among which are, "How they amused themselves a thousand years ago," and "Entrance to a Roman Theatre," the first representing the interior of an Egyptian house, with the family receiving visitors; the second a Roman audience flocking to the theatre. These pictures are pre-Raphaelite in style, full of wonderful detail, accurate no doubt as they could be made after careful archæological study, but interesting more from subject than as works of art. This artist received a second-class prize.

"The Interrupted Prayer," by Bishop; "The Empty Place at the Hearth," by Bles, are both good pictures, the latter with a good deal of feeling and sentiment, representing a father and daughter with an unoccupied chair at the dinner table. "On the Beach at Scheveningen," by Vervier, a semi-marine view, and very accurate and lively picture of the celebrated and much frequented watering-place of the Hague. Israels, of Amsterdam, an artist of much invention and power, exhibited four compositions, the best of which are "The True Support," and "The Last Breath," effective in color, but carelessly manipulated. A third-class prize was awarded to this artist.

The only large pictures were hunting scenes by Kuytenbrouwer, "Stags Fighting," and "Stags after the Fight," the property of the French Emperor—vigorous compositions but roughly treated. "A Sea Piece—Moonlight," by Meyer; "Cows by the River-side," by Roelop; "A Ray of Light in the Shadow," by Scheltema, were fine pictures, the last-named a carefully handled and effective *genre* picture. "The Syndics of the Serge Guild at Leident," by Stroebel, a Dutch interior, and effective picture. "A Dutch Landscape," with cattle, by Tom, is also worthy of mention.

But in the whole Dutch collection there was no picture of great merit. There was a marked lack of historical compositions. The modern Dutch school approximates to the French, having apparently lost the delicacy of touch and harmony of coloring that distinguished the old Dutch painters and individualized their works. The later artists, however, still cherish that important element of art, the proper management of light and shade.

This gallery received a second and third class prize.

It contained some very good specimens of water-color drawings, but none worthy of special enumeration.

PRUSSIA AND NORTHERN GERMANY.

Prussia contributed 98 pictures by 67 artists, and secured three awards: a grand prize to Knaus, a second-class prize to Menzel, and a third-class prize to Achenbach. One-third of the whole collection was from the studios of Dusseldorff, and for sale. Among the best pictures were seven by Knaus, of Weisbaden, of which "A Woman Playing with Two Cats," "An Acrobat Performing in a Barn," surrounded by wondering peasant spectators, were most meritorious works in design, drawing and breadth of coloring. "The Old Schoolmaster's Birthday," by Lasch, of Dusseldorff, and "Marie Antoinette in the Temple, when visited by the Commoners of the National Convention," by Piotrowski, of Koenigsberg, were good compositions, the latter having a fine lamp-light effect. "The Grand Prize," by Meyer, is a picture worthy of notice, as is also "The Banquet of the Generals of Wallestein," by Schoetz, which was the largest composition in this gallery, good in invention, truthful, with fine chiaroscuro. "On the Mountains," by Schenck, was the best animal picture in the collection.

Of the Prussian exhibition it may be said, as of that of Holland, that it is not remarkable for grand designs, but rather distinguished for pleasing domestic scenes, carefully manipulated, and with good effect of light and shade.

It seems hardly worth while to speak separately of the exhibition made by the smaller German states. Hesse was represented by two pictures by Schevesser, one of which, "Forbidden Fruit," of cabinet size, belonging to the French Emperor, represents boys smoking—an excellent design, well handled, in the Dusseldorff style.

Baden sent 19 pictures by the same number of artists, the best of which were : H. Gude, " Montenegrin Mountaineers going to their Chalets in the Spring ;" an effective, carefully handled cabinet picture. F. Keller, "Death of Philip II of Spain," a large figure composition. G. Saal, " Forest of Fontaillebleau by Moonlight."

Wurtemburg sent eleven pictures by eight artists, of which the best was " The Departure of the Monks from the Cloister of Alpirsbach," by Haeberlin, of Stuttgard ; a large picture, broad in color, and well manipulated.

Luxemburg sent four pictures by two artists, of no special merit.

RUSSIA.

Russia exhibited 63 pictures by 42 artists, chiefly from the royal palaces and the St. Petersburg and Moscow Academies of the fine arts. They certainly did great credit to the artists who painted them, and to the commissioners who selected them. In proportion to the number of pictures, no gallery equalled the Russian in excellence. The French were compelled to admit this, but claimed the Russian artists as their pupils, practicing in their schools, and owing their taste and skill to French teaching, example, and influence. Be this as it may, the painters of the best works are unquestionably Russians, and whether they studied at home or abroad is a circumstance not affecting their capacities, whatever influence it may have had on their style.

There were no less than 33 figure compositions in this gallery, many of them large, and displaying a high order of talent. They were all broad in effect, and not frittered away and lost in scattered light.

Of landscapes there were 12, in a number of which figures were introduced. " The Russians Passing the Devil's Bridge," on the St. Gothard road over the Alps, in the Swiss campaign of 1792, by Kotzebue, is a wonderfully effective picture, in which the introduction of a body of soldiers on which the light is concentrated only seems to heighten the grandeur of the scenery in the background. "A Winter's Evening in Finland," by Mestchersky—a lurid sunset, the light just touching the tops of the pine trees, while beneath is the reflected light on the ice-covered rocks, altogether constituting one of the most effective pictures in the whole exhibition.

There were some marine battle-scenes by Bogoliouboff, excellent in composition and drawing, and good in effect.

The most wonderful piece of portrait painting in the whole Exposition was "A Portrait of an Old Lutheran Woman," by Horawsky, which, for microscopic accuracy in color and texture, is only equalled by the two celebrated heads of an old man and an old woman, in the old Pinakothek at Munich.

Judging from the pictures in the Russian gallery, and accepting them as types of Russian art, Russia must be admitted to a front rank in painting. Her pictures are mostly on a large scale, in which poor or careless drawing would be easily detected. Her painters appear to be faithful and careful artists, with no tendency to oddities or tricks, but disposed to confine themselves within the just rules of art—truthful drawing, strong but not glaring coloring, and careful manipulation. Their style may be described as a combination of the French and Dusseldorff schools, with a decided leaning to the first.

A noticeable feature in this gallery was its nationality. Nearly all the pictures, whether figure compositions or landscapes, were Russian in subject, representing incidents in Russian history, or scenes in Russian territory.

SWITZERLAND.

Switzerland exhibited 112 pictures by 58 artists, in a separate building, and received a second-class prize awarded to Vautier. This gallery contained many large pictures of natural scenery, but few figure pieces, and these only mediocre, while there were no marine or *genre* pictures. The principal pictures were: Berthond, "At the Death," an animal piece. Bodmer, "Sheter," a snow scene. Diday, "The Cascade of the Giessbach." Girardet, "Sunrise on the Toccia, Lago Maggiore." Humbert, "The First Autumn Snow on the Mountains," with animals. L. Jackottet, "The Aar and Erlenbach at Haudick." J. Jackottet, "Falls of the Reichenbach." Lugardon, "The Borders of the Lake." Ulrich, "The Rocks of Lazaset at Nice." Veillon, "Souvenir of the Lake of the Four Cantons." Zelger, "Sunrise on Mount Pilatus."

The Swiss landscapes, though large and carefully manipulated, lack effect and contrast; they are mostly dull and sombre pictures, and not what such scenery as Switzerland possesses should inspire. The artists seem to have studied and reproduced nature under her dullest aspects. They lack feeling, and do not seem to comprehend the importance of concentrating the light in a picture. The scenes represented are usually grand in the extreme, but not well treated or effectually managed. The beautiful sunrises and glowing sunsets so often witnessed among the snowy Alps do not appear to have found an indigenous artist sufficiently confident to attempt them.

There were some very creditable animal pictures in this gallery, and some fine steel engravings by Girardet and Weber.

AUSTRIA.

Austria exhibited 89 pictures by 56 artists, and obtained three prizes, viz: one first-class, one second-class, and one third-class. The best and largest picture exhibited was "The Diet of Warsaw, 1773," by Matejiko, which obtained the highest of the prizes just named. It is well and forcibly drawn, rather roughly manipulated, and pervaded by a purple tint which gives it an unreal and garish appearance, exceedingly untrue to nature and offensive to the eye.

The Chevalier F. L. Allemand had two battle pieces in this collection, the property of the Austrian Emperor, of no special merit, but their author secured a second-class prize. Wuerzinger secured a third-class prize for a portrait of Emperor Ferdinand II.

Of the remaining pictures the following are perhaps worthy of mention: "The Convent Soup," by Waldmueller, and "Hungarian Forest," by Schaeffer.

Nearly one-half of the pictures in this gallery were exhibited by the artists themselves, and were for sale, with their prices affixed in the catalogue. The largest part of the remainder were contributed by the Emperor. Judging from this collection, and the public galleries of Vienna, it would appear that painting is not making rapid progress in Austria.

DENMARK.

Denmark exhibited 29 pictures by 19 artists. No prizes. The best works were: Dalogaard, "Churching a Young Mother." Extner, "Puzzled to choose—Card-players." Madam Jerichan, "Shipwreck on the Coast of Jutland." Jacobson, "A Savant of the Middle Ages." Soerensen, "Sunrise at Skagen—Storm," and "Summer Morning on the Shore of Elsenör," two well-handled and vigorous pictures.

SWEDEN.

Sweden contributed 54 pictures by 28 artists, and secured two third-class prizes. The best works were: Berg, "Waterfall in the Province of Bohns." This artist received a third-class prize. Hoeckert, "Fire in the Royal Palace of Stockholm, May 1, 1697;" Malmstroem, "Elves Playing by Moonlight;" Nordenberg, "The Wedding Presents, Souvenirs of the Province of Blekinge;" Wallander, "Young Girls of the Parish of Wingaker;" Fagerlin, "A Declaration of Love," "A Demand in Marriage," and "Jealousy." This artist obtained a third-class prize.

The representations of natural scenery and of domestic life in the Swedish collection were uniformly very good.

NORWAY.

Norway exhibited 45 pictures by 25 artists, and secured a second-class prize. The best were: Gude, "Funeral Procession crossing a Fiord in Norway," and "Return of Whalers." These secured a second-class prize. Eckersberg, "Table Land of Central Norway, the summit of the Yotun in the background; Morning." Tidemand, "Singular Combat of the Olden Times," an excellent picture.

The pictures in this, as in the Swedish collection, although wanting in works of the highest order of art, such as historical compositions, are nevertheless uniformly good. The subjects were mostly natural scenery, marine views and local domestic scenes, all carefully studied and manipulated after the Dusseldorff school, the influence of which is widely perceptible in most of the galleries of northern Europe.

Both Sweden and Norway owe much of the sound development of art within their borders to the artistic taste and discriminating appreciation of their king, Charles XV, who had two landscapes from his own hand in the Exhibition, both of more than average excellence.

ROME AND ITALY.

Rome exhibited 25 pictures by 14 artists, but received no prize. Among them there were few designs, and those with but little invention. In fact the collection was meagre and unimpressive, and inferior in drawing, effect, and color.

Italy exhibited 51 pictures by 42 artists, and secured four awards, one grand prize, one second prize, and two third-class prizes, a larger proportion than given to any other country. It will be interesting to know from the Report of the Fine Art Jury on what principle these awards were made, since the inferiority of this gallery was equally a disappointment and subject of common remark. The landscapes were mediocre, and there were no marine or animal pictures; in fact none exhibiting much invention or power.

The most ambitious work in this gallery was by Gastaldi, "The Defence of Tortona by its Citizens when besieged by Barbarossa," a large picture, good in composition, but lacking breadth of color and chiaroscuro.

There were some remarkably good figure pieces in water-color in this gallery.

SPAIN.

Spain contributed 42 pictures by 35 artists, and obtained four prizes, one first prize, one second, and two third prizes. Here, as in the case of Italy, the attentive visitor to the Exhibition must feel surprised at the awards of the Jury. A great many of the pictures were inferior; the figure compositions lacked action, the landscapes were tame, and there were no marine or animal subjects. There was a marked proportion of religious subjects, as might be expected, but with the exception of those enumerated below, and which were among the best, the others were mostly *genre* and of no particular merit. Rosales, "Isabella the Catholic Dictating her Will," a large figure composition in rough impasto style, not remarkable for good drawing; the coloring after the old Venetian school. First prize. Palmaroli, "Sermon at the Sistine Chapel," showing Michael Angelo's Last Judgment in the back wall, with the Pope, cardinals, and priests in their purple, scarlet, and sombre robes. The figures are small but well designed, and the position and expression of each well studied. The management of light and shadow is excellent, and the manipulation careful. Leon-y-Escesura, a *genre* picture. Gonsalso, "Interior."

PORTUGAL, GREECE, AND OTHER COUNTRIES.

Portugal was represented by 12 pictures by the same number of artists. No prizes were awarded to this gallery. The only picture of any merit was by Lupi, "Tintoretto Painting the Portrait of his Daughter," a large picture broad in effect, and good in tone.

Greece had 14 pictures by the same number of artists. Received no prize. The best picture was "Antigone" by Litras, a moonlight figure scene. This department showed a low state of art in Greece.

Turkey had seven pictures by three artists; only two of them representing Turkish life, and none of them of any merit.

Egypt exhibited no oil paintings, but sent a number of water-color drawings of ancient monuments, more interesting on historical grounds than as works of art.

One picture was catalogued from China, having belonged to a pagoda, and was not an illustration of painting in China at the present day.

The Argentine Confederation sent three pictures, and Brazil three; all poor.

Canada sent six pictures; Cape of Good Hope two; Malta three; Mauritius several; Nova Scotia one; and Victoria one; all mediocre.

GENERAL OBSERVATIONS ON THE PAINTINGS.

Large and varied as was the collection of paintings in the Exhibition, it does not appear to have impressed those connoisseurs and critics in art who had the opportunity of studying the previous French Exposition of 1855, and the English Exhibition of 1862, as an advance upon either of them. The opinion of these authorities has been accepted, if indeed it was not originally anticipated, by the public. It is alleged that a sameness and monotony pervaded nearly all the galleries, indicating that modern art is subsiding to a dead level of conventionalism, unrelieved by originality or genius, and hardly by eccentricity.

"If" (said the *Revue Contemporaine* in an article on painting in the Exhibition) "you put aside England, and certain painters of other countries who exceptionally evince some originality, all modern painting is nothing more than the attenuated fag-end of the school of Bologna. This is evident throughout the whole circle of the Exposition; traceable in the pavilions of Holland and Belgium, in the temples of Bavaria and Switzerland, and patent in the galleries of Russia and Prussia, Italy and Spain. It follows everywhere the great official road, receives profits and recompenses, broods complacently over ephemeral glories, holds the keys of all the academies and of all the royal and imperial treasuries, and has the privilege of handing down to the indifference or mirth of posterity the features of great persons and sovereigns.   *   *   * In other words, painting has become a branch of industry which requires a little more than others a certain education on the part of those who exercise it, but has, in common with ordinary industrial pursuits, its processes and methods, and a marked desire to meet and satisfy the dominant taste. It differs no more in one country or another than do other similar products, cotton stuffs, for example. In France, England, Belgium, and Germany, calico is always calico, fabricated in a common way, from threads of the same kind of filamentous material. The tissue is a little more or less smooth; more or less regular; more or less white;

but it is only calico after all, serving the same purpose and used by the same people. The production is more or less extended, more or less skilful, but it is always difficult, at first sight, to recognize the country of its origin. ·

"That which strikes every intelligent and impartial man who surveys in succession the galleries of the Exposition is this, that with some marked exceptions there is the same desolate uniformity and mediocrity throughout. It is impossible to point out any except faint and shadowy differences. All the paintings may have come from the same studio, and it seems as if they were produced by the same mechanical art that presides over the production of calico. They are painted canvas, to be measured by the yard. They evince great manual aptitude, and expertness, and subtlety, but the types are all the same, and the coloring comes from common practices and the same point of view. A sterile fecundity marks all these productions, and assures for the future a discredit without example in the past, for the reason that never to the same degree as now, even in the periods called barbarous, has painting effected a divorce so complete from art; never has the mechanical part prevailed more over the ideal; never has the intimate and profound intelligence of the human soul and the artistic sense been more systematically banished from the arts of design, and the mercantile idea allowed to dominate so completely in their practice, as now. It was possible at certain epochs to be deprived of the material means of art, and to be without the instruction now so accessible, but never was there such a dearth of the means that may be called spiritual, in opposition to those material. The design may have been clumsy and without proportion; the coloring without graduation of tint or harmony; but never to the point of depriving the human figure or nature of that aureole of poetry which is the divine reflection and highest expression of truth. Under the brush of our painters the image of man alone is produced, man himself rarely appears; nature is photographic or kaleidoscopic; the artist invests neither with life or light. Hence we see that which should be action is only contortion, and that which should be calm is death."

This, perhaps, may be a strong and almost exaggerated statement of the fact; but it is not to be denied that the new and more intimate relations of men and nations, incident on new and easy facilities of intercourse and intercommunication, are working that assimilation in art which they are so rapidly effecting in costume, habits, literature, modes of thought and expression, and even in religion and government. And it must be admitted, that as art is getting every day to be more and more dependent on the people at large for support, in the same proportion will be its tendency to mould itself to the popular taste. But if for the time it lowers itself to meet that taste, or to suit the fashion of the hour, may we not hope that it will rise as that taste improves?

As several times remarked in the preceding pages, the level to which European art is tending is that of France, and the standard by which art is judged is French. If painting has degenerated into a mechanic art, and its products have become merchandise, conforming to commercial laws, it has been through the example and practice of France. That country, by her system of art education, and through the number of her schools of design, which offer easy access to students, has no doubt imposed her tastes and her style, good or bad, on the world of art. She exists to-day as a rich and powerful nation, not through expanse of territory and value of agricultural productions, nor by supplying the world with the ordinary products of manual labor, but through her application of art and taste in design to manufactures and the useful arts. Drawing and painting have become handmaids of the mechanic arts, and if they have suffered from the contact, has not the world been compensated in the general elevation of the popular taste, and has not France been enriched by a practical monopoly of several large classes of manufactures?

A great deficiency in our own country is the almost total absence of schools of design of high standard, without which correct taste and proficiency in the fine arts can neither be created nor fostered. By such schools is meant establishments in which persons, and especially youth with artistic tendencies, may become thoroughly grounded in the elements of art, and have the facilities and instruction necessary to become, in the first place, accurate and firm in drawing, and in the second place, masters of the theory and practice of color. After that, with a few masterpieces of painting distributed in public galleries throughout the country, partly to guide but mainly to stimulate their powers, they may safely be left to the development of their own natural abilities through study, observation, and practice.

SCULPTURE.

In the department of sculpture, still less than in that of painting, can the late Paris Exhibition be called competitive. The range of competition was really circumscribed to two countries, France and Italy. From others there were only isolated specimens. The reason for this is obvious. Works of marble can only be transported for considerable distances at heavy cost and risk, such as few owners or artists care to undertake. Besides, few countries had the space requisite for a proper display of such works, which were obliged to take refuge in the radiating avenues or passage-ways of the building.

French sculpture, like French painting, naturally enough, predominated, and like French painting was, for the most part, of the naturalistic as distinguished from the Italian, classic, or traditional style. In fact, it is in stone what French painting is on canvas; firm and sharp in outline, bold and free in modelling, good in action, and very faithful in anatomy. The French gallery numbered no fewer than 216 figures, and secured 23 out of the 36 prizes. Italy received four prizes; Prussia two; Spain two; and Rome, Switzerland, Greece, Belgium, and Great Britain, each, one, as shown in Appendix B.

3 F A

The criticism to be made on French sculpture is, that it is sometimes meretricious, and occasionally voluptuous, even to coarseness, and that there is a frequent tendency to theatrical extravagance, showing more talent than taste. The realistic tendency was shown by probably the most remarkable piece of sculpture in the French department, a seated figure of Mdlle. Mars, by Thomas, which has been truly described as "a picture in marble, or rather a picture in which the marble is lost in the realization of texture and material. The silk dress flickers in the light and flutters in the wind." The Empress Josephine, by Dubray, is another work of similar character. "Napoleon at St. Helena," by Guillaume, which received one of the four "grand prizes," was probably the piece of sculpture which attracted most attention. It is one of not less than seven busts and statutes of the first Napoleon exhibited by the same sculptor, and represents the fallen Emperor in the garb of the sick-room, seated in an arm chair, with a map of Europe spread on his knees before him, and with his head bowed languidly on his breast as if in mingled weariness and contemplation. The figures on the dressing gown, the folds and texture of the napkin that rests partly over one arm are faithfully reproduced. The head is forcible, but its power is much lessened by the accessories of the figure, which, however, is natural in *pose*.

In the Italian gallery, "Charlotte Corday," by Miglioretti, was probably the best production in the semi-classical style. "Phryne," by Barzaghi, was also pure and excellent, as were also "The Adulteress," by Bernasconti, "Armide" by Bianchi, and "Vanity," by Tantardini.

There were very few works in marble or bronze in the American gallery. Miss Hosmer's "Sleeping Fawn" was unquestionably the best reclining figure in the Exhibition, and commanded appreciative attention. It derived no advantage from being slightly tinted. Ward's "Indian Hunter and his Dog," in bronze, is a bold and vigorous work, with far more powers and action than was shown in any other work in the Exhibition. Thompson's bronze statute of Napoleon I is a reproduction of the conventional great commander, and owed even more attention in the Exhibition to the subject than the execution. Rogers exhibited some of his carefully-studied statuettes, but they were so placed that they did not fall under general notice. Miss Foley, a promising Vermont artist, exhibited some very clever medallions in the Roman department.

Although creditable, the few works exhibited gave no idea of the extent or proficiency of American sculpture, in which the United States has gained a far higher rank than in any of the fine arts. The world, familiar with their names, looked in vain for the works of Powers, Story, Rogers, Rinehardt, Mozier, Brown, and others, who have done so much to vindicate the American name in this department of art.

Peru exhibited a single group, by Suarez, "A Defender of his Country," still in plaster, an animated work, in which, however, there is only the expression of an athlete in the principal figure, instead of lofty and patriotic inspiration and fervor.

ENGRAVING.

Of engraving, whether on steel or wood, the Exposition contained nothing indicative of any real advance in those arts. Indeed, we should rather say that the various works exhibited showed, instead of any advance, a decided falling off in all that is artistically high or admirable. It would be unfair, as well as vain, to particularize where all are of the same quality. Photographs of different degrees of excellence, chromo-lithographs, undoubtedly good, and color-printings, false in taste and clumsy in their methods of execution, take the place of the etcher's and engraver's art. The best steel plates exhibited were half machine-ruled or at best but feeble and inartistic attempts to atone by multiplicity of lines for inferiority in touch and color; while wood-engraving has followed the fashion of the Dalziel mania in England, and gone out of art altogether into a rudeness poorer than even that of the wood-cutting (cutting done with knives) of the worst days before Bewick. Great Britain, the especial country of wood-engraving, sent not even a single decent specimen; the French department was almost equally deficient; Spain and Portugal and Greece sent some childlike and crude beginnings; and Germany contributed only of her worst. To judge of European art, the engraver's art, only from the evidence of the Exposition, we should be constrained to say that it had almost died out, leaving only improved mechanisms, certain facilities for cheapness, greater opportunities for making a mere show just good enough to gratify some untutored appetite for what is strangely miscalled "an illustrated edition."

It is at such a time and under such circumstances that one naturally expects to find new "processes"—inventions intended to supersede the artist's talent and power of hand by some clever mechanism. Of these processes, though several are in use, not one exhibited can be called really successful. Some drawback of rottenness of line in the more delicate and open parts, or of obscurity and muddiness in the darker, spoils always the work as a work of beauty; and all that is obtained is reduction of cost, and sometimes an economy of time. Here again it is idle to particularize where none is markedly eminent.

In conclusion we can only confess that the Exposition would seem to prove that for art we have now substituted machinery, and for the artist the processes of the chemist. With the exception of lithographs, chromo-lithographs, photo-lithographs, and photographs, we have nothing to take the place of the copper engraving (line, aquatint, or mezzotint) or of the wood engraving of past times. Book illustration has notably deteriorated; and even printing (lacking artistic taste to superintend it) is only cheapened and not improved.

Of steel engravings, Marshall exhibited a fine but perhaps over-elaborate portrait of President Lincoln, and the American Bank Note Company a great variety of work, unapproachable in respect of style and finish by anything of the kind exhibited by any other nation.

Of what may be called substitutes for wood and steel engraving, or short easy, and cheap processes for accomplishing some of the results of engravings on wood and steel, several examples were exhibited in the French departments. But none were comparable with that in use, for mechanical engravings, in the Reports of the American Patent Office.

First may be mentioned—

THE GRAPHOTYPE PROCESS,

an American invention, but adopted in England, whereby the drawing is made on a surface of fine chalk, compacted under great hydraulic pressure, with a solution of silica, which hardens the lines, after which the chalk between them is brushed away, and the lines left in relief. From the relief thus produced, it is easy to obtain stereotypes or electrotypes for use in relief printing.

THE COPPER PROCESS.

On a plate of copper, varnished as if for etching, the subject is drawn with an etching point, and bitten in by *aquafortis* exactly in the way that is called "the first biting in."

The plate previously covered with varnish at the back, is then put into a gold bath, (electrogilding,) when the lines on the copper etched and consequently undefended by the varnish and slightly incised, are covered with a thin coating of gold. The plate is now cleaned of the varnish and recovered with varnish applied by a dabber and presents the appearance of a copper-plate inked ready for printing.

It is now cleaned of the varnish, which rests only on the gold incised lines. The plate is then put into an acid bath, and what the engraver calls the "whites" are bitten away leaving the drawing in relief, which is then mounted as an ordinary cast. The object of the gold bath is to protect the lines of the etching against the too uncertain action of the acid; but the process can be effected without this.

THE GILLOT PROCESS.

On a plate of zinc, polished, a transfer is made of an ordinary lithographic drawing, either by pen or pencil. The plate of zinc is covered on the back by a layer of oil varnish and submitted to a bath of sulphuric acid diluted with water. Every part of the plate not covered either by the ink of the transfer or by the varnish is bitten or hollowed out by the acid and leaves in relief the covered parts, which, mounted on wood, form a cast replacing an engraving on wood. This process is very quickly executed, and offers great advantages for illustrations which will not bear delay. It offers also this considerable advantage, that all drawings executed in lithography of all kinds, etchings, engravings on steel, plates of music, etc., etc., from which can be taken a proof on paper, are rapidly transformed into a cast, and can be printed with the test.

The most marked disadvantages of this process are these: 1. Breaking down of the lines in the operation of transfer, however delicately this operation may be performed. This thickens and blots the delicacy of the original work: 2. Difficulty of obtaining the tints when the work is taken directly from lithography in pen and ink. The blotting is very apparent when it is a lithographic drawing.

This process serves for the illustration of a number of journals, *i. e.*, *Le Journal Amusant*, *La vie Parisienne*, *La Lune*, and several others.

THE CONTE PROCESS.

Instead of transferring the drawing on a plate of zinc, the plate is covered with a white water-varnish, sufficiently adherent for a tracing and sufficiently soft to be easily cut away, without scratching the zinc, by a point of wood, ivory, or whalebone. The artist then proceeds as for etching, with this difference of result, that the zinc uncovered is again covered by oil ink, and instead of being incised as in etching, remains in relief in the acid bath, which is the same as in the process Gillot. This process is very convenient for artists, who can judge of their work as if they were drawing on paper by means of a black lead pencil. The lines do not thicken, but remain delicate. The varnish can be removed or put on with a camel's hair pencil, so that any part may be corrected at any time by the artist.

Different processes of the same nature have been tried by different persons who have endeavored to substitute for zinc, copper, steel, and other metals, but without success. However, the principle of all is the same.

Respectfully submitted by

FRANK LESLIE,
U. S. Commissioner to the Paris Universal Exposition of 1867.

APPENDIX A.

FINE ARTS.

GROUP I—UNITED STATES.—LIST OF PAINTINGS, SCULPTURES, ETC., EXHIBITED.

CLASS 1.—PAINTINGS IN OIL.

BAKER, G. A., New York.—1. Portrait of a Child, property of A. M. Cozzens. 2. Portrait of a Lady, property of F. Prentice.

BEARD, W. H., New York.—3. Dancing Bears, property of J. Caldwell.

BIERSTADT, A., New York.—4. The Rocky Mountains, property of J. McHenry.

BOUGHTON, G. H., New York.—5. Winter Twilight, property of R. L. Stuart. 6. The Penitent, property of J. F. Kensett.

CASILEAR, J. W., New York.—7. The Plains of Genesee.

CHURCH, F. E., New York.—8. The Niagara, property of J. Taylor Johnston. 9. The Rainy Season in the Tropics, property of M. O. Roberts.

COLEMAN, S., New York.—10. Landscape view of the Alhambra.

CROPSEY, J. F., New York.—11. Mount Jefferson, New Hampshire, property of R. M. Olyphant. 12. Landscape.

DIX, C. F., New York.—13. Sea Piece.

DURAND, A. B., New York.—14. In the Wood. 15. A Symbol, property of R. M. Olyphant.

ELLIOTT, C. L., New York.—16. A Portrait, property of M. Fletcher Harper.

FAGNANI.—17. A portrait, property of Sir Henry Bulwer.

GIFFORD, S. R., New York.—18. Twilight on Mount Hunter, the property of J. W. Pinchot. 19. An interior of a Dwelling in the Desert, property of M. Knoedler.

GIGNOUX, R., New York.—20. Mount Washington, New Hampshire, property of A. T. Stewart.

GRAY, A. P., New York.—21. The Apple of Discord, property of R. M. Olyphant. 22. The Pride of the Village, property of W. H. Osborn.

HART, J. M., New York.—23. Landscape, River Tunxis, Connecticut.

HEALEY, G. P. A., Chicago, Illinois.—24. Portrait of Lieutenant-General Sherman. 25. Portrait of a Lady, property of W. B. Duncan.

HORNER, W., New York.—26. Confederate Prisoners, property of J. T. Johnston. 27. The Bright Side, property of W. H. Hamilton.

HUBBARD, R. W., New York.—28. View of the Adirondacks, taken on Mount Mansfield, property of Madame H. B. Cromwell. 29. Beginning of Autumn, property of H. G. Marquand.

HUNT, W. M., Boston, Massachusetts.—30. Portrait. 31. Portrait of Abraham Lincoln. 32. Italian Boy. 33. Italian Boy. 34. Dinan, in Brittany. 35. The Quarry.

HUNTINGTON, D., New York.—36. Portrait of M. Gulian Verplanck. 37. The Republican Court in the time of Washington, property of A. T. Stewart.

INNESS, G., Perth Amboy.—38. Sunset in America, property of M. Marcus Spring. 39. Landscape and Animals.

JOHNSON, E., New York.—40. Country Scene in Kentucky, property of Mr. W. H. Derby. 41. Seductive Proposals, property of Major-General John A. Dix. 42. The Violin Player, property of Mr. R. L. Stuart. 43. Sunday Morning, property of Mr. R. M. Hoe.

KENSETT, J. F., New York.—44. Lake George in Autumn, property of Mr. G. F. Olyphant. 45. Views on the Coast of Newport, property of G. F. Olyphant. 46. An Opening in the White Mountains, property of Mr. R. L. Stuart. 47. Morning on the Coast of Massachusetts, property of Mr. S. Gandy.

LAMBDIN, G. C., Philadelphia.—48. The Consecration, 1861, property of Mr. George Whitney. 49. The Last Sleep.

LANGDON, W., New York.—50. The Storm. 51. At Sea.

LAFARGE, J., Newport, Rhode Island.—52. Flowers.

LEUTZE, E., New York.—53. Marie Stuart hearing Mass for the first time at Holyrood after her return from France, property of Mr. John A. Riston.

LEWIS, S. J., Burlington, New Jersey.—54. The Little Fisherman.

MAY, E. C., New York.—55. Lady Jane Grey giving her Tablets to the Governor of the Tower of London on her way to the Scaffold. 56. Lear and Cordelia, ("King Lear," Act IV, Scene 7.) 57. A Portrait.

MAC ENTEE, J., New York.—58. Virginia in 1863, property of M. C. Butler. 59. The End of October, property of M. S. C. Evans. 60. Autumn in the Woods of Ashokan, property of R. M. Hoe.

MIGNOT, L. R.—61. Sources of the Susquehanna, property of H. W. Derby.

MORAN, T., Philadelphia.—62. Autumn on the Conemaugh, Pennsylvania, property of C. L. Sharpless. 63. The Children of the Mountain.

OWEN, G., New York.—64. Landscape of New England, study from nature.

RICHARDS, W. F., Philadelphia.—65. Forest in June, property of R. L. Stuart. 66. A Foggy Day at Nantucket, property of G. Whitney.

WEIR, J. F., New York.—67. The Cannon Foundry, property of R. P. Parrott.

WHISTLER, J. MAC NEIL.—68. The White Girl. 69. Wapping, on the Thames. 70. Old Battersea Bridge. 71. Twilight at Sea.

WHITE, E., New York.—72. Recollections of Siberia, property of R. L. Stuart.

WHITTRIDGE, W., New York.—73. The Old Kentucky Land, property of J. W. Pinchott. 74. The Coast of Rhode Island, property of A. M. Cozzens.

WEBER, P., Philadelphia.—75. Woodlands, Bolton Park, England.

CLASS 2.—OTHER PAINTINGS AND DRAWINGS.

DARLEY, F. O. C., New York.—1. Charge of Cavalry at Fredericksburg, in Virginia, property of W. F. Blodgett. 2. Vignettes for bank notes.

JOHNSON, E., New York.—3. The Wounded Drummer, property of the Century Club.

ROWSE, S. W., Boston.—4. Portrait of Emerson, (crayon.) 5. Portrait of Lowell, (crayon.)

CLASS 3.—SCULPTURE, DIE-SINKING, STONE AND CAMEO ENGRAVING.

HOSMER, Md'lle H. G.—1. The Sleeping Fawn. 2. The Wakened Fawn.

ROGERS, J., New York.—3. Three groups of statuettes.

THOMPSON, L., New York.—4. Statue of Napoleon, property of C. C. D. Pinchot. 5. Bust of W. C. Bryant, property of C. H. Ludington. 6. Bust of the Trapper of the Rocky Mountains.

VOLK, L. W., Chicago, Illinois.—7. Bust of A. Lincoln.

WARD, J. Q. A., New York.—8. The Indian Hunter and his Dog, the property of the Central Park of New York. 9. The Liberated Slave.

CLASS 4.—ARCHITECTURAL DESIGNS AND MODELS.

(No exhibition.)

CLASS 5.—ENGRAVING AND LITHOGRAPHY.

AMERICAN BANK NOTE COMPANY.—1. Specimens of engraving and printing of bank notes.

MARSHALL, W. E.—2. Lincoln, (engraving on steel.) 3. Washington, (engraving on steel.)

NATIONAL BANK NOTE COMPANY.—4. Specimens of engraving of bank notes.

HALPIN, F., New York.—5. President Lincoln, (engraving on steel.)

WISTLER J. MAC NEIL.—6. Twenty-four etchings.

APPENDIX B.

FINE ARTS.—LIST OF AWARDS.

GROUP I.—WORKS OF ART, FIRST SECTION.

Classes 1 and 2 united.

PAINTING AND DESIGN.

GRAND PRIZES.

Cabanel	France.
Gérôme	France.
Ernest Meissonier	France.
Theodore Rousseau	France.
Ussi	Italy.
Guillaume de Kaulbach	Bavaria.
Knaus	Prussia.
Leys	Belgium.

FIRST PRIZES.

Bida	France.
Jules Breton	France.
Charles Daubigny	France.
Français	France.
Fromentia	France.
Jean François Millet	France.
Pils	France.
Joseph Robert Fleury	France.
Calderon	Great Britain.
Horsehelt	Bavaria.
Makejko	Austria.
Piloty	Bavaria.
Rosales	Spain.
Alfred Stevens	Belgium.
Willems	Belgium.

SECOND PRIZES.

Mlle. Rosa Bonheur	France.
Bonnat	France.
Brion	France.
Covot	France.
Delaunay	France.
Jules Dupré	France.
Hamon	France.
Hebert	France.
Jalabert	France.
Yvon	France.
Alma Yadema	Holland.
Church	United States.
Clays	Belgium.
Gude	Norway.
Sigismond L'Allemand	Austria.
Menzel	Prussia.
Morelli	Italy.
Nicol	Great Britain.
Palmaroli	Spain.
Vautier	Switzerland.

4 F A

THIRD PRIZES.

Henry Baron	France.	Bergh	Sweden.
Belly	France.	Fayerlin	Sweden.
Bouguereau	France.	Faruftini	Italy.
Busson	France.	Gisbert	Spain.
Cabut	France.	Gonsalvo	Spain
Comte	France.	Israels	Holland.
De Curzon	France.	Kotzebue	Russia.
Emile Levy	France.	Lenbach	Bavaria.
Puvis de Chavannes	France.	Q. Orchardson	Great Britain.
Velter	France.	Pagliano	Italy.
F. Adam	Bavaria.	Walker	Great Britain.
Andre Achenbach	Prussia.	Wurzinger	Austria.

SECOND SECTION, CLASS 3.—SCULPTURE.

GRAND PRIZES.

Eugene Guillaume	France.	Drake	Prussia.
Perraud	France.	J. Dupré	Italy.

FIRST PRIZES.

Carpeaux	France.	Aime Millet	France.
Gustave Crauk	France.	Ponscarme	France.
Falguière	France.	Jules Thomas	France.
Gumery	France.	Vela	Italy.

SECOND PRIZES.

Paul Dubois	France.	Argenté	Italy.
Fremiet	France.	Blaeser	Prussia.
Gruyère	France.	Caroni	Switzerland.
Mathurin Moreau	France.	Luccardi	Rome.
Ottin	France.	Pescador	Spain.
Salmson	France.	Strazza	Italy.

THIRD PRIZES.

Caïn	France.	Sanson	France.
Cambos	France.	Drossis	Greece.
Cugnot	France.	Pioker	Belgium.
Feugère des Forts	France.	G. Sunol	Spain.
Maillet	France.	J. S. Wyon and A.	
Merley	France.	B. Wyon, (collec-	
Montagny	France.	tive medal)	Great Britain

CLASS 4.—ARCHITECTURE.

GRAND PRIZES.

Ancelet	France.	Waterhouse	Great Britain.
Ferstel	Austria.		

FIRST PRIZES.

Joyau	France.	Late Capt. Fowke	Great Britain.
Lamière	France.	Rosanoff.	Russia.
Thierry	France.	F. Schmitz	Prussia, &c.

SECOND PRIZES.

Boitte	France.	Questel	France.
Deperthes	France.	W. D. Lynn	Great Britain.
Esquié	France.	T. Hanzel	Austria.
Edmond Guillaume	France.	Hlavka	Austria.

THIRD PRIZES.

Ambroise Baudry	France.	E. Barry	Great Britain.
Daumet	France.	Carpentier	Belgium.
Felix Thomas	France.	G. Semper	Switzerland.

CLASS 5.—ENGRAVING AND LITHOGRAPHY.

GRAND PRIZES.

Alphonse François	France.	J. Keller	Prussia, &c.

FIRST PRIZES.

Bertinot	France.	E. Maudel	Prussia.
Achille Martinet	France.		

SECOND PRIZES.

Salmon	France.	N. Barthelmess	Prussia.
Bal	Belgium.	Edouard Girardet	Switzerland.

THIRD PRIZES.

Auguste Blanchard	France.	Jacquemard	France.
Charles Jacques	France.	Rousseaux	France.

PARIS UNIVERSAL EXPOSITION, 1867.
REPORTS OF THE UNITED STATES COMMISSIONERS.

THE FINE ARTS

APPLIED TO

THE USEFUL ARTS.

REPORT BY THE COMMITTEE:

FRANK LESLIE, S. F. B. MORSE, THOMAS W. EVANS,

UNITED STATES COMMISSIONERS.

WASHINGTON:
GOVERNMENT PRINTING OFFICE.
1868.

FINE ARTS APPLIED TO THE USEFUL ARTS.

REPORT OF THE COMMITTEE.

The importance of art education as applied to manufactures is so obvious as to need no enforcement in this age. It is universally recognized that utility and beauty may be combined. This marriage of the fine and the mechanic arts has been encouraged and promoted in Europe for many years, chiefly through government intervention and aid, through establishments like those of the Gobelins and of Sèvres, and through schools of art and of design founded in every principal city. It is doubtful, however, if such an assumption of functions would be consonant with our system of governmental administration; nor would it be desirable, perhaps, even if it were possible. Even in the important matter of rudimentary education the governments of the various states scarcely do more than to take the initiative, and the amplification of our educational system is wisely left to the intelligence and public spirit of our people. Still it is a question, if such an initiation might not be taken by government, in this matter of schools of design, on grounds of public utility.

One of the practical and most important results of the great London Exposition of 1851 was, that it aroused the attention of the people of England to the importance of an extended application of the fine to the useful arts. The display made in the French department showed that, without being in other respects equal to the English, many classes of French manufactures and products were preferred in the markets of the world on account of their greater beauty of form and outline, and greater taste in ornamentation. This led to the establishment in England, through public authority, of a number of schools of design, with advantageous results to the manufacturers of Great Britain, who were gradually enabled to renew their hold on a number of branches of trade, while the general taste, which is an element and reflection of civilization, was encouraged and elevated.

The reports of the English commissioners to the New York exhibition of 1853 are full of references to the fact that in almost every department of manufactures the United States was dependent on foreign designers—French, English, and German. While praising the aptitude and skill of American mechanics, they said that "American manufactures have reached that point at which it has become desirable that originality of thought should be infused into them by means of the instructed designer; for the design once obtained, the very effort to realize it educates the workman." They added that "there was no appearance of an attempt to strike out a national style, although the many peculiar features of the country, the habits of the people, and undoubted originality in the mechanic arts, would lead to the inference that a gradual repudiation of European modes and forms will ultimately take place, and that in art, as applied to the utilities of life, true principles in the education of the people in this respect are alone needed to produce results of a very satisfactory character. At present the co-mingling of totally different styles of decoration in architecture, and in the adoption of European designs for totally different purposes to those for which they were originally intended, are among the least of the errors committed in a vague seeking after novelty."

Although these observations were written 15 years ago, and although American taste has in the interval vastly improved, yet the criticisms they contain still remain substantially true, and the United States still sustains a colonial, not to say provincial, relationship to Europe in the matter of grace and ornamentation in articles and constructions of common use. Chaste and tasteful designs are capable of universal application, not less in lamp-posts and hydrants than in household furniture and jewelry, nor less to the paper-hangings and calicos of the poor than the soft carpetings and silks of the rich. The ingenious mechanic who knows how to devise the implements, instruments, and utensils or other objects required to meet universal or exceptional wants, needs only the aid of the designer to make them pleasing and elegant as well as beautiful.

In this application of the tasteful to the useful, the United States undertook no competitive display in the Exhibition of 1867; yet she exhibited locomotives which, if somewhat meretricious in ornamentation, were nevertheless beautiful in comparison with the ungainly monsters exhibited by other countries. In lightness and tasteful outline, no carriages were comparable with those from the United States, although they were not remarkable in any other respect; and the gas fixtures, chandeliers, and other lighting apparatus exhibited by the Tucker Manufacturing Company were certainly equal to those of any other country. No billiard tables were equal to those of the United States in elegance or excellence.

It is not probably worth while to attempt a discussion of the point whether a "national style" of design is possible or desirable. The principles of design, grace in form and outline, just proportions, harmony both in ornament and coloring, are of universal and kaleidoscopic application, and can hardly be nationalized, although the mode of their application may and ought to vary with the varying circumstances of different people, having different wants and different means of meeting them.

The committée, therefore, think they will better serve the objects of their appointment by submitting drawings of a number of useful objects in the Exhibition, to which the fine arts have been applied with more or less success, as a substitute for descriptions which, from the nature of the case, would scarcely be intelligible. These will be useful in the way of suggestion, if they do not absolutely form models to be followed.

Respectfully submitted:

FRANK LESLIE,
S. F. B. MORSE, } *Committee,*
THOMAS W. EVANS,

U. S. Commissioners to Paris Universal Exposition of 1867.

EXPLANATION OF PLATES.

PLATE I.—Carved Cabinet. Stovesandt, Carlsruhe.

PLATE II.—Ebony Cabinet, mounted in ivory. Angelo Amici, Italy.

PLATE III.—Carved Furniture. Wirth Bros., Brientz, Switzerland.

PLATE IV.—Carved Stand and Table. Leovinson, Berlin.

PLATE V.—Fauteuil, which divides into sofas and easy chairs. Filmer & Co., England.

PLATE VI.—Chimney-piece. Leclercq, Belgium.

PLATE VII.—Chimney-piece. Leclercq, Belgium. Carved Chairs. Leovinson, Berlin.

PLATE VIII.—Cast-iron Fountain. Ducel, Paris.

PLATE IX.—Centre-piece for Flowers, in gold and silver. Sy & Wagner, Prussia.

PLATE X.—Centre-piece. Dziedzinski and Hanusch, Austria.

PLATE XI.—Cast-iron Hat Stand. Crichley & Co., England.

PLATE XII.—Bronze Fender. Raingo Bros., Paris. Bronze and Glass Stand. Lobmeyr, Austria. Cast-iron Umbrella Stand. Crichley & Co., England.

PLATE XIII.—Cast-iron Lamp-posts. Barbezat & Co., Paris. Tripod Lamp. Charpentier, Paris.

PLATE XIV.—Lamp. Schlossmacher & Co., Paris. Candelabrum. Boyer & Sons, Paris. Lamp. Barbezat & Co., Paris.

PLATE XV.—Gas Chandeliers. Best & Hobson, England.

PLATE XVI.—Gas Chandelier. Winfield & Co., England.

PLATE XVII.—Gas Chandeliers. Phillp, England.

PLATE XVIII.—Glass Chandelier. Defries & Sons, England.

PLATE XIX.—Glassware. Dobson, England.

PLATE XX.—Glassware. Dobson, England.

PLATE XXI.—Glassware. Phillips & Pearce, England.

PLATE XXII.—Engraved Glassware. Millar & Co., Scotland.

PLATE XXIII.—Engraved Glassware. Millar & Co., Scotland.

PLATE XXIV.—Table. Royal Porcelain Manufactory of Dresden.

PLATE XXV.—1. Painted Vase. Royal Porcelain Manufactory, Dresden. 2. Painted Vase. Royal Porcelain Works, Prussia. 3. Painted Vase. Imperial Manufactory, Sèvres. 4. Electro-plated Vase. Elkington, England.

PLATE XXVI.—Bronze Clock. Lerolle, Paris. Porcelain Service. Royal Porcelain Manufactory, Berlin.

PLATE XXVII.—1, 2, 3. Water Coolers and Filters. Doulton & Watts, England. 4, 5, 6. Flagons. Sy & Wagner, Prussia. 7. Toilette Bottle in oxydized silver. Rudolphi. 8. Tazza. Boyer & Son, Paris.

PLATE XXVIII.—1, 2. Porcelain and Metal Clock. Boulonnois, Paris. 3. Card Stand in pulverized wood. Latry & Co., Paris.

PLATE XXIX.—Cast-iron Railing. Barbezat & Co., Paris. Gas Bracket. Hollenbach, Austria.

PLATE XXX.—Panel in Silk. Haas & Sons, Austria.

PLATE XXXI.—Marble Chimney-piece. Bródski, Russia.

PLATE XXXII.—Epergne.

PLATE XXXIII.—Bronze Clock.

[As the elaborately executed drawings for the following described plates required several colors to properly represent the various objects they have not been engraved. By direction of the Secretary of State these drawings have been bound in one volume, with a copy of this report, and are deposited in the Congressional Library.]

PLATE XXXIV.—Winfield & Co., Birmingham, England.—Copper-gilt Gas Chandelier Bracket. Gold medal.

PLATE XXXV.—Elkington, London, England.—A, Fork; B, Spoon; C, Epergne, silver chased and rehaussé work ; D, Saltcellar, silver and jet ; E, Vase, silver, gold and glass. Received second medal, (gold.)

PLATE XXXVI.—Heal & Son, London, England.—Bed, citron-wood and blue furniture. Received bronze medal.

PLATE XXXVII.—Ward, London, England.—A, Easy Chair. Silver medal and honorable mention.

Heal & Son, London, England.—B, Table, citron-wood. Bronze medal.

PLATE XXXVIII.—Winfield & Co., Birmingham England.—A, Bracket Lamp, copper gilt; B, Chair, gold and damask. Gold medal.

PLATE XXXIX.—James Green, London, England.—A, Vase, in glass; B, Cup for flowers ; C, Bottle for perfumes. Bronze medal.

PLATE XL.—James Green, London, England.—A, Vase for perfumes; B, Glasses for flowers. Bronze medal.

PLATE XLI.—Minton & Co., Stoke-upon-Trent, England.—A, Vase in porcelain ; B, Small Cup, in same ware ; C, Bottle ; D, Chair in pottery ware. Gold medal, first class.

PLATE XLII.—Barbezat, Paris, France.—A, Dish, imitation Bernard Palissy.

Deck, Paris.—B, Vase, in earthenware.

PLATE XLIII.—Creil & Moutereau, Paris, France.—A, Vase, in earthenware.

Triefus & Ettlinger, Paris.—B, Prayer-book, ivory and gold. Silver medal.

Deck, Paris.—C, Vase, in earthenware.

PLATE XLIV.—Samson, (father and son,) Paris, France.—A, Vase for flowers, wood, imitation of china ware ; B, Plate.

Gonard, Paris.—C, Clock for chimney piece, enamel and gold.

PLATE XLV.—Giroux, Paris, France.—A, Armoire, bronze gilt, velvet pedestal; B, Looking-glass, gold frame.

PLATE XLVI.—Dielh, Paris.—A, Casket; wood, bronze, gilt and marble.

Pull, Paris.—B, Cup; style, Bernard Palissy, (bought by the Emperor.) Bronze medal.

PLATE XLVII.—Tahan, Paris, France.—A, Cup, style antique; B, Clock with Statuette; C, Cup; D, Casket with enamelled paintings.

PLATE XLVIII.—Mazaroz-Ribaillier & Co., Paris and London.—Buffet in carved oak.

PLATE XLIX.—Mazaroz-Ribaillier & Co.—A, Chair, carved oak; B, Chair, gold and blue damask.

Giroux, Paris.—C, Vase in gold and marble. Silver medal, first class.

PLATE L.—W. & J. R. Hunter, London, England.—Armoire in citron and maple wood, medallions in blue and white wedgewood ware. Bronze medal.

PLATE LI.—Leovinson, Berlin and Paris.—A, Writing Desk in carved oak, interior in citron-wood; B, Arm-chair, carved oak, covered in damask. Honorable mention.

PLATE LII.—H. Ullrich, Vienna, Austria.—Vase in colored porcelain. Honorable mention.

PLATE LIII.—H. Ullrich, Vienna, Austria.—Drinking Cup in green glass, with black eagle and medallion iron handles. Honorable mention.

PLATE LIV.—H. Ullrich, Vienna, Austria.—Vase Etagére, in ruby colored carved glass of Bohemia. Honorable mention.

PLATE LV.—Imported by Chanton, Paris, from China.—Bed, iron-wood carved, cane bottom. Silver medal.

PLATE LVI.—Imported by Chanton, Paris, from China.—A, Lantern, from the imperial palace at Pekin. (Bought by the King of the Belgians.) B, Fan in ivory; C, Arm-chair, wood, inlaid with ivory and citron-wood, cane seat. Silver medal.

PLATE LVII.—Imported by Chanton, Paris, from China.—A, Arm-chair, black iron-wood, covered in damask; B, Vase in porcelain mounted on stand of brown iron-wood.

PLATE LVIII.—Leovinson, Berlin, London.—A, Paper Case, carved oak, interior citron wood; B, Chair, carved oak, covered in green damask. Honorable mention.

PLATE LIX.—Adolf Meyr, Austria, Bohemia.—Cup in porcelain and Bohemian glass, mounted on red velvet. Gold medal.

PLATE LX.—J. & L. Lobmeyr, Vienna, Austria.—Cup, glass and bronze. Gold medal.

PLATE LXI.—J. & L. Lobmeyr, Vienna, Austria.—Cup, glass mounted on wood and bronze.

PLATE LXII.—Adolf Meyr, Austria, Bohemia.—A, Vase in white glass; B, Cup, green glass and gold. Gold medal.

PLATE LXIII.—J. & L. Lobmeyr, Austria.—Vase, engraved glass mounted on bronze gilt. Gold medal.

PLATE LXIV.—Wirth Brothers, Brientz, Switzerland.—Cup, in ground glass mounted on dark oak. Silver medal.

PLATE LXV.—Wirth Brothers, Brientz, Switzerland.—Looking-glass, light oak mounting. Silver medal.

PLATE LXVI.—Wirth, Brientz, Switzerland.—Lamp, gray wood and green bronze.

PLATE LXVII.—Schwechten, Berlin, Prussia.—Piano in carved oak. Bronze medal.

Cast-iron Railing. Barbezat & Co., Paris.

I.—Carved Cabinet. Stovesandt, Carlsruhe.

GBS_FOLDOUT

GBS_FOLDOUT

GBS_FOLDOUT

GBS_FOLDOUT

IV.—Carved Stand and Table. Leovinson, Berlin.

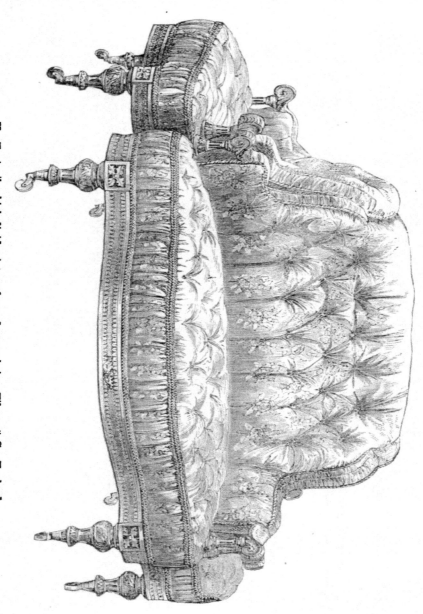

V.—Fauteuil, which divides into sofas and easy chairs. Filmer & Co., England.

**VII.—Chimney-piece. Leclercq, Belgium.
Carved Chairs. Leovinson, Berlin.**

VIII.—Cast-iron Fountain. Ducel, Paris.

IX.—Centre-piece for Flowers in gold or silver. Sy & Wagner, Prussia.

X.—Centre-piece. Dziedzinski and Hanusch, Austria.

XI.—Cast-iron Hat Stand.   Crichley & Co., England.

XII.—Bronze Fender.   Raingo Bros.,   Paris.
Bronze and Glass Stand.   Lobmeyr, Austria.
Cast-iron Umbrella Stand.   Crichley & Co., England.

XIII.—Cast-iron Lamp-posts. Barbezat & Co., Paris.
Tripod Lamp. Charpentier, Paris.

Lamp. Schlossmacher & Co.,   **Candel**
Paris.

XV.—Gas Chandeliers.　　Best & Hobson, England.

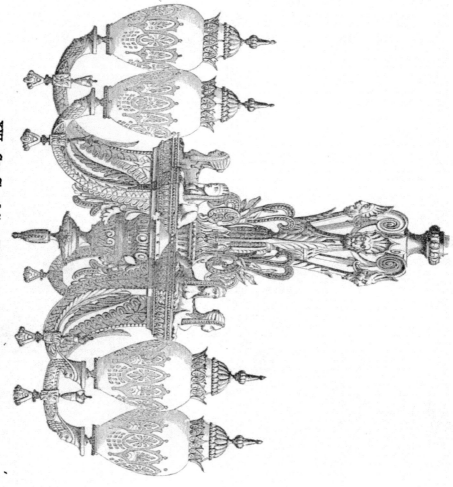

XVI.—Gas Chandelier. Winfield & Co., England.

XVII.—Gas Chandeliers. Phillp, England.

XVIII.—Glass Chandelier.   Defries and Sons, England.

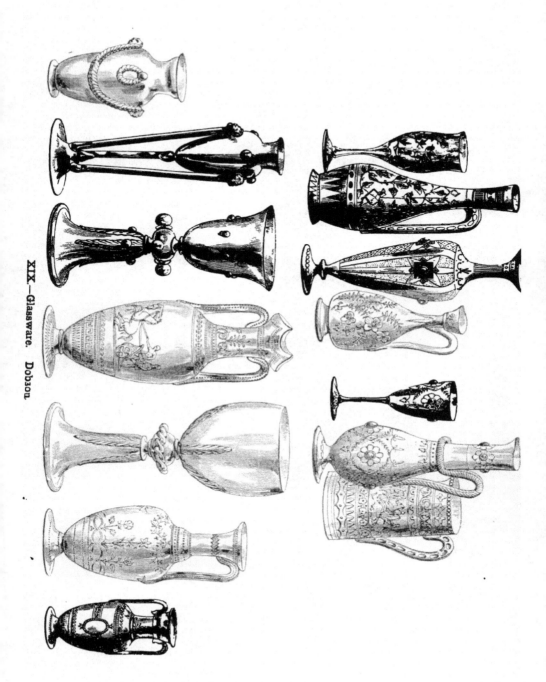

GBS_FOLDOUT

GBS_FOLDOUT

XXI.—Glassware. Phillips & Pearce, England.

XXII.—Engraved Glassware. Millar & Co., Scotland.

XXIII.—Engraved Glassware. Millar & Co., Scotland.

XXV.—1. Painted Vase.   Royal Porcelain Works, Dresden.
2. Painted Vase.   Royal Porcelain Works, Prussia.
3. Painted Vase.   Imperial Manufactory, Sèvres.
4. Electro-plated Vase.   Elkington, England.

XXVI.—Bronze Clock. Lerolle, Paris.
Porcelain Service. Royal Porcelain Manufactory, Berlin.

1. 2. 3. 4. 5. 6. 7. 8.

XXVII.—1, 2, 3. Water Coolers and Filters. Doulton & Watts, England.
4, 5, 6. Flagons. Sy & Wagner, Prussia.
7. Toilet Bottle in oxydized silver. Rudolphi.
8. Tazza. Boyer & Sons, Paris.

1. **2.**

3.

XXVIII.—1 and 2. Porcelain and Metal Clock. Boulonnois, Paris.
3. Card Stand in pulverized wood. Latry & Co., Paris.

XXIX.—Cast-iron Railing. Barbezat & Co., Paris.
Gas Bracket. Hollenbach, Austria.

XXX.—Panel in Silk.   Haas & Sons, Austria.

GBS_FOLDOUT

GBS_FOLDOUT

GBS_FOLDOUT

GBS_FOLDOUT

GBS_FOLDOUT

GBS_FOLDOUT

PARIS UNIVERSAL EXPOSITION, 1867.
REPORTS OF THE UNITED STATES COMMISSIONERS.

EXTRACTS

FROM THE

REPORT OF THE INTERNATIONAL COMMITTEE

ON

WEIGHTS, MEASURES, AND COINS,

WITH A

NOTICE OF THE USE OF THE METRIC SYSTEM
IN THE UNITED STATES

AND ITS

RELATIONS TO OTHER SYSTEMS OF WEIGHTS AND MEASURES.

WASHINGTON:
GOVERNMENT PRINTING OFFICE.
1870.

CONTENTS.

WEIGHTS, MEASURES, AND COINS.

PREFACE.

In the absence of any report by the Commission upon the exhibition of weights and measures, the editor has been induced to prepare a brief notice and to give a translation of the valuable series of tables, published by the committee, of the weights, measures, and coins of the various countries which took part in the display.

Such a notice not only properly forms a portion of the series of reports of the United States Commissioners, but it is especially desirable in connection with them, inasmuch as in many of the reports quantities are expressed in the denominations of the metric system, and also in those of other systems prevailing in various countries, and by means of the tables the relative values of the quantities so expressed may be ascertained without recourse to other publications.

The official report of the committee was addressed by the president, M. L. Mathieu, to his excellency the minister of state and of the finances and vice-president of the Imperial Commission, and was published in Paris.[1] Two copies of this published report were officially transmitted to the Department of State by the Hon. Samuel B. Ruggles, vice-president of the United States Commission, and delegate to the international monetary conference. The report of Commissioner Ruggles upon the proceedings of the sub-committee and of the international monetary conference in relation to the unification of coinage has already been published,[2] and some observations upon the subject, will also be found in the Report upon the Precious Metals, pp. 242-307.

THE EXHIBITION OF WEIGHTS AND MEASURES.

The pavilion for the reception of the collections of weights, measures, and coins of all countries was a circular structure, surmounted by a dome, erected in the center of the central garden of the great Exposition building. It was divided into twenty sections, assigned to each of the different countries or groups of countries.

The visitor who entered this pavilion by the door facing the main

---

[1] Exposition Universelle de 1867, à Paris: Comité des poids et mesures et des monnaies. Rapport et procès-verbaux—Catalogue officiel. Paris: E. Dentu. Small 8°, pp. 163.

[2] Report to the Department of State, by Samuel B. Ruggles, delegate from the United States to the Monetary Conference at Paris, 1867. 4°, 20 pp. No date, (Washington, 1868,) and in 8°. Fortieth Congress, second session, Senate Ex. Doc. No. 14.

avenue of the Exposition building saw the weights and coins of the various countries in the following order, commencing upon the left: France, Holland, Belgium, Prussia, and States of Northern Germany, States of Southern Germany, Austria, Switzerland, Spain, Denmark, Sweden and Norway, Russia, Italy, Turkey, Egypt, China, Japan and the different countries of Asia, Morocco and the different States of Africa, the different countries of South America, Brazil, United States of America, England. This was also the order of succession of the exhibition in the Palace. The names of the different countries were inscribed upon the outside above each of the sections.

The pavilion was divided into two stories. On the ground floor were exhibited:

1. The measures of capacity, at least such of them as were not too large to be admitted in the under part of the glass cases.

2. The weights placed upon a shelf above the measures.

3. The linear measures were ranged vertically facing the plate glass of the case, and were supported by two wires, one on each side, stretched at the height of a metre above the level foundation upon which all the measures rested, with their initial divisions in the same plane. It was thus made easy to compare the length of the metre with all the other measures.

4. The coins systematically arranged and suspended against the interior surface of the glass of the case.

In the first story above the ground floor there were the bank bills, paper money of all kinds, and the postage stamps sent by the different countries. There was also in this part of the building a book-case containing the documents placed at the service of the international committee.

Four clock-dials of different types—Roman, Turkish, Indian, and Chinese—were placed upon the four faces of the pavilion. In the inside of the cupola there was a dial with Arabic figures. All of these dials indicated Paris time. Upon the summit of the dome a terrestrial globe was supported with its axis parallel to the axis of the earth, and it revolved once in twenty-four hours. This motion and the movement of the hands upon the clock dials were all given by a clock movement in the center of the base of the cupola.

Within this building there were exhibited no less than thirty-six different systems of weights based upon thirty-six different units; sixty-seven different measures based upon sixty-two different units; and thirty-five different standards of gold and silver coin, belonging to eighteen different monetary systems, based upon eighteen different units or measures of value.

THE INTERNATIONAL COMMITTEE.

The Committee of Weights, Measures, and Coins, was organized by a decree of the Emperor Napoleon, dated June 7, 1866, with the object of ascertaining the best method of utilizing the general assembling of people of all nations at the Exposition of 1867, for the adoption of a uniform system of weights, measures, and coins. This organization was completed by the decrees of the 14th of February and the 10th of April, 1867.

From the 15th of June, 1866, until the rendering of the report by M. L. Matthieu, the president, in 1867, the committee held twenty-six sessions, and in order to facilitate the work it had divided into three sub-committees, the first of which was charged with the study of the question of weights and measures; the second, with that of coins; and the third, with areometry.

These sub-committees prepared a series of propositions, which, after having been examined by the committee as a whole, were adopted. They also submitted reports as follows:

1. Report upon the uniformity of weights and measures, prepared by M. de Jacobi.

2. Report upon the uniformity of coins, prepared by Baron de Hock.

3. Report upon areometry, by M. de Baumhauer.

The report of the sub-committee upon weights and measures is based upon the following four fundamental propositions:

1. That the decimal system, being in conformity with the system of enumeration universally employed, is the most proper for expressing the multiples and parts of weights, measures, and coins.

2. That the metric system is perfectly fit to be universally adopted, on account of the scientific principles upon which it is established, the homogeneity which exists in the relation of all its parts, and the simplicity and facility of its application in the sciences and arts, in industry and commerce.

3. That the instruments of precision and methods employed for obtaining copies of the original standard have attained such perfection that the accuracy of these copies meets every requirement of industry and commerce, and even the exigencies of modern science.

4. As every economy of labor, both material and intellectual, is equivalent to an actual increase of wealth, the adoption of the metric system, which may be ranked in the same order of ideas as tools and machines, railways, telegraphs, logarithms, &c., particularly commends itself in an economical point of view.

The report, based upon these general propositions, sets forth at length the advantages of the metric system in the different branches of human activity, and the means of extending its use. The following are the chief subdivisions of the report:

Situation of different countries with regard to the adoption of the metric system.

States which have adopted the metric system.

States which have partially adopted the metric system.

Countries in which the systems of weights and measures have nothing in common with the metric system.

Teaching the metric system in primary schools.

Advantages of the use of the metric system in the sciences.

Advantages of the metric system in commercial transactions.

Easy application of the system in the various industries, in mechanics, and in construction.

Utility of the metric system in postal tariffs and in telegraphy.

Taxes upon merchandise duties.

Verification of the accuracy of weights and measures.

Nomenclature.

Inconveniences of a mixed system of transition.

The report concludes by urging the importance of: First. Prescribing the study of the metric system in all schools, and exacting a knowledge of it in all public examinations. Second. Introducing its exclusive use in all scientific publications, in official statistics, in the post-offices, in the custom-houses, in all public works, and in such other executive departments as the governments may find convenient.

The sub-committee upon areometry, under the presidency of E. H. Von Baumbauer, presented the three following propositions as a *résumé* of their report :

1. It is desirable that for all transactions concerning liquids the same systems of graduation should be adopted in all countries.

2. It is desirable that the particular scales used for different liquids should be decimal and based upon the densities or specific volumes.

3. The committee express the wish that the centigrade thermometer and the metric graduation of the barometer should be generally adopted.

The Committee on Weights and Measures enumerate the countries that have fully adopted the metric system of France, and made its use compulsory, as follows: France, Belgium, Holland, Italy, Roman States, Spain, Portugal, Greece, Mexico, Chili, Brazil, New Granada, and other South American States. (To this list the North German Confederation may now be added.) In Brazil the law will not be enforced until 1873.

In Great Britain, since 1864, the use of the metric system has been optional. There are no less than ten different systems of weights authorized by the laws, and there are about twenty different kinds of the bushel.

THE METRIC SYSTEM IN THE UNITED STATES.

The metric-decimal system of France has been legalized and its use made optional in the United States since July 26, 1866, when Congress enacted as follows:

Be it enacted by the Senate and House of Representatives of the United States of America in Congress assembled, That, from and after the passage of this act, it shall be lawful throughout the United States of America to employ the weights and measures of the metric system; and no contract, or dealing, or pleading in any court shall be deemed invalid, or liable to objection, because the weights or measures expressed or referred to therein are weights or measures of the metric system.

SEC. 2. *And be it further enacted*, That the tables in the schedule hereto annexed shall be recognized, in the construction of contracts, and in all legal proceedings, as establishing, in terms of the weights and measures now in use in the United States, the equivalents of the weights and measures expressed therein in terms of the metric system; and said tables may be lawfully used for computing, determining, and expressing in customary weights and measures the weights and measures of the metric system.

Measures of length.

METRIC DENOMINATIONS AND VALUES.		EQUIVALENTS IN DENOMINATIONS IN USE.
Myriametre	10,000 metres.	6.2137 miles.
Kilometre	1,000 metres.	0.62137 mile, or 3,280 feet and 10 inches.
Hectometre	100 metres.	328 feet and 1 inch.
Dekametre	10 metres.	393.7 inches.
Metre	1 metre.	39.37 inches.
Decimetre	$\frac{1}{10}$ of a metre.	3.937 inches.
Centimetre	$\frac{1}{100}$ of a metre.	0.3937 inch.
Millimetre	$\frac{1}{1000}$ of a metre.	0.0394 inch.

Measures of surface.

METRIC DENOMINATIONS AND VALUES.		EQUIVALENTS IN DENOMINATIONS IN USE.
Hectare	10,000 square metres.	2.471 acres.
Are	100 square metres.	119.6 square yards.
Centare	1 square metre.	1550 square inches.

Measures of capacity.

METRIC DENOMINATIONS AND VALUES.			EQUIVALENTS IN DENOMINATIONS IN USE.	
Names.	No. of litres.	Cubic measure.	Dry measure.	Liquid or wine measure.
Kilolitre or stere.	1000	1 cubic metre	1.308 cubic yard	264.17 gallons.
Hectolitre	100	$\frac{1}{10}$ of a cubic metre	2 bushels and 3.35 pecks.	26.417 gallons.
Dekalitre	10	10 cubic decimetres	9.08 quarts.	2.6417 gallons.
Litre	1	1 cubic decimetre	0.908 quart.	1.0567 quart.
Decilitre	$\frac{1}{10}$	$\frac{1}{10}$ of a cubic decimetre	6.1022 cubic inches	0.845 gill.
Centilitre	$\frac{1}{100}$	10 cubic centimetres	0.6102 cubic inch.	0.338 fluid-ounce.
Millilitre	$\frac{1}{1000}$	1 cubic centimetre	0.061 cubic inch	0.27 fluid-drachm.

Weights.

Metric denominations and values.			Equivalents in denominations in use.
Names.	Number of grammes.	Weight of what quantity of water at maximum density.	Avoirdupois weight.
Millier or tonneau	1000000	1 cubic metre	2204.6 pounds.
Quintal	100000	1 hectolitre	220.46 pounds.
Myriagramme.	10000	10 litres	22.046 pounds.
Kilogramme or kilo	1000	1 litre	2.2046 pounds.
Hectogramme	100	1 decilitre	3.5274 ounces.
Dekagramme	10	10 cubic centimetres	0.3527 ounce.
Gramme	1	1 cubic centimetre	15.432 grains.
Decigramme	$\frac{1}{10}$	$\frac{1}{10}$ cubic centimetre	1.5432 grain.
Centigramme	$\frac{1}{100}$	10 cubic millimetres	0.1543 grain.
Milligramme	$\frac{1}{1000}$	1 cubic millimetre	0.0154 grain.

The foregoing schedule and the enactment may be found in the Revised Statutes of the Thirty-ninth Congress, first session, Chap. CCCI, p. 339. They have also been printed in the report of the Committee on Coinage, Weights and Measures, House of Representatives, Report No. 62; in the Report of the Regents of the Smithsonian Institution for 1865, and reprinted in a separate pamphlet in 1868.

Professor H. A. Newton observes: " The numbers are not carried to the highest degree of accuracy, but the amount of the error in them is generally (except in some of the smaller denominations) less than the change due to a difference of two or three degrees of temperature of the standard metre bar."

In addition to the law authorizing the use of the metric system, the Secretary of the Treasury has been directed (1866) to furnish to the governor of each State one set of the standard weights and measures of the metric system for the use of the States respectively; and further, the Postmaster General has been directed to furnish to the post offices exchanging mails with foreign countries, and to such other offices as he shall think expedient, postal balances denominated in grammes, of the metric system, and, until otherwise provided by law, one half ounce avoirdupois is to be deemed and taken for postal purposes as the equivalent of fifteen grammes of the metric weights, and so adopted in progression; and the rates of postage are to be applied accordingly.

For the details of the movement for the adoption of the metric system in the United States, and of the advantages which would attend this adoption, reference may be made to the report of the Committee on Coinage, Weights and Measures, House of Representatives, Thirty-ninth Congress, first session, Report No. 62, May 17, 1866; also, Report of the Committee of the National Academy of Sciences upon Weights, Measures, and Coinage, and the Report upon Weights and Measures by the Hon. John Quincy Adams, Secretary of State, 1821. Other documents relating to the adoption of the metric system in this and other countries are numerous, but among them may be cited particularly the Reports, first and second, of the Standards Commission, British Blue-books; *Rapports et Procès-verbaux, Comité des Poids et Mesures et des Monnaies: Exposition Universelle de* 1867, (already cited;) The Metric System of Weights and Measures, with Tables, &c., prepared by Professor H. A. Newton, of Yale College, published in the Smithsonian Report for 1855. The last-mentioned document contains a very complete and carefully prepared series of tables for the conversion of quantities expressed in the metric system into their equivalents of other systems in use in the United States. A few of these tables have been selected and are here appended for the convenience of the readers of the reports upon the Exposition.

It will be observed that the French orthography is retained throughout. This is done for uniformity, and to avoid the word *gram*. The use of gram for gramme is a constant source of error because of the resemblance of gram to grain, for which last it is constantly mistaken by authors, compositors, proof-readers, and readers generally. As they are both denominations of weight the substitution of one for the other is particularly liable to pass unobserved in reading proof, a difficulty which can be avoided by using the French form of the word.

The French orthography should be retained for the sake of uniformity also. The metric system is destined to be generally international. The names and orthography of all its divisions should be equally so. For this reason alone we should refrain from anglicizing the French names. Examples of error consequent upon the use of the anglicized form of the word gramme are to be found in the Exposition reports, and in congressional documents. The editor regrets that the rule to express all the denominations of the metric system in their appropriate French names was not adopted at the outset of the publication.

TABLES FOR THE CONVERSION OF METRIC WEIGHTS AND MEASURES INTO OTHER DENOMINATIONS.

For the sake of uniformity of these tables the metre is regarded as 39.37 inches, and the kilogramme as 2.2046 avoirdupois pounds.

Scheme of the metric system.

Ratios.	Lengths.	Surfaces.	Volumes.	Weights.
1000000	....	....	....	Millier, or Tonneau.
100000	....	....	....	Quintal.
10000	Myriametre.........	....	Dekastere........	Myriagramme.
1000	Kilometre	....	Kilolitre, or Stere.	Kilogramme, (2.2046 lbs.av.)
100	Hectometre.........	Hectare, (2.471 acres.)	Hectolitre	Hectogramme.
10	Dekametre	....	Dekalitre.........	Dekagramme.
1	Metre, (39.37 in.) ..	Are..............	Litre............	Gramme, (15.4322 grains.)
$\frac{1}{10}$	Decimetre	....	Decilitre	Decigramme.
$\frac{1}{100}$	Centimetre	Centare	Centilitre.........	Centigramme.
$\frac{1}{1000}$	Millimetre.........	....	Millilitre	Milligramme.

Table for converting metres into inches.

Metres.	0.	1.	2.	3.	4.	5.	6.	7.	8.	9.
0	0	39.4	78.7	118.1	157.5	196.8	236.2	275.6	315.0	354.3
10	393.7	433.1	472.4	511.8	551.2	590.5	629.9	669.3	708.7	748.0
20	787.4	826.8	866.1	905.5	944.9	984.2	1023.6	1063.0	1102.4	1141.7
30	1181.1	1220.5	1259.8	1299.2	1338.6	1377.9	1417.3	1456.7	1496.1	1535.4
40	1574.8	1614.2	1653.5	1692.9	1732.3	1771.6	1811.0	1850.4	1889.8	1929.1
50	1968.5	2007.9	2047.2	2086.6	2126.0	2165.3	2204.7	2244.1	2283.5	2322.8
60	2362.2	2401.6	2440.9	2480.3	2519.7	2559.0	2598.4	2637.8	2677.2	2716.5
70	2755.9	2795.3	2834.6	2874.0	2913.4	2952.7	2992.1	3031.5	3070.9	3110.2
80	3149.6	3189.0	3228.3	3267.7	3307.1	3346.4	3385.8	3425.2	3464.6	3503.9
90	3543.3	3582.7	3622.0	3661.4	3700.8	3740.1	3779.5	3818.9	3858.3	3897.6

Table for converting kilogrammes into avoirdupois pounds.

Kilogrammes.	0.	1.	2.	3.	4.	5.	6.	7.	8.	9.
0	0	2.205	4.409	6.614	8.818	11.023	13.228	15.432	17.637	19.841
10	22.046	24.251	26.455	28.660	30.864	33.069	35.274	37.478	39.683	41.887
20	44.092	46.297	48.501	50.706	52.910	55.115	57.320	59.524	61.729	63.933
30	66.138	68.343	70.547	72.752	74.956	77.161	79.366	81.570	83.775	85.979
40	88.184	90.389	92.593	94.798	97.002	99.207	101.412	103.616	105.821	108.025
50	110.230	112.435	114.639	116.844	119.048	121.253	123.458	125.662	127.867	130.071
60	132.276	134.481	136.685	138.890	141.094	143.299	145.504	147.708	149.913	152.117
70	154.322	156.527	158.731	160.936	163.140	165.345	167.550	169.754	171.959	174.163
80	176.368	178.573	180.777	182.982	185.186	187.391	189.596	191.800	194.005	196.209
90	198.414	200.619	202.823	205.028	207.232	209.437	211.642	213.846	216.051	218.255

Table for converting avoirdupois pounds into grammes.

Pounds.	0.	1.	2.	3.	4.	5.	6.	7.	8.	9.
0	0	454	907	1361	1814	2268	2722	3175	3629	4082
10	4536	4990	5443	5897	6350	6804	7258	7711	8165	8618
20	9072	9526	9979	10433	10886	11340	11794	12247	12701	13154
30	13608	14062	14515	14969	15422	15876	16329	16783	17237	17690
40	18144	18597	19051	19505	19958	20412	20865	21319	21773	22226
50	22680	23133	23587	24041	24494	24948	25401	25855	26309	26762
60	27216	27669	28123	28577	29030	29484	29937	30391	30845	31298
70	31752	32205	32659	33113	33566	34020	34473	34927	35381	35834
80	36288	36741	37195	37649	38102	38556	39009	39463	39917	40370
90	40824	41277	41731	42185	42638	43092	43545	43999	44453	44906

Table for converting troy ounces into grammes.

Troy ounces.	0.	1.	2.	3.	4.	5.	6.	7.	8.	9.
0	0	31. 10	62. 21	93. 31	124. 42	155. 52	186. 62	217. 73	248. 83	279. 93
10	311. 04	342. 14	373. 25	404. 35	435. 45	466. 56	497. 66	528. 76	559. 87	590. 97
20	622. 08	653. 18	684. 28	715. 39	746. 49	777. 59	808. 70	839. 80	870. 91	902. 01
30	933. 11	964. 22	995. 32	1026. 43	1057. 53	1088. 63	1119. 74	1150. 84	1181. 94	1213. 05
40	1244. 15	1275. 26	1306. 36	1337. 46	1368. 57	1399. 67	1430. 77	1461. 88	1492. 98	1524. 09
50	1555. 19	1586. 29	1617. 40	1648. 50	1679. 60	1710. 71	1741. 81	1772. 92	1804. 02	1835. 12
60	1866. 23	1897. 33	1928. 44	1959. 54	1990. 64	2021. 75	2052. 85	2083. 95	2115. 05	2146. 16
70	2177. 27	2208. 37	2239. 47	2270. 58	2301. 68	2332. 78	2363. 89	2394. 99	2426. 10	2457. 20
80	2488. 30	2519. 41	2550. 51	2581. 62	2612. 72	2643. 82	2674. 93	2706. 03	2737. 13	2768. 24
90	2799. 34	2830. 45	2861. 55	2892. 65	2923. 76	2954. 86	2985. 96	3017. 07	3048. 17	3079. 28

Table for converting grammes into grains.

Grammes.	0.	1.	2.	3.	4.	5.	6.	7.	8.	9.
0	0	15. 43	30. 86	46. 30	61. 73	77. 16	92. 59	108. 03	123. 46	138. 89
10	154. 32	169. 75	185. 19	200. 62	216. 05	231. 48	246. 92	262. 35	277. 78	293. 21
20	308. 64	324. 08	339. 51	354. 94	370. 37	385. 81	401. 24	416. 67	432. 10	447. 53
30	462. 97	478. 40	493. 83	509. 26	524. 69	540. 13	555. 56	570. 99	586. 42	601. 80
40	617. 29	632. 72	648. 15	663. 58	679. 02	694. 45	709. 88	725. 31	740. 75	756. 18
50	771. 61	787. 04	802. 47	817. 91	833. 34	848. 77	864. 20	879. 64	895. 07	910. 50
60	925. 93	941. 36	956. 80	972. 23	987. 66	1003. 09	1018. 53	1033. 96	1049. 39	1064. 82
70	1080. 25	1095. 69	1111. 12	1126. 55	1141. 98	1157. 42	1172. 85	1188. 28	1203. 71	1219. 14
80	1234. 58	1250. 01	1265. 44	1280. 87	1296. 30	1311. 74	1327. 17	1342. 60	1358. 03	1373. 47
90	1388. 90	1404. 33	1419. 76	1435. 19	1450. 63	1466. 06	1481. 49	1496. 92	1512. 36	1527. 79

Table for converting grains into grammes.

Grains.	0.	1.	2.	3.	4.	5.	6.	7.	8.	9.
0	0	0. 0648	0. 1296	0. 1944	0. 2592	0. 3240	0. 3888	0. 4536	0. 5184	0. 5832
10	0. 6480	0. 7128	0. 7776	0. 8424	0. 9072	0. 9720	1. 0368	1. 1016	1. 1664	1. 2312
20	1. 2960	1. 3608	1. 4256	1. 4904	1. 5552	1. 6200	1. 6848	1. 7496	1. 8144	1. 8792
30	1. 9440	2. 0088	2. 0736	2. 1384	2. 2032	2. 2680	2. 3328	2. 3976	2. 4624	2. 5272
40	2. 5920	2. 6568	2. 7216	2. 7864	2. 8512	2. 9160	2. 9808	3. 0456	3. 1104	3. 1752
50	3. 2400	3. 3048	3. 3696	3. 4344	3. 4992	3. 5640	3. 6288	3. 6936	3. 7584	3. 8232
60	3. 8880	3. 9528	4. 0176	4. 0824	4. 1472	4. 2120	4. 2768	4. 3416	4. 4064	4. 4712
70	4. 5360	4. 6008	4. 6656	4. 7304	4. 7952	4. 8600	4. 9248	4. 9896	5. 0544	5. 1192
80	5. 1840	5. 2488	5. 3136	5. 3784	5. 4432	5. 5080	5. 5728	5. 6376	5. 7024	5. 7672
90	5. 8320	5. 8968	5. 9616	6. 0264	6. 0912	6. 1560	6. 2208	6. 2856	6. 3504	6. 4152

TABLES OF THE WEIGHTS, MEASURES, AND COINS OF VARIOUS COUNTRIES.

The tables which follow, of the weights, measures, and coins in use in various countries, are taken from the official report and catalogue of the international committee, published in the French language. The notes and explanatory statements are translated, but the French names, such as *once, livre, quintal,* &c., of the denominations of weight of other countries, have been retained, though it has not been deemed desirable to give their equivalents in English.

In the report of the committee the value of the principal unit of money for each country is given in francs in explanatory notes at the foot of the page. To these a statement of the value in the dollar, gold standard of the United States, has been added. The value of gold relatively to silver has been assumed to be as 15⅝ to 1; and this may account, in some cases, for the apparent discrepancy in the values as stated in francs and in dollars.

FRANCE.

The metric system was introduced in France by the law of 18 Germinal, year III, but the exclusive use of the new system was not made obligatory until the law of the 4th of July, 1837, which became executory on the 1st day of January, 1840.

WEIGHTS.

Milligramme .... = 0^k.000001.
Centigramme .... = 0^k.00001.
Décigramme .... = 0^k.0001.
Gramme .... = 0^k.001.
Décagramme .... = 0^k.01.
Hectogramme .... = 0^k.1.
Kilogramme .... = 1^k = weight in vacuo of one cubic decimetre of distilled water, at 4° centigrade.
Myriagramme .... = 10^k.
Quintal .... = 100^k.
Tonne .... = 1000^k.

MEASURES.

1. MEASURES OF LENGTH.

Millimètre .... = 0^m.001.
Centimètre .... = 0^m.01.
Décimètre .... = 0^m.1.
Mètre .... = 1^m = the ten-millionth part of the fourth part of the terrestrial meridian.
Décamètre .... = 10^m.
Road measure.. { Hectomètre ... = 100^m.
Kilomètre ..... = 1000^m.
Myriamètre ... = 10000^m.

2. MEASURES OF SURFACE, (SQUARE MEASURE.)

Ordinary measures.

These are the squares of the measures of length.

Land measures.

Centiare .... = 0.01 are.
Are .... = 1 are = one hundred square metres.
Hectare .... = 100 ares.

3. MEASURES OF VOLUME.

Ordinary measures.

These are the cubes of the measures of length.

Measures for wood.

Centistère .... = 0.01 stere.
Décistère........................ = 0.1 stere.
Stère............................ = 1 stere = one cubic metre.
Décastère .... = 10 steres.

Measures of capacity.

Millilitre .... = $0^l.001$.
Centilitre .... = $0^l.01$.
Décilitre .... = $0^l.1$.
Litre............................. = 1^l = one cubic decimetre.
Décalitre .... = 10^l.
Hectolitre .... = 100^l.
Kilolitre .... = 1000^l.

COINS.[1]

Metal.	Name.	Fineness.	Weight in grammes.	Value in francs.
Gold.........................	Piece of 100 francs...................	$\frac{900}{1000}$	32. 25806	100
	Piece of 50 francs..................	... do....	16. 12903	50
	Piece of 20 francs..................	... do....	6. 45161	20
	Piece of 10 francs..................	... do....	3. 22580	10
	Piece of 5 francs..................	... do....	1. 61290	5
Silver......................	Piece of 5 francs..................	... do....	25	5
	Piece of 2 francs..................	$\frac{835}{1000}$	10	2
	Piece of 1 franc..................	... do....	5	1
	Piece of 50 centimes	... do....	2. 5	0. 50
	Piece of 20 centimes	... do....	1	0. 20
Bronze......................	Piece of 10 centimes	....	10	0. 10
	Piece of 5 centimes	....	5	0. 05
	Piece of 2 centimes	....	2	0. 02
	Piece of 1 centime	....	1	0. 01

[1] According to the regulations of the Monetary Convention of the 23d of December, 1865, concluded between France, Italy, Switzerland, and Belgium, for the weights, fineness, and the values of the coins of gold and of silver. The value of the 20-franc gold piece in dollars (U. S.) is $3.8591; of the silver 5-franc, $0.9726.

NETHERLANDS.

The system of weights and measures of Holland is the same as the metric system. The old names only have been retained.

WEIGHTS.

Korrel	(*grain*)	= decigramme.
Wigtje	(*gramme*)	= gramme.
Lood	(*gros*)	= decagramme.
Ons	(*once*)	= hectogramme.
NEDERLANDSCH POND (¹)	(*livre néerlandaise*)	= KILOGRAMME.
Centnaar	(*quintal*)	= 100 kilogrammes.

MEASURES.

1. MEASURES OF LENGTH.

Streep	(*ligne*)	= millimetre.
Duim	(*pouce*)	= centimetre.
Palm	(*palme*)	= decimetre.
EL	(*aune*)	= METRE.
Roede	(*perche*)	= decametre.
		= hectometre.
Mijl	(*mille*)	= kilometre.

2. MEASURES OF SURFACE.

VIERKANTE ROEDE	(*perche carrée*)	= ARE.
Bunder	(*bonnier*)	= hectare.

3. MEASURES OF VOLUME.

Dry measures.	Liquids.		Value.
	Vingerhoed (*dé*)		= centilitre.
Maatje (*mesurette*)	Maatje (*mesurette*)		= decilitre.
KOP (*litron*)	KAN (*pot*)		= LITRE.
Schepel (*boisseau*)			= decalitre.
Mudde (*rasière*)	Vat (*baril*)		= hectolitre.
Scheepston or wisse (*tonne*)			= kilolitre.
Last (*charge*)			= 30 hectolitres.

(¹) In pharmacy the Nederlandsch pond = 375 grammes, and is divided into 12 ounces; each ounce into 8 drams; each dram into 3 scruples; and each scruple into 20 grains.

COINS. (¹)

Metal.	Name.	Fineness.	Weight in grammes.	Value in gulden.
Gold'.	Double ducat	$\frac{983}{1000}$	6. 988	
	Ducat	do	3. 494	
	Double William	$\frac{900}{1000}$	13. 458	
	William	do	6. 729	
	Half William	do	3. 3645	
Silver	Rijksdaalder	$\frac{945}{1000}$	25	2. 5
	Gulden, (florin)	do	10	1
	Half gulden, (½ florin)	do	5	0. 5
Billon	25 cents, (Pays-Bas)	$\frac{640}{1000}$	3. 575	0. 25
	10 cents, (Pays-Bas)	do	1. 400	0. 10
	5 cents, (Pays-Bas)	do	0. 685	0. 05
	¼ gulden, (Colonies)	$\frac{720}{1000}$	3. 180	0. 25
	1/10 gulden, (Colonies)	do	1. 250	0. 10
	1/20 gulden, (Colonies)	do	0. 610	0. 05
Copper	Cent, (Pays-Bas)	Pure copper.	3. 845	0. 01
	½ cent, (Pays-Bas)	do	1. 922	0. 005
	2½ cents, (Colonies)	do	12. 500	0. 025
	Cent, (Colonies)	do	4. 800	0. 01
	½ cent, (Colonies)	do	2. 300	0. 005

BELGIUM.

WEIGHTS AND MEASURES.

THE METRIC SYSTEM.

COINS. (²)

Metal.	Name.	Fineness.	Weight in grammes.	Value in francs.
Gold	Piece of 20 francs	$\frac{900}{1000}$	6. 45161	20
Silver	Piece of 5 francs	do	25	5
	Piece of 2 francs	$\frac{835}{1000}$	10	2
	Piece of 1 franc	do	5	1
	Piece of 50 centimes	do	2. 5	0. 50
	Piece of 20 centimes	do	1	0. 20
Nickel	Piece of 20 centimes			0. 20
	Piece of 10 centimes			0. 10
	Piece of 5 centimes			0. 05
Bronze	Piece of 2 centimes			0. 02
	Piece of 1 centime			0. 01

(¹) The gold coins are only merchandise, not being current. The "William" is worth about 10 florins. The florin = 2.1164 francs, or $0.4084.

(²) According to the rules established by the Monetary Convention of the 23d of December, 1865, between France, Italy, Belgium, and Switzerland, for coins of gold and of silver.

PRUSSIA.

WEIGHTS.

ORDINARY WEIGHTS.

Zent			$= \frac{1}{3000}$	pfund.
Quentchen	(quentin)	$= 10$ zent	$= \frac{1}{300}$	pfund.
Loth	($\frac{1}{2}$ once)	$= 10$ quentchen	$= \frac{1}{30}$	pfund.
Pfund	(livre)	$= 30$ loths	$= 500$	grammes.
Centner	(quintal)		$= 100$	pfund.
Tonne	(tonne)	$= 3$ centner	$= 300$	pfund.

FOR PRECIOUS METALS.

Gold.

Gran	(grain)	$= \frac{1}{188}$	of the mark.
Karat	(karat)	$= \frac{1}{24}$	of the mark.
Mark	(mark).. $= \frac{1}{2}$ of the old Prussian pound..	$= 233.855$	grammes.

Silver.

Gran	(grain)	$= \frac{1}{288}$	of the mark.
Loth	($\frac{1}{2}$ once)	$= \frac{1}{16}$	of the mark.

PHARMACY.

Gran	(grain)	$= \frac{1}{5760}$	of the pfund.
Skrupel	(scrupule)	$= 20$ gran	$= \frac{1}{288}$	of the pfund.
Drachme	(drachme)	$= 3$ skrupel	$= \frac{1}{96}$	of the pfund.
Unze	(once)	$= 8$ drachmen	$= \frac{1}{12}$	of the pfund.
Pfund	(livre)	$= 12$ unzen	$= 350.783$	grammes.

MEASURES.

1. MEASURES OF LENGTH.

In common use.				In surveying.		
Linie ...(ligne)		$= \frac{1}{1728}$ ruthe.	Decimal linie	$= \frac{1}{10000}$ ruthe.
Zoll .... (pouce)	$= 12$ linien	$= \frac{1}{144}$ ruthe.	Decimal zoll	$= \frac{1}{100}$ ruthe.
Fuss....(pied)	$= 12$ zoll	$= \frac{1}{12}$ ruthe.	Decimal fuss	$= \frac{1}{10}$ ruthe.
Elle ....(aune)	$= 2\frac{1}{3}$ fuss	$= \frac{17}{36}$ ruthe.		
Faden ..(toise)	$= 6$ fuss	$= \frac{1}{2}$ ruthe.		

Ruthe .... (perche) .... $= 3.76624$ metres.

Meile.......... (mille) .... $= 2000$ ruthen.

2. MEASURES OF SURFACE.

Ordinary measures.

Squares of the measures of length.

Land measures.

Morgen (arpent) $= 180$ square ruthen.

3. MEASURES OF VOLUME.

Solids.

Cubes of the measures of length. For measuring wood the *klafter* (cord) is used. This $= 108$ cubic feet $= 3.3389$ cubic metres.

Dry measures.

Metze............. ....(*mesure*) .... $=$ 3 viertel .... $= \frac{1}{16}$ scheffel.
Viertel ....(*quart*) .... $= \frac{1}{48}$ scheffel.
SCHEFFEL............(*boisseau*) .... $=$ 16 metzen .... $= 1\frac{7}{9}$ cubic fuss.
Tonne ....(*tonne*) .... $=$ 4 scheffel.
Malter.............(*muid*) .... $=$ 3 fass .... $=$ 12 scheffel.
Wispel .... $=$ 2 malter .... $=$ 24 scheffel.
Last.............. ..(*charge*) .... $=$ 60 scheffel.

Liquids.

For wines, brandy, beer.

QUART ....(*quart*)... $=$ 64 cubic zoll.

Anker............. ...... $=$ 30 quart.	Tonne .... $=$ 100 quart.
Eimer.........(*seau*)...... $=$ 60 quart.	Fass .... $=$ 200 quart.
Ohm..........(*muid*)...... $=$ 120 quart.	Kufen........... ...... $=$ 400 quart.
Oxhoft...(*muid et demi*).... $=$ 180 quart.	Gebräude .... $=$ 1800 quart.

COINS.

Metal.	Name.	Fineness.	Weight in Prussian pounds, (500 gr.)	Value in thalers.
Gold.....................	Krone, (crown)	$\frac{900}{1000}$	0. 0222	....
	Halbe krone	do....	0. 0111	....
Silver................. .	Doppelte thaler, (double thaler)...........	do....	0. 074074	2
	Thaler ([1])	do....	0. 037037	1
	Sechstel-thaler stücke, (5 gros)	$\frac{520}{1000}$	0. 010684	$\frac{1}{6}$
Billon.....................	Stücke von 2½ silbergroschen, (2½ gros)....	$\frac{375}{1000}$	0. 0064412	$\frac{1}{12}$
	Silbergroschen, (1 gros)	$\frac{220}{1000}$	0. 0043522	$\frac{1}{30}$
	Halbe silbergroschen, (½ gros).............	do....	....	$\frac{1}{60}$
	4 pfennig, (denier).........................	....	....	$\frac{1}{90}$
	3 pfennig	....	....	$\frac{1}{120}$
	2 pfennig	....	....	$\frac{1}{180}$
	1 pfennig	....	....	$\frac{1}{360}$

([1]) The thaler $=$ 3 fr. 64 c., or $0.7204. The crown $=$ 34 fr. 39 c., or $6.64615.

BAVARIA.

WEIGHTS.

ORDINARY WEIGHTS.

Quentchen ....(*gros*)............... =	.... =	$\frac{1}{128}$ pfund.
Loth............($\frac{1}{4}$ *once*)............. =	4 quentchen .... =	$\frac{1}{32}$ pfund.
Halber vierling....($\frac{1}{2}$ *quart*).......... =	4 loth .... =	$\frac{1}{8}$ pfund.
Vierling...........(*quart*)............. =	2 halbe vierling .... =	$\frac{1}{4}$ pfund.
PFUND (¹) ....(*livre*) .... =	.... =	500 grammes.
Centner...........(*quintal*) .... =	.... =	100 pfund.

PHARMACY.

Gran.............(*grain*) .... =	.... =	$\frac{1}{5760}$ pfund.
Skrupel ....(*scrupule*)........... =	20 gran............... =	$\frac{1}{288}$ pfund.
Drachme .... =	3 skrupel............. =	$\frac{1}{96}$ pfund.
Loth .... =	4 drachmen........... =	$\frac{1}{24}$ pfund.
Unze.............(*once*)............. =	2 loth .... =	$\frac{1}{12}$ pfund.
MEDICINAL PFUND.(*livre médicinale*).... =	12 unzen .... =	360 grammes.

MEASURES.

1. MEASURES OF LENGTH.

Linie ....(*ligne*) .... = $\frac{1}{144}$ fuss.
Zoll ....(*pouce*) .... = 12 linien = $\frac{1}{12}$ fuss.
Fuss (²).......(*pied*) .... = 12 zoll.
Elle ....(*aune*) ...., = 2 fuss, 10¼ zoll, sub-divided $\begin{cases} \frac{1}{2}, \frac{1}{4}, \frac{1}{8}, \frac{1}{16}, \frac{1}{32}. \\ \frac{1}{3}, \frac{1}{6}, \frac{1}{12}, \frac{1}{24}. \end{cases}$
Klafter ....(*toise*)........... = 6 fuss.
Ruthe ....(*perche*) .... = 10 fuss.
Meile (³) ......(*mille*) .... = $\frac{1}{15}$ degree of the equator = 25421.6 fuss.

2. MEASURES OF SURFACE.

Ordinary measures.

Squares of the measures of length.

Land measures.

TAGEWERK (⁴) ....(*journal*) .... = 40000 fuss square, (decimal divisions.)

3. MEASURES OF VOLUME.

Solids.

Cubes of the measures of length. (⁵)

Dry measures.

Maas ....(*mesure*)........................ ...... = $\frac{3}{104}$ metze.

Metze ....(*minot*)..subdivided into $\begin{cases} \frac{1}{4} \text{ metze.....} = 1 \text{ viertel.} \\ \frac{1}{16} \text{ metze.....} = 1 \text{ maasel.} \\ \frac{1}{32} \text{ metze.....} = 1 \text{ dreissiger.} \\ \frac{1}{64} \text{ metze.....} = 1 \text{ half-dreissiger.} \end{cases}$

Scheffel ....(*boisseau*)..... .... = 6 metzen.

(¹) The pfund (livre or pound) of 500 grammes is the pound of the Zollverein. The commercial pound = 560 grammes.

(²) The fuss (foot) = 0.291859172 metre at the temperature of +13° R. Engineers and architects divide he foot decimally.

(³) The meile = 7419.50 metres.

(⁴) The tagewerk = 34.0727 ares.

(⁵) The cubic foot = 0.024861 cubic metre. For wood they use the klafter.(cord) = 126 cubic feet 3.1325 cubic metres.

Liquid measures.

MAAS ..... (*mesure*) .... = 43 cubic feet.
EIMER..... (*seau*) ..... = 64 maas (legal) = 60 maas (commercial) for certain liquids.

COINS. (¹)

Metal.	Name.	Fineness.	Weight in grammes.	Value in florins.
Gold	Krone, (crown).........................	$\frac{900}{1000}$	11. 11111	16
	Halbe krone, (half-crown)....................	... do....	5. 555556	8
Silver..............	Vereins 2-thaler stücke, (piece of 2 thalers)	... do....	37. 037037	3½
	Vereins thaler	... do....	18. 518519	1½
	2-gulden stücke, (piece of 2 florins)	... do....	21. 164021	2
	Gulden, (²) (florin)......................	... do....	10. 582011	1
	½-gulden stücke, (half-florin).....................	... do....	5. 291005	½
	Viertel gulden	$\frac{520}{1000}$	4. 578758	¼
Billon..............	6-kreutzer stücke	....	2. 463054	$\frac{1}{10}$
	3-kreutzer stücke	....	1. 231527	$\frac{1}{20}$
	1-kreutzer stücke	....	0. 841751	$\frac{1}{60}$
Bronze..............	Zwei pfennige, (2 pf.).......................	....	2. 500	$\frac{1}{120}$
	Pfennig	....	1. 250	$\frac{1}{240}$
	Heller	....	6. 625	

WURTEMBERG.

Since the year 1806 Wurtemberg has had a uniform and well-defined system of weights and measures. At that date the standards were selected very nearly equal to the old measures in use before that time. A law of 1859, in force in 1860, introduced the Zollverein pound of 500 grammes as the unit of weight, and made its use obligatory in commerce for the greater portion of merchandise.

WEIGHTS.

ORDINARY WEIGHTS.

Richtfennig .... = $\frac{1}{512}$ pfund.
Quentchen ....(*gros*) .... = 4 richtfennig...... = $\frac{1}{128}$ pfund.
Loth...............(½ *once*)................. = 4 quentchen....... = $\frac{1}{32}$ pfund.
PFUND (³)..........(*livre*)................... = 32 loth .... = 500 grammes.
Centner ....(*quintal*)....................................... = 100 pfund.

SPECIAL WEIGHTS.

Precious metals.

MARK (⁴).........(*marc*) .... = 233.855 grammes

Pharmacy.

MEDICINAL PFUND(*livre*) .... = 357.6337 grammes.

(¹) Since the Monetary Convention of the 24th of January, 1857, the fineness and the weights of the pieces of gold and silver have been the same in all the states of Southern Germany.
(²) The florin = 2 fr. 13 c., or $0.4117.
(³) The Zollverein pound. The division into grammes is optional. It is used for weighing coin.
(⁴) The mark is divided for weighing gold into 24 carats of 12 grains, and for silver into 16 carats of 18 grains.

MEASURES.

1. MEASURES OF LENGTH.

Linie	*(ligne)*		$=$	$\frac{1}{100}$ fuss.
Zoll	*(pouce)*	$=$ 10 linien	$=$	$\frac{1}{10}$ fuss.
Fuss (¹)	*(pied)*	$=$ 10 zoll	$=$	1 fuss·
Elle	*(aune)*	(Divisions $\frac{1}{2}$, $\frac{1}{4}$, $\frac{1}{8}$)	$=$	2.144 fuss.
Ruthe	*(perche)*		$=$	10 fuss.
Meile	*(mille)*	$=$ 260 ruthen	$=$	2600 fuss.

2. MEASURES OF SURFACE.

Ordinary measures.

Squares of the measures of length.

Land measures.

MORGEN .... *(journal)* .... $=$ 38400 square fuss.   (Divisions: $\frac{1}{4}$ morgen and $\frac{1}{2}$ morgen.)

3. MEASURES OF VOLUME.

Ordinary measures.

Cubes of the measures of length.

Dry materials. (²)

Ecklein			$=$	$\frac{1}{32}$ simri.
Vierling	*(quarteron)*	$=$ 8 ecklein	$=$	$\frac{1}{4}$ simri.
SIMRI (³)		$=$ 4 vierling	$=$	$942\frac{1}{4}$ cubic zoll.
Scheffel	*(boisseau)*		$=$	8 simri.

Liquids.

Schoppen	*(chopine)*		$=$	$\frac{1}{4}$ maas.
MAAS (⁴)	*(mesure)*	$=$ 4 schoppen	$=$	$78\frac{1}{4}$ cubic zoll.
Imi			$=$	10 maas.
Eimer	*(seau)*	$=$ 16 imi	$=$	160 maas.
Fuder	*(foudre)*	$=$ 6 eimer	$=$	960 maas.

(¹) The fuss of Wurtemberg $=$ 0.2864903 metre.

(²) There are other measures of volume: For fire-wood, the klafter, (cord,) of which the three dimensions are 6, 6, and 4 feet of 0.2864903 metre, and the volume 144 cubic feet; for hay and straw the measure of 512 cubic feet; for sand, 8 eight cubic feet; for lime, 3.960 cubic feet; for coal, 29 cubic feet, &c.

(³) The simri $=$ 22.1533 litres.

(⁴) This maas is the Hellaichmaas; it equals $78\frac{1}{4}$ cubic inches $=$ 1.837 litre. There is also the Trübaichmaas, which is 7·160 greater; and the Schenkmaas, which is 1·10 smaller than the Hellaichmaas.

COINS. [1]

Metal.	German names.	Fineness.	Value in florins.
Gold.............	Vierfache jubiläums dukaten.(*quadruple ducat du jubilé*)...	$\frac{986}{1000}$	23
	Dukaten....................... (*ducat*)	....do ...	$5\frac{3}{4}$
Silver.............	Doppelte vereins thaler........(*double V. thaler*)	$\frac{900}{1000}$	$3\frac{1}{2}$
	2-gulden stücke ....(*double florin*)	....do ...	2
	Vereins thaler.................(*V. thaler*).....................	....do ...	$1\frac{3}{4}$
	Gulden........................ (*florin*).........................	....do ...	1
	½-gulden stücke...............(½ *florin*).....................	....do ...	$\frac{1}{2}$
Billon.............	6-kreutzer stücke.............(*piece of 6 kreutzer*)	$\frac{333}{1000}$ 1·3	$\frac{1}{10}$
	3-kreutzer stücke.............(*piece of 3 kreutzer*)	....do ...	$\frac{1}{20}$
	1-kreutzer stücke.............(*piece of 1 kreutzer*)...........	$\frac{166}{1000}$ 2·3	$\frac{1}{60}$
Copper.............	½-kreutzer stücke.............(½ *kreutzer*)		$\frac{1}{120}$
	¼-kreutzer stücke.............(¼ *kreutzer*)		$\frac{1}{240}$

BADEN.

WEIGHTS.

ORDINARY WEIGHTS.

Richttheile	....	(*trait*)	....	= $\frac{1}{131072}$	pfund.
Gränchen	....	(*petit grain*)...	= 4 richttheile ..	= $\frac{1}{32768}$	pfund.
Gran	....	(*grain*)	= 4 gränchen....	= $\frac{1}{8192}$	pfund.
Karat	....	....	= 4 grans	= $\frac{1}{2048}$	pfund.
Pfennig	....	(*denier*)	= 4 karat	= $\frac{1}{512}$	pfund.
Quentchen	....	....	= 4 pfennig	= $\frac{1}{128}$	pfund.
Loth	....	(½ *once*).......	= 4 quentchen ..	= $\frac{1}{32}$	pfund.
Unze	....	(*once*)	= 2 loth.........	= $\frac{1}{16}$	pfund.
Vierling	....	(*quart*)	= 4 unze	= $\frac{1}{4}$	pfund.
Mark	....	(*marc*)	= 2 vierling	= $\frac{1}{2}$	pfund.
As	....	....		= $\frac{1}{10000}$	pfund.
Dekas	....	....		= $\frac{1}{1000}$	pfund.
Centas	....	....		= $\frac{1}{100}$	pfund.
Zehntlinge	....	....		= $\frac{1}{10}$	pfund.
PFUND [2]	....	(*livre*)	....	= 500	grammes.
Stein	....	(*pierre*)	....	= 10	pfund.
Centner	....	(*quintal*)	....	= 100	pfund.

PHARMACY.

Gran	....	(*grain*)	....	= $\frac{1}{5760}$	pfund.
Skrupel	....	(*scrupule*).....	= 20 gran	= $\frac{1}{288}$	pfund.
Drachme	....	(*drachme*)	= 3 skrupel ...	= $\frac{1}{96}$	pfund.
Unze	....	(*once*)	= 8 drachmen .	= $\frac{1}{12}$	pfund.
MEDICINAL PFUND	....	(*livre*)	= 12 unzen....	= $\frac{5}{6}$	zollpfund.

[1] System of the Monetary Convention of the 24th of January, 1857, concluded between the states of Southern Germany. The zwei-guldon stücke = $0.8252.

[2] The pound of the Zollverein, called the *zoll-pfund*.

MEASURES.

1. MEASURES OF LENGTH.

Punkt	(point)			$= \frac{1}{1000}$ fuss.
Linie	(ligne)	$= 10$ punkte		$= \frac{1}{100}$ fuss.
Zoll	(pouce)	$= 10$ linien		$= \frac{1}{10}$ fuss.
Fuss (1)	(pied)	$= 10$ zoll		$= 1$ fuss
Elle	(aune)	$= 2$ fuss		$= 2$ fuss.
Klafter	(toise)	$= 3$ ellen		$= 6$ fuss.
Ruthe	(perche)			$= 10$ fuss.
Wegstunde	(lieue)			$= 14814.8148$ fuss.
Meile	(mille)			$= 29629.6296$ fuss.

2. MEASURES OF SURFACE.

Ordinary measures.

Squares of the measures of length. (2)

LAND MEASURES.

Morgen(3) ... (journal) ... $= 400$ square perches, (subdivided into 4 parts.)

3. MEASURES OF VOLUME.

Solids.

Cubes of the measures of length.(4)

Dry measures.

Becher	(gobelet)		$=$	$\frac{1}{10}$ messlein.
MESSLEIN(5)	(petite mesure)		$=$	1 messlein.
Sester	(setier)		$=$	10 messlein.
Malter	(muid)	$= 10$ sester	$=$	100 messlein.
Zuber	(tonneau)	$= 10$ malter	$=$	1,000 messlein.

Liquids.

Glas	(verre)		$=$	$\frac{1}{10}$ maas.
MAAS (6)	(mesure)		$=$	1 maas.
Stütze			$=$	10 maas.
Ohm	(muid)	$= 10$ stützen	$=$	100 maas.
Fuder	(foudre)	$= 10$ ohm	$=$	1,000 maas

COINS. (7)

Metal.	Name.	Fineness.	Value in florins.
Silver	Vereins thaler	$\frac{900}{1000}$	$1\frac{1}{2}$
	Gulden, (florin)	...do	1
	½-gulden stücke	...do	$\frac{1}{2}$
Billon	3-kreutzer stücke, (piece of 3 krentzer)		$\frac{1}{20}$
Bronze	Kreutzer		$\frac{1}{60}$
	½-kreutzer, (½ kreutzer)		$\frac{1}{120}$

(1) The foot of Baden $= 0.30$ centimetre.

(2) The square foot $= 0.09$ square metre.

(3) The morgen $= 36$ ares.

(4) The cubic foot $= 0.027$ cubic metre. For wood, the klafter (corde) $= 3.888$ steres.

(5) The messlein $= 1.5$ litre.

(6) The maas $= 1.5$ litre.

(7) Monetary Convention of the 24th January, 1857, between the states of Southern Germany. The gulden $= \$0.4117$.

HESSE.

WEIGHTS.

ORDINARY WEIGHTS.

Quentchen	(*gros*)		=	$\frac{1}{128}$ pfund.
Loth	(*demi-once*)	= 4 quentchen	=	$\frac{1}{32}$ pfund.
PFUND (1)	(*livre*)	= 32 loth	=	500 grammes.
Centner	(*quintal*)		=	100 pfund.

MEASURES.

1. MEASURES OF LENGTH.

Linie	(*ligne*)		=	$\frac{1}{100}$ fuss.
Zoll	(*pouce*)	= 10 linien	=	$\frac{1}{10}$ fuss.
Fuss (2)	(*pied*)	= 10 zoll	=	1 fuss.
Elle	(*aune*)		=	$2\frac{3}{8}$ fuss.
Klafter	(*toise*)		=	10 fuss.
Wegstunde	(*lieue*)	= 2000 klafter	=	20000 fuss.
Meile	(*mille*)	= 3000 klafter	=	30000 fuss.

2. MEASURES OF SURFACE.

Ordinary measures.

Squares of the measures of length.(3)

Land measures.

Morgen (4) ...... (*journal*) .... = 400 square klafter, divided into four parts.

3. MEASURES OF VOLUME.

Solids.

Cubes of the measures of length.(5)

Dry measures.

MÄSSCHEN (6)	(*petite mesure*)		=	1 mässchen.
Gescheid			=	4 mässchen.
Kumpf	(*écuelle*)	= 4 gescheid	=	16 mässchen.
Simmer		= 4 kumpf	=	64 mässchen.
Malter	(*muid*)	= 4 simmer	=	256 mässchen.

Liquids.

SCHOPPEN (7)	(*chopine*)		=	1 schoppen.
Mass	(*mesure*)		=	4 schoppen.
Viertel	(*quart*)	= 4 mass	=	16 schoppen.
Ohm	(*muid*)	= 20 viertel	=	320 schoppen.

(1) Pound of the Zollverein.   (2) The foot = 0.25 metre.
(3) The square foot = 0.0625 square metre.   (4) The morgen = 25 ares.
(5) The cubic foot = 0.015625 cubic metre; for wood the stecken = 100 cubic feet = 1.5625 cubic metre; for lime the kalkbütte = 10 cubic feet = 0.156 cubic metre.
(6) The mässchen = $\frac{1}{2}$ litre.   (7) The schoppen = $\frac{1}{2}$ litre.

COINS.[1]

Metal.	Name.	Fineness.	Value in florins.
Silver	Verins 2-thaler stücke, (piece of 2 thaler)..........	$\frac{900}{1000}$	$3\frac{1}{4}$
	2-gulden stücke, (piece of 2 florins)....................	..... do....	2
	Vereins thaler	..... do....	$1\frac{1}{2}$
	Gulden, (florin)	..... do....	1
	$\frac{1}{4}$-gulden stüke, ($\frac{1}{4}$ florin)	..... do....	$\frac{1}{4}$
Billon	6-kreutzer stücke		$\frac{1}{10}$
	3-kreutzer stücke		$\frac{1}{20}$
	1-kreutzer stücke		$\frac{1}{60}$
Copper	1 pfennig		$\frac{1}{240}$

AUSTRIA.

WEIGHTS.

ORDINARY WEIGHTS.

Quentchen.... ..... *(gros)* .... $=$ 4 pfennig........ $= \frac{1}{128}$ pfund.
Loth............... *(demi-once)* .... $=$ 4 quentchen .... $= \frac{1}{32}$ pfund.
PFUND [2] .... *(livre)* .... $=$ 32 loth.
Stein .... *(pierre)* .... $=$ 20 pfund.
Centner .... *(quintal)* .... $=$ 100 pfund.
Lägel .... $=$ 125 pfund.
Saum.... .... *(charge)* .... $=$ 250 pfund (for the steel of Styria.)
Saum............... *(charge)* .... $=$ 275 pfund.
Karch............... *(charge)* .... $=$ 400 pfund.
Last.... .... *(charge)* .... $=$ 2000 pfund (for ships.)
Last........... .... *(charge)* .... $=$ 40000 pfund (ordniary.)

PRECIOUS METALS.

Pfennig............ ...... *(grain)* .... (subdiv. of the pfennig $\frac{1}{2}$, $\frac{1}{4}$) $= \frac{1}{256}$ mark.
Quentchen......... *(gros)* .... $=$ 4 pfennig................. $= \frac{1}{64}$ mark.
Loth............... *($\frac{1}{2}$-once)* ..... $=$ 4 quentchen............. $= \frac{1}{16}$ mark.
MARK .... *(marc)* .... $=$ 280.644 grammes.
Karat .... $=$ 0.206085 grammes divided into $\frac{1}{2}$, $\frac{1}{4}$, $\frac{1}{8}$, $\frac{1}{16}$, $\frac{1}{32}$, $\frac{1}{64}$ (for jewels.)

PHARMACY.

Gran .... *(grain)* .... $= \frac{1}{5760}$ pfund.
Skrupel .... *(scrupule)* .... $=$ 20 gran .... $= \frac{1}{288}$ pfund.
Drachme .... *(drachme)* .... $=$ 3 skrupel................. $= \frac{1}{96}$ pfund.
Unze .... *(once)* .... $=$ 8 drachmen .... $= \frac{1}{12}$ pfund.
PFUND [3]...... ....*(livre)* .... $=$ 12 unzen .... $=$ 24 loth.

---

(1) Monetary Convention of January 24, 1857, concluded between the states of Southern Germany.

(2) The Austrian pound $=$ 56.012 grammes; but in the customs and in the collection of taxes and for coins, they make use of the pound of the Zollverein of 500 grammes, divided decimally in 0.500, 0.200, 0.100, 0.050, 0.020, 0.005, 0.002, and 0.001.

(3) The pound of pharmacy $=$ 420.009 grammes.

MEASURES.

1. MEASURES OF LENGTH.

Punkt		$= \frac{1}{1728}$ fuss.
Linie	*(ligne)*	$= \frac{1}{144}$ fuss.
Zoll	*(pouce)*	$= \frac{1}{12}$ fuss.
Fuss ([1])	*(pied)*	$= 1$ fuss.
Elle	*(aune)*	$= 2.465$ fuss.
Klafter	*(toise)*	$= 6$ fuss.
Decimal-linie		$= \frac{1}{1000}$ ruthe.
Decimal-zoll		$= \frac{1}{100}$ ruthe.
Decimal-fuss		$= \frac{1}{10}$ ruthe.
Ruthe	*(perche)*	$= 12$ fuss.
Meile	*(mille)*	$= 24000$ fuss.

2. MEASURES OF SURFACE.

Ordinary measures.

Squares of the measures of length. ([2])

Land measures.

Metzen	$= \frac{1}{6}$ joch.
JOCH ([3])	$= 57600$ square fuss.

3. MEASURES OF VOLUME.

Solids.

Cubes of the measures of length. ([4])

Dry measures.

Becher	$= \frac{1}{128}$ metzen.
Kleines massel	$= \frac{1}{64}$ metzen.
Grosses massel	$= \frac{1}{32}$ metzen.
Halbachtel, müller massel	$= \frac{1}{16}$ metzen.
Achtel	$= \frac{1}{8}$ metzen.
Viertel	$= \frac{1}{4}$ metzen.
Halbe metzen	$= \frac{1}{2}$ metzen.
METZEN *(minot)*	$= 1.9471$ cubic foot.
Muth *(muid)*	$= 30$ metzen.

Liquids.

Seidel *(chopine)*		$= \frac{1}{100}$ eimer.
Gross-seidel *(grande chopine)*	$= 1\frac{1}{2}$ seidel	
Halbe maas *($\frac{1}{2}$ mesure)*	$= 4$ seidel	$= \frac{1}{80}$ eimer.
Maas *(mesure)*	$= 2$ halbe maas	$= \frac{1}{40}$ eimer.
EIMER ([5]) *(seau)*	$= 40$ maas	$= 1.792$ cubic feet.
Fass *(tonneau)*	$= 10$ eimer (for wine)..	$= 2$ eimer, (for beer.)
Dreiling		$= 24$ eimer.
Fuder *(foudre)*		$= 32$ eimer.

([1]) The Austrian foot = 0.316111 metre.
([2]) The square foot = 0.099926 square metre.
([3]) The joch = 57.5545 ares.
([4]) The cubic foot = 0.031588 cubic metre. For coke and coal they use the stübich = 2 metzen; for lime, the kalk-müthel = $2\frac{1}{2}$ metzen; for wood, the cord = $\frac{1}{2}$ kubic-klafter.
([5]) The eimer = 56.60524 litres.

COINS.

Metal.	Name.	Fineness.	Weight in Austrian pounds of 500 grammes.	Value in gulden, (florins.)
Gold	Vierfacher dukaten, (quadruple ducat)	$\frac{986\frac{1}{2}}{1000}$		
	Dukaten, (ducat)	do		
	Krone,[1] (couronne)	$\frac{900}{1000}$	0.0222	
	Halbe krone, ($\frac{1}{2}$-couronne)	do	0.0111	
Silver	Maria-Theresien thaler	$\frac{833\frac{1}{3}}{1000}$		
	3-gulden stücke	$\frac{900}{1000}$	0.74	3
	Vereinsthaler	do	0.037	1$\frac{1}{2}$
	2-gulden stücke	do	0.049	2
	Gulden,[2] (florin)	do	0.024	1
Billon	$\frac{1}{4}$-gulden stücke, ($\frac{1}{4}$ florin)	$\frac{520}{1000}$	0.010	0.25
	10-kreutzer stücke, (piece of 10 kreutzer)	$\frac{500}{1000}$	0.004	0.10
	5-kreutzer stücke, (piece of 5 kreutzer)	$\frac{375}{1000}$	0.002	0.05
Bronze	4-kreutzer stücke, (piece of 4 kreutzer)		0.026	0.04
	Kreutzer		0.006	0.01
	$\frac{1}{2}$-kreutzer stücke, ($\frac{1}{2}$ kreutzer)		0.003	0.005

SWITZERLAND.

In consequence of the law of the 23d of March, 1851, in force since the 31st of December, 1856, Switzerland has had a single system of weights and measures replacing the multiplicity of the old systems.

WEIGHTS.

ORDINARY WEIGHTS.

Loth = $\frac{1}{32}$ livre.   Milligramme.
Once = $\frac{1}{16}$ livre.   Centigramme.
      $\frac{1}{8}$ livre.   Gramme.
      $\frac{1}{4}$ livre.   Hectogramme.
      $\frac{1}{2}$ livre.

LIVRE   = 500 grammes.
Quintal = 100 livres.

PHARMACY.

Grain .... = $\frac{1}{5760}$ livre.
Scrupule .... = 20 grains .... = $\frac{1}{288}$ livre.
Drachme .... = 3 scrupules .... = $\frac{1}{96}$ livre.
Once .... = 8 drachmes .... = $\frac{1}{12}$ livre.
LIVRE .... = $\frac{3}{4}$ of the ordinary livre .... = 375 grammes = 12 onces

MEASURES.

1. MEASURES OF LENGTH.

Trait .... = $\frac{1}{1000}$ pied, (foot.)
Ligne .... = 10 traits .... = $\frac{1}{100}$ pied.
Pouce .... = 10 lignes................................... = $\frac{1}{10}$

[1] The crown = 34 fr. 44c., or $6.6462.   [2] The florin = 2 fr. 47c., or $0.4803.

Pied	$= 10$ pouces	$=$	30 centimetres.
Brache.........		$=$	2 pieds.
Aune.......	$= 2$ braches	$=$	4 pieds.
Toise	$= 3$ aunes	$=$	6 pieds.
Perche		$=$	10 pieds.
Lieue		$=$	16,000 pieds.

2. MEASURES OF SURFACE.

Ordinary measures.

Squares of the measures of length.

Land measures.

Arpent .... $= 400$ pieds square.

3. MEASURES OF VOLUME.

Solids.

Cubes of the measures of length. (¹)

Dry measures.

Emine .... $= \frac{1}{10}$ quarteron.
QUARTERON, (²)(*boisseau*) $=$ volume of 30 livres of water distilled at 4° C. $= 1\frac{8}{8}$ cubic pied.
Sac .... $= 10$ quarterons.

Liquids.

POT (³) .... $=$ volume of 3 livres of water distilled at 4° C. $= \frac{1}{18}$ cubic pied.
Setier .... $= 25$ pots.
Muid.. $= 100$ pots.

COINS.

System of the Monetary Convention of the 23d December, 1865. Gold pieces are not coined in Switzerland. The pieces of 2 francs and of 1 franc, issued in Switzerland in accordance with the law of the 31st January, 1860, are to be current until the 1st of January, 1868.

(¹) For measuring firewood the *toise moule* is used. This is the volume of a solid having for its anterior and posterior faces the area of a square toise. It is left to the cantons to decide upon the length of the sticks, but it must be expressed in legal measures.

(²) The quarteron $= 15$ litres. It is divided in practice in 1-4 and 1-16 quarteron. There are also double quarterons.

(³) The pot $= 1.50$ litre. It is divided into $\frac{1}{2}$, $\frac{1}{4}$, $\frac{1}{8}$ pot.

SPAIN.

The metric system has been legally adopted in all its modifications since the 1st of January, 1859. The denominations of the system have only received the Spanish terminations "metro," "litro," &c.

WEIGHTS AND MEASURES.

THE METRIC SYSTEM.

COINS.[1]

Metal.	Name.	Fineness.	Weight in grammes.	Value in escudos.
Gold	Doblon.....................	$\frac{900}{1000}$	{ 8. 3870 3. 3548 1. 6774	10 4 2
Silver.....................	Duro.....................	.....do	25. 960	2
	Escudo [2]	.....do	12. 980	1
	Peseta	$\frac{810}{1000}$	5. 192	0. 40
	Media peseta	.....do	2. 596	0. 20
	Real	.....do	1. 298	0. 10
Bronze [3]	Medio real	....	12. 500	0. 05
	Cuartilla	....	6. 250	0. 025
	Décima	....	2. 500	0. 010
	Media décima.....................	....	1. 250	0. 005

PHILIPPINE ISLANDS.

Metal.	Name.	Fineness.	Value in escudos.
Gold	Doblon	$\frac{875}{1000}$	{ 4 2 1
Silver.....................	50 centésimos.....................	$\frac{900}{1000}$	0. 50
	20 centésimos.....................	.....do ...	0. 20
	10 centésimos.....................	.....do ...	0. 10

DENMARK.

WEIGHTS.

ORDINARY WEIGHTS.

Ort ....(denier)	=		$\frac{1}{1000}$	pund.
Quint ....(gros) .... = 10 ort.....................	=		$\frac{1}{100}$	pund.
PUND [4]..........(livre) .... = 100 ort.....................	=		500	grammes.
Centner ....(quintal)	=		100	pund.

[1] The monetary system established by the law of June 26, 1864. It has the double standard, the relation of the value of gold to that of silver being 15.476.

[2] The escudo = 2.631 fr., or $0.504975.   [3] 95 copper, 4 tin, 1 zinc.

[4] The pound of 500 grammes was introduced in 1852.

PRECIOUS METALS.

Ort ....(*denier*)... $=$ $\frac{1}{256}$ mark.
Quintin ....(*gros*) .... $=$ 4 ort.................... $=$ $\frac{1}{64}$ mark.
Lod ....(¼ *once*) .... $=$ 4 quintin................. $=$ $\frac{1}{16}$ mark.
MARK............(*marc*) .... $=$ 16 lod..................... $=$ 233. 855 grammes.

PHARMACY.

Gran ....(*grain*) .... $=$ $\frac{1}{5760}$ pund.
Drachme ....(*drachme*) .... $=$ 60 gran .... $=$ $\frac{1}{96}$ pund.
Unze ....(*once*)........... $=$ 8 drachmen................ $=$ $\frac{1}{12}$ pund.
PUND (^{1})........(*livre*) .... $=$ 12 unzen .... $=$ 375   grammes.

MEASURES.

1. MEASURES OF LENGTH.

Linie.............(*ligne*) .... $=$ $\frac{1}{144}$ fod.
Tomme ....(*pouce*) .... $=$ 12 linien.................... $=$ $\frac{1}{12}$ fod.
FOD (^{2})........ (*pied*) .... $=$ 12 tomme .... 1 fod.
Alen .... (*aune*) .... 2 fod.
Favn .... (*toise*) .... $=$ 3 alen .... $=$ 6 fod.
Rode .... (*perche*) .... $=$ 5 alen .... $=$ 10 fod.
Miil .... (*mille*) .... $=$ 12000 alen................ $=$ 24000 fod.

2. MEASURES OF SURFACE.

Ordinary measures.

Squares of the measures of length.

Land measures.

TÖNDELAND.... (*arpent*)... $=$ 14000 square alen.

3. MEASURES OF VOLUME.

Ordinary measures.

Cubes of the measures of length.

Dry measures.

TÖNDE (^{3})..... (*tonneau*) .... $=$ 144 pots .... $=$ 4.50 cubic fod.
½ tönde .... (½ *tonneau*).. $=$ ½ tönde.
¼ tönde... $=$ ¼ tönde.
Skjæppe .... $=$ $\frac{1}{8}$ tönde.
½ skjæppe.................... $=$ 2   fjerdingkar.......... $=$ $\frac{1}{16}$ tönde.
Fjerdingkar .... $=$ 2   ottingkar .... $=$ $\frac{1}{32}$ tönde.
½ fjerdingkar... $=$ $\frac{1}{64}$ tönde.
Halvotting... $=$ $\frac{1}{128}$ tönde.

Liquids.

Pœgel.. $=$ ¼ pot.
POT (^{4})........ (*pot*).. $=$ $\frac{1}{32}$ cubic fod.
Kande... $=$ 2 pots.
Tönde .... (*tonneau*) .... $=$ 4.25 cubic fod, (tönde of beer.)
                              $=$ 120 pot, (tönde of tar.)
Anker........ (*ancre à vin*)...... .... $=$ 39 pots.

(^{1}) The pound of pharmacy is ¾ of the commercial pound.
(^{2}) The Danish fod (foot) $=$ 0.3139 metre.
(^{3}) For coal and charcoal, the tönde $=$ 5. 5 cubic feet.
(^{4}) The pot $=$ 0. 966 litre.

COINS.

Metal.	Name.	Fineness.	Value in rigsdaler.
Gold	Dobbelt Christian d'or	$\frac{8\,9\,0}{1000}$	$14\frac{1}{3}$
.	Frédéric d'or	....do ...	$7\frac{1}{4}$
	Species ducat	$\frac{8\,7\,9}{1000}$	$4\frac{1}{12}$
Silver..............................	Dobbelt daler	$\frac{8\,7\,5}{1000}$	2
	Rigsdaler (¹)...................................	....do ...	1
	Halvdaler	....do ...	$\frac{1}{2}$
	Mark.......................................	$\frac{6\,0\,0}{1000}$	$\frac{1}{6}$
Silver billon	20 cents	$\frac{6\,2\,5}{1000}$	....
	10 cents	....do ...	....
	5 cents	....do ...	....
	3 cents	....do ...	....
	4 skilling...................................	$\frac{2\,5\,0}{1000}$	....
Bronze..............................	Skilling	....	$\frac{1}{71}$
	½ skilling	....	$\frac{1}{36}$
	Cent.......................................	....	$\frac{1}{192}$

SWEDEN.

The ordinances of January 31, 1853, and of November 10, 1865, prescribe the use of the decimal system based upon the old units.

WEIGHTS.

ORDINARY WEIGHTS.

Korn .... (*grain*)....... .... = $\frac{1}{10000}$ skalpund.
Ort .... = 100 korns......... = $\frac{1}{100}$ skalpund.
SKALPUND (²).......(*livre*) ... = 100 orts .... = $\frac{1}{61.022}$ of the weight of a cubic foot of distilled water at 150° C.
Centner ....(*quintal*) . = 100 skalpund..... = 100 skalpunds.
Last...............(*nylast*) .. = 100 quintaux ..... = 10000 skalpunds.

There also exist the weights:

1, 2, 10, 20, 50 korns.
1, 2, 10, 20, 50 orts.
1, 2, 10, 20, 50 skalpunds.

PHARMACY.	POSTAL.
0. 10 korn.	1. 5 ort.
0. 20 korn.	3. 0 orts.
0. 50 korn.	

(¹) The rigsdaler is worth 2 fr. 85 c., or $0,5463. The division into hundredths is customary in the colonies only.

(²) The skalpund = 0.42354 gramme.

MEASURES.

1. MEASURES OF LENGTH.

Linie ....(*ligne*)................................. $=$ $\frac{1}{100}$ fot.
Tum ....(*pouce*).... ...... $= 10$ linien .... $=$ $\frac{1}{10}$ fot.
Fot(1) ......(*pied*)........... $= 10$ tum.......... $= \frac{1}{59.33064}$ of the length of a pendulum beating seconds at Stockholm.

Stang ......(*perche*)........... $= 10$ fots.......... $=$ 10 fot.
Ref ....(*corde*)........... $= 10$ stang .... $=$ 100 fot.

2. MEASURES OF SURFACE.

Ordinary measures.

Squares of the measures of length.

Land measures.

Ref, square .... $=$ 10000 square fot.

3. MEASURES OF VOLUME.

Ordinary measures.

Cubes of the measures of length.

Dry measures.

10 cubic tums .... $\frac{1}{100}$ cubic fot.
20 cubic tums.. $\frac{1}{50}$ cubic fot.
50 cubic tums.. $\frac{1}{20}$ cubic fot.
1 kann $=$ 100 cubic tums................................. $\frac{1}{10}$ cubic fot.
2 kann $=$ 200 cubic tums................................. $\frac{1}{5}$ cubic fot.
3 kann $=$ 300 cubic tums................................. $\frac{3}{10}$ cubic fot.
5 kann $=$ 500 cubic tums................................. $\frac{1}{2}$ cubic fot.
1 CUBIC FOT (2) .... 1 cubic fot.
2 cubic fot .... 2 cubic fot.
3 cubic fot .... 3 cubic fot.
5 cubic fot... 5 cubic fot.

Liquids.

1 cubic tum.. $=$ $\frac{1}{1000}$ cubic fot.
2 cubic tum.. $=$ $\frac{1}{500}$ cubic fot.
5 cubic tum.. $=$ $\frac{1}{200}$ cubic fot.
10 cubic tum... $=$ $\frac{1}{100}$ cubic fot.
20 cubic tum... $=$ $\frac{1}{50}$ cubic fot.
50 cubic tum... $=$ $\frac{1}{20}$ cubic fot.
1 kann .... $=$ $\frac{1}{10}$ cubic fot.
2 kann .... $=$ $\frac{1}{5}$ cubic fot.
5 kann .... $=$ $\frac{1}{2}$ cubic fot.
1 CUBIC FOT $=$ 10 kann................................. $=$ 1 cubic fot.

---

(1) The fot (foot) = 0.2969 metre.   (2) The cubic fot (cubic foot) = 26.17188 litres.

3 W M C

COINS.

Metal.	Name.	Fineness.	Weight in grammes.	Value.
Gold....................	4 ducat	$\frac{976}{1000}$	....	4 ducat.
	2 ducat	....do	....	2 ducat.
	1 ducat (¹)	....do	3. 482	1 ducat.
Silver...................	4 riksdaler	$\frac{750}{1000}$	....	4 riksdaler.
	2 riksdaler	....do	....	2 riksdaler.
	1 riksdaler (²)	....do	33. 925	1 riksdaler.
	0.50 riksdaler	....do	....	0. 50 riksdaler.
	0.25 riksdaler	....do	....	0. 25 riksdaler.
	0.10 riksdaler	....do	....	0. 10 riksdaler.
Bronze..................	0.05 riksdaler	....	....	0. 05 riksdaler.
	0.02 riksdaler	....	....	0.02 riksdaler.
	Ore....................................	....	....	0. 01 riksdaler.
	½ ore	....	....	0. 005 riksdaler.

NORWAY.

The following are according to the law of July 28, 1824 :

WEIGHTS.

ORDINARY WEIGHTS.

Gran ..... $= \frac{1}{131072}$ pund.
Es .... $= 16$ gran .... $= \frac{1}{8192}$ pund.
Ort .... $= 16$ es .... $= \frac{1}{512}$ pund.
Qvintin....... .... $= 4$ ort.................... $= \frac{1}{128}$ pund.
Lod, (half-ounce) .... $= 4$ qvintin .... $= \frac{1}{32}$ pund.
Mark .... $= 16$ lod..................... $= \frac{1}{2}$ pund.
PUND, (³) (pound)............... .... $= 2$ mark $=$ Weight in vacuo of $\frac{1}{62}$ of a Norwegian cubic foot of distilled water at 4°C.
Bismerpund .... $= 12$ pund.
Lispund .... $= 16$ pund.
Vog .... $= 3$ bismerpund .... $= 36$ pund.
Skippund.......................... $= 20$ lispund .... $= 320$ pund.

MONETARY WEIGHTS. (⁴)

PHARMACY.

Gran......... (*grain*) .... $= \frac{1}{37160}$ pund.
Skrupel.................... $= 20$ gran .... $= \frac{1}{288}$ pund.
Drachme.................. $= 3$ skrupel . ... $= \frac{1}{96}$ pund.
Unze .... $= 8$ drachmen .... $= \frac{1}{12}$ pund.
PUND (⁵) .... $= 12$ unzen........................... $= 1$ pund.

(¹) The ducat $= 11$ fr. 66c.   (²) The riksdaler $= 5$ fr. 61c., or $1,1023.
(³) The pound of commerce $= 498.114$ grammes.
(⁴) The monetary pound is equal to the monetary pound of Cologne ; it is divided, like the ordinary pound, into marks, lods, qvintins, orts, es, and grans. It is equal to 123144. 5 grans of the commercial weight, and $= 467.99$ grammes.
(⁵) The pharmaceutical pound is equal to that of Nuremberg; it is worth 94162.25 grains of the commercial pound $= 357.85$ grammes. The law of May 12, 1866, prescribes the use in medicine of the metric system of France.

MEASURES.

1. Measures of length.

Line .... $= \frac{1}{144}$ fod.
Inch.. $= 12$ lines............................... $= \frac{1}{12}$ fod.
$\frac{1}{16}$ alen .... $= 1\frac{1}{2}$ inch............................ $= \frac{1}{8}$ fod.
$\frac{1}{4}$ alen .... $= 6$ inches............................ $= \frac{1}{2}$ fod.
Fod..................(*pied*)............ $= 0.31374$ metre.   This is $\frac{1}{18}$ of the length
of a pendulum beating
seconds in vacuo at 45°
north latitude, and at
the sea level.

Alen ....(*aune*) .... $=$ 2 fods.
Favn ....(*toise*) .... $= 3$ alen .... $=$ 6 fods.
Rode..................(*perche*) .... $= 5$ alen .... $=$ 10 fods.
Grenzmiil ....(*lieue*) .... $=$ 30000 fods.
Miil..................(*mille*) .... $=$ 36000 fods.

2. Measures of surface.

Ordinary measures.

Squares of the measures of length.

Land measures.

TÖNDELAND ([1]) .....(*arpent*).. $= 14000$ square alens.

3. Measures of volume.

Solids.—Ordinary measures.

Cubes of the measures of length.

Dry measures. ([2])

Ottingkar...............(*demi-quarteron*) ..... $= \frac{1}{8}$ skjeppe .... $= 2.25$ pot.
Fjerdingkar ....(*quarteron*)........... $= \frac{1}{4}$ skjeppe .... $= 4.50$ pot.
Notting .... $= \frac{1}{3}$ skjeppe .... $= 6$ pot.
Sœtting .... $= \frac{1}{2}$ skjeppe .... $= 9$ pot.
Skjeppe ....(*boisseau*) .... $= 18$ pot.
Fjerding...............(*quartant*)... $= 36$ pot.
Tönde..................(*tonne*) .... $= 4$ quartants $= 144$ pot $= 4\frac{1}{2}$ cubic fod.

Liquids.

Pögel..................(*chopine*).. $= \frac{1}{4}$ pot.
Pot ([2]).. $= 54$ cubic inches.
Kande.. $= 2$ pot.
Viertel .... $= \frac{1}{16}$ fisktönde ..... $= 7.5$ pot.
Anker .... $= \frac{1}{4}$ am............. $= 40$ pot.
Fisktönde..............(*tonne de poisson*) .... $= 3$ anker .... $= 120$ pot.
Am .... $= \frac{2}{3}$ oxhoved....... $= 160$ pot.
Oxhoved ....(*muid*)............ $= 2$ fisktönde....... $= 240$ pot.

([1]) The Töndeland $= 5512.5875$ metres.
([2]) The unit for dry and liquid materials is the *pot*, which is equal to 0.9651 of the litre.

COINS.

Metal.	Name.	Fineness.	Weight.	Value in speciedaler.
Silver....................	Speciedaler (¹)	$\frac{876}{1000}$	28. 949	1
	½ speciedaler...........................	.....do	14. 474	½
	1 ort	$\frac{878}{1000}$	5. 970	$\frac{3}{10}$
	12 skilling	.....do	2. 890	$\frac{1}{10}$
Bronze..............	4 skilling...............................	....	....	$\frac{1}{30}$
	2 skilling..............................	....	....	$\frac{1}{60}$
	1 skilling (²)	....	....	$\frac{1}{120}$
	½ skilling	....	....	$\frac{1}{240}$

RUSSIA.

At the commencement of the last century, considering the commercial relations between Russia and England, Peter the Great decreed that the unit of length, the sagène, should be equal to seven English feet, and should be divided into three archines.

WEIGHTS.

ORDINARY WEIGHTS.

Dolis........... .... $= \frac{1}{9169}$ founte.

Zolotnick........... .... $= 96$ dolis .... $= \frac{1}{96}$ founte.

Loth .... $= 3$ zolotnicks .... $= \frac{1}{32}$ founte.

FOUNTE (³) ....(livre)........... $=$ weight in vacuo of 25.019 duime of distilled water at the temperature of $16\frac{2}{3}°$ C.

Poude.......... .... $=$　40 founte.

Berkovetz .... $=$ 10 poudes .... $=$　400 founte.

PHARMACY.

Grains................... .... $= \frac{1}{8160}$ founte.

Scrupule.................. ... $= 20$ grains .... $= \frac{1}{288}$ founte.

Drachme .... $=$ 3 scrupules ...... $= \frac{1}{96}$ founte.

Once.... .... $=$ 4 drachmes ...... $= \frac{1}{12}$ founte.

FOUNTE ....(livre) .... $= 84$ zolotnicks ..... $= \frac{7}{8}$ of the commercial pound.

MEASURES.

1. MEASURES OF LENGTH.

Linia...........(ligne) .... $= \frac{1}{1008}$ sagène.

Duime ....(pouce)........ $=$ 12 linia...................... $= \frac{1}{84}$ sagène.

Verchok .... $= \frac{1}{48}$ sagène.

Tchetverke .... $=$ 4 verchok .... $= \frac{1}{12}$ sagène.

Foute.........(pied) .... $=$ 12 duimes .... $= \frac{1}{7}$ sagène.

Archine.......................... $=$ 4 tchetverk.............. $= \frac{1}{3}$ sagène.

SAGÈNE (⁴)..................... $=$ 3 archines................ $=$ 7 English feet.

Vieersta........(verste)........ $=$ 500 sagènes................ $=$ 3500 English feet.

(¹) The speciedaler $= 5.616$ fr., or $1.0929.

(²) The skilling $= 0.0468$ fr.

(³) The founte, (Russian pound,) avoirdupois and monetary pound $= 0.40952$ kilogramme.

(⁴) The sagène $= 2.13356$ metres.

2. MEASURES OF SURFACE.

Ordinary measures.

Squares of the measures of length.

Land measures.

Deciatina .... = 2400s quare sagènes.

3. MEASURES OF VOLUME.

Ordinary measures.—Solids.

Cubes of the measures of length.

Dry measures.

Garnetz .... = ⅛ tchetverik.
Tchetverkas .... = 2 garnetz .... = ¼ tchetverik.
Polou-tchetverik .... = 2 tchetverkas .... = ½ tchetverik.
TCHETVERIK (¹) .... = volume of 64 litres of distilled water at 16¾°
           C., and weighed in vacuo.
Osminas .... = 4 tchetverik .... = 4 tchetverik'
Tchetvert .... = 2 osminas .... = 8 tchetverik.

In the official catalogue of the Russian Section, the measures of volume are given as follows:

Laste .... = 12 tchetvertes .... = 25. 166 hectolitres.
Tchetverte .... = 2 osminas .... = 2. 097 hectolitres.
Osmina .... = 4 tchetverik .... = 1. 049 hectolitre.
Tchetverik .... = 8 garnetz .... = 2. 621 hectolitres.
Garnetz .... = 3. 277 litres.
Vedro .... = { 8 stoffs .... = 1. 229 decalitre.
           { 10 krouchkas .... = 12. 290 litres.
Krouchka .... = 10 tcharkas .... = 1. 229 litre.
Tonneau .... = 40 vedros .... = 4. 916 hectolitres.

Liquids.

Tcharka .... (*petit verre*) .... = $\frac{1}{100}$ vedro.
Quars .... = $\frac{1}{3\frac{1}{3}}$ vedro.
Polou-krouchka (½ *cruche*) .... = 2½ krouchka .... = $\frac{1}{20}$ vedro.
Polou-stoff .... (⅓ *pot*) .... = 2 quars .... = $\frac{1}{16}$ vedro.
Krouchka ...... (*cruche*) .... = 10 tcharkas .... = $\frac{1}{10}$ vedro.
Stoff .... (*pot*) .... = 2 demi-stoffs .... = ⅛ vedro.
Polou-vedro .... (½ *seau*) .... = 4 stoffs .... = ½ vedro.
VEDRO (²) ..... (*seau*) .... = volume of 750. 57 cubic inches (*duime*) of
           distilled water = volume of 30 pounds
           weighed in vacuo at 16¾° C.

(¹) The tchetverik = 26.238 litres.   (²) The vedro = 12.299 litres.

COINS.

Metal.	Name.	Fineness.	Weight.		Value in rubles.
			Zolotnik.	Dolis.	
Gold	Polou-impérial	$\frac{880}{1000}$	1	51	5
Silver	Rouble (¹)	$\frac{833}{1000}$	4	83½	1
	Poltinnik	....do....	2	41	0.50
	Tchetvertak	....do....	1	22	0.25
	Abassis	....do....		94¼	0.20
	Zloty-Polski	$\frac{830}{1000}$		70	0.15
	Grivenik	....do....		48	0.10
	Piatak	....do....		24	0.05
Bronze					0.05
					0.03
					0.02
	Kopek				0.01
	Deneschika				0.005
	Polushka				0.0025

ITALY.

WEIGHTS AND MEASURES.

The metric system was adopted on the 11th of September, 1845, and the decree has been in force since January 1, 1846.

COINS.

The same as the French. The franc is called *lira*.

TURKEY.

There exist in Turkey a great many systems of weights and measures varying with the provinces and the nature of the substances to be measured. The following is in use at Constantinople. It has recently been decided by the Minister of Public Works that the unit of length, the archine or pik, shall hereafter have a length of 75 centimetres exactly. But there is no law obliging the public to use this official unit:

WEIGHTS.

Karat	= 32 subdivisions	= $\frac{1}{6400}$ occa
Denke	= 4 karats	= $\frac{1}{1600}$ occa
Dirhème ......(*drachme*)	= 4 denkes	= $\frac{1}{400}$ occa
¼ occa	= 100 dirhèmes	= ¼ occa
Occa (²)	= 400 dirhèmes	= 1 occa
Kantar ......(*quintal*)		= 44 occa
Tchéki (³)	= 4 kantars	= 176 occa

(¹) The rouble = 4 fr. 04 c., or $0.7779.

(²) The occa = 1285.56 grammes. The prototype of the occa, in platina, was constructed in 1847 by M. Dehevil.

(³) There are many other tchékis in use for different commodities; the above is used in the wood trade.

MEASURES.

1. MEASURES OF LENGTH.

For construction and topography.

Nocktat......(*point*) .... $= \frac{1}{3456}$ archine.
Hatt ....(*ligne*) .... $= 12$ nocktats .... $= \frac{1}{288}$ archine.
Parmack.....(*pouce*) .... $= 12$ hatts .... $= \frac{1}{24}$ archine.
ARCHINE (¹) .... $= 24$ parmacks.

For measuring piece goods.

Guirah .... $= \frac{1}{16}$ endazé.
Ouroub .... $= 2$ guirahs .... $= \frac{1}{8}$ endazé.
ENDAZÉ (²) .... $= 8$ ouroubs .... $= 1$ endazé.

2. MEASURES OF SURFACE.

The squares of the measures of length.

3. MEASURES OF VOLUME.

For cereals.

Chinik............ .... $= \frac{1}{4}$ kilé.
Kilé (³) .... $= 4$ chiniks .... $=$ volume of 21 occas.

COINS.

Metal.	Name.	Fineness.	Weight.		Value in piasters, (kourouche.)
			Dirhèmes.	*Karats.*	
Gold....................	Becheyuslyk..........	$\frac{916.50}{1000}$	11	4	500
	Ykiyuzlyk............	....do	5	10	250
	Yusilykmedjid........	....do	2	4	100
	Ellyklyk.............	....do	1	2	50
	Yirmibechlyk.........	....do	0	9	25
Silver.................	Yrmilykmedjid	$\frac{830}{1000}$	7	8	20
	Onlyk...............	....do	3	12	10
	Bechelyk	....do	1	14	5
	Ikylyk	....do	0	12	2
	Kourouche, (⁴) (*piastre*)	....do	0	6	1
	Yaremlyk............	....do	0	3	$\frac{1}{2}$
Bronze....................	40 paras.............	....	6	10 $\frac{39}{92}$	1
	20 paras	....	3	5 $\frac{18}{92}$	$\frac{1}{2}$
	10 paras	....	1	10 $\frac{21}{92}$	$\frac{1}{4}$
	5 paras	....	0	13 $\frac{10}{92}$	$\frac{1}{8}$
	1 para	....	0	2 $\frac{31}{92}$	$\frac{1}{40}$

(¹) The archine $= 0.757738$ metre. $\}$ The prototypes of these measures of length were made in 1841
(²) The endazé $= 0.6528$ metre. $\}$ by M. Dieu Antoine.

(³) The kilé is without relation to the unit of length. It only contains 21 occas of grain. The original is at Constantinople. Merchandise, other than grain, wine, oil, and liquids in general, are weighed rather than measured.

(⁴) The kourouche (piaster) $= 0.225$ fr., or $0.04325.

EGYPT.

WEIGHTS.

DIVISION OF THE DARHIM.

Kamha...........................(*grain*) .... $=$ 1 kamha.
Kérat... $=$ 4 kamha.
DARHIM ([1]) ....(*drachme*) .... $=$ 24 kérat.
Mitkal, (used for weighing precious substances.)

NEW WEIGHTS.

$\frac{1}{64}$ occa... $=$ 6. 25 darhim.
$\frac{1}{32}$ occa .... $=$ 12. 50 darhim.
$\frac{1}{16}$ occa .... $=$ 25   darhim.
$\frac{1}{8}$ occa .... $=$ 50   darhim.
$\frac{1}{4}$ occa .... $=$ 100 darhim.
$\frac{1}{2}$ occa .... $=$ 200 darhim.
1 OCCA ([2]) .... $=$ 400 darhim.

OLD WEIGHTS.

$\frac{1}{24}$ rottoli .... $=$ 6   darhim.
$\frac{1}{12}$ oukich .... $=$ 12 darhim.
$\frac{1}{6}$ oukich .... $=$ 24 darhim.
$\frac{1}{4}$ oukich .... $=$ 36 darhim.
$\frac{1}{2}$ oukich .... $=$ 72 darhim.
1 ROTTOLI ([3]) .... $=$ 144 darhim.

MEASURES.

1. MEASURES OF LENGTH.

Kérat .... $=$ $\frac{1}{24}$ deraâ.
Abdat...........(*palme*)................. $=$ 4 kerats .... $=$ $\frac{1}{6}$ deraâ.
Kadam.........(*pied*)................. $=$ 3 abdats .... $=$ $\frac{1}{2}$ deraâ.
DERAÂ ([4]) ....(*pic*)............. $=$ 2 kadam .... $=$ 1 deraâ.
Cassaba of the cadastre................. $=$ 3 metres 55 centimetres.

Road measure. ([5])

Bââh..............(*brasse*) .... $=$ 2¼ deraâ.
Cassaba........................... $=$ 2 bââh .... $=$ 5 deraâ.
Mili..........(*mille*) .... $=$ 500 cassabas .. .. $=$ 2,500 deraâ.
Farsâk ....(*lieue*)......... $=$ 3 mili . ... $=$ 7,500 deraâ.
Baride ....(*étape*) .... $=$ 4 farsâk .... $=$ 30,000 deraâ.
Safar-yome ....(*journée de marche*)... $=$ 2 baride .... $=$ 60,000 deraâ.

([1]) The darhim (drachme) $=$ 3.0934 grammes.
([2]) The occa $=$ 1.23739 kilogramme.
([3]) The rottoli $=$ 0.445458 kilogramme.
([4]) There are six kinds of deraâ, all sub-divided into abdats and karats ; they are :
    Deraâ Nili (pik of the Nile) $=$ 0.2545 metre.
    Deraâ Baladi (pik indigène) $=$ 0.5682 metre.
    Deraâ Istambouli (pik or archine of Constantinople) $=$ 0.6691 metre.
    Deraâ Hendazeh (of the merchants, or yard) $=$ 0.6479 metre.
    Deraâ Melmari (for construction) $=$ 0.75 metre.
    Deraâ itinéraire $=$ 0.7389 metre.
([5]) The degree of the meridian is divided into 2¼ safar-yome, (day's march,) into 20 farsâk, (leagues,) or into 60 mili.

2. Measures of surface.

Land measures.

Square of the deraâ meimari and of the cassaba of the cadastre.

FADDAN (¹) $\dots\dots = 333\frac{1}{3}$ square cassabas.
Kérat $\dots\dots = \frac{1}{24}$ faddan.

3. Measures of volume.

Dry measures.

Kérat		$= \frac{1}{32}$ kaddah.
Kharroubah	2 kérat	$= \frac{1}{16}$ kaddah.
Thoumn-kaddah	2 kharroubah	$= \frac{1}{8}$ kaddah.
Roubb-kaddah	2 thoumn-kaddah	$= \frac{1}{4}$ kaddah.
Nisf-kaddah	2 roubb-kaddah	$= \frac{1}{2}$ kaddah.
KADDAH (²)	2 nisf-kaddah	$= 1$ kaddah.
Maluah	2 kaddah	$= 2$ kaddah.
Roubouh	2 maluah	$= 4$ kaddah.
Kélé	2 roubouh	$= 8$ kaddah.
Ouebeh	2 kélé	$= 16$ kaddah.
Ardebb	6 ouebeh	$= 96$ kaddah.

Liquids.

Guirbeh (³) $\dots\dots$ (voie d'eau) $\dots\dots = 66.666$ litres.

COINS.

Metal.	Name.	Fineness.	Weight in grammes.	Value in piasters.
Gold	Livre, (⁴) or Egyptian guinea	$\frac{875}{1000}$	8. 53927	100
	Half livre	do	4. 26964	50
	$\frac{1}{20}$ livre, (quite rare)	do	0. 41891	5
Silver	Tallari, (Egyptian dollar)	$\frac{8333}{10000}$	27. 84126	20
	Half tallari	do	13. 92063	10
	Quarter tallari	do	6. 96031	5
	Piaster, (⁵)	$\frac{750}{1000}$	1. 35339	1
	Half piaster	do	0. 67664	0. 50
	Quarter piaster	do	0. 33832	0. 25
Copper	Piaster		51. 814	1
	Piece of 20 paras		25. 907	0. 50
	Piece of 10 paras		12. 95339	0. 25
	Piece of 5 paras		6. 47669	0. 125

JAPAN.

WEIGHTS.

The exhibition by Japan in the pavilion for weights, measures, and coins included a balance (*tembihanakari*) for weighing gold and silver coins, with a series of nineteen weights, of which the smallest weighed

---

(¹) The faddan = 420. 8383 square metres = $333\frac{1}{3}$ square cassabas ; it is divided into 24 kérats.
(²) The kaddah = 1.9112 litre
(³) The guirbeh = 66.666 litres.
(⁴) The livre is worth 25.923 fr., or $4.969.
(⁵) The piaster is worth 0.259 fr., or $0.0495.

about two decigrammes, and the others of the series were formed by doubling each term successively; also, a series of five *tigoueri*, (like steel-yards,) graduated decimally. The two largest are intended for ordinary traffic; the two smallest, in ivory, are used in pharmacy.

MEASURES.

1. MEASURES OF LENGTH.

The exhibition of scales of lineal measure contained six specimens made of bamboo, well finished.

No. 1. The kofkoudassi, divided into twenty parts, with ten subdivisions; used for measuring cloth, silk, &c.

No. 2. The kanedassi; used for measuring stones and metals.

No. 3. The kanedassi, equal to three-quarters of No. 2, divided into fifteen parts, each of which is subdivided into ten; used in building.

No. 4. The korkoudassi, equivalent to the half of No. 1, divided into ten parts, each decimally divided.

No. 5. A kanedassi about thirty centimetres long, equal to two-thirds of No. 3, and also used in building.

No. 6. A kanisaci, square, in copper, with the sides graduated; used by builders, cabinet-makers, and carpenters.

2. MEASURES OF VOLUME.

Itchighau .... = 1.68 litre (about). .... = $\frac{1}{100}$ ittomassé.
Gonghau .... = 5 itchigau .... = $\frac{1}{20}$ ittomassé.
Ischiomassé .... = 2 gonghau .... = $\frac{1}{10}$ ittomassé.
Ittomassé .... = 10 ischiomassé .... = 1 ittomassé.

These measures are parallelopipeds of wood, with a diagonal rod of iron serving as a handle; used for all kinds of merchandise.

COINS.

Metal.	Name.	Form.	Mean value in francs. [1]	Observations.
Gold and silver..................	Oban	Oval	250	Not in circulation.
Gold	Hoodji-Koban........	..do	33. 60	Do.
Gold	Koban	..do	16. 60	Do.
Silver and copper...............	Tchooging	Ingot	4. 80	Do.
Silver and copper...............	Manneging	..do	....	Do.
Gold and silver	Nibou	Square....	4. 80	Current coins.
Gold	Itzibou	..do	2. 40	Do.
Silver	Itzibou.............	..do	2. 40	Do.
Gold and Silver	Nishou	..do	1. 20	Do.
Silver	Ischiou	..do	0. 60	Do.
Brass	Tempo..............	Oval [2] ...	0. 15	Do.
Copper.........................	Djunimon...........	..do	$\frac{0.15}{8}$	Do.
Copper.........................	Hatchimon	..do	$\frac{0.15}{12}$	Do.
Copper or iron	Simon	..do	$\frac{0.15}{24}$	Do.
Copper or iron	Itchimon	..do	$\frac{0.15}{96}$	Do.

[1] This valuation varies daily with the demand.   [2] With a hole in the center.

PORTUGAL.

WEIGHTS AND MEASURES.

The metrical system, by law of the 29th of July, 1864.

COINS. (¹)

Metal.	Name.	Fineness. $\frac{11}{12}$	Weight.	Value in reis.
			Grammes.	
Gold	Coroa, (²) (*crown*)	$\frac{916}{1000}$	17. 733.	10000
	½ coroa	do	8. 867	5000
	⅕ coroa	do	3. 547	2000
	¹⁄₁₀ coroa	do	1. 774	1000
Silver	5 tostaos	do	12. 5	500
	2 tostaos	do	5	200
	1 tostão, (*teston*)	do	2. 5	100
	½ tostão	do	1. 25	50
Copper				20
				10
				5

BRAZIL.

WEIGHTS AND MEASURES.

The metric system was made obligatory by a law which will not be put into execution until the first of January, 1873. The old system is still in use.

COINS. (³)

Metal.	Name.	Fineness.	Weight in grammes.
Gold	Piece of 10000 reis	$\frac{916}{1000}$	8. 963
	Piece of 5000 reis	do	4. 486
Silver	Piece of 2000 reis	do	25. 495
	Piece of 1000 reis	do	12. 747
	Piece of 500 reis	do	6. 373
	Piece of 200 reis	do	2. 549
Copper	Piece of 20 reis		
	Piece of 10 reis		

(¹) The silver coins are only used for change. 4500 reis = 1 pound sterling. 180 reis = 1 franc.
(²) The coroa, (crown = 10000 rëis,) legal = $5.8257.
(³) 350 reis = 1 franc. 20 milreis, legal = $10. 9235.

UNITED STATES OF AMERICA.

The metric system has been optional in the United States since the law of 1866.

WEIGHTS.

AVOIRDUPOIS WEIGHT.

Dram ....(*drachme*)		$= \frac{1}{256}$ pound.
Ounce .... (*once*)....	$= 16$ drams	$= \frac{1}{16}$ pound.
POUND (¹) ....(*livre*)	$= 16$ ounces	$= 1$ pound.
Quarter..................... (*quart*)	$= 25$ pounds	$=$ 25 pounds.
Hundredweight (cwt.) ....(*quintal*)	$= 4$ quarters	$=$ 100 pounds.
Ton ....(*tonne*)	$= 20$ cwt..........	$=$ 2000 pounds.

TROY WEIGHT.

Grain ....(*grain*)		$= \frac{1}{5760}$ pound.
Pennyweight ....(*denier poids*)....	$= 20$ grains	$= \frac{1}{240}$ pound.
Ounce .... (*once*)	$= 24$ pennyweights	$= \frac{1}{12}$ pound.
POUND (²) .... (*livre*)	$= 12$ ounces	$=$ 1 pound.

APOTHECARIES' WEIGHT.

Grain .... (*grain*)		$= \frac{1}{5760}$ pound.
Scruple ....(*scrupule*)	$= 20$ grains	$= \frac{1}{288}$ pound.
Drachm............... (*drachme*)	$= 3$ scruples......	$= \frac{1}{96}$ pound.
Ounce .... (*once*)	$= 8$ drams	$= \frac{1}{12}$ pound.
POUND.................... (*livre*)	$= 12$ ounces	$=$ pound Troy.

MEASURES.

MEASURES OF LENGTH.

Line .... (*ligne*)		$= \frac{1}{360}$ yard.
Inch .... (*pouce*)	$= 10$ lines	$= \frac{1}{36}$ yard.
Hand .... (*palme*)	$= 4$ inches	$= \frac{1}{9}$ yard.
Foot .... (*pied*)	$= 3$ hands........	$= \frac{1}{3}$ yard.
YARD (³) ....(*aune*)	$= 3$ feet..........	$=$ 1 yard.
Fathom .... (*brasse*)...........	$= 2$ yards	$=$ 2 yards.
Rod .... (*perche*)....	$= 5\frac{1}{2}$ yards........	$=$ 5½ yards.
Road measure. { Furlong ......(*stade*)	$= 40$ rods..........	$=$ 220 yards.
Statute mile.. (*mille*)	$= 8$ furlongs	$=$ 1760 yards.
League ......(*lieue*)	$= 3$ miles	$=$ 4280 yards.

2. MEASURES OF SURFACE.

Ordinary measures.

Squares of the measures of length.

Land measures.

Rood of land..	$=$ 40 square rods.
Acre (⁴)...	$=$ 4 roods.

(¹) The avoirdupois pound = 453.59265 grammes. [The standard avoirdupois pound of the United States, as determined by Mr. Hassler, is the weight of 27.7015 cubic inches of distilled water.] This weight is in general use for most articles of merchandise.

(²) The Troy pound = 373.242 grammes.

(³) The yard = 0.91438348 metre.

(⁴) The acre = 0.40467 hectare.

3. MEASURES OF VOLUME.

Ordinary measures.

Cubes of the measures of length.

Dry measures.

Pint ....(*pinte*)... = $\frac{1}{64}$ bushel.
Quart....................(*quart*)..................... = 2 pints.......... = $\frac{1}{32}$ bushel.
Peck.....................(*picotin*) .... = 8 quarts .... = $\frac{1}{4}$ bushel.
BUSHEL (¹)...............(*boisseau*).............. = 4 pecks .... = 1 bushel.

Liquids.

Gill... = $\frac{1}{32}$ gallon.
Pint ....(*pinte*)............. = 4 gills .... = $\frac{1}{8}$ gallon.
Quart ....(*quart*) .... = 2 pints.............. = $\frac{1}{4}$ gallon.
GALLON (²)............................ = 4 quarts .... = 1 gallon.
Barrel..............(*baril*)............ = 31½ gallons............. = 31½ gallons.
Hogshead ....(*poinçon*) .... = 2 barrels............... = 63 gallons.

COINS.

Metal.	Name.	Fineness.	Weight in grains Troy.	Value in dollars.
Gold	Double eagle, (*double aigle*)............	$\frac{900}{1000}$	516	20
	Eagle	... do....	258	10
	Half eagle, (½ *aigle*).....................	... do....	129	5
	3 dollars	... do....	77. 4	3
	2½ dollars	... do....	64. 5	2. 50
	Dollar (³)...............................	... do....	25. 8	1
Silver....................	Dollar	... do....	412. 50	1
	Half-dollar (½ *dollar*).................	... do....	192	0. 50
	Quarter-dollar	... do....	96	0. 25
	Dime	... do....	38. 40	0. 10
	Half-dime (½ *dime*).....................	... do....	19. 20	0. 05
	3 cents.................................	... do....	11. 52	0. 03
Nickel	5 cents			0. 05
	3 cents			0. 03
Bronze....................	2 cents.................................			0. 02
	1 cent..................................			0. 01

GREAT BRITAIN AND IRELAND.

The use of the metric system has been optional in England since 1864.

WEIGHTS.

TROY WEIGHT.

Grain ....(*grain*) .... = $\frac{1}{5760}$ pound.
Pennyweight .... = 24 grains .... = $\frac{1}{240}$ pound.
Ounce ....(*once*) .... = 20 pennyweight.... = $\frac{1}{12}$ pound.
POUND (⁴).......... (*livre*)............. = 12 ounces.......... = 1 pound.

(¹) The bushel = 35.2432 litres. (²) The gallon = 3.786 litres.
(³) The dollar = 5 fr. 17c. It weighs 25.8 grains Troy = 1.67 grammes. Its weight exceeds that of the piece of five francs of France by 57.5 milligrammes.
(⁴) The pound troy = 373.241948 grammes.

AVOIRDUPOIS WEIGHT.

Dram ....(*drachme*)........	=		$\frac{1}{256}$ pound.
Ounce ....(*once*)	= 16 drams ...,........	=	$\frac{1}{16}$ pound.
POUND ....(*livre*)	= 16 ounces	=	1 pound.
Stone	= 14 pounds	=	14 pounds.
Quarter ....(*quart*)	= 2 stones	=	28 pounds.
Hundredweight(cwt).(*quintal*)	= 4 quarters	=	112 pounds.
Ton ....(*tonneau*)	= 20 cwt	=	2240 pounds.

APOTHECARIES' WEIGHT.

Grain............(*grain*)		=	$\frac{1}{5760}$ pound.
Scruple ....(*scruple*)	= 20 grains	=	$\frac{1}{288}$ pound.
Dram............(*drachme*)...........	= 3 scruples	=	$\frac{1}{96}$ pound.
Ounce ....(*once*)	= 8 drams	=	$\frac{1}{12}$ pound.
POUND (1)............(*livre*)	= 12 ounces	=	1 pound Troy.

MEASURES.

1. MEASURES OF LENGTH.

Inch ....(*pouce*)	=		$\frac{1}{36}$ yard.
Foot ....(*pied*)	= 12 inches	=	$\frac{1}{3}$ yard.
YARD (2)	= 3 feet...........	=	1 yard.
Fathom...........(*toise*)	= 2 yards........	=	2 yards.
Pole ....(*perche*)	= $5\frac{1}{2}$ yards..........	=	$5\frac{1}{2}$ yards.
Chain ....(*chaîne d'arpenteur*)..	= 100 links	=	22 yards.
Furlong ....(*stade*)	= 40 poles.............	=	220 yards.
Mile ....(*mille*).............	= 8 furlongs............	=	1760 yards.

2. MEASURES OF SURFACE.

Ordinary measures.

Squares of the measures of length.

Land measures.

Rood................	=	1210 square yards.
Acre (3)...............	=	4840 square yards.

3. MEASURES OF VOLUME.

Ordinary measures.

Cubes of the measures of length. (4)

Dry measures.

GALLON (5) ....(*gallon*)	=		1 gallon.
Peck ....(*picotin*)	= 2 gallons	=	2 gallons.
Bushel...........(*boisseau*)	= 4 pecks	=	8 gallons.
Sack ....(*sac*)	= 3 bushels	=	24 gallons.
Coomb	= 4 bushels	=	32 gallons.
Quarter ....(*quart*)	= 2 coombs	=	64 gallons.
Load ....(*charge*)	= 5 quarters	=	320 gallons
Last ....(*lest*)...	= 2 loads	=	640 gallons.

(1) The pound avoirdupois = 453.592645 grammes.   (2) The yard = 0.91438348 metre.
(3) The acre = 0.404671 hectare.   (4) For fire-wood, the cord = 128 cubic feet.
(5) The gallon = 4.5434 58 litres.

Liquids.

Gill .... (*gill*) .... = $\frac{1}{32}$ gallon.
Pint .... (*pinte*) .... = 4 gills .... = $\frac{1}{8}$ gallon.
Quart .... ($\frac{1}{4}$ *gallon*) .... = 2 pints .... = $\frac{1}{4}$ gallon.
GALLON .... (*gallon*) .... = 4 quarts .... = 1 gallon.
Firkin .... (*quarteau*) .... = 9 gallons .... = 9 gallons.
Kilderkin .... ($\frac{1}{2}$ *baril*) .... = 18 gallons .... = 18 gallons.
Barrel .... (*baril*) .... = 36 gallons .... = 36 gallons.
Hogshead .... (*muid*) .... = 54 gallons .... = 54 gallons.
Pipe .... (*pipe*) .... = 2 hogsheads .... = 108 gallons.
Ton .... (*tonne*) .... = 2 pipes .... = 216 gallons.

COINS. [1]

Metal.	Name.	Fineness.	Weight in grammes.	Value in pounds.
Gold	5 pounds, (5 *livres*)	$\frac{916}{1000}$	....	5
	Sovereign, [2] (*souverain*)	....do	7. 981	1
	Half sovereign, ($\frac{1}{2}$ *souverain*)	....do	3. 995	$\frac{1}{2}$
Silver	Crown, (*couronne*)	$\frac{925}{1000}$	28. 250	$\frac{1}{4}$
	Half crown, ($\frac{1}{2}$ *couronne*)	....do	14. 125	$\frac{1}{8}$
	Florin	....do	11. 300	$\frac{1}{10}$
	Shilling [3]	....do	5. 650	$\frac{1}{20}$
	Six-pence	....do	2. 825	$\frac{1}{40}$
	Groat, (4 pence)	....do	1. 883	$\frac{1}{60}$
	Three-pence	....do	1. 412	$\frac{1}{80}$
	Two-pence	....do	0. 941	$\frac{1}{120}$
	One-penny	....do	0. 470	$\frac{1}{240}$
Bronze	Penny			$\frac{1}{240}$
	Half-penny			$\frac{1}{480}$
	Farthing			$\frac{1}{960}$

[1] The standard is the pound (£) sterling.
[2] The pound sterling or sovereign, legal = $4. 8666, by United States custom-house valuation $4.84.
[3] The shilling, legal = $0. 2261.

PARIS UNIVERSAL EXPOSITION, 1867.
REPORTS OF THE UNITED STATES COMMISSIONERS.

BIBLIOGRAPHY

OF THE

PARIS UNIVERSAL EXPOSITION OF 1867,

BY

WILLIAM P. BLAKE.

COMMISSIONER FROM THE STATE OF CALIFORNIA.

WASHINGTON:
GOVERNMENT PRINTING OFFICE.
1870.

BIBLIOGRAPHY.

An observant visitor at the Paris Exposition could not fail to be impressed with the fact that a vast amount of valuable statistical information was there gathered together. Each country (with few exceptions) brought forward the latest attainable commercial and industrial statistics, so that in making the circuit of the building one could collect the materials for a statistical summary of the extent, resources, and industrial condition of the principal nations. The publications of this nature made in the form of catalogues, with introductions to the various classes, published in special brochures, and, in many cases, (as for example Russia, Brazil, Australia,) in volumes of good size, have more than a transient importance, and when brought together and grouped with the special technical reports by the commissioners, juries, and by various experts, form a collection which gives a most accurate and complete picture of the present condition of the industries of the world.

The value and importance of such a collection is evident. Many of the publications so made were collected and sent in by the Commissioner General at the close of the Exposition, and the Editor has since endeavored to add to them so as to form a complete collection of the books published in connection with the Exposition, to be deposited in the library of the State Department.

In forming this collection the bibliographical list has been prepared with the object of facilitating the collection of the publications, and also with the conviction that it would prove useful to investigators in various departments of industry and science, and to the reader of the Exposition reports. It is also hoped that it may have some influence inducing the formation of special collections of such books in connection with some of our libraries.

The value of the statistical publications made for distribution upon the occasion of great international exhibitions would be greatly enhanced if some comprehensive plan of collection and presentation of the statistics could be adopted by each of the participating countries. It would be well if a general and simultaneous decennial census could be agreed upon, or a simultaneous summing up of the yearly statistical returns of every country, and that a great international exhibition should immediately follow, and be accompanied by carefully prepared statistical abstracts for distribution and exchange.

The idea of forming a bibliographical list of publications connected with, or immediately resulting from the Paris Exposition of 1867, is not a novel one. M. Léon Morillot, of Paris, published a list in 1869, in the bulletin of the " Association Internationale pour le développement du commerce et des expositions," and I am indebted to that list as the foundation upon which mine has been prepared and for titles of some of the works I have not been able to procure.   There are doubtless many publications upon the Exposition unknown to me, and which do not therefore appear in the list.   An effort has been made to include all published up to January, 1870, and if time had permitted, the titles of the principal reports upon the previous great exhibitions would have been included.

The list now presented contains over three hundred titles, and shows in a most striking manner the value of great international exhibitions in stimulating the preparation and publication of industrial works.   It is chiefly by such publications that the great object of the exhibitions, the general elevation and advancement of art and industry, is effected. They supplement the exhibitions.   Their influence is not circumscribed by the walls of the exhibition buildings nor confined to the country in which the exhibitions are made.   They reach every center of human industry, and carry the fruits of the exhibitions to families and workshops in all countries.

AMERICA.—UNITED STATES.

The preliminary publications, calling the attention of the people of the United States to the Exposition and setting forth from time to time the progress made, appeared at intervals until the time of opening in 1867. These publications, in the order of their issue, were entitled as follows:

FRENCH UNIVERSAL EXPOSITION for 1867, to open April 1, 1867, close October 31, 1867.   Official correspondence on the subject, published by the Department of State for the information of citizens of the United States, containing general regulations, classification of articles, &c. Washington, Government Printing Office, 1865.   Quarto pamphlet, 28 pp.   A second edition of this was printed with additions, 42 pp.

THE FRENCH UNIVERSAL EXPOSITION of 1867.   Interesting letters from United States Commissioner Beckwith, and other papers, containing valuable advice and information for exhibitors.   Government Printing Office, 1865.   Quarto pamphlet, 7 pp.

The following circulars and documents in octavo form appeared in regular succession and included the letters published in the former circulars :

MESSAGE from the President of the United States, December 11, 1865, transmitting a report from the Secretary of State, concerning the Universal Exposition to be held at Paris in the year 1867.   8º.   14 pp.

SUPPLEMENTAL circular relative to the Paris Universal Exposition of 1867 : Proceedings of the Chamber of Commerce of the State of New York.   Washington, Government Printing Office, 1866.   8º.   14 pp.

SPEECH of Hon. N. P. Banks, of Massachusetts, upon the representation of the United States at the Exposition of the World's Industry, Paris, 1867. Washington, D. C., Mansfield & Martin, publishers, 1866. 8°. 24 pp.

SECOND SUPPLEMENTAL PAMPHLET, Paris Universal Exposition of 1867: Details of organization. Washington, Government Printing Office, 1866. 8°. 64 pp.

THIRD SUPPLEMENTAL CIRCULAR respecting the Paris Exposition of 1867 : Importance of prompt action, &c. J. C. Derby, general agent for the United States. Washington, Government Printing Office, 1866. 8°. 71 pp.

PRESIDENT'S MESSAGE. Senate Ex. Doc. No. 5, 39th Congress, 2d session : Message of the President of the United States, communicating, in compliance with a resolution of the Senate of the 19th December, 1866, information in respect to the progress made in collecting the products, and the weights, measures, and coins of the United States, for exhibition at the Universal Exposition at Paris in April next. 8°. 52 pp.

The following were published after the opening of the Exposition :

OFFICIAL CATALOGUE of the products of the United States of America exhibited at Paris, 1867, with statistical notices. Catalogue in English. Catalogue Français. Deutscher Catalog. (Printed in the three languages together, with the exception of the statistical notices, which were in French.) Paris, Imprimerie centrale des chemins de fer. A. Chaix et cie, Rue Bergère 20, 1867. 12°. (There were three editions printed,) 3d edition, 160 pp.

MINERALS of the United States of America. Catalogue compiled by H. F. Q. d'Aligny. Paris, Brière, 257 Rue Saint Honoré, 1867. 8°. 91 pp.

COLORADO in the United States of America. Schedule of ores contributed by sundry persons to the Paris Universal Exposition of 1867, with some information about the region and its resources, by J. P. Whitney, commissioner from the Territory. London, Cassell, Petter, and Galpin, 1867. Large 8°. 62 pp. The French edition; Paris, Berger, 1867. 71 pp. (Large octavo, wide margin, editions published separately in English, French, and in German.)

ALABAMA.—The State of Alabama, (United States of America,) its mineral, agricultural, and manufacturing resources, by Hiram Haines. Paris, Simon Raçon, 1867. 8°. 120 pp. Also an edition in French.

MUNITIONS OF WAR.—Report to the government of the United States on the munitions of war exhibited at the Paris Universal Exhibition, 1867, by Ch. B. Norton and W. J. Valentine, United States Commissioners. New York, Office of Army and Navy Journal, 39 Park Row. London, Spon, 48 Charing Cross, 1868. 8°. 286 pp.

IRON AND STEEL.—The production of iron and steel in its economic and social relations, by Abram S. Hewitt, United States Commissioner to the Universal Exposition at Paris, 1867. Washington, 1868. 8°. pp. iv to 104. (House document, ordered to be printed March 28, 1868.)

SILK.—Report on silk and silk manufactures, by Elliot C. Cowdin, esq., commissioner to the Paris Exposition of 1867. (Transmitted to the House of Representatives March 25, 1868, and ordered to be printed.)

SILK.—Report to the Department of State on silk and silk manufactures, by Elliot C. Cowdin, United States Commissioner. [Author's edition.] 8°. 114 pp. Washington, 1868.

ÉTATS-UNIS D'AMÉRIQUE.—Exposition Universelle de 1867. États-Unis d'Amérique. Le Névada Oriental: Géographie, ressources, climat, et état social. Rapport adressé au comité local pour l'Exposition de Paris, par Myron Angel. 1re édition de cinq cents exemplaires. 9 Rue de Fleurus, juillet 1867. 12°. 164 pp. Paris, Imprimerie générale de Ch. Lahure. (Printed in French and in English.)

NEW YORK.—Norton, C. B. Report of the New York State commissioner to the Paris Exposition of 1867.

NEW YORK.—Second report of the New York State commissioner to the Paris Exposition of 1867. Printed by order of the New York State senate. 8°. 125 pp. With plates. Albany, 1868.

CALIFORNIA.—Report of the commissioner to the Paris Exposition, 1867. Notes upon the Universal Exposition at Paris, 1867, by William P. Blake, commissioner for the State of California to the Universal Exposition at Paris, 1867, and delegate of the State board of agriculture. 8°. 100 pp. Sacramento, D. W. Gelwicks, State printer, 1868.

STATE OF ILLINOIS and the Universal Exposition of 1867 at Paris, France. Report of John P. Reynolds, delegate from the Illinois State Agricultural Society and commissioner for the State. 8°. viii–134 pp. Springfield, State Journal office, 1868.

The following list gives the titles of the final official reports, alphabetically arranged according to the subjects. The reports all bear the imprint of the Government Printing Office and the year of publication. This imprint is omitted in the list, but the exact date of publication is supplied. The copies of reports not separately issued as above have been grouped together and bound in six volumes, under the general title of "Reports of the United States Commissioners to the Paris Universal Exposition of 1867; published under the direction of the Secretary of State, by authority of the Senate of the United States."

A list of the reports, in the order in which they are grouped in volumes, will be found at the end of Volume I and of Volume VI.

LIST OF THE REPORTS, BY THEIR TITLES, ARRANGED ALPHABETICALLY ACCORDING TO THE SUBJECTS.

ARTS.—Machinery and processes of the industrial arts and apparatus of the exact sciences, by Frederick A. P. Barnard, LL. D., United States Commissioner.—pp. ix, 669. August 4, 1869. (In volume iii.)

ASPHALT AND BITUMEN.—Report on asphalt and bitumen, as applied to the construction of streets and sidewalks in Paris; also to terraces, roofs, &c., and to various products in the Exposition of 1867; with observations upon macadamized streets and roads, by Arthur Beckwith, Civil Engineer.—pp. 31. January 15, 1869. (In volume iv.)

BEET-SUGAR.—The manufacture of beet-sugar and alcohol and the cultivation of sugar-beet, by Henry F. Q. D'Aligny, United States Commissioner.—pp. 90. November 3, 1869. (In volume v.)

BÉTON-COIGNET.—Report on Béton-Coignet, its fabrication and uses—construction of sewers, water-pipes, tanks, foundations, walls, arches, buildings, floors, terraces; marine experiments, &c., by Leonard F. Beckwith, Civil Engineer.—pp. 21. January 15, 1869. (In volume iv.)

BIBLIOGRAPHY.—Bibliography of the Paris Universal Exposition of 1867, by William P. Blake, Commissioner of the State of California to the Paris Exposition of 1867. June, 1870. (In volume i.)

BUILDINGS.—Report upon buildings, building materials, and methods of building, by James H. Bowen, United States Commissioner.—pp. 96. September 28, 1869. (In volume iv.)

CEREALS.—Report on cereals: The quantities of cereals produced in different countries compared, by Samuel B. Ruggles, Vice-President of the United States Commission. The quality and characteristics of the cereals exhibited, by George S. Hazard, United States Commissioner.—pp. 26. September 28, 1869. (In volume v.)

CHEMISTRY.—The progress and condition of several departments of industrial chemistry, by J. Lawrence Smith, United States Commissioner.—pp. ix, 146. September 7, 1869. (In volume ii.)

CIVIL ENGINEERING.—Civil engineering and public works, by William P. Blake, Commissioner of the State of California.—pp. 49. March 5, 1870. (In volume iv.)

CLOTHING.—Report on clothing and woven fabrics; being classes twenty-seven to thirty-nine of group four. By Paran Stevens, United States Commissioner. In press, April, 1870. (In volume vi.)

COAL.—Report on the manufacture of pressed or agglomerated coal, by Henry F. Q. D'Aligny, United States Commissioner.—pp. 19. October 8, 1869. (In volume v.)

COTTON.—Report upon cotton, by E. R. Mudge, United States Commissioner, with a supplemental report by B. F. Nourse, Honorary Commissioner.—pp. ii, 115. June 28, 1869. (In volume vi.)

EDUCATION.—Report on education, by J. W. Hoyt, United States Commissioner.—pp. 398. June, 1870. (In volume vi.)

Report on school-houses and the means of promoting popular education, by J. R. Freese, United States Commissioner.—pp. 13. October 8, 1869. (In volume v.)

ENGINEERING.—Report upon steam-engineering, as illustrated by the Paris Universal Exposition, 1867, by William S. Auchincloss, Honorary Commissioner.—pp. 72. August 2, 1869. (In volume iv.)

FINE ARTS.—Report on the fine arts, by Frank Leslie, United States Commissioner.—pp. 43. February 6, 1869. (In volume i.)

The fine arts applied to the useful arts—report by the committee, Frank Leslie, S. F. B. Morse, Thomas W. Evans, United States Commissioners.—pp. 8, with 33 leaves of wood engravings. February 6, 1869. (In volume i.)

FOOD.—Report on the preparation of food, by W. E. Johnston, M. D., Honorary Commissioner.—pp. 19. October 8, 1869. (In volume v.)

GENERAL SURVEY.—General survey of the Exposition, with a report on the character and condition of the United States Section.—pp. 325. January 7, 1869. (In volume i.)

GOLD AND SILVER.—(See *Precious metals.*)

INTRODUCTION.—Introduction, with selections from the correspondence of United States Commissioner General Beckwith and others, showing the organization and administration of the United States Section.—pp. 184. May, 1870. (In volume i.)

IRON AND STEEL.—The production of iron and steel, in its economic and social relations, by Abram S. Hewitt, United States Commissioner, 1868.—pp. 183. January 7, 1869. (In volume ii.)

MINING.—Report on mining and the mechanical preparation of ores, by Henry F. Q. D'Aligny, United States Commissioner, and Alfred Huet, F. Geyler, and C. Lepainteur, Civil and Mining Engineers, Paris, France. February 19, 1869. (In volume iv.)

MUNITIONS OF WAR.—Report on the munitions of war, by Charles B. Norton and W. J. Valentine, United States Commissioners.—pp. 213. January 7, 1869. (In volume v.)

MUSICAL INSTRUMENTS.—Report upon musical instruments, by Paran Stevens, United States Commissioner.—pp. 18. June 21, 1869. (In volume v.)

ORES, MECHANICAL PREPARATION OF.—(See *Mining.*)

PHOTOGRAPHY.—Photographs and photographic apparatus, by Henry F. Q. D'Aligny, United States Commissioner.—pp. 19. October 8, 1869. (In volume v.)

PRECIOUS METALS.—Report upon the precious metals, being statistical notices of the principal gold and silver producing regions of the world represented at the Paris Universal Exposition, by William P. Blake,

Commissioner of the State of California.—pp. viii, 369. March 11, 1869. (In volume ii.)

SCHOOL-HOUSES.—(See *Education.*)

SILK.—Report on silk and silk manufactures, by Elliot C. Cowdin, United States Commissioner.—pp. 51. January 7, 1869. (In volume vi.)

SURGERY.—Report on instruments and apparatus of medicine, surgery, and hygiene, surgical dentistry and the materials which it employs, anatomical preparations, ambulance tents and carriages, and military sanitary institutions in Europe, by Thomas W. Evans, M. D., United States Commissioner.—pp. 70. January 28, 1869. (In volume v.)

TELEGRAPHY.—Examination of the telegraphic apparatus and the processes in telegraphy, by Samuel F. B. Morse, LL. D., United States Commissioner.—pp. 166. November 20, 1869. (In volume iv.)
Outline of the history of the Atlantic cables, by H. F. Q. D'Aligny, United States Commissioner.—pp. 13. October 8, 1869. (In volume v.)

UNITED STATES SECTION, REPORT ON.—(See *General survey,* &c.)

VINE.—Report upon the culture and products of the vine, by Marshall P. Wilder, Alexander Thompson, William J. Flagg, Patrick Barry, committee.—pp. 28. October 8, 1869. (In volume v.)

WOOL.—Report upon wool and manufactures of wool, by E. R. Mudge, United States Commissioner, assisted by John L. Hayes, Secretary of the National Association of Wool Manufacturers.—pp. 143. January 7, 1869. (In volume vi.)

WEIGHTS, MEASURES, AND COINS.—Extracts from the report of the International Committee on Weights, Measures, and Coins, with a notice of the introduction of the metrical system in the United States and its relations to other systems of weights and measures.—pp. 47. June, 1870. (In volume i.)

AMERICA.—CENTRAL AND SOUTH AMERICA.

RÉPUBLIQUES de l'Amérique centrale et méridionale. Notices et catalogues. (Bolivia, Argentine Confederation, Costa Rica, Ecuador, Hayti, Nicaragua, New Granada, Paraguay, Peru, Salvador, Uruguay, Venezuela.) Paris, Bouchard-Huzard, 5 Rue de l'Éperon, 1867.

ÉTATS AMÉRICAINS. Leurs produits, leur commerce, en vue de l'Exposition Universelle de Paris, par Tenré. Paris, Henri Plon, 1867.

CHILI.—Notice statistique sur le Chili. Montereau, Zanote.

ÉQUATEUR.—République de l'Équateur. Notice et catalogue. Paris, Bouchard-Huzard, 5 Rue de l'Éperon, 1867.

CHILI.—Notice statistique sur le Chili et catalogue des minéraux envoyés à l'Exposition Universelle. Paris, Poitevin, 2 Rue Damiette, 1867. 8° 83 pp.

LA CONFÉDÉRATION ARGENTINE, 48 pp.

LA RÉPUBLIQUE ARGENTINE, &c. Rapport adressé au gouvernement de sa Majesté britannique, par M. Frances Clair Ford, secrétaire de la légation britannique à Buenos Ayres, &c. 8°. 79 pp. (Translated from the British Blue Book.)

RÉPUBLIQUE DE NICARAGUA.—(Single sheet of statistics.)

AUSTRIA.

KATALOG der österreichischen Abtheilung, herausgegeben vom k. k. Central-Comité für die Pariser Ausstellung. Wien.

CATALOGUE spécial du royaume de Hongrie. Paris, Aug. Marc, 22 Rue de Verneuil, 1867.

KATALOG der vom k. k. Ministerium für Cultus und Unterricht ausgestellten Unterrichtsgegenstände, herausgegeben vom k. k. Central-Comité für die Pariser Ausstellung. Wien.

OFFICIELLER AUSSTELLUNGSBERICHT, herausgegeben durch das k. k. österreichische Central-Comité. Ausstellung zu Paris, 1867. Wien, Wilhelm Braumüller, 1867–1869. In seven volumes 8°., including one. volume of illustrations.

MÜNZEN, MASSE, UND GEWICHTE.—Die internationale Münz- Mass und Gewichts- Commission der Pariser Ausstellung von 1867. Vortrag gehalten in der Wochenversammlung des Vereins vom 21. December, 1867, von F. Bömches. Wien, Verlag des Verfassers, 1868.

FORÊTS.—Les richesses forestières de l'Autriche et leur exportation. Explication relative aux objets faisant partie de l'exposition forestière de l'Autriche à Paris, par Joseph Wessely. Vienne, 1867.

VERZEICHNISS der Anmeldungen für die Welt-Ausstellung zu Paris im Jahre 1867. Im Auftrage des k. k. Central-Comités für die Agricultur- Kunst- und Industrie-Ausstellung zu Paris, zusammengestellt von Dr. E. Hornig. Wien, 1866.

MINISTÈRE DE LA GUERRE.—Notice sur les objets formant l'exposition collective du ministère de la guerre I. R. d'Autriche à l'Exposition de Paris. Paris, Aug. Marc, 1867.

KURZE MITTHEILUNGEN über Berg- und Hüttenwesens-Maschinen und Baugegenstände auf der allgemeinen Industrie-Ausstellung zu Paris, 1867, in 114 selbständigen, durch Holzschnitte illustrirten, Artikeln, von Peter Ritter von Rittinger. Wien, Druck und Verlag der k. k. Hof- und Staats-Druckerei, 1867.

L'INSTITUT GÉOLOGIQUE impérial et royal d'Autriche, par F. de Hauer. Vienne, Geitler, 1867.

LA SLAVONIE. Sa production et son commerce. Aperçu rédigé à l'occasion de l'Exposition Universelle à Paris, par Felix Lay. Essegg, 1867.

AUTRICHE.—Notice sur les objets formant l'exposition collective du ministère de la guerre I.-R. d'Autriche, à l'Exposition Internationale de Paris, 1867. 8º. 54 pp. Paris, Typographie Auguste Marc, 1867.

BADEN.

LES EXPOSANTS du grand-duché de Bade et leurs produits. Publication de la Commission Grand-ducale. Carlsruhe, Fr. Müller, 1867.

DIE BETHEILIGUNG des Grossherzogthums Baden an der Universal-Ausstellung zu Paris, (1867,) herausgegeben von der badischen Ausstellung-Commission. Carlsruhe, F. Müller, 1867.

BAVARIA.

CATALOGUE de l'exposition des beaux-arts. (Notices.) Paris, Kugelmann, 13 Rue Grange-Batelière, 1867.

DIE INDUSTRIE und Landwirthschaft Bayerns auf der internationalen Ausstellung zu Paris im Jahre 1867. Mittheilungen und Aufschlüsse über die bayerischen Aussteller und deren Producte. München, C. Wolf und Sohn, 1867.

L'INDUSTRIE de la Bavière à l'Exposition Universelle de Paris en 1867. Notes détaillées sur les exposants et leurs produits. Paris, Kugelmann, 1867.

KÖNIGREICH BAYERN. Statistische Mittheilungen. München, M. Pössenbacher' sche Buchdruckerei.

DIE BAYERISCHE Landwirthschaft auf der Ausstellung zu Paris, 1867. München, Pössenbacher' sche Buchdruckerei.

EXPOSÉ de la fondation, du développement et de l'activité de l'École Royale des Beaux-Arts et des Métiers à Nuremberg, servant d'explication aux travaux de l'école destinés à l'Exposition de Paris en 1867. Nuremberg, Bieling (Dietz.)

BELGIUM.

CATALOGUE des produits industriels et des œuvres d'art. Bruxelles, Bruyant-Christophe. 12º. 685 pp. Bruxelles, Imprimerie Bruylaut-Christophe & Cie. 33 Rue Blaes, 1867.

NOTICE sur les constructions élevées dans la partie du parc reservée à la Belgique, par Du Pré, membre délégué à Paris de la Commission Belge. Bruxelles, Van Dooren, 25 Chaussée de Wavre, 1867.

RÉSUMÉ DE STATISTIQUE BELGE, d'après les documents officiels, par Faider, délégué de la Belgique et des Pays-Bas au jury spécial institué pour un Nouvel Ordre de Récompenses. Bruxelles, Bruyant-Christophe, 1867.

MATÉRIEL et procédés des exploitations rurales et forestières, par Leclerc, inspecteur général de l'agriculture et des chemins vicinaux en Belgique. A volume containing 7 plates and 59 figures interspersed in the text. Paris, Librairie agricole, 26 Rue Jacob. Bruxelles, Librairie polytechnique, 9 Rue de la Madeleine, 1868.

Du Pré.—Note sur la transmission télodynamique, inventée par Hirn, par Du Pré, ingénieur-en-chef honoraire des ponts et chaussées belges. Bruxelles, Van Dooren, 1869.

BRAZIL.

L'Empire du Brazil à l'Exposition Universelle de 1867 à Paris. 8°. 200 pp. Rio Janeiro, Laemmert, 61 B., Rue des Invalides, 1867.

Brazil.—The empire of Brazil at the Paris International Exhibition of 1867. Printed by E. & H. Laemmert. 8°. 197 pp. Rio de Janeiro, 61 B., Rua dos Invalidos, 1867. (An edition in French, also, as above.)

Relatorio sobre á exposição universal de 1867, redigido pelo secretario da commissão brazileira Julio Constancio de Villeneuve. 2 volumes. Paris, Claye, 7 Rue Saint Benoît, 1868.

Travaux au sujet des produits du Brésil, qui sont à l'Exposition Universelle de Paris en 1867, par José de Saldanha da Gama. Paris, Brière, 257 Rue Saint-Honoré, 1867. 29 pp.

Breve noticia sobre a collecção das madeiras do Brazil apresentada na exposição internacional de 1867, pelos Srs. Freire Allemão, Custudio Alves Serrao, Ladislau Netto e J. de Saldanha da Gama. Rio de Janeiro, 1867.

Classement botanique des plantes alimentaires du Brésil, par José de Saldanha da Gama. Paris, Martinet, 2 Rue Mignon, 1867.

Blumenau.—La colonie de Blumenau. (A small pamphlet.) Paris, Berger, 5 Impasse des Filles-Dieu, 1867.

DENMARK.

Le Danemark à l'Exposition Universelle de 1867, publié par la Commission danoise; la partie historique, par Valdemar Schmidt. Paris, Reinwald, 15, Rue des Saints-Pères, 1867.

EGYPT.

L'Égypte à l'Exposition Universelle de 1867, par Charles Edmond, commissaire général de l'exposition vice-royale d'Égypte; ouvrage orné du portrait de S. A. le vice-roi d'Égypte, et de trois belles planches: Temple, "Selamlik," "Okel." Paris, Dentu, 1867.

Aperçu de l'histoire ancienne d'Égypte, pour l'intelligence des monuments exposés dans le temple égyptien du parc, par Mariette-Bey. Paris, Dentu.

FRANCE.

Catalogue général, publié par la Commission Impériale. Three editions, one in one volume, another in two volumes, and the last in nine numbers. 8°. 1538–xxviii pp. Paris, Dentu. No date, [1867.]—

Londres, Johnson & Sons, Castle street, Holborn, 1867.—The edition in two volumes, as well as that in nine numbers, contain interesting introductions to each of the classes. Edition in English in one octavo volume, Johnson & Sons.

CATALOGUE GÉNÉRAL, publié par la Commission Impériale. Histoire du travail et des monuments historiques, 2 volumes. Paris, Dentu. Londres, Johnson & Sons.

CATALOGUE GÉNÉRAL, publié par la Commission Impériale. Annexe agricole; Billancourt, instruments et specimens de culture. Paris, Dentu, 1867.

CATALOGUE SPÉCIAL de l'exposition d'horticulture. Paris, Dentu, 1867.

CATALOGUE RAISONNÉ des collections exposées par l'administration des forêts de France. Paris, Imprimerie impériale, 1867.

CATALOGUE ET NOTICES des missions protestantes et évangeliques. Paris, Dentu, 1867. Large 8°. 191 pp.

DIXIÈME GROUPE.—L'enquête du dixième groupe : catalogue analytique des documents, mémoires, et rapports exposés hors classe dans le dixième groupe et relatifs aux institutions publiques et privées, créées par l'état, les départements, les communes et les particuliers, pour améliorer la condition physique et morale de la population. Paris, Dentu, 1867. 8°. 283 pp.

POIDS ET MESURES.—Rapports, procès-verbaux et catalogue du comité des poids et mesures et des monnaies.—pp. 163. Paris, Dentu.

EXPOSANTS RÉCOMPENSÉS.—Catalogue officiel des exposants récompensés par le jury international. Paris Dentu, [1867.] 8°. About 500 pages ; each group separately paged.

RAPPORTS DU JURY INTERNATIONAL, publiés sous la direction de M. Michel Chevalier, 13 volumes. 8°. Paris, Paul Dupont, 1868.

The majority of the different reports contained in this collection have been printed separately, and form as many interesting monographs.

The following is a list of the titles of the reports in the series in their order of succession, from the Introduction, in Volume I, to the end of Volume XIII :

RAPPORTS DU JURY INTERNATIONAL.

INTRODUCTION, par M. Michel Chevalier. Tome premier, pp. i–dxcvi. A general view of the whole range of human industry and art as presented at the Exposition, followed by a list of authors of reports in the series, and a complete table of contents for the thirteen volumes.

PEINTURE, dessins, sculpture, architecture, gravure et lithographie, par M. Ernest Chesneau. (Group I, Classes 1–5.) Rapports du jury international, tome 1, pp. 1–133.

HISTOIRE DU TRAVAIL: Rapport de M. E. du Sommerard, commissaire délégué. Rapports du jury international, tome 1, pp. 137–246.

MONUMENTS et spécimens d'architecture élevés dans le parc du Champ de Mars. Rapport spécial de M. A. de Saint-Yves. Rapports du jury international, tome 1, pp. 247–354.

NOUVEL ORDRE DE RECOMPENSES, institué en faveur des établissements et des localités qui ont développé la bonne harmonie entre les personnes coopérant aux mêmes travaux et qui ont assuré aux ouvriers le bien-être matériel, intellectuel et moral. Rapport par M. Alfred Leroux, vice-président du corps législatif, membre de la Commission Impériale. Rapports du jury international, tome 1, pp. 355–534.

PRODUITS D'IMPRIMERIE et de librairie, par M. Paul Boiteau. (Group II, Class 6.) Rapports du jury international, tome 2, pp. 4–99.

OBJETS DE PAPETERIE: Matériel des arts, de la peinture et du dessin: Section I. Papeterie, par M. Roulhac. Section II. Papiers. Succédanés des chiffons, par M. Anselme Payen. Section III. Matériels des arts, de la peinture et du dessin, par M. Roulhac. (Group II, Class 7.) Rapports du jury international, tome 2, pp. 103–139.

APPLICATION DU DESSIN et la plastique aux arts industriels: Section I. Procédés et enseignement de l'art industriel, par M. Baltard. Section II. Applications de l'art à l'industrie, par M. Edmund Taigny. Section III. Gravures sur pierres dures, par M. Barre. (Group II, Class 8.) Rapports du jury international, tome 2, pp. 145–189.

ÉPREUVES et appareils de photographie, par M. Davanne. (Group II, Class 9.) Rapport du jury international, tome 2, pp. 193–234.

INSTRUMENTS DE MUSIQUE, par M. Fétis. (Group II, Class 10.) Rapports du jury international, tome 2, pp. 289–318.

APPAREILS et instruments de l'art médical, ambulances civiles et militaires: Section I. Hygiène et médecine, par MM. les Docteurs A. Tardieu et Sir John Oliffe. Section II. Instruments de chirurgie, par M. le Docteur Nélaton. Section III. Appareils et ouvrages de gymnastique, par M. le Docteur Demarquay. Section IV. Appareils orthopédiques, prosthèse chirurgicale, bandages, secours aux blessés, par M. le Docteur Tillaux. Section V. Voitures et tentes d'ambulance, par M. le Docteur Thomas W. Evans. Section VI. Chirurgie dentaire, par M. le Docteur Thomas W. Evans. (Group II, Class 2.) Rapports du jury international, tome 2, pp. 321–411.

INSTRUMENTS de précision et matériel de l'enseignement des sciences: Section I. Observations générales, par M. Lissajous. Section II. Appareils d'électricité, de magnétisme et de physique mécanique, par M. Privat-Deschanel. Section III. Instruments d'astronomie, de géodésie, de topographie, de marine, d'optique et d'acoustique, par M. Lissajous. Section IV. Poids et mesures, monnaies, par M. de Lapparent. Section V. Appareils densimétriques, par M. L. H. de Baumhauer. Section VI.

Instruments de mathématiques et modèles pour l'enseignement des sciences, par M. Ed. Grateau. Section VII. Modèles d'anatomie, par M. Tillaux. (Group II, Class 12.) Rapports du jury international, tome 2, pp. 415–553.

CARTES ET APPAREILS de géographie, de géologie et de cosmographie : Section I. Cartes topographiques, hydrographiques et géographiques, plans en relief, par M. le Colonel Ferri Pisani. Section II. Cartes marines, par M. Daroudeau. Section III. Cartes géologiques, (première partie,) par M. Edmond Fuchs ; cartes géologiques, (seconde partie,) · par M. Daubrée. (Group II, Class 13.) Rapports du jury international, tome 2, pp. 557–653.

MEUBLES DE LUXE : Section I. Considérations sur l'art dans ses applications à l'industrie, par M. E. Guichard. Section II. Meubles de luxe, par MM. Diéterle et Pollen. (Group III, Class 14.) Rapports du jury international, tome 3, pp. 5–35.

OUVRAGES de tapissier et de décorateur, par MM. Jules Diéterle et Digby Wyatt. (Group III, Class 15.) Rapports du jury international, tome 3, pp. 39–56.

CRISTAUX, verrerie de luxe et vitraux : Section I. Verrerie, par MM. E. Peligot et G. Bontemps. Section II. Vitraux, par MM. Bontemps et Bœswilwald. (Group III, Class 16.) Rapports du jury international, tome 3, pp. 59–97.

PORCELAINES, faïences et autres poteries de luxe : Section I. Terres cuites et grès, par M. Chandelon. Section II. Faïences fines, faïences · décoratives et porcelaines tendres, par M. Aimé Gerard. Section III. Porcelaines dures, par M. F. Dommartin. (Group III, Class 17.) Rapports du jury international, tome 3, pp. 103–181.

TAPIS, tapisseries et tissus d'ameublement : Section I. Tapis et tapisseries, par M. Badin. Section II. Tapis d'un usage ordinaire, par M. W. Chocqueel. Section III. Tissus d'ameublement, par M. Carlhian. Section IV. Toiles cirées, par M. Persoz fils. (Group III, Class 18.) Rapports du jury international, tome 3, pp. 187–218.

PAPIERS PEINTS, par M. Aldrophe. (Group III, Class 19.) Rapports du jury international, tome 3, pp. 221–236.

COUTELLERIE, par M. Dubocq. (Group III, Class 20.) Rapports du jury international, tome 3, pp. 239–256.

ORFÉVRERIE : Section I. Orfévrerie, par M. Paul Christofle. Section II. Émaux et damasquine, par M. Philippe Delaroche. (Group III, Class 21.) Rapports du jury international, tome 3, pp. 259–280.

BRONZES D'ART, fontes d'art diverses, objets en métaux repoussés, par M. Barbedienne. (Group III, Class 22.) Rapports du jury international, tome 3, pp. 283–313.

HORLOGERIE, par M. Bréguet. (Group III, Class 23.) Rapports du jury international, tome 3, pp. 317–336.

APPAREILS et procédés de chauffage et d'éclairage : Section I. Appareils d'économie domestique, par M. Müller. Section II. Chauffage et ventilation, par M. Louis Ser. Section III. Lampes servant à l'éclairage au moyen des huiles animales, végétales, ou minérales accessoires de l'éclairage, par M. Henri Péligot. Section IV. Allumettes, par M. Henri Péligot. (Group III, Class 24.) Rapports du jury international, tome 3, pp. 339–400.

PARFUMERIE, par M. Barreswil. (Group III, Class 25.) Rapports du jury international, tome 3, pp. 403–423.

OBJETS de maroquinerie, de tabletterie, et de vannerie: Section I. Reliure, par M. Paul Boiteau. Section II. Objets divers de maroquinerie, de tabletterie, et de vannerie, par M. Louis Aucoc. Group III, Class 26.) Rapports du jury international, tome 3, pp. 427–500.

FILS ET TISSUS du coton : Section I. Filature du coton, par M. Mimerel fils. Section II. Industrie cotonnière, tissage, par M. Gustave Roy. Section III. Tissus de coton imprimés, par M. Jules Kœchlin. (Group IV, Class 27.) Rapports du jury international, tome 4, pp. 6–65.

FILS ET TISSUS de lin, de chanvre, &c. : Section I. Lins et chanvres, par M. Casse. Section II. Tissus de fibres végétales, équivalents du lin et du chanvre, jute, China-grass, et textiles divers, par M. A. F. Legentil. (Group IV, Class 28.) Rapports du jury international, tome 4, pp. 69–104.

FILS ET TISSUS de laine peignée: Section I. Laines peignées et fils de laine peignée, par M. J. E. Charles Seydoux. Section II. Tissus de pure laine peignée, tissus de laine mélangée d'autres matières, et étoffes de fantaisie en laine cardée légèrement foulée, par M. Larsonnier. (Group IV, Class 29.) Rapports du jury international, tome 4, pp. 108–137.

FILS ET TISSUS de laine cardée : Section I. Filature de la laine cardée, par M. Balsan. Section II. Industrie drapière, par M. Vauquelin. (Group IV, Class 30.) Rapports du jury international, tome 4, pp. 141–158.

SOIES ET TISSUS DE SOIE : Section I. Soies, par M. Jules Raimbert. Section II. Tissus de soie, par M. Alphonse Payen. Section III. Rubans, par M. Girodon. (Group IV, Class 31.) Rapports du jury international, tome 4, pp. 162–230.

CHÂLES, par M. David Gerson. (Group IV, Class 32.) Rapports du jury international, tome 4, pp. 223–230.

DENTELLES, tulles, broderies et passementeries : Section I. Dentelles, par M. Félix Aubry. Section II. Tissus de soie et de coton unis, par M. Delhaye. Section III. Broderies, par M. Rondelet. Section IV. Passementerie, par M. Louvet. Section V. Broderies et passementeries orientales, M. de Launay. (Group IV, Class 33.) Rapports du jury international, tome 4, pp. 233–280.

ARTICLES de bonneterie et de lingerie, objets accessoires du vêtement: Section I. Bonneterie, par M. Tailbonis. Section II. Lingerie confectionée pour hommes: chemises, flanelles, cols-cravates, et faux cols, par M. Hayem aîné. Section III. Industrie des corsets, par M. E. Deschamps. Section IV. Parapluies et ombrelles, cannes, fouets, et cravaches, par M. Duvelleroy. Section V. Fabrication des éventails, par M. Duvelleroy. Section VI. Gants et bretelles, par M. Carcenac. Section VII. Boutons, par M. Trélon. (Group IV, Class 34.) Rapports du jury international, tome 4, pp. 283–344.

HABILLEMENTS des deux sexes: Section I. Vêtements d'homme et de femme, par M. Auguste Dusautoy. Section II. Fleurs et plumes, chapeaux de paille, modes et coiffures de femme, par M. Charles Petit. Section III. Ouvrages en cheveux, par M. Maxime Gaussen. Section IV. Chaussures, par M. Maxime Gaussen. Section V. Chapellerie, par M. Laville. (Group IV, Class 35.) Rapports du jury international, tome 4, pp. 347–408.

JOAILLERIE et bijouterie, par MM. Fossin et Beaugraud. (Group IV, Class 36.) Rapports du jury international, tome 4, pp. 411–438.

ARMES PORTATIVES: Section I. Armes de guerre portatives, par M. le Baron Treuille de Beaulieu. Section II. Les armes, par M. Challeton de Brughat. Section III. Armes de tous les temps, par M. Henri Berthoud. (Group IV, Class 37.) Rapports du jury international, tome 4, pp. 442–498.

ARTICLES de voyage et de campement, par M. Teston. (Group IV, Class 38.) Rapports du jury international, tome 4, pp. 501–526.

BIMBELOTERIE, par M. Jules Delbruck. (Group IV, Class 39.) Rapports du jury international, tome 4, pp. 529–538.

PRODUITS de l'exploitation des mines et de la métallurgie: Section I. Substances minérales, par M. Daubrée. Section II. Combustibles artificiels, par Edmond Fuchs. Section III. Acier, par M. Goldenberg. Section IV. L'acier en 1867. Section V. Fontes et fers, par MM. Edmond Fuchs et P. Worms de Romilly. Section VI. Fers et aciers ouvrés, par M. Martelet. Section VII. Cuivres bruts et affinés, par M. J. Martelet. Section VIII. Exploitation et traitement des minerais de plomb, par M. L. E. Rivot. Section IX. Zinc, par M. Edmond Fuchs. Section X. Platine, oxygène, silicium et bore, glucinium, par M. Sainte-Claire Deville. Section XI. Métaux rares, par M. Sainte-Claire Deville. Section XII. Métaux divers, par M. Petitgand. Section XIII. Observations générales sur l'état du travail des mines, par M. Petitgand. (Group V, Class 40.) Rapports du jury international, tome 5, pp. 5–700.

PRODUITS des expositions et des industries forestières: Section I. Par M. Émile Fournier. Section II. Matières tannantes, par M. Cavaré fils. (Group V, Class 41.) Rapports du jury international, tome 6, pp. 3–102.

2 B

PRODUITS de la chasse, de la pêche, et des cueillettes: Section I. Spécimens et collections d'animaux de toute sorte, par M. Ad. Focillon ;
produits de la chasse : fourrures et pelleteries, poils, crins, plumes, du-
vets, cornes, dents, ivoire ou écaille, musc, &c., par M. Servant. Section II. Gommes, résines, et gommes résines, par M. I. M. Da Sylva
Couthinho.   Section IV. Blanc de baleine, stéarinerie, par M. J. Lawrence Smith.   (Group V, Class 42.)   Rapports du jury international,
tome 6, pp. 105–182.

PRODUITS agricoles (non alimentaires) de facile conservation : Section
I. Production de coton, par M. Augel Dolfus.   Section II. Essais de
culture du coton en France, par M. Focillon. Section III. Lins et chanvres, par M. Moll.   Section IV. Laines, par M. Louis Moll.   Section V.
Cocons, par M. Robinet.   Section VI. L'histoire naturelle médicale a
l'Exposition Universelle, par M. Chatin.   Section VII. Houblons, par M.
Victor Borie.   Section VIII. Tabacs, par M. Barral.   Section IX.   Four-
rages, par M. Barral.   Section X. Les expositions agricoles collectives,
par M. Jules Lestiboudois.   Section XI. Exposition collective de l'Algérie, par M. Thémistocle Lestiboudois.   Section XII. État de l'agriculture et de l'industrie dans le Levant, par M. Marie de Launay.   Section
XIII. Produits agricoles, non alimentaires, de l'Amérique Méridionale,
par M. Martin de Moussy.   Section XIV. L'Amérique Centrale et
l'Amérique Méridionale à l'Exposition Universelle de 1867, par M. V.
Martin de Moussy.   Section XV. Notice sur les Îles Hawaï, par M.
William Martin.   (Group V, Class 43.)   Rapports du jury international, tome 6, pp. 185–568.

PRODUITS chimiques et pharmaceutiques: Section I. Produits chimiques
pour la grande industrie, par M. Balard.   Section II. La méthode des
vases clos et ses applications, par M. Berthelot. Section III. Savons et
industrie savonnière, par M. Fourcade.   Section IV. Industrie stéarique; bougies, parafine, par M. Alphonse Fourcade.   Section V. Produits de l'industrie du caoutchouc et de la gutta-percha, par M. G. Gerard.   Section VI. Découverte des nouvelles couleurs dérivées de la
houille, par M. Balard.   Section VII. Matières colorantes derivées de
la houille, par MM. A. W. Hofmann, Georges de Laire et Charles
Girard.   Section VIII. Produits pharmaceutiques, par MM. Fumouge
et Barreswil.   (Group V, Class 44.)   Rapports du jury international,
tome 7, pp. 7–318.

SPÉCIMENS des procédés chimiques de blanchiment, de teinture, d'impression et d'apprêt: Section I. Considérations générales sur l'industrie
du blanchiment, de la teinture, de l'impression et de l'apprêt, sur les
matières textiles, (laines, soie, coton, lin et chanvre,) par M. Aimé Bou-
tarel.   Section II. Teintures et impressions, par M. J. Persoz fils.
(Group V, Class 45.)   Rapports du jury international, tome 7, pp.
321–361.

CUIRS ET PEAUX, par M. Fauler.   (Group V, Class 46.)   Rapports du
jury international, tome 7, pp. 365–379.

MATÉRIEL et procédés de l'exploitation des mines et de la métallurgie : Section I. Sondages, par M. Gernaert. Section II. Détails des travaux récents de sondage, par M. Ch. Laurent-Degousée. Section III. Sondages du Sahara Oriental de la Province de Constantine, par M. Dubocq. Section IV. Travaux de captage des eaux minérales, établissements thermaux, par M. Jules François. Section V. Matériel et procédés de l'exploitation des mines, par M. Callon. Section VI. Procédés métallurgiques, par M. Lan. Section VII. Foyers fumivores, par M. Ed. Grateau. Section VIII. Galvanoplastie, par M. de Jacobi. Section IX. Applications en grand de la galvanoplastie et de l'électrométallurgie, par M. Oudry. (Group VI, Class 47.) Rapports du jury international, tome 8, pp. 5–171.

MATÉRIEL et procédés des exploitations rurales et forestières : Section I. Matériel et procédés des exploitations rurales, par M. Boitel. Section II. Machines locomobiles et machines routières, par M. Tresca. Section III. Matériel et procédés des exploitations forestières, par M. Serval. Section IV. Matières fertilisantes d'origine organique ou minérale, par M. Le Baron Justus de Liebig. (Traduit de l'allemand, par M. Michel Rempp.) Section V. Assainissement des fosses et conversion des vidanges en engrais, par M. Dumas. Section VI. État de l'industrie des engrais, par M. Paul Boiteau. (Group VI, Class 48.) Rapports du jury international, tome 8, pp. 175–258.

ENGINS et instruments de la chasse, de la pêche et des cueillettes : Section I. Articles de pêche, cannes, lignes, moulinets, hameçons, appats, &c., filets de mer et d'eau douce, machine à fabriquer les filets, par M. le Docteur A. Gillet de Grandmont. Section II. Matériel et procédés de pisciculture fluviale, par M. Coumes. Section III. Appareils plongeurs et scaphandres, par M. le Docteur A. Gillet de Grandmont. (Group VI, Class 49.) Rapports du jury international, tome 8, pp. 262–292.

MATÉRIEL et procédés des usines agricoles et des industries élémentaires : Section I. Outillage pour la fabrication du sucre de betterave, par M. le Baron Thénard. Section II. Pétrisseurs mécaniques, par M. Lebandy. Section III. Matériel de la chocolaterie, par M. le Baron Thénard. Section IV. Fabrication de la glace, par M. Arnould Thénard. (Group VI, Class 50.) Rapports du jury international, tome 8, pp. 295–377.

MATÉRIEL des arts chimiques de la pharmacie et de la tannerie : Section 1. Industrie stéarique, par M. Motard. Section II. Usines a gaz, par M. Eugène Pelouze. Section III. Matériel de la pharmacie, par M. Amédée Vée. Section IV. Préparation des tabacs, par M. Cavaré fils. Section V. Matériel et outillage mécanique de la tannerie et de la mégisserie, par M. A. Perrault. Section VI. Produits refractaires, par M. Chandelon. Section VII. Matériaux et appareils des usines a gaz, par M. Lawrence Smith, (Group VI, Class 51.) Rapports du jury international, tome 8, pp. 382–476.

MOTEURS, générateurs et appareils mécaniques spécialement adaptés aux besoins de l'Exposition: Section I. Service mécanique et service hydraulique, par MM. Jacqmin et Cheysson. Section II. Manutention et appareils de levage employés au déchargement et au chargement des colis, par M. E. Hangard. Section III. Distribution du gaz au palais et dans le parc du Champ de Mars, par M. Guérard. Section IV. Ventilation du palais, par M. le Vicomte d'Ussel. (Group VI, Class 52.) Rapports du jury international, tome 8, pp. 479–597.

MACHINES et appareils de mécanique générale: Section I. Pièces détachées de machines, paliers, embrayages, déclics, appareils de graissage, compteurs, dynamomètres, modèles et dessins de machines, par M. P. Worms de Romilly. Section II. Appareils fumivores, compteurs, appareils de jaugeage, pompes, presses, &c., par M. Lebleu. Section 111. Machines servant à élever les fardeaux, grues, monte-charges, crics, courroies, par M. P. Worms de Romilly. Section IV. Moteurs hydrauliques, par M. P. Worms de Romilly. Section V. Machines à vapeur, chaudières, générateurs, &c., par M. P. Luuyt. Section VI. Machines à gaz, à air chaud, à ammoniaque, moteurs électriques, moulins à vent, &c., par M. Lebleu. (Group VI, Class 53.) Rapports du jury international, tome 9, pp. 5–107.

MACHINES-OUTILS et procédés de la confection des objets de mobilier et d'habitation: Section I. Machines-outils, par M. Tresca. Section II. Machines-outils servant spécialement au travail des bois, par MM. Tresca et Lecœuvre. Section III. Machines servant au travail des matières argileuses, par M. Tresca. (Group VI, Class 54.) Rapports du jury international, tome 9, pp. 110–162.

MATÉRIEL et procédés de la filature, par MM. Michel Alcan et Édouard Simon. (Group VI, Class 55.) Rapports du jury international, tome 9, pp. 165–193.

MATÉRIEL du tissage et des apprêts, par MM. Michel Alcan et Édouard Simon. (Group VI, Class 56.) Rapports du jury international, tome 9, pp. 195–222.

MATÉRIEL et procédés de la couture et de la confection des vêtements, par M. Henry F. Q. d'Aligny. (Group VI, Class 57.) Rapports du jury international, tome 9, pp. 225–255.

MATÉRIEL et procédés de la papeterie, des teintures et des impressions, par MM. Doumerc, Laboulaye et Normand. (Group VI, Class 59.) Rapports du jury international, tome 9, pp. 261–294.

MACHINES, instruments et procédés usités dans divers travaux, par MM. Charles Callon et Férd. Kohn. (Group VI, Class 60.) Rapports du jury international, tome 9, pp. 297–312.

CARROSSERIE ET CHARRONNAGE, par MM. L. Binder et C. Lavollée. (Group VI, Class 61.) Rapports du jury international, tome 9, pp. 318–327.

BOURRELERIE ET SELLERIE, par M. Noisette. (Group VI, Class 62.) Rapports du jury international, tome 9, pp. 332–335.

MATÉRIEL des chemins de fer: Section I. Chemins de fer; exposé économique, par MM. Eugène Flachat et de Goldschmidt. Section II. Voie et matériel fixe de la voie, par MM. Eugène Flachat et de Goldschmidt. Section III. Locomotives, par M. Couche. Section IV. Matériel roulant, voitures et wagons, par M. Henry Mathieu. Section V. Signaux optiques et acoustiques, par M. J. Morandière. Section VI. Modèles, plans et dessins de gares, de stations, de remises et de dépendances de l'exploitation des chemins de fer, par M. J. Morandière. (Group VI, Class 63.) Rapports du jury international, tome 9, pp. 340–532.

MATÉRIEL et procédés de la télégraphie: Section I. Application de l'électricité à la télégraphie, par M. Ed. Becquerel. Section II. Pose du cable transatlantique, par M. de Vougy. Section III. Applications de l'électricité considerée au point de vue dynamique, par M. Ed. Becquerel. (Group VI, Class 64.) Rapports du jury international, tome 10, pp. 5–42.

MATÉRIEL et procédés du génie civil, des travaux publics et de l'agriculture: Section I. Matériaux de construction, par M. Delesse. Section II. Terres cuites et poteries, par M. E. Baude. Section III. Matériel des travaux du génie civil et de l'architecture, par M. Viollet-Le-Duc. Section IV. Routes et ponts, navigation intérieure, fondations et opérations diverses, par M. E. Baude. Section V. Percement du Mont Cenis, par M. Edmond Huet. Section VI. Percement de l'Isthme de Suez, par M. E. Baude. Section VII. Alimentation en eau et assainissement des villes, par M. Edmond Huet. Section VIII. Emploi agricole des eaux d'égout, par M. Mille. Section IX. Travaux maritimes, par M. Charles Marin. Section X. Phares, par M. Léonce Reynaud. (Group VI, Class 65.) Rapports du jury international, tome 10, pp. 45–354.

MARINE: Section I. Cales et bassins de Radout, docks flottants, &c., par M. Pasquier-Vauvillers. Section II. Marine commerciale, M. de Fréminville. Section III. Marine militaire, par M. A. de Fréminville. Section IV. Balisage, par M. Dumoustier. Section V. Sauvetage, par M. Dumoustier. Section VI. Arsenaux et établissements de la marine militaire, par M. Pasquier-Vauvillers. (Group VI, Class 66.) Rapports du jury international, tome 10, pp. 358–481.

CÉRÉALES et autres produits farineux comestibles avec leurs dérivés: Section 1. Les céréales alimentaires, par M. Gustave Heuzé. Section II. Céréales et autres produits farineux comestibles en Orient, par M. Ohannès-Effendi Tuyssuzian. Section III. Notice sur les principales productions du Mexique, par M. Thomas. Section IV. Levûre pressée allemande, par M. Anselme Payen. Section V. Pâtes d'Italie, gluten granulé et couscous des Arabes, par M. Payen. (Group VII, Class 67.) Rapports du jury international, tome 11, pp. 5–81.

PRODUITS de la boulangerie et de la pâtisserie, par MM. A. Husson et L. Foubert. (Group VII, Class 68.) Rapports du jury international, tome 11, pp. 85–102.

CORPS GRAS alimentaires, laitages et œufs: Section I. Les huiles, par M. J. A. Barral. Section II. Corps gras alimentaires, laitages et œufs, par M. Poggiale. (Group VII, Class 69.) Rapports du jury international, tome 11, pp. 106–148.

VIANDES ET POISSONS, par MM. Payen et Martin de Moussy. (Group VII, Class 70.) Rapports du jury international, tome 11, pp. 151–182.

LÉGUMES ET FRUITS: Section I. Fruits et légumes à l'état frais, par M. Pépin. Section II. Conserves de légumes, par M. L. Bignon. Section III. Légumes et fruits secs, par M. le Docteur Wittmack. Section IV. Oranges, citrons et raisins secs, par M. le Marquis d'Arcicolar. (Group VII, Class 71.) Rapports du jury international, tome 11, pp. 185–250.

CONDIMENTS et stimulants, sucres et produits de la confiserie: Section I. Moutarde, par M. Éric Baker. Section II. Thés, par M. Éric Baker. Section III. Café, succédanés du café, cacao et chocolat, coca et maté, par M. Ménier. Section IV. État de l'industrie du sucre, par M. B. Dureau. Section V. Confiserie, par M. Jacquin. (Group VII, Class 72.) Rapports du jury international, tome 11, pp. 253–333.

BOISSONS FERMENTÉES: Section I. Vins, par M. Teissonière. Section II. Production des vins en Amérique et dans les Colonies Anglaises, par M. Émile Chédieu. Section III. Bière, par M. Anselme Payen. Section IV. Eaux-de-vie et alcools, boissons spiritueuses; genièvre, rhum, tafia, kirsch, &c., par M. Gustave Claudon. Section V. Liqueurs aromatisées alcooliques, par M. Champoiseau. (Group VII, Class 73.) Rapports du jury international, tome 11, pp. 337–419.

SPÉCIMENS d'exploitations rurales et d'usines agricoles: Section I. Considérations générales sur l'agriculture, sur ses progrès et ses besoins, par M. Eugène Tisserand. Section II. Charrures, semoirs, distributeurs d'engrais, moteurs à vapeur, manéges, moulins à bras, à eau, à vapeur; hache-paille, presses et pressoirs, machines à élever l'eau, par M. J. A. Grandvoinnet. Section III. Principaux instruments et travaux divers de l'agriculture, par M. Aureliano. Section IV. Constructions rustiques par M. Albert Le Play. Section V. Travaux divers de l'agriculture, par M. Lesage. Section VI. Desséchement du lac Fucino et mise en culture du terrain conquis, par M. Ed. Grateau. (Group VIII, Class 74.) Rapports du jury international, tome 12, pp. 5–196.

CHEVAUX, anes, mulets, &c., maréchalerie: Section I. Exposition chevaline, par M. Rouy. Section II. Les chevaux étrangers, par M. Basile de Kopteff. Section III. Anes et mulets, par M. Ed. Prillieux. Section IV. Chameaux, par M. Ed. Prillieux. Section V. La maréchalerie, par M. Bouley. (Group VIII, Class 75.) Rapports du jury international, tome 12, pp. 199–264.

BŒUFS BUFFLES, par M. André Sanson. (Group VIII, Class 76.) Rapports du jury international, tome 12, pp. 267–307.

MOUTONS ET CHÈVRES, par M. Magne. (Group VIII, Class 77.) Rapports du jury international, tome 12, pp. 312–332.

PORCS ET LAPINS : Section I. Porcs, par M. Reynal. Section II. Lapins, par M. J. Laverrière. (Group VIII, Class 78.) Rapports du jury international, tome 12, pp. 335–344.

OISEAUX DE BASSE-COUR, par M. Florent Prévost. (Group VIII, Class 79.) Rapports du jury international, tome 12, pp. 347–352.

RACES CANINES, par M. Pierre Pichot. (Group VIII, Class 80.) Rapports du jury international, tome 12, pp. 355–399.

INSECTES UTILES : Section I. Les insectes utiles, par M. Émile Blanchard. Section II. Sériciculture, par M. de Quatrefages. (Group VIII, Class 81.) Rapports du jury international, tome 12, pp. 403–450.

POISSONS, crustacés et mollusques, par M. de Champeaux. (Group VIII, Class 82.) Rapports du jury international, tome 12, pp. 453–466.

SERRES et matériel de l'horticulture : Section I. Expositions d'horticulture, par M. L. Bouchard-Huzard. Section II. Parcs et matériel de l'horticulture, par M. J. Darcel. (Group IX, Class 83.) Rapports du jury international, tome 12, pp. 472–521.

FLEURS et plantes d'ornement de pleine terre, par M. Verlot. (Group IX, Class 84.) Rapports du jury international, tome 12, pp. 527–544.

PLANTES POTAGÈRES, par M. Courtois-Gérard. (Group IX, Class 85.) Rapports du jury international, tome 12, pp. 550–565.

ARBRES fruitiers et fruits : Section I. Arbres fruitiers et fruits, par M. Alphonse de Galbert. Section II. La viticulture et ses produits, par M. le Docteur Jules Guyot. (Group IX, Class 86.) Rapports du jury international, tome 12, pp. 569–618.

GRAINES et plantes forestières; procédés divers de repeuplement des forêts, par MM. Frédéric Moreau et Eugène de Gayffier. (Group IX, Class 87.) Rapports du jury international, tome 12, pp. 621–642.

PLANTES DE SERRES, par M. Édouard Morren. (Group IX, Class 88.) Rapports du jury international, tome 12, pp. 645–711.

MATÉRIEL et méthodes de l'enseignement des enfants; bibliothèques et matériel de l'enseignement donné aux adultes dans la famille, l'atelier, le commerce ou la corporation : Section I. Considérations sur le Groupe X, par M. Charles Robert. Section II. Introduction aux rapports des Classes 89 et 90, par M. Philibert Pompée. Section III. Crèches et asiles, par M. Philibert Pompée. Section IV. Écoles primaires, plans, mobilier et matériel des maisons d'école, par M. Ch. Barbier. Section V. Méthodes de lecture, d'écriture, d'arithmétique et de système métrique, par M. Ch. Barbier. Section VI. Enseignement spécial des aveugles, des sourds-muets et des idiots, par M. P. A. Dufau. Section VII. Résultats de l'instruction primaire, par M. Ph. Pompée. Section VIII. Écoles d'adultes, par M. Ph. Pompée. Section IX. Enseignement secondaire spécial, par M. Ch. Sauvestre. Section X. Enseignement secondaire des adultes, cours polytechniques, par M. Ph. Pompée. Section XI. Globes, cartes, appareils pour l'enseignement de la géographie, par M. le Baron de Watteville. Section XII. Enseignement de la musique, par M. Laurent de Rille. Section XIII. De l'enseignement du dessin en 1867, par M. Édouard Brongniart. Section XIV. Enseignement technique : agriculture, industrie, marine et commerce, par M. Ph. Pompée. Section XV. Almanachs, aides-memoires et autres publications utiles destinées au colportage, par M. le Comte Sérurier. Section XVI. Bibliothèques, par M. de Mofras. Section XVII. Collections diverses, par M. Philibert Pompée. (Group X, Classes 89 and 90.) Rapports du Jury International, tome 13, pp. 5–772.

MEUBLES, vêtements et aliments de toute origine, distingués par les qualités utiles, unies au bon marché : Chapitre I. Introduction, par M. A. Cochin. Chapitre II. Mobilier, par M. A. Selière. Chapitre III. Papiers peints, par M. Moreno-Henriquès. Chapitre IV. Tissus de coton, par M. A. Selière. Chapitre V. Tissus de lin, chanvre, jute et coton, par M. Fr. Ducuing. Chapitre VI. Tissus de laine peignée non foulée, tissus mélangés de coton, tissus de fil et de coton, châles, par M. Bouffard. Chapitre VII. Draps, par MM. V. Darroux et Moreno-Henriquès. Chapitre VIII. Bonneterie, tricots à la main, tissus à mailles, confections en tissus à mailles, ganterie de tricots, articles de filets, &c., par M. Moreno-Henriquès. Chapitre IX. Effilochages de laine, par MM. V. Darroux et Moreno-Henriquès. Chapitre X. Confection de vêtements pour hommes, femmes et enfants, par MM. Moreno-Henriquès et V. Darroux. Chapitre XI. Chaussures à bon marché, par M. V. Darroux. Chapitre XII. Industries accessoires, par M. Fréd. Jourdain. (Group X, Class 91.) Rapports du jury international, tome 13, pp. 775–853.

SPÉCIMENS des costumes populaires des diverses contrées, par M. Armand-Dumaresq. (Group X, Class 92.) Rapports du jury international, tome 13, pp. 857–878.

HABITATIONS caractérisées par le bon marché uni aux conditions d'hygiène et de bien-être, par M. E. Degrand, ingénieur des ponts et chaussées, et M. le Docteur J. Faucher (de Berlin.) (Group X, Class 93.) Rapports du jury international, tome 13, pp. 881–952.

PRODUITS de toute sorte, fabriqués par des ouvriers chefs de métiers, par M. A. Saint-Yves, ingénieur des ponts et chaussées, et M. Auguste Vitu, rédacteur-en-chef de l'Etendard. (Group X, Class 94.) Rapports du jury international, tome 13, pp. 355–984.

INSTRUMENTS et procédés de travail spéciaux aux ouvriers chefs de métiers, par M. Darimon, deputé au corps legislatif, et M. Van Blarenberghe, ingénieur-en-chef des ponts et chaussées. (Group X, Class 95.) Rapports du jury international, tome 13, pp. 987–1012.

RAPPORTS des délégations ouvrières, publiés par la commission d'encouragement des études des ouvriers à l'Exposition Universelle de 1867, 3 volumes. 4°. Paris, Morel, 13 Rue Bonaparte.

RAPPORTS addressés à s. exc. le ministre de l'instruction publique, par les membres de la commission chargée d'examiner les travaux d'élèves et les moyens d'enseignement exposés au ministère et au Champ de Mars. Paris, Imprimerie impériale, 1867.

RAPPORTS sur les progrès accomplis pendant les vingt dernières années, dans les sciences et dans les lettres, publiés par ordre de s. exc. le ministre de l'instruction publique.

Twenty-eight reports, forming 28 large volumes in octavo, have appeared as follows :

Analyse mathématique, par M. Bertrand, de l'Institut. Mécanique appliquée, par MM. Combes, de l'Institut, Phillips et Collignon. Minéralogie, par M. Delafosse, de l'Institut. Instruction publique, par M. Jourdain, de l'Institut. Hygiène navale, par M. Leroy de Méricourt. Hygiène militaire, par M. Michel Lévy. Médecine vétérinaire, par M. Magne. Archéologie, par M. Maury, de l'Institut. Astronomie, par M. Delaunay, de l'Institut. Anthropologie, par M. de Quatrefages, de l'Institut. Électricité, magnétisme, capillarité, par M. Quet. Zoologie, par M. Milne-Edwards, de l'Institut. Chirurgie, par MM. Velpau et Nélaton, de l'Institut. Thermodynamique, par M. Bertin. Hygiène civile, par M. Bouchardat. Physiologie, par M. Claude Bernard, de l'Institut. Géologie expérimentale, par M. Daubrée, de l'Institut. Médecine, par MM. Béclard et Axenfeld. Progrès des études relatives à l'Égypte et à l'Orient, sous la direction de M. Guigniaut, de l'Institut. Études historiques, par MM. Geffroy, Zeller et Thiénot. Paléontologie, par M. d'Archiac, de l'Institut. État des lettres, par MM. de Sacy, de l'Institut, Paul Féval et E. Thierry. Philosophie, par M. Ravaisson, de l'Institut. Études classiques et du moyen-âge, sous la direction de M. Guigniaut, de l'Institut. Théorie de la chaleur, par M. Desains, de l'Institut. Botanique physiologique, par M. Duchartre. Botanique phytographique, par M. Brongniart, de l'Institut. Stratigraphie, par M. Élie de Beaumont, de l'Institut.

The following named ten volumes remained to be published in 1867, some or all of which have since appeared :

La géologie, (phénomènes éruptifs,) par MM. Ch. Sainte-Claire Deville, de l'Institut, et Fouqué. La chimie, par M. Dumas, de l'Institut. L'optique, par M. Jamin, de l'Institut. La géométrie, par M. Chasles, de l'Institut. Le droit des gens, par M. de La Guéronnière, sénateur. La législation civile et pénale, par M. Duvergier, président de section au conseil d'état. L'histoire du droit, par M. Giraud, de l'Institut. Le droit public et administratif, par M. Boulatignier, conseiller d'état. L'économie politique, par M. Michel Chevalier, de l'Institut. L'épigraphie grecque et latine, par MM. Léon Renier, de l'Institut, et Frescher.

CONFÉRENCES PÉDAGOGIQUES faites à la Sorbonne aux instituteurs primaires venus à Paris pour l'Exposition Universelle de 1867. Organization pédagogique des écoles; législation scolaire; maisons d'école; hygiène; matières de l'enseignement, 3 volumes. Paris, Hachette, 1868.

CONSTRUCTIONS CIVILES.—Rapports du jury international sur les travaux publics et les constructions civiles, réunis par ordre de s. exc. M. de Forçade la Roquette, ministre de l'agriculture, du commerce et des travaux publics. Paris, P. Dupont, 1868.

NOUVEL ORDRE DE RÉCOMPENSES.—Rapport sur le Nouvel Ordre de Récompenses institué en faveur des établissements et des localités qui ont dévelopé la bonne harmonie entre les personnes coopérant aux mêmes travaux et qui ont assuré aux ouvriers le bien-être matériel, intellectuel et moral, par Alfred Le Roux, vice-président du corps législatif, membre de la Commission Imperiale et du jury spécial. Paris, Paul Dupont, 1867.

PHARES ET BALISES, (extrait du catalogue du ministère de l'agriculture, du commerce et des travaux publics.) Notices. Paris, 1867.

TRAVAUX PUBLICS.—Notices sur les modèles, cartes, et dessins relatifs aux travaux publics, réunis par les soins du ministère de l'agriculture, du commerce et des travaux publics. Paris, Thunot & c^{ie}, 26 Rue Racine, 1867.

CARTES ET DESSINS.—Notices sur les collections, cartes et dessins relatifs au service du corps impérial des mines, réunis par les soins du ministère de l'agriculture, du commerce, et des travaux publics. Paris, Paul Dupont, 1869.

RAPPORTS du comité départemental de la Seine-Inférieure sur l'Exposition Universelle de 1867. Verrerie, terre réfractaire, horlogerie, chronométrie, indiennes, rouenneries, teintures, apprêts, laines cardées, draperie, chimie industrielle et agricole, cuirs et peaux, instruments agricoles, produits agricoles et béstiaux, industrie cotonnière. Rouen, Lapierre & c^{ie}, 1 Rue Saint-Étienne des Tonneliers, 1867.

L'INDUSTRIE LINIÈRE à l'Exposition Universelle de 1867 dans ses rapports avec les intérêts du département des Côtes-du-Nord. (Extrait des annales du comité linier du littoral.) Saint-Brienne, L. Prud'homme, 1867.

L'INDUSTRIE SUCRIÈRE de l'arrondissement de Valenciennes à l'Exposition Universelle de 1867. Rapport dressé par ordre du comité des fabricants de sucre des arrondissements de Valenciennes et d'Avesnes, suivi de notes sur la fabrication du sucre dans l'arrondissement d'Avesnes, &c., par Mariage. Valenciennes. Lemaître, 14 Rue du Quesnoy, 1867.

ÉTUDES SUR L'EXPOSITION de 1867, annales et archives de l'industrie au XIXme siècle; nouvelle technologie des arts et metiers, des manufactures, de l'agriculture, des mines, &c., description générale ency-clopédique, méthodique et raisonnée de l'état actuel des arts, des sciences, de l'industrie et de l'agriculture chez toutes les nations ; recueil de travaux historiques, techniques, théoriques et pratiques, par les rédacteurs de Annales du Génie Civil, 8 volumes, (No. 8 à 15 des Annales du Génie Civil,) plus 2 atlas. Paris, E. Lacroix, 54 Rue des Saints-Pères.

PROMENADE à l'exposition scolaire de 1867, souvenir de la visite des instituteurs, par Charles Defodon. Ouvrage contenant des gravures, plans et vignettes. Paris, Hachette, 1868.

L'EXPOSITION UNIVERSELLE de 1867 illustrée. Publication interna-tionale autorisée par la Commission Impériale. Moitié texte, moitié gravures. 2 volumes. 4°. Each 484 pp. Paris, 1867.

LES MERVEILLES de l'Exposition de 1867 par Jules Mesnard. 2 vol-umes. 4°. Illustrated, Paris.

REVUE DE L'EXPOSITION. Noblet, éditeur. From the Revue Univer-selle des Mines, &c.

HISTOIRE GÉNÉRALE de l'Exposition Universelle de 1867, par Aymar-Bression. 1 volume. Paris, 41 Rue du Cardinal Fesch, 1868.

MAGASIN PITTORESQUE.—A series of articles on the Exposition, pub-lished during the years 1867 and 1868, in volumes 35 and 36 of the collection. Paris, 29 Quai des Grands-Augustins.

PROMENADES PRÉHISTORIQUES à l'Exposition Universelle, par G. de Mortilet, directeur des matériaux pour l'histoire primitive de l'homme. Paris, Reinwald, 15 Rue des Saints-Pères, 1867.

VISITES d'un ingénieur à l'Exposition Universelle de 1867. Notes et croquis, chiffres, et faits utiles, par C. A. Oppermann. 1 volume et un atlas. Paris, Baudry, 15 Rue des Saints-Pères, 1867.

LA PRODUCTION ANIMALE ET VÉGÉTALE.—Études faites à l'Exposition Universelle de 1867. Société impériale d'acclimatation. Paris, au siége de la Société, 19 Rue de Lille, et chez Dentu, 1867.

AISNE.—L'Exposition Universelle de 1867, étudiée au point de vue des intérêts du département de l'Aisne. Laon, Coquet & Stenger, 22 Rue Sérurier, 1868.

ALLIER.—Le département de l'Allier à l'Exposition Universelle de 1867, par Lavergne, avec une introduction par le Marquis de Montlaur, membre du conseil général de l'Allier. Moulins, Desrosiers, 1868.

ARTS TEXTILES.—Études sur les arts textiles à l'Exposition Universelle de 1867, comprenant les perfectionnements récents apportés à la fila-ture, au retordage, &c., du coton, du chanvre, du lin, de la laine, de la soie, du jute, du china-grass, &c.; à la fabrication des cordages ; au tissage des étoffes à fils serrés et à mailles unies et façonnées et aux apprêts des fils et des étoffes, par Michel Alcan. 1 volume, 8°. pp. 424, et 1 atlas en 4°. Paris, Baudry, 15 Rue des Saints-Pères, 1868.

LES PLANTES TEXTILES, études faites à l'Exposition Universelle de 1867, par Carcenac. Paris, Société d'acclimatation, 19 Rue de Lille, 1867.

MÉMOIRE sur les épreuves des arcs métalliques de la galerie des machines faites par ordre de la Commission Impériale, par G. Eiffel, ingénieur-constructeur.

L'EXPOSITION UNIVERSELLE de 1867. Guide de l'exposant et du visiteur, avec les documents officiels, un plan et une vue de l'Exposition, par Henri de Parville. Paris, Hachette, 1866.

LA CLEF de l'Exposition Universelle de 1867. Paris, Alcan-Lévy, 62 Boulevard de Clichy.

GUIDE-LIVRET international de l'Exposition Universelle de 1867, published in five languages, (French, English, German, Italian, and Spanish.) Paris, Lebigre-Duquesne, 16 Rue Hautefeuille, 1867.

NOTICE DESCRIPTIVE de l'exposition ethnographique de la société d'ethnographie. Paris, Amyot, 1867.

LAVOLLÉE.—Les expositions de l'industrie et l'Exposition Universelle de 1867, par Lavollée. Conférences populaires faites à l'asile impérial de Vincennes. Paris, Hachette.

SUEZ.—Exposition de la compagnie universelle du canal maritime de Suez. Catalogues followed by a notice of the works and with explanatory additions. Paris, Auguste Vallée, 15 Rue Bréda, 1867.

LES VITRAUX à l'Exposition Universelle de 1867, par Édouard Didron. Paris, Didron, 23 Rue Saint-Dominique-Saint-Germain, 1868.

LE CREUSOT: Son industrie, sa population. Memorandum delivered to the special jury for the New Order of Rewards. Paris, Chaix, 20 Rue Bergere, 1867.

DÉPARTEMENT DE LA GUERRE ET DE LA MARINE.—Rapport officiel de la commission chargée de l'examen de différentes collections de matériel ressortissant aux départements de la guerre et de la marine, admises à l'Exposition Universelle de 1867. Paris, Paul Dupont, 1869.

RAPPORT DE L'ADMINISTRATION de la Commission Impériale sur l'Exposition Universelle de 1867; containing an abstract of the administrative and financial operations; some considerations on the future prospects of the exhibitions; the general list of the collaborators of the Imperial Commission; statistical tables, documents and plates. 1 volume in octavo. Imprimerie impériale.

COLLECTION DES DOCUMENTS officiels, publiés par la Commission Impériale de l'Exposition de 1867. 1 volume. 4°. Imprimerie impériale.

CÉRUSE.—Notice sur la fabrique de céruse de MM. Théodore Lefebvre & cie, à Lille, (Nord.) Lille, Horemans, 1867. 8°. 14 pp.

ENGINS DE SAUVETAGE pour les naufragés. Étude sur les canons et les fusils ports-amarres. (Extrait du rapport addressé par le jury, à s. exc. M. Rouher, ministre d'état.) Paris, Paul Dupont, 1867. 8°. 55 pp.

ASSOCIATION internationale pour le développement du commerce et des expositions. Bulletin. Tome premier; année 1868. 8°. 121 pp. Paris, au siége provisoire de l'association, 19 juillet 1869. (Contains the "Bibliographie de l'Exposition Universelle," by Léon Morillot.)

TURGAN. Études sur l'Exposition Universelle, 1867. Large 8°. viii, 224 pp. Paris, Michel Lévy frères, 2 Rue Vivienne et 15 Boulevard des Italiens.

LE VER À SOIE du chêne à l'Exposition Universelle de 1867. Insectes utiles vivants, par Camille Personnat. pp. 1–14. Paris, Librairie agricole de la Maison Rustique, 1868.

LES INSECTES UTILES (vers à soie et arbeilles) et les insectes nuisibles, par Maurice Girard. 8°. pp. 1–40. Paris, Librairie agricole, 1867.

LA MUSIQUE à l'Exposition Universelle de 1867, par le Marquis de Pontécoulant. 8°. lxxii–238 pp. Paris, au bureau du journal L'Art Musical, 1868.

FRENCH COLONIES.

CATALOGUE spécial, accompagné de notices sur les produits agricoles et industriels de l'Algérie. Paris, Challamel aîné, 30 Rue des Boulangers, et 27 Rue de Bellechasse, 1867.

COLONIES FRANÇAISES.—Catalogue des produits des colonies françaises précédé d'une notice statistique. Paris, Challamel aîné, 27 Rue de Bellechasse, 1867.

GREAT BRITAIN.

OFFICIAL CATALOGUE. English version, translated from the proof-sheets of the French catalogue published by the Imperial Commission. London, Johnson & Sons, 3 Castle street, Holborn, 1867.

CATALOGUE of the British section, containing a list of the exhibitors of the United Kingdom and its colonies, and the objects which they exhibit. (In English, French, German, and Italian.) London, New Street square, 1867.

ILLUSTRATIONS OF PRINTING.—Catalogue of the illustrations of printing, executed in the United Kingdom in 1866, exhibited in Paris in 1867. London, Spottiswoode & Co., New Street square.

FINE ARTS official catalogue, with descriptive notices furnished by the artists and owners of many of the most important pictures and works of art exhibited. London, Johnson & Sons, 3 Castle street, Holborn, 1867.

FINE ARTS DIVISION, comprising the objects illustrating the history of labor before 1800, and list of the contributors of the various works of art; also a catalogue of the pictures, sculptures, mosaics, &c., in four languages. London, Spottiswoode.

REPORTS on the Paris Universal Exposition, 1867. Presented to both houses of Parliament by command of her Majesty. London, printed by George E. Eyre and William Spottiswoode, printers to the Queen's most excellent Majesty, for her Majesty's stationery office, (20,406.) In six volumes. 8°. 1868.

REPORTS OF ARTISANS, selected by a committee appointed by the council of the Society of Arts to visit the Paris Universal Exposition, 1867. 8°. London. Published for the Society for the Encouragement of Arts, Manufactures, and Commerce, 1867. In two parts. Part I: vii, 476 pp. Part II: xvi, 213 pp.

ENGLISH COLONIES.

NOVA SCOTIA.—Catalogue of the Nova Scotian department, with introduction and appendices. Paris, Gustave Bossange, 25 Quai Voltaire, 1867.

NEW SOUTH WALES.—Catalogue of the natural and industrial products of New South Wales, forwarded to the Paris Universal Exposition of 1867. Sydney, Thomas Richards.

VICTORIA.—Catalogue des produits de la colonie de Victoria, (Australie,) figurant à l'Exposition Universelle de 1867, précédé d'une introduction ou notes sommaires sur la colonie et l'exploitation de ses mines. Londres, Spottiswoode et c^ie, New Street square, 1867.

SOUTH AUSTRALIA.—Catalogue of contributions to the Paris Universal Exposition, held in Paris, 1867.—pp. 31. Adelaide, W. C. Cox, Victoria square, 1866.

QUEENSLAND.—Catalogue of the natural and industrial products of Queensland.—pp. 44. London, Ed. Stanford, 6 Charing Cross, S. W., 1867.

INDIAN DEPARTMENT.—Catalogue of the articles forwarded from India. London, Spottiswoode, 1867.

NATAL.—Catalogue of contributions from the colony of Natal, by Peniston. London, Jarrold & Sons, 12 Paternoster Row, 1867.

CAPE OF GOOD HOPE.—Catalogue of the articles contributed to the Paris Exposition of 1867, by the Cape of Good Hope, with an introductory sketch of the colony, its institutions, physical features and products, by J. B. Currey. London, Jarrold.

BRITISH GUIANA.—Catalogue of contributions transmitted from British Guiana to the Paris Universal Exposition of 1867. London, Ed. Stanford, 6 Charing Cross, S. W., 1867.

GUIANA.—Exposition Universelle de Paris, 1867. Catalogue des produits exposés par la Guyane anglaise, publié par la comité de correspondance de la societé royale d'agriculture et du commerce. cii–52 pp. 2 maps, 1 tab. 8°. Londres, E. Standford, 1867.

CANADA.—Esquisse géologique du Canada, suivie d'un catalogue descriptif de la collection de cartes et coupes géologiques, livres imprimés, roches, fossiles et minéraux économiques envoyés à l'Exposition Universelle de 1867. Paris, Bossange, 25 Quai Voltaire, 1867.

CANADA.—Catalogue des végétaux ligneux du Canada, par l'abbé Ovide Brunet. Québec, Darveau, Rue La Montagne, 1867.

GREECE.

CATALOGUE définitif rédigé par la commission centrale. Athènes, 1867.

ΠΕΡΙΛΗΠΤΙΚΗ. περιγραφη των εις την παγκοσμιων εχθεσιντων Παρισιων του ετους 1867, αποσταλησομενων ελληνικων μαρμαρων και ορυκτων. Αθηναι, 1866.

GRAND-DUCHY OF HESSE.

SPECIAL-KATALOG für das Grossherzogthum Hessen, herausgegeben von der grossherzoglich-hessischen Commission. Darmstadt, Buchdruckerei von Heinrich Brill, 1867.

ITALY.

ESPOSIZIONE universale del 1867 a Parigi. Parte prima : Atti ufficiali della r. commissione italiana. Firenze, Barbèra, 1867. 4°. pp. 334.

RELAZIONE del r. commissario al ministro di agricoltura, industria e commercio ed elenco dei premiati della sezione italiana. Firenze, Barbèra, 66 Via Faenza, 1868.

L'ITALIE ÉCONOMIQUE en 1867 avec un aperçu des industries italiennes à l'Exposition Universelle de Paris. Florence, Barbèra, 1867.

ELENCO dei premiati della sezione italiana. Firenze, Barbèra, 66 Via Faenza, 1868.

LES PRODUITS de l'agriculture du Piémont, de la Lombardie et de la Vénétie, par Gaëtan Cantoni. Paris, Librairie agricole, 26 Rue Jacob, 1867.

RELAZIONI dei giurati italiani sulla Esposizione Universale del 1867. Firenze, G. Pellas, 1868.

ESPOSIZIONE Universale del 1867. Sui prodotti greggi e l'avorati delle industrie estrattive. Relazione di Giulio Curioni, giurati al esposizione di Firenze nel 1861, di Londra nel 1862, e di Parigi, nel 1867. Firenze, Stabilimento di Gius Pellas, 1869. 8°. 1 pl. 159 pp.

SYSTÈME AGUDIO. Locomoteur avec adhérence au moyen du rail central. Turin, 1867.

CENNI SUGLI ultimi perfezionamenti delle macchine a vapore, locomotive preceduti da alcuni principi generali di termodinamica di Leonardo Carpi. Paris, Paul Dupont, 1867.

MOROCCO.

NOTICE sur le Maroc, par Auguste Beaumier. Brochure lithographiée par Valeur, 35 Rue d'Argenteuil. Paris.

NETHERLANDS.

CATALOGUE SPÉCIAL.—Édition officielle de la commission royale des Pays-Bas. Preceded by a statistical notice of the kingdom. Harlem, Enschedé et fils, 1867.

PAPAL STATES.

ÉTATS-PONTIFICAUX.—(Catalogue pour l'Exposition Universelle de 1867 à Paris.) Paris, Adrien Le Clerc, 29 Rue Cassette 1867.

ELENCO GENERALE ragionato di tutti gli oggetti spediti dal Governo Pontificio alla Esposizione Universale di Parigi nell' anno 1867, par mezzo del ministero del commercio, belle arti, industria, agricoltura e lavori publici. Roma, Tipografia della rev. cam. apostolica, 1867.

APERÇU GÉNÉRAL sur les catacombes de Rome et description du modèle d'une catacombe exposé à Paris en 1867, par J. B. de Rossi. Paris, Hachette, 1867.

LA MÉTÉOROLOGIE et le météographe du P. Secchi à l'Exposition Universelle. Paris, Gauthiers-Villars, 55 Quai des Grands-Augustins. 1867.

PORTUGAL.

CATALOGUE SPÉCIAL de la section portugaise. Paris, Dentu, et Paul Dupont, 1867.

CATALOGUE DESCRIPTIF de la collection des minéraux utiles, accompagné d'une notice sur l'industrie minérale du pays, par J. A. C. das Neves Cabral. Paris, Paul Dupont, 1867.

NOTICE sur le Portugal, par J. J. Roderigues de Freitas, (junior.) Paris, Paul Dupont, 1867.

EAUX MINÉRALES.—Renseignements sur les eaux minérales portugaises. Paris, Dentu, 1867.

NOTICE ABRÉGÉE de l'imprimerie nationale de Lisbonne, suivie du catalogue des produits qu'elle présenta dans l'Exposition Universelle de Paris en 1867. Texte français et portugais. Lisbonne, 1867.

PRUSSIA AND NORTHERN GERMANY.

AMTLICHER SPECIAL-KATALOG der Ausstellung Preussens und der Norddeutschen Staaten.—12°. pp. 337. Berlin, Bernstein, 1867.

PRUSSE et États de l'Allemagne du Nord. Catalogue spécial. Paris, Dentu, 1867.

KATALOG der auf der Pariser allgemeinen Ausstellung in 1867 in dem preussischen Schulhause aufgelegten Lehrmittel. Paris, Renou et Maulde, 1867.

KATALOG für die Sammlung der Bergwerks-und Steinbruchs-Produkte Preussens auf der Industrie-und Kunst-Ausstellung zu Paris, im

Jahre 1867. Im Auftrage Sr. Excellenz des Herrn Ministers für Handel, Gewerbe und öffentliche Arbeiten, Grafen von Itzenplitz, verfasst von Dr. Hermann Wedding. Berlin, gedruckt in der königlichen Staatsdruckerei. 8°. 76 pp.

STATISTIQUE AGRICOLE, industrielle et commerciale de la Prusse. (Brochure extraite du Moniteur Prussien.) Berlin, R. de Decker, 75 Wilhelmstrasse, 1867.

BERICHT über die allgemeine Ausstellung zu Paris, erstattet von den für Preussen und die Norddeutschen Staaten ernannten Mitgliedern der internationalen Jury. Berlin, 1868.

BERICHT über den landwirthschaftlichen Theil der Pariser Welt-Ausstellung von 1867, herausgegeben im Auftrage des königlich-preussischen Ministeriums für die landwirthschaftlichen Angelegenheiten, von Salviati. Berlin, 1868.

DIE LANDWIRTHSCHAFTLICHEN Maschinen und Geräthe auf der Welt-Ausstellung zu Paris 1867. Bericht erstattet dem königlich preussischen Ministerium für die landwirthschaftlichen Angelegenheiten, von Emil Perels. Berlin, 1867.

L'ÉCOLE PRIMAIRE prussienne à l'Exposition de Paris. 1867.

BERICHT über literarische Leistungen im Königreiche Sachsen lebender Schriftsteller während der Jahre 1847–1867, von Oswald Marbach. Leipzig, Giesecke und Devrient, 1867.

INSTRUCTION PUBLIQUE.—Exposé de l'état de l'instruction publique dans le royaume de Saxe. Dresde, Meinhold, 1867.

VERZEICHNISS UND BESCHREIBUNG der von der königlich preussischen Direction der Niederschlesisch-Märkischen Eisenbahn im Jahre 1867 zur Ausstellung nach Paris gesandten Gegenstände. (Brochure.) Von verschiedenen verfassern. xvi, 376 pp., 17 l, 11 pl. 4°. Berlin, Wiegardt & Kempel, 1868.

DIE ERZEUGNISSE des Pflanzen- und Thierreichs und das Rüstzeug des Landwirths mit Ausnahme der Maschinen und Geräthe auf der Welt-Ausstellung zu Paris 1867. Bericht, erstattet dem königlich-preussischen Ministerium für die landwirthschaftlichen Angelegenheiten.

ACIÉRIE de Fried. Krupp, à Essen, (Prusse rhénane.) Paris, Wittersheim, 1867.

ROUMANIA.

ROUMANIE.—Notice sur la Roumanie, principalement au point de vue de son économie rurale, industrielle et commerciale, avec une carte de la principauté de Roumanie. Paris, Franck, 67 Rue Richelieu, 1867.

KURTEA.—L'église du monastère épiscopal de Kurtea d'Argis en Valachie, with four plates and twenty-five wood engravings. Vienne, Ch. Gérold fils, 1867. (There is a German edition.)

3 B

RUSSIA.

CATALOGUE SPÉCIAL de la section russe à l'Exposition Universelle de Paris en 1867. Publié par la Commission Impériale de Russie. 8°. xii, 289 pp. Paris, Imprimerie générale de Ch. Lahure, 9 Rue de Fleurus.

APERÇU STATISTIQUE des forces productives de la Russie, par De Buschen, membre du comité central de statistique de Saint-Petersbourg. Annexe au catalogue spécial de la section russe de l'Exposition Universelle de Paris en 1867. 8°. 368 pp. Paris, Imprimerie générale de Ch. Lahure, 9 Rue de Fleurus, 1867.

MINISTÈRE DE LA GUERRE. Service de l'équipement et du campement des troupes et des hôpitaux militaires. Catalogue de la collection présentée à l'Exposition Universelle de Paris, en 1867. Moscou, W. Gautier.

MÉMOIRE EXPLICATIF de la collection des substances préparées dans le laboratoire de l'institut agricole de Saint-Petersbourg pour l'Exposition de Paris, 1867. Paris, Librairie agricole, 26 Rue Jacob.

SIAM.

NOTICE sur le royaume de Siam, publiée par A. A. Gréhan, Phra Siam Dhuranwraks, consul de S. M. le suprème roi de Siam et son commissaire général près l'Exposition Universelle. Paris, Simon Raçon, 1 Rue d'Erfurth, 1867. Deux éditions. The second edition is ornamented with a portrait of H. M. the king of Siam, drawn by Riou, and with nine photo-lithographs representing the exhibition by Siam.

SWEDEN AND NORWAY.

EXPOSITION d'horticulture de la Suède. Paris, P. Dupont, 1867. (Small pamphlet.)

LA SUÈDE : Son développement moral, industriel, et commercial, d'après des documents officiels, par C. E. Ljungberg. Avec une carte et trente-et-un tableaux. Traduit par L. de Lilliehöok. 8°. vii, 177 pp. Paris, Imprimerie de Dubuisson & Cie., 5 Rue Coq-Héron, 1867. (With map.)

APERÇU de la végétation et des plantes cultivées de la Suéde, par N. J. Anderson. Stockholm, Norstedt et fils, 1867.

FERS ET ACIERS.—Quelques renseignements sur la fabrication des fers et aciers de la Suède, ainsi que sur les autres objets des Classes 40 et 47, à l'occasion de l'Exposition Universelle de Paris en 1867, par L. Rinman, ingénieur et métallurgiste de la Suède. Paris, 1866.

HISTOIRE DU TRAVAIL.—Notice sommaire sur l'histoire du travail dans le royaume de Norvége. Paris, P. Dupont, 1867.

NORVÉGE.—Notice statistique sur le royaume de Norvége. Paris, Dupont, 1867.

LES PÊCHES de la Norvége, par Herman Baars, [special commissioner from Norway, for the fisheries and navigation, at the Universal Exposition of 1867, at Paris.] Paris, J. Bonaventure, 55 Quai des Grands-Augustins, 1867.

SPAIN.

CATALOGO GENERAL de la seccion española, publicado por la comision regia de España. Paris, Lahure, 1867.

CATALOGUE GÉNÉRAL de la section espagnole, publié par la commission royale d'Espagne. Paris, Lahure, 1867.

L'ESPAGNE à l'Exposition Universelle de 1867, par Léon Droux. Paris, Dentu, 1868.

TRAVAUX PUBLICS.—Notice sur l'état des travaux publics en Espagne et sur la législation spéciale qui les régit. Madrid, Rivadeneyra, 3 Rue du Duc d'Ossuna, 1867. 8°. pp. 135.

NOTICE sur les documents appartenant à la collection paléographique de Jean do Troy Ortolano, présentés à l'Exposition Universelle de 1867, à Paris. Paris, Lahure, 1867.

SWITZERLAND.

RAPPORTS sur la participation de la Suisse à l'Exposition Universelle de 1867, avec catalogue. Berne, C. J. Wyss, 1868.

LES INSTITUTIONS OUVRIÈRES de la Suisse, par Gustave Moynier. Mémoire rédigé à la demande de la commission centrale de la Confédération Suisse pour l'Exposition Universelle de Paris. Genève, Cherbuliez. Paris, Cherbuliez, 33 Rue de Seine, 1867.

TUNIS.

TRADUCTION LITTÉRALE du travail, publié en Arabe pour M. le baron Jules de Lesseps, commissaire général de Tunis, du Maroc, de la Chine et du Japon, par Soliman Al-Haraïri. Paris, Victor Goupy, 5 Rue Garancière, 1866. The Arab text: Paris, Jousset, Clet et cie, 8 Rue de Furstenberg.

TUNIS.—Notices abrégées sur la régence de Tunis, par Ch. Cubisol, 1866. Pamphlet lithographed by Valeur, 35 Rue d'Argenteuil, Paris.

TURKEY.

LA TURQUIE à l'Exposition Universelle de 1867, ouvrage publié par les soins et sous la direction de s. exc. Salaheddin-Bey, commissaire impérial ottoman pour l'Exposition Universelle. Paris, Hachette, 1867.

RAPPORTS de la commission scientifique impériale ottomane, réuni à Paris sous la présidence de s. exc. Djemil Pacha. (These reports were expected to be published in twelve volumes.) Paris, Lahure.

WURTEMBERG.

CATALOGUE DESCRIPTIF du royaume de Wurtemberg. Stuttgart, Metzler, 1867.

WURTEMBERG.—Descriptive catalogue of the products of the kingdom of Wurtemberg. Published by authority of the Royal Wurtemberg Commission. 12°. 128 pp. Stuttgart, J. B. Metzler, printer, 1867.

CATALOGUE pour l'exposition spéciale des poids, mesures, et monnaies du royaume de Wurtemberg. Stuttgart, Grüninger, 1867.

AGRICULTURE.—Exposition des produits de l'agriculture wurtembergeoise, organisée par le conseil central de l'agriculture du royaume de Wurtemberg à Stuttgart. 1867.

RAPPORT sur l'économie politique et sociale du royaume de Wurtemberg, publié sous la direction du conseil royal pour l'industrie et le commerce, par Mæhrlen. Stuttgart, Grüninger, 1868.

DAS BESONDERE PREISGERICHT und die neugeschaffenen Preise für die Pflege der Eintracht in Fabriken und Ortschaften, und die Sicherung des Wohlstandes, der Sittlichkeit und Intelligenz in den Arbeiter-Kreisen. ·Officieller Bericht von Alfred Le Roux, in's Deutsche übertragen auf Veranlassung und unter Revision des süddeutschen Mitgliedes und Referenten im Preisgericht Dr. F. V. Steinbeis. Stuttgart, Grüninger, 1868.

LIST OF THE REPORTS

OF

THE UNITED STATES COMMISSIONERS

IN THE ORDER OF SUCCESSION IN THE VOLUMES.

VOLUME I.

VOLUME II.

VOLUME III.

MACHINERY AND PROCESSES OF THE INDUSTRIAL ARTS, AND APPARATUS OF THE EXACT SCIENCES. BY FREDERICK A. P. BARNARD, LL. D.

VOLUME IV.

EXAMINATION OF THE TELEGRAPHIC APPARATUS AND THE PROCESSES IN TELEGRAPHY. BY SAMUEL F. B. MORSE, LL. D.

STEAM ENGINEERING AS ILLUSTRATED BY THE PARIS UNIVERSAL EXPOSITION. BY WILLIAM S. AUCHINCLOSS.

ENGINEERING AND PUBLIC WORKS. BY WILLIAM P. BLAKE.

BÉTON-COIGNET; ITS FABRICATION AND USES, ETC. BY LEONARD F. BECKWITH.

ASPHALT AND BITUMEN AS APPLIED IN CONSTRUCTION, ETC. BY ARTHUR BECKWITH.

BUILDINGS, BUILDING MATERIALS, AND METHODS OF BUILDING. BY JAMES H. BOWEN.

MINING AND THE MECHANICAL PREPARATION OF ORES. BY HENRY F. Q. D'ALIGNY, AND MESSRS. HUET, GEYLER, AND LEPAINTEUR.

VOLUME V.

QUANTITIES OF CEREALS PRODUCED IN DIFFERENT COUNTRIES COMPARED. BY SAMUEL B. RUGGLES.

THE QUALITY AND CHARACTERISTICS OF THE CEREAL PRODUCTS EXHIBITED. BY GEORGE S. HAZARD.

REPORT ON THE PREPARATION OF FOOD. BY W. E. JOHNSTON, M. D.

GENERAL ALPHABETICAL INDEX

TO THE

REPORTS OF THE UNITED STATES COMMISSIONERS

TO THE

UNIVERSAL EXPOSITION AT PARIS, 1867.

NOTE.—As each report is paged separately, it is necessary to specify in the references not only the volume but the report. This has been done by giving, first the number of the volume in Roman numerals, next an abbreviation of the title of the report, followed by the number of the page or pages in Arabic numerals. When a subject is discussed in two or more of the reports, reference is made to each report, and, generally, in the order of the volumes.

The sequence of reports in each of the volumes may be ascertained by consulting the full list of the reports or the table of contents in each volume. The running titles of the right-hand pages will further aid in finding a report sought for.

A.

C.

2

E.

I.

3

M.

Q.

R.

S.

T.

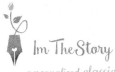

Im The Story

personalised classic books

JANE
IN
WONDERLAND

LEWIS
CARROLL

"Beautiful gift.. lovely finish.
My Niece loves it, so precious!"

Helen R Brumfieldon

⭐⭐⭐⭐⭐

UNIQUE GIFT

FOR KIDS, PARTNERS
AND FRIENDS

Timeless books such as:

Kids

Alice in Wonderland • The Jungle Book • The Wonderful Wizard of Oz
Peter and Wendy • Robin Hood • The Prince and The Pauper
The Railway Children • Treasure Island • A Christmas Carol

Adults

Romeo and Juliet • Dracula

Highly Customizable

Change Books Title

Replace Characters Names with yours

Upload Photo for inside page

Add Inscriptions

Visit
Im The Story .com
and order yours today!

CPSIA information can be obtained
at www.ICGtesting.com
Printed in the USA
BVHW041248160819
556068BV00019B/1699/P